UNIVERSITY PRESS

Great Clarendon Street, Oxford ox2 6DP

Oxford University Press is a department of the University of Oxford.
It furthers the University's objective of excellence in research, scholarship,
and education by publishing worldwide in

Oxford New York

Auckland Cape Town Dar es Salaam Hong Kong Karachi
Kuala Lumpur Madrid Melbourne Mexico City Nairobi
New Delhi Shanghai Taipei Toronto

With offices in

Argentina Austria Brazil Chile Czech Republic France Greece
Guatemala Hungary Italy Japan Poland Portugal Singapore
South Korea Switzerland Thailand Turkey Ukraine Vietnam

Oxford is a registered trade mark of Oxford University Press
in the UK and in certain other countries

Published in the United States
by Oxford University Press Inc., New York

First edition 2001
Second edition 2010
Reprinted 2010 (twice), 2011 (with corrections), 2012, 2013

British Library Cataloguing in Publication Data
Data available

Library of Congress Cataloging in Publication Data
Data available

Typeset by SPI Publisher Services, Pondicherry, India

ISBN 978–0–19–957336–3 (hbk)
     978–0–19–957337–0 (pbk)

Preface

Nine years have passed since the first edition of *Optical properties of solids* was published, and in these years I have received many helpful comments and suggestions about how to improve the text. By and large, the comments from students have been concerned with sections that need further clarification, while those from colleagues have been about adding new topics. This second edition gives me the opportunity to make both types of improvements.

Science move on, and, even in the relatively short time since the first edition was published, some completely new subjects have arisen, while others have grown in importance. There are also other topics that should have been included in the first edition, but were omitted. It is not possible to cover everything in a book of this length, and in the end I have settled on the following list of new topics for the second edition:

Electro-optics and magneto-optics New sections on induced birefringence, optical chirality, and electro-optics have been added, namely Sections 2.5.2, 2.6, and 11.3.4.

Spintronics Three new sections have been added—Sections 3.3.7, 5.3.4, and 6.4.5—to cover the physics of optical spin injection in semiconductors.

Cathodoluminescence This topic is covered in Section 5.4.4.

Quantum dots Section 6.8 has been substantially expanded to reflect the prominence of quantum dots in current semiconductor research and device development.

Plasmonics The discussion of bulk plasmons in Section 7.5 has been improved, and a new subsection on surface plasmons has been added.

Negative refraction Section 7.6 gives a brief discussion of this phenomenon.

Carbon nanostructures Graphene, nanotubes, and bucky balls are discussed in Section 8.5.

Diamond NV centres Section 9.2.2 has been added to reflect the growing interest in diamond NV centres for quantum information processing.

Solid-state lighting A discussion of white light LEDs has been added to Section 9.5.

This choice undoubtedly reflects my personal opinions on the present state of the subject, but it also based on the suggestions that I have received from colleagues. With some ingenuity, it has been possible to work all of this new material into the chapter structure of the first edition outlined in Fig. 1. Note, however, that the title of Chapter 6 has been changed from 'Quantum wells' to 'Quantum confinement' to reflect the greater emphasis on quantum dots.

In addition to these new topics, I have made improvements throughout the whole text, and have tried to correct any errors or misleading remarks that were present in the first edition. All of the chapters have been updated, with new examples and exercises added where appropriate. In some cases new data have been included. The most significant improvements have been made to the sections on the Kramers–Kronig relationships (2.3), birefringence (2.5.1), and the quantum-confined Stark effect (6.5). It is inevitable that some errors will still persist in this second edition, and new ones occur. A web page with the *errata* will be posted as these errors are discovered.

M.F.
Sheffield
January 2010

Preface to the First Edition

This book is about the way light interacts with solids. The beautiful colours of gemstones have been valued in all societies, and metals have been used for making mirrors for thousands of years. However, the scientific explanations for these phenomena have only been given in relatively recent times. Nowadays, we build on this understanding and make use of rubies and sapphires in high power solid-state lasers. Meanwhile, the arrival of inorganic and organic semiconductors has created the modern opto-electronics industry. The onward march of science and technology therefore keeps this perennial subject alive and active.

The book is designed for final year undergraduates and first year graduate students in physics. At the same time, I hope that some of the topics will be of interest to students and researchers of other disciplines such as engineering or materials science. It evolved from a final year undergraduate course in condensed matter physics given as part of the Master of Physics degree at Oxford University. In preparing the course I became aware that the discussion of optical phenomena in most of the general solid-state physics texts was relatively brief. My aim in writing was therefore to supplement the standard texts and to introduce new subjects that have come to the fore in the last 10–20 years.

Practically all textbooks on this subject are built around a number of core topics such as interband transitions, excitons, free electron reflectivity, and phonon polaritons. This book is no exception. These core topics form the backbone for our understanding of the optical physics, and pave the way for the introduction of more modern topics. Much of this core material is well covered in the standard texts, but it can still benefit from the inclusion of more recent experimental data. This

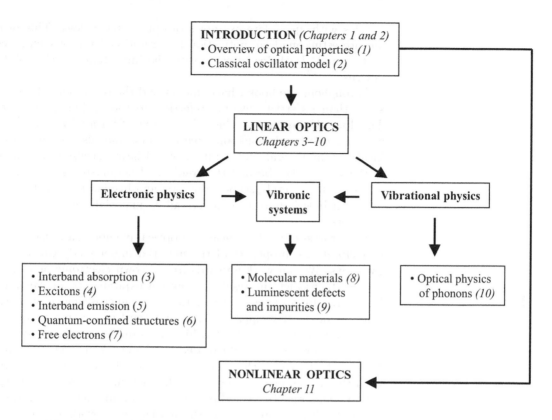

Fig. 1 Scheme of the topics covered in this book. The numbers in brackets refer to chapters.

is made possible through the ever-improving purity of optical materials and the now widespread use of laser spectroscopy.

The overall plan of the subject material is summarized in Fig. 1. The flow diagram shows that some of the chapters can be read more or less independently of the others, on the assumption that the introductory material in Chapters 1 and 2 has been fully assimilated. I say 'more or less' here because it does not really make sense, for example, to try to understand nonlinear optics without a firm grasp of linear optics. The rest of the chapters have been arranged into groups, with their order following a certain logical progression. For example, knowledge of interband absorption is required to understand quantum wells, and is also needed to explain certain details in the reflectivity spectra of metals. Similarly, molecular materials provide an intuitive introduction to the concept of configuration diagrams, which are required for the understanding of colour centres and luminescent impurities.

The inclusion of recent developments in the subject has been one of the main priorities motivating this work. The chapters on semiconductor quantum wells, molecular materials, and nonlinear optics will not be found in most of the standard texts. Other new topics such as the Bose–Einstein condensation of excitons are included alongside traditional subject material. Furthermore, it is my deliberate intention to illustrate the physics with up-to-date examples of optical technology. This provides an interesting modern motivation for traditional topics such as colour centres and also helps to emphasize the importance of the solid-state devices.

Throughout the book I have understood the term 'optical' in a wider sense than its strict meaning referring to the visible spectral region. This has allowed me to include discussions of infrared phenomena such as those due to phonons and free carriers, and also the properties of insulators and metals in the ultraviolet. I have likewise taken the scope of the word 'solid' beyond the traditional emphasis on crystalline materials such as metals, semiconductors, and insulators. This has allowed me to include important non-crystalline materials such as glasses and polymers.

The process of relating measured optical phenomena to the electronic and vibrational properties of the material under study can proceed in two ways. We can work forwards from known electronic or vibrational physics to predict the results of optical experiments, or we can work backwards from experimental data to the microscopic properties. An example of the first approach is to use the free electron theory to explain why metals reflect light, while an example of the second is to use absorption or emission data to deduce the electron level structure of a crystal. Textbooks such as this one inevitably tend to work forwards from the microscopic properties to the measured data, even though an experimental scientist would probably be working in the other direction.

The book presupposes that the reader has a working knowledge of solid-state physics at the level appropriate to a third-year undergraduate, such as that found in H.M. Rosenberg's *The solid state* (Oxford University Press, third edn, 1988). This puts the treatment at about the same level as, or at a slightly higher level than, that given in the *Introduction to solid state physics* by Charles Kittel. The book also necessarily presupposes a reasonable knowledge of electromagnetism and quantum theory. Classical and quantum arguments are used interchangeably throughout, and the reader will need to revise their own favourite texts on these subjects if any of the material is unfamiliar. Four appendices are included to provide a succinct summary of the principal results from band theory, electromagnetism, and quantum theory that have been presupposed.

The text has been written in a tutorial style, with worked examples in most chapters. A collection of exercises is provided at the end of each chapter, with solutions at the end of the book. The exercises follow the presentation of the material in the chapter, and the more challenging ones are identified with an asterisk. A solutions manual is available on request for instructors from the Oxford University Press web page.

M.F.
Sheffield
January 2001

Acknowledgements

I would like to acknowledge the many people who have helped in various ways in the production of both editions of this book. Pride of place goes to Sonke Adlung and his staff at Oxford University Press—especially Anja Tschoertner, Richard Lawrence, Emma Lonie, and April Warman—for bringing the books to fruition, and to Julie Harris for assistance with the LaTeX typesetting. I would also like to offer special thanks to Dr Geoff Brooker of Oxford University for critical reading of the whole of the first edition and for a major input to the revised section on plasmons (Section 7.5) in this present edition.

Numerous colleagues have helped to clarify my understanding of certain specialist topics and have made comments on parts of the text. Among these I would like to thank especially: Prof. Arturo Lousa from the Universidad de Barcelona, for comments on several chapters and permission to use exercises from his course; Prof. David Smith from the University of Vermont, for comments on the theory of dispersion; Prof. Richard Harley of the University of Southampton and Dr Odilon Couto Jr of the University of Sheffield for suggestions about optical spin injection; Prof. Jeremy Allam of the University of Surrey, for providing material on carbon nanotubes; my former colleagues Dr Simon Martin and Dr Paul Lane at the University of Sheffield, for their critical reading of the chapter on molecular materials in the first edition; Dr Friedemann Reinhard from the Universität Stuttgart, and Victor Acosta and Prof. Dmitry Budker from the University of California, for critical reading of the section on diamond NV centres; and Dr Oleg Shchekin from Philips Lumileds Lighting, for comments on white light LEDs. I am, of course, also very grateful to the students who have used the text and offered advice on how to improve it.

The figures are a major part of this book, and I would like to express my thanks to the publishers who have permitted the reproduction of diagrams in both editions. I would also like to thank a large number of colleagues who have provided original or unpublished data. In particular, I would like to thank: Dr Steve Collins for Fig. 2.12(b); Prof. Robert Taylor for providing unpublished data for use in Figs 5.3, 6.16, and 6.23; Dr Adam Ashmore for taking the data in Figs 5.6 and 5.13; Prof. Gero von Plessen and Dr Andrew Tomlinson for the data presented in Fig. 4.5; Prof. Mark Hopkinson for Fig. 6.21(a); Prof. Maurice Skolnick for Fig. 6.22; Dr Tim Richardson and Mark Sugden for Fig. 7.17; Prof. Frank Hegmann and Dr Aaron Slepkov for Fig. 8.11; Prof. David Lidzey for Fig. 8.19; Dr Fedor Jelezko, Philipp Neumann, Dr Friedemann Reinhard, and Prof. Jörg Wrachtrup for Fig. 9.6; Prof. Dmitry Budker and Victor Acosta for help with Fig. 9.7(b); Prof. Richard Warburton for Fig. 11.12; and Prof. Steve Blundell for the periodic table and list of fundamental constants on the inside covers.

Finally, I would like to thank the Royal Society for supporting me as a University Research Fellow during the writing of most of the first edition, and the University of Sheffield for supporting me in the remaining years.

1.3   The complex refractive index and dielectric constant

In the previous section we mentioned that the absorption and refraction of a medium can be described by a single quantity called the **complex refractive index**. This is usually given the symbol $\tilde{n}$ and is defined through the equation:

$$\tilde{n} = n + i\kappa. \tag{1.14}$$

The real part of $\tilde{n}$, namely n, is the same as the normal refractive index defined in eqn. 1.2. The imaginary part of $\tilde{n}$, namely κ, is called the **extinction coefficient**. As we shall see below, κ is directly related to the absorption coefficient α of the medium.

The relationship between α and κ can be derived by considering the propagation of plane electromagnetic waves through a medium with a complex refractive index. If the wave is propagating in the z direction, the spatial and time dependence of the electric field is given by (see eqn A.32 in Appendix A):

$$\mathcal{E}(z,t) = \mathcal{E}_0\, e^{i(kz - \omega t)}, \tag{1.15}$$

where k is the wave vector of the light and ω is the angular frequency. $\mathcal{E}_0$ is the amplitude at $z = 0$. In a non-absorbing medium of refractive index n, the wavelength of the light is reduced by a factor n compared to the free-space wavelength λ. k and ω are therefore related to each other through:

$$k = \frac{2\pi}{(\lambda/n)} = \frac{n\omega}{c}. \tag{1.16}$$

This can be generalized to the case of an absorbing medium by allowing the refractive index to be complex:

$$k = \tilde{n}\frac{\omega}{c} = (n + i\kappa)\frac{\omega}{c}. \tag{1.17}$$

On substituting eqn 1.17 into eqn 1.15, we obtain:

$$\begin{aligned}
\mathcal{E}(z,t) &= \mathcal{E}_0\, e^{i(\omega\tilde{n}z/c - \omega t)} \\
&= \mathcal{E}_0\, e^{-\kappa\omega z/c}\, e^{i(\omega n z/c - \omega t)}.
\end{aligned} \tag{1.18}$$

This shows that a non-zero extinction coefficient leads to an exponential decay of the wave in the medium. At the same time, the real part of $\tilde{n}$ still determines the phase velocity of the wave front, as in the standard definition of the refractive index given in eqn 1.2.

The optical intensity of a light wave is proportional to the square of the electric field, namely $I \propto \mathcal{E}\mathcal{E}^*$ (cf. eqn A.44). We can therefore deduce from eqn 1.18 that the intensity falls off exponentially in the medium with a decay constant equal to $2 \times (\kappa\omega/c)$. On comparing this to Beer's law given in eqn 1.4 we conclude that:

$$\alpha = \frac{2\kappa\omega}{c} = \frac{4\pi\kappa}{\lambda}, \tag{1.19}$$

where λ is the vacuum wavelength of the light. This shows that κ is directly proportional to the absorption coefficient.

We can relate the refractive index of a medium to its relative dielectric constant ϵ_r by using the standard result derived from Maxwell's equations (cf. eqn A.31 in Appendix A):

$$n = \sqrt{\epsilon_r} \, .\qquad(1.20)$$

This shows us that if n is complex, then ϵ_r must also be complex. We therefore define the **complex relative dielectric constant** $\tilde{\epsilon}_r$ according to:

$$\tilde{\epsilon}_r = \epsilon_1 + i\epsilon_2 \, .\qquad(1.21)$$

By analogy with eqn 1.20, we see that $\tilde{n}$ and $\tilde{\epsilon}_r$ are related to each other through:

$$\tilde{n}^2 = \tilde{\epsilon}_r \, .\qquad(1.22)$$

We can now work out explicit relationships between the real and imaginary parts of $\tilde{n}$ and $\tilde{\epsilon}_r$ by combining eqns 1.14, 1.21, and 1.22. These are:

$$
\begin{aligned}
\epsilon_1 &= n^2 - \kappa^2 & (1.23)\\
\epsilon_2 &= 2n\kappa \, , & (1.24)
\end{aligned}
$$

and

$$
\begin{aligned}
n &= \frac{1}{\sqrt{2}}\left(\epsilon_1 + (\epsilon_1^2 + \epsilon_2^2)^{1/2}\right)^{1/2} & (1.25)\\
\kappa &= \frac{1}{\sqrt{2}}\left(-\epsilon_1 + (\epsilon_1^2 + \epsilon_2^2)^{1/2}\right)^{1/2} \, . & (1.26)
\end{aligned}
$$

This analysis shows us that $\tilde{n}$ and $\tilde{\epsilon}_r$ are not independent variables: if we know ϵ_1 and ϵ_2 we can calculate n and κ, and vice versa. Note that, if the absorption of the medium is weak, the imaginary part of $\tilde{n}$ is small. In this limit with $n \gg \kappa$, eqns 1.23 and 1.24 imply:

$$
\begin{aligned}
n &= \sqrt{\epsilon_1} & (1.27)\\
\kappa &= \frac{\epsilon_2}{2n} \, . & (1.28)
\end{aligned}
$$

These equations show us that the refractive index is basically determined by the real part of the dielectric constant, while the absorption is mainly determined by the imaginary part. This generalization is obviously not valid if the medium has a very large absorption coefficient.

The microscopic models that we shall be developing throughout the book usually enable us to calculate $\tilde{\epsilon}_r$ rather than $\tilde{n}$. The measurable optical properties can then be obtained by converting ϵ_1 and ϵ_2 to n and κ through eqns 1.25 and 1.26. The refractive index is given directly by n, while the absorption coefficient can be worked out from κ by using eqn 1.19. The reflectivity depends on both n and κ and is given by:

$$R = \left|\frac{\tilde{n}-1}{\tilde{n}+1}\right|^2 = \frac{(n-1)^2 + \kappa^2}{(n+1)^2 + \kappa^2} \, .\qquad(1.29)$$

The relative dielectric constant is also called the **relative permittivity**.

This formula is derived in eqn A.54. It gives the coefficient of reflection between the medium and the air (or vacuum) at normal incidence.

In a transparent material such as glass in the visible region of the spectrum, the absorption coefficient is very small. Equations 1.19 and 1.24 then tell us that κ and ϵ_2 are negligible, and hence that both $\tilde{n}$ and $\tilde{\epsilon}_r$ may be taken as real numbers. This is why tables of the properties of transparent optical materials generally list only the real parts of the refractive index and dielectric constant. On the other hand, if there is significant absorption, then we shall need to know both the real and imaginary parts of $\tilde{n}$ and $\tilde{\epsilon}_r$.

In the remainder of this book we shall take it as assumed that both the refractive index and the dielectric constant are complex quantities. We shall therefore drop the tilde notation on n and ϵ_r from now on, except where it is explicitly needed to avoid ambiguity. It will usually be obvious from the context whether we are dealing with real or complex quantities.

Example 1.2

The complex refractive index of germanium at 400 nm is given by $\tilde{n} = 4.141 + \mathrm{i}\, 2.215$. Calculate for germanium at 400 nm: (a) the phase velocity of light, (b) the absorption coefficient, and (c) the reflectivity.

Solution
(a) The velocity of light is given by eqn 1.2, where n is the real part of $\tilde{n}$. Hence we obtain:

$$v = \frac{c}{n} = \frac{2.998 \times 10^8}{4.141}\, \mathrm{m\,s^{-1}} = 7.24 \times 10^7\, \mathrm{m\,s^{-1}}.$$

(b) The absorption coefficient is given by eqn 1.19. By inserting $\kappa = 2.215$ and $\lambda = 400$ nm, we obtain:

$$\alpha = \frac{4\pi \times 2.215}{400 \times 10^{-9}}\, \mathrm{m^{-1}} = 6.96 \times 10^7\, \mathrm{m^{-1}}.$$

(c) The reflectivity is given by eqn 1.29. On inserting $n = 4.141$ and $\kappa = 2.215$ into this, we obtain:

$$R = \frac{(4.141 - 1)^2 + 2.215^2}{(4.141 + 1)^2 + 2.215^2} = 47.1\,\%.$$

Example 1.3

We shall see in Chapter 10 that the Reststrahl absorption is caused by the interaction between the light and the optical phonons.

Salt (NaCl) absorbs very strongly at infrared wavelengths in the 'Reststrahl' band. The complex dielectric constant at 60 μm is given by $\tilde{\epsilon}_r = -16.8 + \mathrm{i}\, 91.4$. Calculate the absorption coefficient and the reflectivity at this wavelength.

Solution

We must first work out the complex refractive index by using eqns 1.25 and 1.26. This gives:

$$n = \frac{1}{\sqrt{2}} \left(-16.8 + ((-16.8)^2 + 91.4^2)^{1/2} \right)^{1/2} = 6.17$$

and

$$\kappa = \frac{1}{\sqrt{2}} \left(+16.8 + ((-16.8)^2 + 91.4^2)^{1/2} \right)^{1/2} = 7.41 \,.$$

We then insert these values into eqns 1.19 and 1.29 to obtain the required results:

$$\alpha = \frac{4\pi \times 7.41}{60 \times 10^{-6}} \text{ m}^{-1} = 1.55 \times 10^6 \text{ m}^{-1} \,,$$

and

$$R = \frac{(6.17 - 1)^2 + 7.41^2}{(6.17 + 1)^2 + 7.41^2} = 76.8 \, \% \,.$$

1.4   Optical materials

We shall be studying the optical properties of many different types of solid-state materials throughout this book. The materials can be loosely classified into five general categories:

- Crystalline insulators and semiconductors
- Glasses
- Metals
- Molecular materials
- Doped glasses and insulators.

Before delving into the details, we give here a brief overview of the main optical properties of these materials. This will serve as an introduction to the optical physics that will be covered in the following chapters.

1.4.1   Crystalline insulators and semiconductors

Figure 1.4(a) shows the transmission spectrum of crystalline sapphire (Al_2O_3) from the infrared to the ultraviolet spectral region. The spectrum for sapphire shows the main features observed in all insulators, although of course the details will vary considerably from material to material. The principal optical properties can be summarized as follows.

(1) Sapphire has a high transmission in the wavelength range 0.2–6 µm. This defines the **transparency range** of the crystal. The transparency region of sapphire includes the whole of the visible spectrum, which explains why it appears colourless and transparent to the human eye.

Sapphire gemstones tend to be blue. This is caused by the presence of chromium, titanium, and iron impurities in the Al_2O_3 crystal. Pure synthetic Al_2O_3 crystals are colourless.

Table 1.2 Approximate transparency range, band gap wavelength λ_g, and refractive index n of a number of common semiconductors. n is measured at $10\,\mu m$. Data from Driscoll & Vaughan (1978), Kaye & Laby (1986), and Madelung (1996).

Crystal	Range (μm)	λ_g (μm)	n
Ge	1.8 – 23	1.8	4.00
Si	1.2 – 15	1.1	3.42
GaAs	1.0 – 20	0.87	3.16
CdTe	0.9 – 14	0.83	2.67
CdSe	0.75 – 24	0.71	2.50
ZnSe	0.45 – 20	0.44	2.41
ZnS	0.4 – 14	0.33	2.20

Table 1.3 Refractive index, n, of synthetic fused silica versus wavelength. Data from Kaye & Laby (1986).

Wavelength (nm)	n
213.9	1.53430
239.9	1.51336
275.3	1.49591
334.2	1.47977
404.7	1.46962
467.8	1.46429
508.6	1.46186
546.1	1.46008
632.8	1.45702
706.5	1.45515
780.0	1.45367
1060	1.44968
1395	1.44583
1530	1.44427
1970	1.43853
2325	1.43293

which means that the whole of the transparency range lies outside the visible spectrum. Hence no visible light is transmitted through the crystal, and it has a dark metallic appearance to the eye.

Table 1.2 lists the transparency range and refractive index of several semiconductors. The data show that the lower limit of the transmission range coincides closely with the wavelength of the fundamental band gap. This happens because the band gap determines the lowest energy for interband transitions, as will be explained in Chapter 3. Note that the refractive index increases as the band gap wavelength gets larger. This is a consequence of the Kramers–Kronig relationships between the real and imaginary parts of the complex refractive index. (See Section 2.3.)

The upper limit of the transmission range is determined by the lattice absorption, as for insulators, and also by free carrier absorption. Free carriers are present in semiconductors at room temperature through the thermal excitation of electrons across the band gap or due to the presence of impurities. This causes infrared absorption, as will be explained in Section 7.4. Insulators have very small free carrier densities due to their large band gaps.

One very important aspect of the optical properties of semiconductors is that a subset of them, namely those with direct band gaps, luminesce strongly when electrons are promoted to the conduction band. This is the physical basis for the light-emitting devices used in the optoelectronics industry. The physical processes behind the luminescence will be explained in Chapter 5. The main point is that the wavelength of the luminescence coincides with the band gap of the semiconductor. In Chapter 6 we shall see how quantum size effects in low-dimensional semiconductors can be used to shift the effective band gap to higher energy. This is a highly desirable feature, because it provides a way to 'tune' the emission wavelength by controlled variation of the parameters during the crystal growth.

1.4.2   Glasses

Glasses are extremely important optical materials. They have been used for centuries in prisms and lenses for optical instruments, in addition to their common usage in windows and glassware. In more recent times they have found new applications in optical fibre technology. With the exception of stained glasses, they are usually made to be transparent in the visible spectrum. They are not crystalline solids, and therefore do not exhibit the optical anisotropy that is characteristic of some crystals.

Most types of glass are made by fusing sand (silica: SiO_2) with other chemicals. Pure fused silica is an insulator, and shows all the characteristic features of insulators discussed in the previous section. It is transparent in the visible region, but absorbs in the ultraviolet due to the electronic transitions of the SiO_2 molecules, and in the infrared due to vibrational absorption. The transparency range thus goes from around 200 nm in the ultraviolet to beyond 2000 nm in the infrared.

The properties of fused silica will be described in more detail in Sec-

Table 1.4 Composition, refractive index and ultraviolet transmission of some common glasses. The letters after the names give the abbreviations used to identify the glass type. The composition figures are the percentage by mass. The refractive index is measured at 546.1 nm, and the transmission is for a 1 cm plate at 310 nm. Data from Driscoll & Vaughan (1978), and Lide (1996).

Name	SiO_2	B_2O_3	Al_2O_3	Na_2O	K_2O	CaO	BaO	PbO	P_2O_5	n	T
Fused silica	100									1.460	0.91
Crown (K)	74			9	11	6				1.513	0.4
Borosilicate crown (BK)	70	10		8	8	1	3			1.519	0.35
Phosphate crown (PK)		3	10		12	5			70	1.527	0.46
Light flint (LF)	53			5	8			34		1.585	0.008
Flint (F)	47			2	7			44		1.607	–
Dense flint (SF)	33				5			62		1.746	–

tion 2.2.3. Fused silica is used extensively in the fibre-optics industry, as the principal material from which many fibres are made. It has been refined to such an extent that the absorption and scattering losses are so small that light can travel many kilometres down the fibre before being fully attenuated.

The refractive index of silica in the transparency range is tabulated against the wavelength in Table 1.3. This variation of the refractive index with wavelength is called dispersion. Note that it is not a very large effect: n changes by less than 1% over the whole visible spectral region. Note also that the dispersion is largest at the shortest wavelengths near the fundamental absorption edge. Dispersion is present in all optical materials, as will be explained in Section 2.4.

Chemicals are commonly added to silica during the fusion process to produce a whole range of other types of glasses. The presence of these additives can alter the refractive index and the transmission range. Table 1.4 lists the composition of a number of common glasses together with their refractive index and ultraviolet transmission. It is apparent that the additives have the effect of increasing the refractive index, at the expense of decreasing the ultraviolet transmission. A high refractive index is desirable for cut-glass products, since it increases the reflectivity (see Exercise 1.2), and hence gives the glassware a more shiny appearance. Note that the glass with the highest refractive index listed in Table 1.4 is the 'dense flint' type. This glass contains a large amount of lead, which explains why cut glass with a high degree of sparkle is rather heavy.

Coloured glass can be made by adding semiconductors with band gaps in the visible spectral region during the fusion process. The properties of these coloured glasses will be discussed further in Section 1.4.5 below.

Pure SiO_2 has a very large band gap of about 10 eV, which corresponds to a wavelength of 120 nm. The additives reduce the energy of the fundamental absorption edge, although not out of the ultraviolet spectral region. This means that the glasses are still transparent at visible wavelengths, but have reduced transmission in the ultraviolet. The reduction of the band gap increases the refractive index through the Kramers–Kronig relationship. (See Section 2.3.)

1.4.3   Metals

The characteristic optical feature of metals is that they are shiny. This is why metals like silver and aluminium have been used for making mirrors for centuries. The shiny appearance is a consequence of their very high

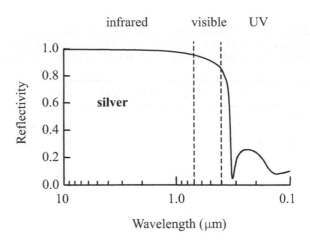

Fig. 1.5 Reflectivity of silver from the infrared to the ultraviolet. Data from Lide (1996).

reflection coefficients. We shall see in Chapter 7 that the high reflectivity is caused by the interaction of the light with the free electrons that are present in the metal.

Figure 1.5 shows the reflectivity of silver from the infrared spectral region to the ultraviolet. We see that the reflectivity is very close to 100% in the infrared, and stays above 80% throughout the whole visible spectral region. The reflectivity then drops sharply in the ultraviolet. This general behaviour is observed in all metals. There is strong reflection for all frequencies below a characteristic cut-off frequency called the plasma frequency. The plasma frequency usually corresponds to a wavelength in the ultraviolet spectral region, and so metals reflect infrared and visible wavelengths, but transmit ultraviolet wavelengths. This effect is called the ultraviolet transmission of metals.

Some metals have characteristic colours. Copper, for example, has a reddish colour, while gold is yellowish. These colours are caused by interband electronic transitions that occur in addition to the free carrier effects that cause the reflection. This point will be explained in Section 7.3.2 of Chapter 7.

1.4.4   Molecular materials

The term 'molecular material' could in principle cover the solid phase of any molecule. However, the crystalline phases of inorganic molecules such as NaCl or GaAs are classified as insulators or semiconductors in this book, while simple organic molecules such as methane (CH_4) tend to be gases or liquids at room temperature. We therefore restrict our attention here to large organic molecules.

Some organic compounds form crystals in the condensed phase, but many others are amorphous. The solids are held together by the relatively weak van der Waals interactions between the molecules, which are themselves held together by strong covalent bonds. The optical properties of the solid therefore tend to be very similar to those of the individual molecules.

Organic compounds can be generally classified into either saturated or conjugated systems. This classification depends on the type of bonding in the molecule, and will be explained in more detail in Chapter 8.

In saturated compounds, the valence electrons are incorporated into strong, localized bonds between neighbouring atoms. This means that all the electrons are tightly held in their bonds, and can only respond at high frequencies in the ultraviolet spectral range. Saturated compounds are therefore usually colourless and do not absorb in the visible region. Their properties are generally similar to those of the glasses discussed in Section 1.4.2: they absorb in the infrared and ultraviolet spectral regions due to vibrational and electronic transitions respectively, and are transparent at visible frequencies. Plastics such as poly-methyl-methacrylate (commonly known as 'perspex' or 'plexiglass') or poly-ethylene (polythene) are typical examples.

Conjugated molecules, by contrast, have much more interesting optical properties. The electrons from the p-like atomic states of the carbon atoms form large delocalized orbitals called π orbitals which spread out across the whole molecule. The standard example of a conjugated molecule is benzene (C_6H_6), in which the π electrons form a ring-like orbital above and below the plane of the carbon and hydrogen atoms. Further examples include the other aromatic hydrocarbons, dye molecules, and conjugated polymers.

π electrons are less tightly bound than the electrons in saturated molecules, and interact with light at lower frequencies. In benzene the absorption edge is in the ultraviolet spectral region at 260 nm, but with other molecules the transition energy is shifted down to visible frequencies. The molecules with visible absorption also tend to emit strongly at visible frequencies. This makes them of high technological interest for applications as light-emitting devices. These are the solid state counterparts of the organic dyes that have been used in liquid lasers for several decades.

The optical processes that occur in π conjugated materials will be described in Chapter 8. By way of example, Fig. 1.6 shows the absorption spectrum of the technologically important polyfluorene-based polymer called 'F8'. Thin film samples of this material are typically prepared by spin coating the molecules onto a glass slide. The data in Fig. 1.6 show that the polymer is transparent throughout most of the visible spectral region, but absorbs strongly at ultraviolet wavelengths. The broad absorption band which peaks at 380 nm is caused by vibrational–electronic transitions to the first singlet excited state of the molecule. This band extends slightly into the blue spectral region, and gives the material a pale yellow colour.

Conjugated polymers such as F8 luminesce strongly when electrons are promoted into the excited states of the molecule. The luminescence is Stokes-shifted to lower energy compared to the absorption, and typically occurs in the middle of the visible spectral region. An attractive feature of these organic materials is that the emission wavelength can be 'tuned'

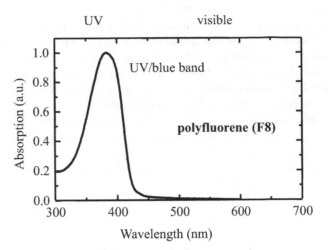

Fig. 1.6 Absorption spectrum of the polyfluorene-based polymer F8 [poly(9,9-dioctylfluorene)]. After Buckley et al. (2001), © Excerpta Medica Inc., reprinted with permission.

by small alterations to the chemical structure of the molecular units within the polymers. We shall see in Section 8.4 how this property has been used to develop organic light-emitting devices to cover the full range of the visible spectral region.

1.4.5 Doped glasses and insulators

We have already mentioned in Section 1.4.2 that coloured glass can be made by adding appropriately chosen semiconductors to silica during the fusion process. This is a typical example of how a colourless material such as fused silica can take on new properties by controlled doping with optically-active substances.

The colour of doped glass can be controlled in two different ways.

(1) The most obvious way is by variation of the composition of the dopant. For example, the glass might be doped with the alloy semiconductor $Cd_xZn_{1-x}Se$ during the fusion process, with the value of x determined by the ZnSe:CdSe ratio in the original melt. The band gap of the alloy can be 'tuned' through the visible spectrum region by varying x, and this determines the short wavelength transmission cut-off for the glass.

(2) The size of the semiconductor crystallites within the glass can be very small, and this can also have an effect on the colour produced. Normally, the optical properties of a material are independent of the size of the crystal, but this ceases to be the case if the dimensions are comparable to the electron wavelength. The 'quantum size effect' increases the energy of the electrons and hence shifts the effective band gap to higher energy. This point will be explained further in Section 6.8 of Chapter 6.

The principle of doping optically-active atoms into colourless hosts is employed extensively in the crystals used for solid-state lasers. A typical

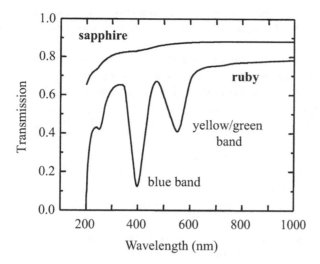

Fig. 1.7 Transmission spectrum of ruby (Al_2O_3 with 0.05% Cr^{3+}) compared to sapphire (pure Al_2O_3). The thicknesses of the two crystals were 6.1 mm and 3.0 mm respectively. After McCarthy (1967), reprinted with permission.

example is the ruby crystal. Rubies consist of Cr^{3+} ions doped into Al_2O_3 (sapphire). In the natural crystals, the Cr^{3+} ions are present as impurities, but in synthetic crystals, the dopants are deliberately introduced in controlled quantities during the crystal growth process.

Figure 1.7 compares the transmission spectra of synthetic ruby (Al_2O_3 with 0.05% Cr^{3+}) to that of synthetic sapphire (pure Al_2O_3). It is apparent that the presence of the chromium ions produces two strong absorption bands, one in the blue spectral region and the other in the yellow/green region. These two absorption bands give rubies their characteristic red colour. The other obvious difference between the two transmission curves is that the overall transmission of the ruby is lower. This is caused in part by the increased scattering of light by the impurities in the crystal.

The optical properties of crystals like ruby will be covered in Chapter 9. We shall see there that the broadening of the discrete transition lines of the isolated dopant ions into absorption bands is caused by vibronic coupling between the valence electrons of the dopant and the phonons in the host crystal. We shall also see how the centre wavelength of the bands is affected by the crystal-field effect, that is, the interaction between the dopant ions and electric field of the host crystal. These properties are very important in the design of solid-state lasers and phosphors.

1.5 Characteristic optical physics in the solid state

The previous section has given a brief overview of the optical properties of several different classes of solid-state materials. It is natural to ask whether any of these properties are exclusive to the solid state. In other words, how do the optical properties of a solid differ from those of its constituent atoms or molecules? This question is essentially the same as asking what the difference is between solid-state and atomic or molecular physics.

The answer clearly depends on the type of material that we are considering. In some materials there will be a whole range of new effects associated with the solid state, while with others, the differences may not be so great. Molecular materials are an example of the second type. We would expect the absorption spectra of a solid film and that of an equivalent dilute solution to be fairly similar. This happens because the forces between the molecules in the condensed phase are relatively weak compared to the forces within the molecule itself. The appeal of the solid state in this case is the high number density of molecules that are present, and the possibility of incorporating them into solid-state electronic devices.

With many other materials, however, there will be substantial differences between the condensed phase and the gaseous or liquid state. It is obviously not possible to give a full catalogue of these effects in an introductory chapter such as this one. Instead, we highlight here five aspects that make the physics of the solid state interesting and different, namely

- Crystal symmetry
- Electronic bands
- Vibronic bands
- The density of states
- Delocalized states and collective excitations.

There are many others, of course, but these themes occur over and over again and are therefore worth considering briefly in themselves before we start going into the details.

1.5.1 Crystal symmetry

Most of the materials that we shall be studying occur as crystals. Crystals have long-range **translational order**, and can be categorized into 32 classes according to their **point group symmetry**. The point group symmetry refers to the group of symmetry operations that leaves the crystal invariant. Examples of these include rotations about particular axes, reflections about planes, and inversion about points in the unit cell. Some crystal classes such as the cubic ones possess a very high degree of symmetry. Others have much lower symmetry.

The link between the measurable properties and the point group symmetry of a crystal can be made through **Neumann's principle**. This states that:

> *Any macroscopic physical property must have at least the symmetry of the crystal structure.*

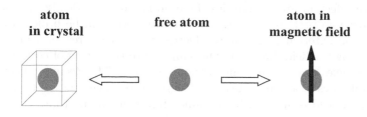

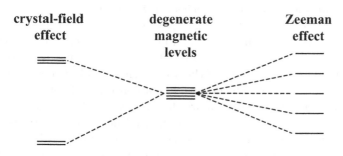

Fig. 1.8 Splitting of the magnetic levels of a free atom by the crystal-field effect. In the free atoms, the magnetic levels are degenerate. We must apply a magnetic field to split them by the Zeeman effect. However, the magnetic levels can be split even without applying an external magnetic field in a crystal. The details of the way the levels split are determined by the symmetry class of the crystal.

For example, if a crystal has four-fold rotational symmetry about a particular axis, then we must get the same result in any experiment we might perform in the four equivalent orientations.

It is instructive to compare the properties of a crystal to those of the atoms from which it has been formed. A gas of atoms has no translational order. Therefore we expect to find new effects in the solid state that reflect its translational symmetry. The formation of electronic bands and delocalized states discussed in Sections 1.5.2 and 1.5.5 below are examples of this. At the same time, the point group symmetry of a crystal is lower than that of the individual atoms, which have the highest possible symmetry due to their spherical invariance. We therefore expect to find other effects in the solid state that relate to the lowering of the symmetry on going from free atoms to the particular point group of the crystal class. Two specific examples of this are discussed briefly here, namely **optical anisotropy** and the **lifting of degeneracies**.

A crystal is said to be anisotropic if its properties are not the same in all directions. Anisotropy is only found in the solid state, because gases and liquids do not have any preferred directions. The degree of anisotropy found in a crystal depends strongly on the point group symmetry that it possesses. In cubic crystals, for example, the optical properties must be the same along the x, y, and z axes because they are physically indistinguishable. On the other hand, in a uniaxial crystal, the properties along the optic axis will be different from those along the axes at right angles to it. The optical anisotropy is manifested by the property of birefringence which is discussed in Section 2.5.1. It is also important for the description of the nonlinear optical coefficients of crystals discussed in Chapter 11.

The lifting of degeneracies by reduction of the symmetry is a well-known effect in atomic physics. Free atoms are spherically symmetric and have no preferred directions. The symmetry can be broken by applying an external magnetic or electric field which creates a preferred axis along the field direction. This can lead to the lifting of certain level degeneracies that are present in the free atoms. The Zeeman effect, for example, describes the splitting of degenerate magnetic levels when a magnetic field is applied. If the same atom is introduced into a crystal, it will find itself in an environment with a point group symmetry determined by the lattice. This symmetry is lower than that of the free atom, and therefore some level degeneracies can be lifted.

This point is illustrated schematically in Fig. 1.8, which shows how the magnetic levels of a free atom can be split by the crystal-field effect in an analogous way to the Zeeman effect. The splitting is caused by the interaction of the orbitals of the atoms with the electric fields of the crystalline environment. The details do not concern us here. The important point is that the splittings are determined by the symmetry class of the crystal and do not require an external field. Optical transitions between these crystal-field split levels often occur in the visible spectral region, and cause the material to have very interesting properties that are not found in the free atoms. These effects will be explored in more detail in Chapter 9.

Before closing this section on crystal symmetry, it is worth pointing out that many important solid-state materials do not possess long-range translational symmetry. Glass is an obvious example. Other examples include thin molecular films such as light-emitting polymers sputtered onto substrates, and amorphous silicon. The optical properties of these materials may be very similar to those of their constituent atoms or molecules. Their importance is usually related to the convenience of the solid phase rather than to new optical properties that relate to the solid-state physics.

1.5.2   Electronic bands

The atoms in a solid are packed very close to each other, with the inter-atomic separation approximately equal to the size of the atoms. Hence the outer orbitals of the atoms overlap and interact strongly with each other. This broadens the discrete levels of the free atoms into bands, as illustrated schematically in Fig. 1.9.

The electron states within the bands are delocalized and possess the translational invariance of the crystal. **Bloch's theorem** states that the wave functions should be written in the form:

$$\psi_{\boldsymbol{k}}(\boldsymbol{r}) = u_{\boldsymbol{k}}(\boldsymbol{r}) \exp(\mathrm{i}\boldsymbol{k} \cdot \boldsymbol{r}) \,, \tag{1.30}$$

where $u_{\boldsymbol{k}}(\boldsymbol{r})$ is a function that has the periodicity of the lattice. The Bloch states described by eqn 1.30 are modulated plane waves. Each electronic band has a different function $u_{\boldsymbol{k}}(\boldsymbol{r})$ which retains some of the atomic character of the states from which the band was derived.

Optical transitions can occur between the electronic bands if they are allowed by the selection rules. This 'interband' absorption is possible over a continuous range of photon energies determined by the lower and upper energy limits of the bands. This contrasts with the absorption spectra of free atoms, which consist of discrete lines. The observation of broad bands of absorption rather than discrete lines is one of the characteristic features of the solid state.

Interband transitions will be discussed at length in a number of chapters in this book, most notably Chapters 3 and 5. The absorption strength is usually very high because of the very large density of absorbing atoms in the solid. This means that we can produce sizeable optical effects in very thin samples, allowing us to make the compact optical devices that form the basis of the modern opto-electronics industry.

1.5.3 Vibronic bands

The electronic states of the atoms or molecules in a solid may be strongly coupled to the vibrational modes of the crystal through the vibronic interaction. A typical example of where this effect occurs is the doped insulator crystals introduced in Section 1.4.5. The vibronic coupling broadens the discrete electronic states of the isolated dopant atoms into bands. This has the effect of broadening the absorption and emission lines of the atoms into continuous bands. These vibronic effects will be described in more detail in Chapter 9.

It is important to realize that the reason for the formation of the vibronic bands is different to that for the electronic bands considered in the previous section. In the case of vibronic bands, the continuum of states arises from the coupling of discrete electronic states to a continuous spectrum of vibrational modes (i.e. phonons). This contrasts with the electronic bands, where the continuum arises from interactions between electronic states of neighbouring atoms.

Vibronic effects are also observed in molecular materials. This is an interesting case which highlights the difference between the solid state and the liquid or gaseous phase. The absorption spectra of simple free molecules also show vibrational–electronic bands, but the transition frequencies are discrete because both the electronic energies and the vibrational energies are discrete. In molecular solids, by contrast, the vibrational frequencies are continuous, and this causes continuous absorption and emission spectra.

1.5.4 The density of states

The concept of the **density of states** is an inevitable corollary of band formation in solids. The electronic and vibrational states of free molecules and atoms have discrete energies, but this is not the case in a solid: both the electronic states and the phonon modes have a continuous range of energies. This continuum of states leads to continuous absorption and emission bands, as has already been stressed in the previous

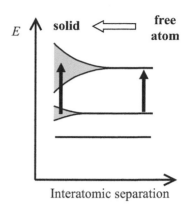

Fig. 1.9 Schematic diagram of the formation of electronic bands in a solid from the condensation of free atoms. As the atoms are brought closer together to form the solid, their outer orbitals begin to overlap with each other. These overlapping orbitals interact strongly, and broad bands are formed. The inner core orbitals do not overlap and so remain discrete even in the solid state. Optical transitions between the bands can occur, and this causes absorption over a continuous range of frequencies rather than discrete lines.

two sections.

The number of states within a given energy range of a band is conveniently expressed in terms of the density of states function $g(E)$. This is defined as:

$$\text{Number of states in the range } E \to (E + dE) = g(E)\, dE . \qquad (1.31)$$

$g(E)$ is usually worked out by first calculating the density of states in momentum space $g(k)$, and then using the relationship between $g(E)$ and $g(k)$, namely:

$$g(E) = g(k)\frac{dk}{dE} . \qquad (1.32)$$

This can be evaluated from knowledge of the E–k relationship for the electrons or phonons. Knowledge of $g(E)$ is crucial for calculating the absorption and emission spectra due to interband transitions and also for calculating the shape of vibronic bands.

1.5.5 Delocalized states and collective excitations

The fact that the atoms in a solid are very close together means that it is possible for the electron states to spread over many atoms. The wave functions of these delocalized states possess the underlying translational symmetry of the crystal. The Bloch waves described by eqn 1.30 are a typical example. The delocalized electron waves move freely throughout the whole crystal and interact with each other in a way that is not possible in atoms. The delocalization also allows collective excitations of the whole crystal rather than individual atoms. Two examples that we shall consider in this book are the excitons formed from delocalized electrons and holes in a semiconductor, and the plasmons formed from free electrons in metals and doped semiconductors. The excitonic effects will be discussed in Chapter 4, while plasmons are covered in Section 7.5. The collective excitations may be observed in the optical spectra, and have no obvious counterpart in the spectra of free atoms.

Other wave-like excitations of the crystal are delocalized in the same way as the electrons. In the case of the lattice vibrations, the delocalized excitations are described by the phonon modes. We have already mentioned above that the phonon frequencies are continuous, which contrasts with the discrete vibrational frequencies of molecules. Some optical effects related to phonons have direct analogies with the vibrational phenomena observed in isolated molecules, but others are peculiar to the solid state. Examples of the former are Raman scattering and infrared absorption. Examples of the latter include the phonon-assisted interband transitions in semiconductors with indirect band gaps (cf. Section 3.4), and the broadening of the discrete levels of impurity atoms into continuous vibronic bands by interactions with phonons as discussed in Chapter 9.

The delocalized states of a crystal are described by quantum numbers such as $\boldsymbol{k}$ and $\boldsymbol{q}$ which have the dimensions of inverse length. These quantum numbers follow from the translational invariance, and are therefore a fundamental manifestation of the crystal symmetry. To all intents and purposes, k and q behave like the wave vectors of the excitations, and they will be treated as such whenever we encounter them in derivations. However, it should be borne in mind that this is really a consequence of the deep underlying symmetry which is unique to the solid state.

1.6 Microscopic models

In the following chapters we shall be developing many microscopic models to explain the optical phenomena that are observed in the solid state. The types of models will obviously vary considerably, but they can all be classified into one of the following three general categories:

- Classical
- Semi-classical
- Fully quantum.

These approaches get progressively more difficult, and so we usually apply them in the order listed above.

In the classical approach we treat both the medium and the light according to classical physics. The dipole oscillator model described in Chapter 2 is a typical example. This model is the basic starting point for understanding the general optical properties of a medium, and in particular for describing the main effects due to free electrons (Chapter 7) and phonons (Chapter 10). We shall also use it as a starting point for the discussion of nonlinear optics in Chapter 11. It would be a mistake to undervalue the classical approach in this modern day and age. The value of more sophisticated models will only be appreciated fully once the classical physics has been properly understood.

In semi-classical models we apply quantum mechanics to the atoms, but treat the light as a classical electromagnetic wave. The treatment of interband absorption in Chapter 3 is a typical example. The absorption coefficient is calculated using Fermi's golden rule, which requires knowledge of the wave functions of the quantized levels of the atoms, but treats the light–matter interaction as that between a quantized atom and a classical electric field. This semi-classical approach is used extensively throughout the book. Appendix B summarizes the main results that will be needed.

The final approach is the full quantum treatment. This is the realm of **quantum optics**, where both the atoms and the light are treated quantum mechanically. We use this approach implicitly whenever we refer to the light as a beam of photons and draw Feynman diagrams to represent the interaction processes that are occurring. This might give the impression that the explanations we are giving are fully quantum because we speak in terms of photons interacting with atoms. However, in the equations used to describe the processes, the light is treated classically and only the atoms are quantized. The quantitative description is therefore only semi-classical. The use of the fully quantum approach at the quantitative level is beyond the scope of the present book.

Chapter summary

- The propagation of light though a medium is quantified by the complex refractive index $\tilde{n}$. The real part of $\tilde{n}$ determines the velocity of light in the medium, while the imaginary part determines the absorption coefficient. Beer's law (eqn 1.4) shows that the intensity of light in an absorbing medium decays exponentially.

- Reflection occurs at the interface between two optical materials with different refractive indices. The coefficient of reflectivity can be calculated from the complex refractive index by using eqn 1.29.

- The transmission of a sample is determined by the reflectivities of the surfaces and the absorption coefficient. If the light is incoherent, the transmission coefficient of a plate is given by eqn 1.6.

- The complex refractive index is related to the complex dielectric constant through eqn 1.22. Microscopic models of optical materials normally calculate $\tilde{\epsilon}_r$ rather than $\tilde{n}$, and the measurable optical coefficients are evaluated by working out the real and imaginary parts of $\tilde{n}$ from eqns 1.23–1.28.

- Luminescent materials re-emit light by spontaneous emission after absorbing photons. The frequency shift between the emission and absorption is called the Stokes shift.

- Scattering causes an exponential attenuation of the optical beam. The scattering is called elastic if the frequency is unchanged, and inelastic otherwise.

- The optical spectra of solid-state materials usually consist of broad bands rather than sharp lines. The bands arise either from electronic interactions between neighbouring atoms or from vibronic coupling to the phonon modes.

- Insulators and glasses have vibrational absorption at infrared wavelengths and electronic absorption in the ultraviolet spectral region. They are transparent and colourless in the visible spectral region between these two absorption bands. In semiconductors and molecular materials the electronic absorption usually occurs at lower frequencies in the near infrared or visible spectral region.

- The free carriers present in metals make them highly reflective in the infrared and visible spectral regions. The colouration of some metals is caused by electronic interband absorption.

- The addition of optically active dopants to a colourless host crystal or glass produces the characteristic colours of coloured glasses and gemstones.
- Crystals have both translational symmetry and point group symmetry. The consequences of the point group symmetry for the optical properties are determined by Neumann's principle.

Further reading

A good general discussion of the optical properties of materials can be found in Hecht (2001). A more advanced treatment may be found in Born and Wolf (1999). The discussion of the optical properties of different materials given in Section 1.4 will be expanded in subsequent chapters, where suitable further reading will be suggested. The exception is the discussion of glasses, which are only covered briefly elsewhere in this book. A more detailed discussion of the optical properties of different types of glass may be found in Krause (2005) or Bach and Neuroth (1995).

The relationship between the optical properties and the complex refractive index and dielectric constant is discussed in most texts on electromagnetism, for example, Bleaney and Bleaney (1976), or Lorrain, Corson, and Lorrain (2000). This material is also covered in Born and Wolf (1999).

A classic discussion of the effects of the point group symmetry on the physical properties of crystals is given in Nye (1985).

Exercises

(1.1) Crown glass has a refractive index of 1.51 in the visible spectral region. Calculate the reflectivity of the air–glass interface, and the transmission of a typical glass window.

(1.2) Use the data in Table 1.4 to calculate the ratio of the reflectivities of fused silica and dense flint glass.

(1.3) The complex dielectric constant of the semiconductor cadmium telluride is given by $\tilde{\epsilon}_r = 8.92 + i\,2.29$ at 500 nm. Calculate for CdTe at this wavelength: the phase velocity of light, the absorption coefficient and the reflectivity.

(1.4) The detectors used in optical fibre networks operating at 850 nm are usually made from silicon, which has an absorption coefficient of $1.3 \times 10^5 \, m^{-1}$ at this wavelength. The detectors have coatings on the front surface that makes the reflectivity at the design wavelength negligibly small. Calculate the thickness of the active region of a photodiode designed to absorb 90% of the light.

(1.5) GaAs has a refractive index of 3.68 and an absorption coefficient of $1.3 \times 10^6 \, m^{-1}$ at 800 nm. Calculate the transmission coefficient and optical density of a GaAs plate of thickness 2 μm.

(1.6) Sea water has a refractive index of 1.33 and absorbs 99.8% of red light of wavelength 700 nm in a depth of 10 m. What is its complex dielectric constant at this wavelength?

(1.7) How would you expect the absorption coefficient of a yellow colour-glass filter to vary with wavelength?

(1.8) A beam of light is incident on a parallel-sided plate of thickness l as shown in Fig. 1.10. We assume that the plate is 'thick', so that l exceeds the coherence length of the light and interference effects do not have to be considered. Let R_1 and R_2 be respectively the reflectivities of the front and back surfaces, and α be the absorption coefficient of the medium.
(a) By adding the intensities of the beams transmitted after multiple reflections to the intensity of the beam transmitted after the first pass, show

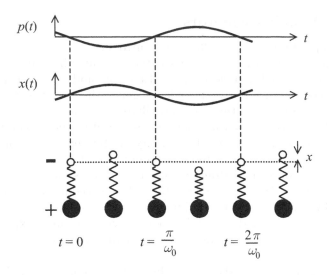

Fig. 2.2 Oscillations of a classical dipole consisting of a heavy positive charge and a light negative charge bound together by a spring. $x(t)$ is the time-dependent displacement of the negative charge from its equilibrium position. The natural vibrations of the dipole about the equilibrium position at angular frequency ω_0 generate a time-dependent dipole moment $p(t)$ as indicated in the top part of the figure.

as

$$\boldsymbol{p} = q(\boldsymbol{r}_+ - \boldsymbol{r}_-)\,. \tag{2.3}$$

Hence the positive nucleus and negative electron form a dipole with magnitude equal to $e|\boldsymbol{r}_N - \boldsymbol{r}_e|$.

During the oscillations of the atomic dipole, the nucleus remains more or less stationary due to its heavy mass, while the electron oscillates backwards and forwards at angular frequency ω_0. Hence the oscillations produce a time-varying dipole in addition to any permanent dipole the atom might have. The magnitude of the time-varying dipole is given by:

$$p(t) = -ex(t)\,, \tag{2.4}$$

where $x(t)$ is the time-varying displacement of the electron from its equilibrium position. This connection between the electron displacement and the time-dependent atomic dipole is illustrated in the top half of Fig. 2.2. The oscillating dipole behaves like a tiny aerial and radiates electromagnetic waves at angular frequency ω_0, in accordance with the theory of classical Hertzian dipoles. Hence the atom is expected to radiate light at its resonant frequency whenever sufficient energy is imparted to excite the oscillations.

We assume here that the forces exerted by the electric fields are very small compared to the binding forces that hold the electrons to the nucleus. This approximation may not be valid if we are using a very powerful laser beam to excite the medium. If this were the case, then we would be working in the regime of nonlinear optics. These effects are considered in Chapter 11.

We can also use the dipole model to understand how the atom interacts with an external electromagnetic wave at angular frequency ω. The AC electric field exerts forces on the electron and the nucleus and drives oscillations of the system at frequency ω. If ω coincides with one of the natural frequencies of the atom, then we have a resonance phenomenon. This induces very large amplitude oscillations, and transfers energy from the external wave to the atom. The atom can therefore absorb energy from the light wave if $\omega = \omega_0$. The absorption strength is characterized by the absorption coefficient α, and the intensity of the wave will decay exponentially according to Beer's law (eqn 1.4).

We now know from quantum theory that what actually happens during absorption is that the atom jumps to an excited state by absorbing a photon. This can only occur if $\hbar\omega = E_2 - E_1$, where E_1 and E_2 are the quantized energies of the initial and final states. Once it has been excited, the atom can return to the ground state by a series of radiationless transitions, in which case the energy from the absorbed photon is ultimately converted into heat. Alternatively, it can luminesce by re-emitting a photon at some later time. The re-radiated photons are incoherent with each other and are emitted in all directions rather than in the specific direction of the incoming wave. Hence there is a net decrease in the energy flow in the beam direction, which is equivalent to absorption.

If ω does not coincide with any of the resonant frequencies, then the atoms will not absorb, and the medium will be transparent. In this situation the light wave drives non-resonant oscillations of the atoms at its own frequency ω. The oscillations of the atoms follow those of the driving wave, but with a phase lag. This phase lag is a standard feature of forced oscillators and is caused by damping. (See Exercise 2.2.) The oscillating atoms all re-radiate instantaneously, but the phase lag acquired in the process accumulates through the medium and retards the propagation of the wave front. This implies that the propagation velocity is smaller than in free space. The reduction of the velocity in the medium is characterized by the refractive index defined in eqn 1.2.

The slowing of the wave due to the non-resonant interactions can be considered as a repeated scattering process. The scattering is both coherent and elastic, and each atom behaves like a Huygens point source. The scattered light interferes constructively in the forward direction, and destructively in all other directions, so that the direction of the beam is unchanged by the repetitive scattering process. However, each scattering event introduces a phase lag which causes a slowing of the propagation of the phase front through the medium.

2.1.2   Vibrational oscillators

An optical medium may contain other types of dipole oscillators in addition to those originating from the bound electrons within the atoms. If the medium is ionic, it will contain oppositely charged ions. Vibrations of these charged atoms from their equilibrium positions within the crystal lattice will produce an oscillating dipole moment, in exactly the same way as the oscillations of the electrons within the individual atoms that we considered above. Therefore, we must also consider the optical effects due to these vibrational oscillators when we consider the interaction of light with an ionic optical medium.

The optical effects of vibrational oscillators are well known in molecular physics. Figure 2.3 gives a schematic illustration of a classical polar molecule. This consists of two charged atoms bound together in a stable configuration, with the spring representing the molecular bond between them. The charged atoms can vibrate about their equilibrium

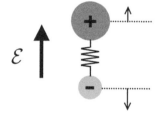

Fig. 2.3 Classical model of a polar molecule. The atoms are positively and negatively charged, and can vibrate about their equilibrium separation. These vibrations produce an oscillating electric dipole which will radiate electromagnetic waves at the resonant frequency. Alternatively, the molecule will interact with the electric field $\mathcal{E}$ of a light wave through the forces exerted on the charged atoms.

positions and induce an oscillating electric dipole in an analogous way
to the bound electrons in the atoms. We see immediately from eqn 2.2
that the vibrations will occur at lower frequencies because the reduced
mass is larger. The vibrations therefore occur at infrared frequencies
with $\omega_0/2\pi \sim 10^{12}$–$10^{13}$ Hz. These molecular vibrations are associated
with strong absorption lines in the infrared spectral region.

The interaction between the vibrations of the molecule and the light
wave occurs through the forces exerted on the atoms by the electric
field. It is obvious that this can only happen if the atoms inside the
molecule are charged. This is why we specified that the molecule was
polar in the preceding paragraph. A polar molecule is one in which
the charge cloud of the electrons that form the bond sits closer to one
of the atoms than to the other. Ionic molecules like the alkali halides
(e.g. Na^+Cl^-) clearly fall into this category, while purely covalent ones
such as the elemental molecules (e.g. O_2) do not. Many other molecules
fall somewhere between these two limits. Water (H_2O) is a well-known
example. Oxygen has a greater electron affinity than hydrogen, and so
the valence electrons in the O–H bond sit closer to the oxygen atoms.
The two hydrogen atoms therefore possess a small positive charge which
is balanced by a negative charge of twice the magnitude on the oxygen
atom.

In a crystalline solid formed from the condensation of polar molecules,
the atoms are arranged in an alternating sequence of positive and negative ions. The ions can vibrate about their equilibrium positions, and
this produces oscillating dipole waves. These oscillations are associated
with **lattice vibrations**, and they occur at frequencies in the infrared
spectral region. We shall consider the optical properties related to the
lattice vibrations in detail in Chapter 10. We shall see there that the
light–matter interaction is associated with the excitation of **phonons**,
which are quantized lattice waves. At this stage, we simply note that
the lattice vibrations of a polar crystal give rise to strong optical effects
in the infrared spectral region. These effects occur in addition to those
due to the bound electrons of the atoms that comprise the crystal. In
practice we can treat these two types of dipoles separately because the
resonances occur at very different frequencies. Therefore the resonant
effects of the bound electrons are negligible at the frequencies of the
lattice vibrations, and vice versa. This point will be considered in more
detail in Section 2.2.2.

2.1.3   Free electron oscillators

The electronic and vibrational dipoles considered above are both examples of bound oscillators. Metals and doped semiconductors, by contrast,
contain significant numbers of **free electrons**. As the name implies,
these are electrons that are not bound to any atoms, and therefore do
not experience any restoring forces when they are displaced. This implies
that the spring constant in eqn 2.2 is zero, and hence that the natural
resonant frequency $\omega_0 = 0$.

The free electron model of metals is attributed to Paul Drude, and so the application of the dipole oscillator model to free electron systems is generally called the **Drude–Lorentz model**. The dipole oscillator model is perfectly valid, except that we must set $\omega_0 = 0$ throughout. The optical properties of free electron systems will be discussed in Chapter 7.

2.2   The dipole oscillator model

In the previous section we introduced the general assumptions of the dipole oscillator model. We now want to use the model to calculate the frequency dependence of the refractive index and absorption coefficient. This will provide a simple explanation for the dispersion of the refractive index in optical materials. It will also illustrate the general point, which will be developed further in Section 2.3, that the phenomena of absorption and refraction are related to each other.

2.2.1   The Lorentz oscillator

We consider the interaction between a light wave and an atom with a single resonant frequency ω_0 due to the bound electrons, as given by eqn 2.2. We model the displacement of the atomic dipoles as damped harmonic oscillators. The inclusion of damping is a consequence of the fact that the oscillating dipoles can lose their energy by collisional processes. In solids, this would typically occur through an interaction with a phonon which has been thermally excited in the crystal. As we shall see, the damping term has the effect of reducing the peak absorption coefficient and broadening the absorption line.

The electric field of the light wave induces forced oscillations of the atomic dipole through the driving forces exerted on the electrons. We make the reasonable assumption that the nuclear mass is much greater than the electron mass (i.e. $m_N \gg m_0$), so that we can ignore the motion of the nucleus. The displacement x of the electron is governed by an equation of motion of the form:

$$m_0 \frac{d^2x}{dt^2} + m_0\gamma\frac{dx}{dt} + m_0\omega_0^2 x = -e\mathcal{E},\tag{2.5}$$

where γ is the damping rate, e is the magnitude of the electric charge of the electron, and $\mathcal{E}$ is the electric field of the light wave. The terms on the left-hand side represent the acceleration, the damping and the restoring force respectively. The damping is modelled by a frictional force which is proportional to the velocity and impedes the motion. The term on the right-hand side represents the driving force due to the AC electric field of the light wave.

We consider the interaction of the atom with a monochromatic light wave of angular frequency ω. The time dependence of the electric field is given by

$$\mathcal{E}(t) = \mathcal{E}_0 \cos(\omega t + \Phi) = \mathcal{E}_0 \,\Re\left(e^{-i(\omega t + \Phi)}\right),\tag{2.6}$$

We know from experimental observations that atoms must have many natural resonant frequencies to account for the multiplicity of lines in the absorption and emission spectra. However, the salient features of the physical behaviour are well illustrated by a singly resonant system, and the inclusion of multiple resonances complicates the discussion without adding much to the physical understanding at this stage. We therefore postpone the discussion of the effects of multiple resonances to Section 2.2.2.

In some conventions, the time dependence is written as $e^{j\omega t}$ instead of $e^{-i\omega t}$. This makes no physical difference. Consistency can be obtained by replacing $-i$ with $+j$ throughout.

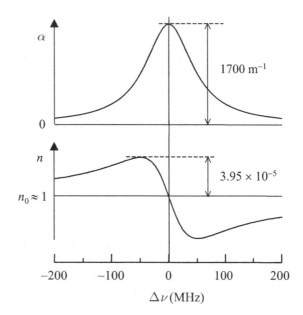

Fig. 2.5 Absorption coefficient and refractive index of the atomic gas considered in Example 2.1. n_0 represents the off-resonant refractive index, which is approximately equal to unity.

This gives $\Delta\omega = \pm\gamma/2$. We see from Fig. 2.4 that $\Delta\omega = -\gamma/2$ corresponds to the local maximum, while $\Delta\omega = +\gamma/2$ corresponds to the local minimum. Therefore the peak in the refractive index occurs 50 MHz below the line centre.

(c) From part (b) we know that the local maximum in the refractive index occurs when $\Delta\omega = -\gamma/2$. We see from eqns 1.27 and 2.20 that the refractive index at this frequency is given by:

$$n_{\max} = \sqrt{\epsilon_1} = \left(\epsilon_\infty + \frac{Ne^2}{2\epsilon_0 m_0 \omega_0 \gamma}\right)^{1/2} = n_0\left(1 + \frac{7.90 \times 10^{-5}}{n_0^2}\right)^{1/2},$$

where $n_0 = \sqrt{\epsilon_\infty}$ is the off-resonant refractive index. We are dealing with a low-density gas, and so it is justified to take $n_0 \approx 1$ here. This implies that the peak value of the resonant contribution to the refractive index is 3.95×10^{-5}.

The full frequency dependence of the absorption and refractive index near this absorption line is plotted in Fig. 2.5.

2.2.2   Multiple resonances

In general, an optical medium will have many characteristic resonant frequencies. We already discussed in Section 2.1 how we expect to observe separate resonances due to the lattice vibrations and to the oscillations of the bound electrons within the atoms. Furthermore, a particular medium may have many resonances of each type. We can treat these multiple resonances without difficulty in our model provided they occur at distinct frequencies.

In writing eqn 2.12 we split the polarization of the medium into a resonant part and a non-resonant part. We then discussed the resonant part in detail, without specifying very accurately what we meant by the 'non-resonant' term. We simply stated that P was proportional to $\mathcal{E}$ through the susceptibility χ. In reality, the non-resonant polarization of the medium must originate from the polarizability of the atoms in exactly the same way as the resonant part. Equation 2.19 tells us that the dielectric constant decreases each time we go through an absorption line. The contributions that enter the background electric susceptibility χ in eqn 2.12 thus arise from the polarization due to all the other oscillators at higher frequencies.

We can understand this point better by making it more quantitative. The contribution to the polarization of a particular oscillator is given by eqn 2.10. In a medium with many electronic oscillators of different frequencies, the total polarization will therefore be given by

$$P = \left(\frac{Ne^2}{m_0} \sum_j \frac{1}{(\omega_j^2 - \omega^2 - \mathrm{i}\gamma_j\omega)} \right) \mathcal{E} \,, \qquad (2.22)$$

where ω_j and γ_j are the angular frequency and damping coefficient of a particular resonance line. We then substitute this into eqn 2.11, and recall the definition of ϵ_r given in eqn 2.13, to obtain:

$$\epsilon_\mathrm{r}(\omega) = 1 + \frac{Ne^2}{\epsilon_0 m_0} \sum_j \frac{1}{(\omega_j^2 - \omega^2 - \mathrm{i}\gamma_j\omega)} \,. \qquad (2.23)$$

This equation takes account of all the transitions in the medium and can be used to calculate the full frequency dependence of the dielectric constant.

The refractive index and absorption coefficient calculated from eqn 2.23 are plotted against frequency in Fig. 2.6. The figure has been calculated for a hypothetical solid with three well-separated resonances with ω_j equal to 4×10^{13} rad/s, 4×10^{15} rad/s, and 1×10^{17} rad/s respectively. The width of each absorption line has been set to 10% of the centre frequency by appropriate choice of the damping coefficient. The resonance in the infrared is included to represent the vibrational absorption. In a real solid, we would have to adapt the model appropriately to account for the different reduced mass and effective charge of the vibrational oscillator.

We can understand this figure by starting at the highest frequencies and gradually working our way down to the lower frequencies. At the very highest frequencies, the electrons are too sluggish to respond to the driving field. The medium therefore has no polarization, and the dielectric constant is unity. As we reduce the frequency, we first run into the transitions of the inner electrons in the X-ray/vacuum-ultraviolet spectral region, and then the transitions of the outer electrons in the ultraviolet and visible. We then have a region with no transitions until we finally reach the vibrational frequencies in the infrared. Each time we

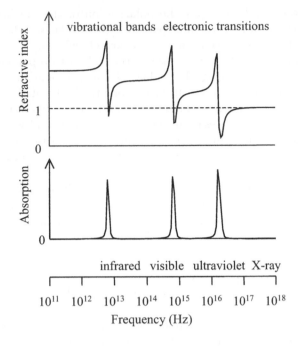

Fig. 2.6 Schematic diagram of the frequency dependence of the refractive index and absorption of a hypothetical solid from the infrared to the X-ray spectral region. The solid is assumed to have three resonant frequencies with $\omega_j = 4 \times 10^{13}$ rad/s, 4×10^{15} rad/s, and 1×10^{17} rad/s respectively. The width of each absorption line has been set to 10% of the centre frequency by appropriate choice of γ_j.

An astute reader will have noticed that the peak absorption coefficient for the three transition lines shown in Fig. 2.6 decreases slightly with decreasing frequency. This happens because n is larger at the lower frequencies. The transitions all have the same peak ϵ_2, but we can see from eqn 1.24 that κ must be slightly smaller if n is larger.

go through one of these resonances, we see the characteristic frequency dependence of the Lorentz oscillator, with a peak in the absorption spectrum and a 'wiggle' in the refractive index. In between the resonances the medium is transparent: the absorption coefficient is zero and the refractive index is almost constant.

The value of the refractive index in the transparent regions gradually increases as we go through more and more resonance lines on decreasing the frequency. This increase of the refractive index is caused by the fact that $\epsilon_{\mathrm{st}} > \epsilon_\infty$ (cf. eqn 2.19), which implies that n is larger below an absorption line than above it. By reference to Fig. 2.6, we now see that we have to understand 'static' and '∞' as relative to a particular resonance. The variation of n with frequency due to the resonances is the origin of the dispersion found in optical materials even when they are transparent. This point will be discussed further in Section 2.4.

The dipole oscillator model predicts that each oscillator contributes a term given by eqn 2.10. This leads to a series of absorption lines of the same strength. However, experimental data shows that the absorption strength actually varies considerably between different atomic transitions. With the benefit of hindsight, we know that this is caused by the variation of the quantum mechanical transition probability. (See Appendix B.) In classical physics, however, there is no explanation, and we just assign a phenomenological **oscillator strength** f_j to each transition, rewriting eqn 2.23 as:

$$\epsilon_{\mathrm{r}}(\omega) = 1 + \frac{Ne^2}{\epsilon_0 m_0} \sum_j \frac{f_j}{\left(\omega_{0j}^2 - \omega^2 - \mathrm{i}\gamma_j\omega\right)} \,. \tag{2.24}$$

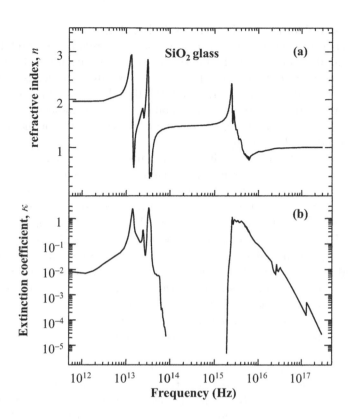

Fig. 2.7 (a) Refractive index and (b) extinction coefficient of fused silica (SiO_2) glass from the infrared to the x-ray spectral region. After Palik (1985).

It can be shown from quantum mechanics that we must have $\sum_j f_j = 1$ for each electron. Since the classical model predicts $f_j = 1$ for each oscillator, we then interpret this by saying that a particular electron is involved in several transitions at the same time, and the absorption strength is being divided between these transitions.

2.2.3 Comparison with experimental data

The schematic behaviour shown in Fig. 2.6 can be compared to experimental data on a typical optical material. Figure 2.7 shows the frequency dependence of the refractive index and extinction coefficient of fused silica (SiO_2) glass from the infrared to the X-ray spectral region. The general characteristics indicated by Fig. 2.6 are clearly observed, with strong absorption in the infrared and ultraviolet, and a broad region of low absorption in between. The data confirms that $n \gg \kappa$ except near the peaks of the absorption. This means that the approximation whereby we associate the frequency dependence of n with that of ϵ_1, and that of κ with ϵ_2 (eqns 1.27 and 1.28), is valid at most frequencies.

The general behaviour shown in Fig. 2.7 is typical of optical materials which are transparent in the visible spectral region. We already noted in Sections 1.4.1 and 1.4.2 that the transmission range of colourless materials is determined by the electronic absorption in the ultraviolet and the vibrational absorption in the infrared. This is demonstrated by the transmission data for sapphire shown in Fig. 1.4(a).

Fused silica is a glass, and hence does not have a regular crystal lattice. The infrared absorption is therefore caused by excitation of vibrational quanta in the SiO_2 molecules themselves. Two distinct peaks are observed at 1.4×10^{13} Hz (21 µm) and 3.3×10^{13} Hz (9.1 µm) respectively. These correspond to different vibrational modes of the molecule. The detailed modelling of these absorption bands by the oscillator model will be discussed in Chapter 10.

The ultraviolet absorption in silica is caused by interband electronic transitions. SiO_2 has a fundamental band gap of about 10 eV, and interband transitions are possible whenever the photon energy exceeds this value. Hence we observe an absorption threshold in the ultraviolet at 2×10^{15} Hz (150 nm). The interband absorption peaks at around 3×10^{15} Hz with an extremely high absorption coefficient of $\sim 10^8$ m^{-1}, and then gradually falls off to higher frequency. Subsidiary peaks are observed at $\sim 3 \times 10^{16}$ Hz and 1.3×10^{17} Hz. These are caused by transitions of the inner core electrons of the silicon and oxygen atoms. The fact that the electronic absorption consists of a continuous band rather than a discrete line makes it hard to model accurately as a Lorentz oscillator. We shall discuss the quantum theory of the interband absorption in Chapter 3.

The refractive index of glass has resonances in the infrared and the ultraviolet which correspond respectively to the vibrational and interband absorption bands. In the far infrared region below the vibrational resonance, the refractive index is ~ 2, while in the hard ultraviolet and X-ray region it approaches unity. In the transparency region between the vibrational and interband absorption, the refractive index has a value of ~ 1.5. Closer inspection of Fig. 2.7 shows that the refractive index actually increases with frequency in this transparency region, rising from a value of 1.40 at 8×10^{13} Hz (3.5 µm) to 1.55 at 1.5×10^{15} Hz (200 nm). This dispersion originates from the low-frequency wings of the ultraviolet absorption and the high-frequency wings of the infrared absorption, and will be discussed in more detail in Section 2.4.

The data in Fig. 2.7 show that the refractive index falls below unity at a number of frequencies. This implies that the phase velocity of the light is greater than c, which might seem to imply a contradiction with relativity. However, this overlooks the fact that a signal must be transmitted as a wave packet rather than as a monochromatic wave. In a dispersive medium, a wave packet will propagate at the **group velocity** v_g given by:

$$v_g = \frac{d\omega}{dk},\qquad(2.25)$$

rather than at the phase velocity $v = \omega/k = c/n$. The relationship between v_g and v is:

$$v_g = v\left(1 + \frac{\omega}{n}\frac{dn}{d\omega}\right)^{-1} = v\left(1 - \frac{\lambda}{n}\frac{dn}{d\lambda}\right)^{-1},\qquad(2.26)$$

where λ is the vacuum wavelength of the light. The derivation of this

It is apparent from Fig. 2.6 that $dn/d\omega$ will be negative at some frequencies close to a resonance line. Equation 2.26 then implies that $v_g > v$, and so we could again run into a problem with relativity. However, the medium is highly absorbing in these frequency regions, and this means that the signal travels with yet another velocity called the signal velocity. This is always less than c.

result is left as an exercise to the reader. (See Exercise 2.7.) We shall see in Section 2.4 that $dn/d\omega$ is positive in most materials at optical frequencies. This then implies that v_g is always less than v, and if we were to try to transmit a signal in a spectral region where $v > c$, we would always find that v_g is less than c. The proof of this for a simple Lorentz oscillator is considered in Exercise 2.8.

2.2.4 Local field corrections

The calculation of the dielectric constant given in eqn 2.24 is valid in a rarefied gas with a low density of atoms. However, in a dense optical medium such as a solid, there is another factor that we must consider. The individual atomic dipoles respond to the **local field** that they experience. This may not necessarily be the same as the external field, because the dipoles themselves generate electric fields that will be felt by all the other dipoles. The actual local field experienced by an atom therefore takes the form:

$$\mathcal{E}_{local} = \mathcal{E} + \mathcal{E}_{other\ dipoles}, \qquad (2.27)$$

where $\mathcal{E}$ and $\mathcal{E}_{other\ dipoles}$ represent the fields due to the external field and the other dipoles respectively. We should have been using $\mathcal{E}_{local}$ instead of $\mathcal{E}$ all along throughout the calculation given in Sections 2.2.1 and 2.2.2.

The calculation of the correction field due to the other dipoles in the medium is actually a rather complicated one. An approximate solution due to Lorentz can be derived if we assume that all the dipoles are parallel to the applied field and are arranged on a cubic lattice. The calculation works by separating the contribution from the nearby dipoles and that from the rest of the sample, as indicated in Fig. 2.8. The division is effected by an imaginary spherical surface with a radius large enough to make it sensible to average the material outside it. The problem is then reduced to summing the field of the dipoles inside the sphere at the one in the middle, and then calculating the effect of a uniformly polarized dielectric outside the sphere. The final result is:

$$\mathcal{E}_{other\ dipoles} = \frac{P}{3\epsilon_0}, \qquad (2.28)$$

where P is the polarization of the dielectric outside the sphere. The derivation of this result is the subject of Exercise 2.9. By using the result of eqn 2.28 in eqn 2.27 we find that:

$$\mathcal{E}_{local} = \mathcal{E} + \frac{P}{3\epsilon_0}. \qquad (2.29)$$

The macroscopic polarization P will be given by

$$P = N\epsilon_0\chi_a\mathcal{E}_{local}, \qquad (2.30)$$

where χ_a is the electric susceptibility per atom. χ_a is defined by:

$$p = \epsilon_0\chi_a\mathcal{E}_{local}, \qquad (2.31)$$

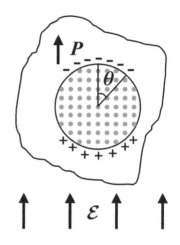

Fig. 2.8 Model used to calculate the local field by the Lorentz correction. An imaginary spherical surface drawn around a particular atom divides the medium into nearby dipoles and distant dipoles. The field at the centre of the sphere due to the nearby dipoles is summed exactly, while the field due to the distant dipoles is calculated by treating the material outside the sphere as a uniformly polarized dielectric.

p being the induced dipole moment per atom. This is analogous to the usual definition of the macroscopic susceptibility given in eqn A.1, except that it is now applied to individual atoms interacting with the local field. We can see from eqn 2.10 that χ_a is given by

$$\chi_a = \frac{e^2}{\epsilon_0 m_0} \frac{1}{(\omega_0^2 - \omega^2 - i\gamma\omega)} , \qquad (2.32)$$

if there is just a single resonance. This is modified to

$$\chi_a = \frac{e^2}{\epsilon_0 m_0} \sum_j \frac{f_j}{(\omega_j^2 - \omega^2 - i\gamma_j\omega)} , \qquad (2.33)$$

if there are multiple resonances (cf. eqn 2.24).

We can combine eqns 2.29 and 2.30 with eqns 2.11 and 2.13 by writing

$$\boldsymbol{P} = N\epsilon_0\chi_a \left(\boldsymbol{\mathcal{E}} + \frac{\boldsymbol{P}}{3\epsilon_0}\right) = (\epsilon_r - 1)\epsilon_0\boldsymbol{\mathcal{E}} . \qquad (2.34)$$

We put all this together to find that:

$$\frac{\epsilon_r - 1}{\epsilon_r + 2} = \frac{N\chi_a}{3} . \qquad (2.35)$$

This result is known as the **Clausius–Mossotti relationship**. The relationship works well in gases and liquids. It is also valid for those crystals in which the Lorentz correction given in eqn 2.29 gives an accurate account of the local field effects, namely cubic crystals.

2.3   The Kramers–Kronig relationships

The discussion of the dipole oscillator shows that the refractive index and the absorption coefficient are not independent parameters but are related to each other. This is a consequence of the fact that they are derived from the real and imaginary parts of a single parameter, namely the complex refractive index. If we invoke the law of causality (that an effect may not precede its cause) and apply complex number analysis, we can derive general relationships between the real and imaginary parts of the refractive index. These are known as the **Kramers–Kronig relationships** and may be stated as follows:

The derivation of the Kramers–Kronig relationships may be found, for example, in Dressel and Grüner (2002). The principal part of an integral that has a divergence at c within the integration range $a \to b$ is defined as:

$$P\int_a^b f(x)\,dx =$$

$$\lim_{\delta\to 0}\left(\int_a^{c-\delta} f(x)\,dx + \int_{c+\delta}^b f(x)\,dx\right) .$$

$$n(\omega) - 1 = \frac{2}{\pi} P\int_0^\infty \frac{\omega'\kappa(\omega')}{\omega'^2 - \omega^2}\,d\omega' \qquad (2.36)$$

$$\kappa(\omega) = -\frac{2}{\pi\omega} P\int_0^\infty \frac{\omega'^2[n(\omega') - 1]}{\omega'^2 - \omega^2}\,d\omega', \qquad (2.37)$$

where P indicates that we take the principal part of the integral.

The Kramers–Kronig relationships allow us to calculate n from κ, and vice versa. This can be very useful in practice, because it would allow us, for example, to measure the frequency dependence of the optical absorption and then calculate the dispersion without needing to make a separate measurement of n.

As examples of the way the Kramers–Kronig relationships are used, we consider here two points raised in Section 1.4 of Chapter 1.

(1) In the discussion of the data in Table 1.2 in Section 1.4.1, we noted that the refractive index of a semiconductor in its transparency range tends to increase with the band gap wavelength. Example 2.2 below shows that the refractive index at frequencies well below an absorption band centred at ω_0 has a contribution that varies as $1/\omega_0$. In the case of a semiconductor, we can put $\omega_0 \approx E_g/\hbar = 2\pi c/\lambda_g$, which then implies that the refractive index should increase linearly with λ_g. Figure 2.9 plots the refractive index data from Table 1.2 against the band gap wavelength. It is apparent that the relationship between n and λ_g is indeed approximately linear, which confirms the basic point that the higher frequency absorption in larger band gap materials gives a smaller contribution to the refractive index through the ω'^2 term in the denominator of eqn 2.36. The fact that the graph does not extrapolate to $n = 1$ at $\lambda_g = 0$ indicates that the narrow absorption band assumption of Example 2.2 should not be taken too far, and that there are other contributions to the refractive index that have not been considered here.

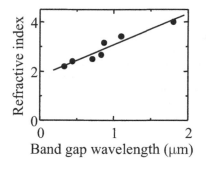

Fig. 2.9 Refractive index measured at $10\,\mu m$ against band gap wavelength for several different semiconductors. The data are taken from Table 1.2.

(2) In the discussion of the data in Table 1.4 in Section 1.4.2, we noted that the addition of compounds such as PbO to SiO_2 glass tends to reduce the frequency of the fundamental absorption edge and increase the refractive index in the visible spectral region. We can explain this by using a very simple model of the glass in which we assume that the material has an absorption band that runs from photon energy E_1 to E_2. If we assume that the absorption has a constant value of α_0 throughout the band, a Kramers–Kronig analysis shows that the refractive index at a photon energy E below the absorption edge at E_1 is given by (see Exercise 2.12):

$$n(E) = 1 + \frac{c\hbar\alpha_0}{2\pi E} \ln \frac{(E_2 - E)(E_1 + E)}{(E_2 + E)(E_1 - E)}. \quad (2.38)$$

For pure SiO_2, the value of E_1 is about $10\,eV$, which corresponds to a wavelength of $120\,nm$. The additives reduce E_1, which has the effect of increasing n through the relatively small $(E_1 - E)$ term in the denominator.

Example 2.2

A solid has a single absorption band of width γ centred at frequency ω_0 such that the extinction coefficient may be written in the form:

$$\kappa(\omega) = \kappa_0, \quad \omega_0 - \gamma/2 \le \omega \le \omega_0 + \gamma/2,$$
$$= 0 \quad \text{otherwise}.$$

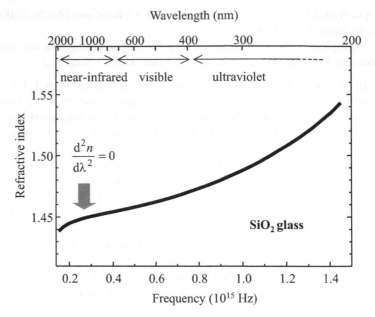

Fig. 2.10 Refractive index of SiO_2 glass in the near infrared, visible, and ultraviolet spectral regions. The thick arrow indicates the frequency at which the group velocity dispersion is zero. After Palik (1985).

Calculate the refractive index at low frequencies on the assumption that $\omega_0 \gg \gamma$.

Solution

We are asked to compute the refractive index at low frequencies, which means that we can assume $\omega_0 \gg \omega$, and therefore put $\omega = 0$ in the Kramers–Kronig relationships. On substituting into eqn 2.36 and using the fact that $\kappa(\omega)$ is zero at all frequencies apart from those within the absorption band, we can write:

$$n(0) = 1 + \frac{2}{\pi} \int_{\omega_0 - \gamma/2}^{\omega_0 + \gamma/2} \frac{\kappa_0}{\omega'} \, d\omega'.$$

Since $\omega_0 \gg \gamma$, we can take $\omega' = \text{constant} = \omega_0$ in the denominator of the integrand, and then obtain:

$$n(0) = 1 + \frac{2}{\pi} \frac{\kappa_0}{\omega_0} \times \gamma = 1 + \frac{2\gamma\kappa_0}{\pi\omega_0}.$$

2.4   Dispersion

Figure 2.10 plots the refractive index data from Fig. 2.7 in more detail. The data show that the refractive index increases with frequency in the near infrared and visible spectral regions. We have seen in Section 2.2.3 that this dispersion originates mainly from the interband absorption in the ultraviolet spectral region. At visible frequencies the absorption from these transitions is negligible and the glass is transparent. However, the ultraviolet absorption still affects the refractive index through the extreme wings of the Lorentzian line. In the near infrared, the dispersion is also affected by the high-frequency wings of the vibrational absorption at lower frequency.

A material in which the refractive index increases with frequency is said to have **normal** dispersion, while one in which the contrary occurs is said to have **anomalous** dispersion. A number of empirical formulae to describe the normal dispersion of glasses have been developed over the years. (See Exercise 2.13.)

The dispersion of the refractive index of glasses such as silica can be used to separate different wavelengths of light with a prism, as shown in Fig. 2.11. The blue light is refracted more because of the higher index of refraction, and is therefore deviated through a larger angle by the prism. (See Exercise 2.14.) This effect is used in prism spectrometers.

One of the effects of dispersion is that light of different frequencies takes a different amount of time to propagate through a material. A pulse of light of duration t_{p} must necessarily contain a spread of frequencies given approximately by

$$\Delta \nu \approx \frac{1}{t_{\mathrm{p}}} \qquad (2.39)$$

in order to satisfy the 'uncertainty principle' $\Delta\nu\Delta t \sim 1$. Dispersion will therefore cause the pulse to broaden in time as it propagates through the medium, due to the different velocities of the frequency components of the pulse. This can become a serious problem when attempting to transmit very short pulses through a long length of an optical material, for example in a high speed optical fibre telecommunications system.

We mentioned in Section 2.2.3 that a pulse of light travels with the group velocity v_{g}. The important parameter for pulse spreading due to dispersion is therefore the **group velocity dispersion** (GVD). It is shown in Exercise 2.15 that the temporal broadening of a pulse due to group velocity dispersion is proportional to the second derivative of the refractive index with respect to the vacuum wavelength, and it is therefore useful to define a **material dispersion parameter** as follows:

$$D = -\frac{\lambda}{c} \frac{\mathrm{d}^2 n}{\mathrm{d}\lambda^2}\,. \qquad (2.40)$$

This is usually quoted in units of $\mathrm{ps\,nm^{-1}\,km^{-1}}$, so that the temporal broadening $\Delta\tau$ of a pulse is given by:

$$\Delta\tau \ (\mathrm{ps}) = |D|\,\Delta\lambda\,L\,, \qquad (2.41)$$

where $\Delta\lambda$ is the spectral width of the pulse measured in nm and L is the length of the medium in km.

The Lorentz model indicates that, within the normal dispersion region, $\mathrm{d}^2 n/\mathrm{d}\lambda^2$ is negative for frequencies below an absorption line and positive above it. On applying this to the data in Fig. 2.10, we have a positive material dispersion parameter in the infrared due to the vibrational absorption and a negative one in the visible due to the interband

The use of the words 'normal' and 'anomalous' is somewhat misleading here. The dipole oscillator model shows us that all materials have anomalous dispersion at some frequencies. The phraseology was adopted before measurements of the refractive index had been made over a wide range of frequencies and the origin of dispersion had been properly understood.

Fig. 2.11 Separation of white light into different colours by dispersion in a glass prism.

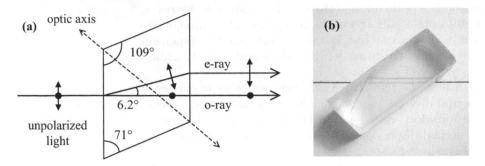

Fig. 2.12 (a) Double refraction in a calcite crystal. The shape of the crystal and the orientation of the optic axis is determined by the natural cleavage planes of calcite. An unpolarized incident light ray is split into two spatially separated, orthogonally polarized rays. The • symbol for the o-ray indicates that it is polarized with its field pointing out of the page. (b) An example of double refraction in a birefringent crystal. The single line underneath the crystal appears double, due to the separation of the o- and e-rays. Photo courtesy of S. Collins.

absorption in the ultraviolet. These two effects cancel at a wavelength in the near infrared, when the curvature of the graph of n against ω changes from negative to positive. This region of zero group velocity dispersion occurs around 1.3 µm in silica optical fibres, and is identified in Fig. 2.10. Short pulses can be transmitted down the fibre with very small temporal broadening at this wavelength, and so it is one of the preferred wavelengths for optical fibre communication systems.

2.5   Optical anisotropy

For liquids and gases, it is reasonable to assume that the optical properties are isotropic, i.e. the same in all directions. This will also be the case for glasses and amorphous materials, which have no preferred physical axes. However, it will not, in general, be a valid assumption for crystals, which have well-defined axes arising from their structure. In this section we shall discuss the effects of anisotropy on optical materials, beginning with the natural anisotropy found in crystals, and then moving on to induced anisotropy caused by strain and external fields.

2.5.1   Natural anisotropy: birefringence

The atoms in a solid are locked into a crystalline lattice with well-defined axes. In general, we cannot assume that the optical properties along the different crystalline axes are equivalent. For example, the separation of the atoms might not be the same in all directions. This would lead to different vibrational frequencies, and hence a change in the refractive index between the relevant directions. Alternatively, the molecules locked into the lattice might preferentially absorb certain polarizations of light.

One of the most clear manifestations of optical anisotropy is the phenomenon of **birefringence**, which can be observed in transparent, anisotropic crystals. We can describe the properties of a birefringent crystal by generalizing the relationship between the polarization and the applied electric field. If the electric field is applied along an arbitrary direction relative to the crystalline axes, we must write a tensor equation to relate P to $\mathcal{E}$:

$$P = \epsilon_0 \chi \mathcal{E} \qquad (2.42)$$

where χ represents the **susceptibility tensor**. Written explicitly in terms of the components, we have:

$$P_i = \epsilon_0 \sum_j \chi_{ij} \mathcal{E}_j . \qquad (2.43)$$

This can be conveniently expressed in terms of matrices as:

$$\begin{pmatrix} P_x \\ P_y \\ P_z \end{pmatrix} = \epsilon_0 \begin{pmatrix} \chi_{11} & \chi_{12} & \chi_{13} \\ \chi_{21} & \chi_{22} & \chi_{23} \\ \chi_{31} & \chi_{32} & \chi_{33} \end{pmatrix} \begin{pmatrix} \mathcal{E}_x \\ \mathcal{E}_y \\ \mathcal{E}_z \end{pmatrix} . \qquad (2.44)$$

We can simplify the form of χ by choosing the Cartesian coordinates x, y, and z to correspond to the principal axes of the crystal. In this case, the off-diagonal components are zero, and the susceptibility tensor takes the form:

$$\chi = \begin{pmatrix} \chi_{11} & 0 & 0 \\ 0 & \chi_{22} & 0 \\ 0 & 0 & \chi_{33} \end{pmatrix} . \qquad (2.45)$$

The relationships between the components are determined by the crystal symmetry.

- In cubic crystals, the x, y, and z axes are indistinguishable. They therefore have $\chi_{11} = \chi_{22} = \chi_{33}$, and their optical properties are isotropic.

- Crystals with tetragonal, hexagonal, or trigonal (rhombohedral) symmetry are called **uniaxial** crystals. These crystals possess a single **optic axis**, which is usually taken to be the z axis. In hexagonal crystals, for example, the optic axis is defined by the direction normal to the plane of the hexagons. The optical properties are the same along the x and y directions, but not along the z direction. This implies that $\chi_{11} = \chi_{22} \neq \chi_{33}$. Some examples of uniaxial crystals are listed in Table 2.1.

- Crystals with orthorhombic, monoclinic, or triclinic symmetry are called **biaxial** crystals. They have two optic axes, and all three diagonal components of the susceptibility tensor are different. Mica is an important example of a biaxial crystal.

One very striking demonstration of birefringence is the phenomenon of **double refraction**. In this effect an unpolarized light ray is separated into two rays which emerge displaced from each other, as shown in Fig. 2.12. These two rays are called 'ordinary' and 'extraordinary', and are orthogonally polarized to each other. It is apparent from Fig 2.12

Equation 2.42 should be contrasted with the usual *scalar* relationship between P and $\mathcal{E}$ namely (cf. eqn A.1):

$$P = \epsilon_0 \chi \mathcal{E} ,$$

which only applies to isotropic materials.

Crystals with cubic symmetry are only isotropic as regards their *linear* optical properties. We shall see in Chapter 11 that cubic crystals can actually have anisotropic *nonlinear* optical properties. This is possible because the nonlinear properties are described by *second*-rank tensors (e.g. $\chi_{ijk}^{(2)}$), rather than the *first*-rank tensors (e.g. χ_{ij}) used for the linear optical properties. See Nye (1985) for further details.

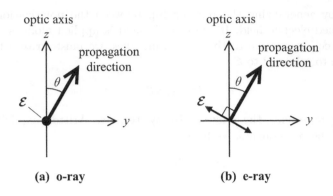

Fig. 2.13 Electric field vector of a ray propagating in a uniaxial crystal with its optic axis along the z direction. The propagation direction is determined by the direction of the Poynting vector. The ray makes an angle of θ with respect to the optic axis. The x and y axes are chosen so that the beam is propagating in the y, z plane. The polarization can be resolved into: (a) a component along the x axis and (b) a component at an angle of $90° - \theta$ to the optic axis. (a) is the o-ray and (b) is the e-ray.

that the e-ray does not obey Snell's law of refraction, which is why it is called 'extraordinary'.

The phenomenon of double refraction can be explained by assuming that the crystal has different refractive indices for the orthogonal polarizations of the ordinary and extraordinary rays. These two refractive indices are usually labelled n_o and n_e respectively. Consider the propagation of a beam of unpolarized light that enters a uniaxial crystal at an angle θ to the optic axis, which is taken to lie along the z axis. The optical properties are isotropic in the x, y plane, and so we can choose the axes so that the beam is propagating in the y, z plane without loss of generality, as shown in Fig. 2.13. This allows us to split the polarization of the light into two orthogonal components, one of which is polarized along the x axis, and the other polarized at an angle of $(90° - \theta)$ to the optic axis. The former is the o(rdinary)-ray, and the latter is the e(xtraordinary)-ray. Now the refractive index will be different for light which is polarized along the z axis or in the x, y plane. Therefore the o-ray, which has no polarization component along the z axis, experiences a different refractive index to the e-ray, which has a component along z. The two rays will thus be refracted differently: hence double refraction. On the other hand, if the beam propagates along the optic axis so that $\theta = 0$, the $\mathcal{E}$-vector of the light will always fall in the x, y plane. In this case, no double refraction will be observed because the x and y directions are equivalent.

Double refraction was first observed in natural uniaxial crystals such as calcite ('Iceland Spar'). Table 2.1 lists the refractive indices for the o- and e-rays of calcite, together with those of several other uniaxial crystals. The birefringent crystals are classified as being either positive or negative depending on whether n_e is larger or smaller than n_o.

Uniaxial birefringent crystals find widespread application in making optical components to control the polarization state of light. Figure 2.14 illustrates the operating principle of a **Glan–Foucault polarizing prism**. The polarizer consists of two identical birefringent prisms mounted with an air gap between them, with their optic axes in the plane of the front surface. Unpolarized light at the input surface of the prism is split into the o-ray and e-ray, which are then incident at the intermediate

Table 2.1 Refractive indices of some common uniaxial crystals at 589.3 nm. Data from Driscoll & Vaughan (1978).

Crystal	Chemical structure	Symmetry class	Type	n_o	n_e
Ice	H_2O	trigonal	positive	1.309	1.313
Quartz	SiO_2	trigonal	positive	1.544	1.553
Beryl	$Be_3Al_2(SiO_3)_6$	hexagonal	negative	1.581	1.575
Sodium nitrate	$NaNO_3$	trigonal	negative	1.584	1.336
Calcite	$CaCO_3$	trigonal	negative	1.658	1.486
Tourmaline	complex silicate	trigonal	negative	1.669	1.638
Sapphire	Al_2O_3	trigonal	negative	1.768	1.760
Zircon	$ZrSiO_4$	tetragonal	positive	1.923	1.968
Rutile	TiO_2	tetragonal	positive	2.616	2.903

air interface. For the case of positive uniaxial crystals, the apex angle θ of the prisms is chosen so that the o-ray suffers total internal reflection, but not the e-ray. (See Exercise 2.17.) The light that emerges at the output of the polarizer therefore only consists of the e-ray, and is linearly polarized. The Glan–Foucault prism therefore turns unpolarized light into linearly polarized light. The **Glan–Thompson polarizing prism** is a variation of the Glan–Foucault prism, with optical cement in the gap between the two birefringent prisms. This improves the acceptance angle of the polarizer (i.e the tolerance to the angle of incidence of the input beam) at the expense of reducing the optical damage threshold. This makes the Glan–Thompson prism more useful for general purpose optics, but less useful when working with high power lasers.

Figure 2.15 illustrates another important optical component that exploits birefringence, namely the **retarder plate**. The retarder is made by cutting a uniaxial birefringent crystal so that the optic axis lies in the plane of the input surface of the plate. Figure 2.15 shows a retarder plate with a linearly polarized input beam. As the beam propagates through the crystal, it can be resolved into an o-ray and e-ray, as shown in Fig. 2.15(b). The two rays experience different refractive indices, and therefore propagate with different phase velocities. This introduces a relative phase difference (or 'retardation') between the o- and e-rays. The magnitude $\Delta\phi$ of the phase difference is given by:

$$\Delta\phi = \frac{2\pi|n_o - n_e|d}{\lambda} = \frac{2\pi|\Delta n|d}{\lambda}, \tag{2.46}$$

where d is the thickness of the retarder plate, λ is the vacuum wavelength of the light, and Δn is the difference between the ordinary and extraordinary refractive indices. The thickness of the plate is usually chosen so that $\Delta\phi$ equals either $\pi/2$ or π. For the case of $\Delta\phi = \pi/2$, the phase difference is equivalent to a quarter of a wave, and so the retarder is called a **quarter-wave plate**. For similar reasons, a plate with $\Delta\phi = \pi$ is called a **half-wave plate**. A quarter-wave plate will turn linearly-polarized light into circularly-polarized light, and vice versa, while a half-wave plate will rotate the polarization of linearly polarized light.

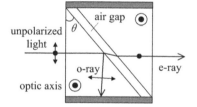

Fig. 2.14 Glan–Foucault polarizing prism. The optic axes of the crystals point vertically out of the page. In the case shown here, the crystals are assumed to have positive birefringence. The role of the o- and e-rays are reversed if the crystals have negative birefringence.

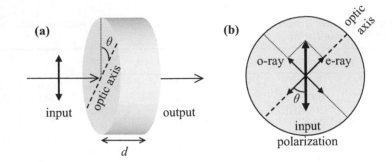

Fig. 2.15 (a) A birefringent optical retarder plate with linear input polarization. The optic axis of the birefringent crystal lies in the plane of the input surface. (b) Input surface of the retarder, showing the resolution of the input polarization into the o- and e-rays. θ is the angle between the input polarization and the optic axis, and d is the thickness of the plate.

(See Exercise 2.18.)

Further discussion of the detailed effects of birefringence may be found in most optics textbooks. The main purpose of discussing birefringence here is to illustrate the phenomenon of optical anisotropy and make the point that it arises from the underlying symmetry of the crystal structure. This is a very standard example of an optical effect that occurs in crystalline solids and is not found in gases or liquids.

Example 2.3

The optic axis of a uniaxial crystal lies along the z axis. The refractive index for light polarized in the z direction is n_e, while that for light polarized in the x, y plane is n_o. Write down the dielectric constant tensor defined through the tensor relationship

$$\boldsymbol{D} = \epsilon_0 \epsilon_r \boldsymbol{\mathcal{E}} \, .$$

Solution

We make use of eqns 2.11 and 2.42 to write:

$$\begin{aligned} \boldsymbol{D} &= \epsilon_0 \boldsymbol{\mathcal{E}} + \boldsymbol{P} \\ &= \epsilon_0 \boldsymbol{\mathcal{E}} + \epsilon_0 \chi \boldsymbol{\mathcal{E}} \\ &= \epsilon_0 \left(1 + \chi \right) \boldsymbol{\mathcal{E}} \equiv \epsilon_0 \epsilon_r \boldsymbol{\mathcal{E}} \, . \end{aligned} \tag{2.47}$$

Hence we see that:

$$\epsilon_r = 1 + \chi \, . \tag{2.48}$$

The susceptibility tensor is given by eqn 2.45, and hence the dielectric constant tensor will take the form:

$$\epsilon_r = \begin{pmatrix} 1 + \chi_{11} & 0 & 0 \\ 0 & 1 + \chi_{22} & 0 \\ 0 & 0 & 1 + \chi_{33} \end{pmatrix} \, . \tag{2.49}$$

In a uniaxial crystal with the optic axis along the z direction, we must have $\chi_{11} = \chi_{22} \neq \chi_{33}$.

We now further assume that the crystal is transparent, so that the dielectric constant is just equal to the square of the refractive index (cf. eqns 1.27 and 1.28 with $\kappa = 0$). If we had a linearly polarized light beam with the electric field directed along the x or y directions, we would measure a refractive index of n_{o}. This tells us that

$$1 + \chi_{11} = 1 + \chi_{22} = n_{\text{o}}^2 \, .$$

On the other hand, if $\mathcal{E}$ is along the z axis, we would measure a refractive index of n_{e}, which implies that

$$1 + \chi_{33} = n_{\text{e}}^2 \, .$$

Therefore the dielectric constant tensor must be:

$$\epsilon_{\mathbf{r}} = \begin{pmatrix} n_{\text{o}}^2 & 0 & 0 \\ 0 & n_{\text{o}}^2 & 0 \\ 0 & 0 & n_{\text{e}}^2 \end{pmatrix} \, . \tag{2.50}$$

Example 2.4

Calculate the thickness of a quartz half-wave plate designed for a wavelength of 589 nm.

Solution

The thickness can be calculated by substituting the refractive index data from Table 2.1 into eqn 2.46. With $n_{\text{o}} = 1.544$ and $n_{\text{e}} = 1.553$, we have $|\Delta n| = 0.009$. In a half-wave plate we require $\Delta \phi = \pi$, and so we can find d by solving:

$$\Delta \phi = \pi = \frac{2\pi |\Delta n| d}{\lambda} \, .$$

This gives $d = \lambda / 2 |\Delta n| = 0.033$ mm.

A quartz plate of thickness 0.033 nm would be very fragile, and so optical companies frequently design their wave plates to have a retardation of $(2\pi m + \Delta \phi)$, where m is an integer. This makes no difference at the design wavelength, and allows more practical thicknesses to be used.

2.5.2 Induced optical anisotropy

Isotropic materials such as liquids, gases, and glasses are not birefringent. However, the application of external perturbations can break the symmetry, thereby producing birefringence. This gives rise to a range of induced optical phenomena associated with strain and electric fields. Note that the application of a magnetic field induces optical activity rather than birefringence, and so it must be treated differently. (See Section 2.6 below.)

The most obvious way to break the symmetry of an isotropic medium is to compress it in one direction. The resulting strain-induced birefringence is called the **photo-elastic effect**. The effect can readily be observed by placing a piece of stressed glass between crossed polarizers.

In the absence of strain, the glass should have no effect on the polarization of the light, and so there should be no transmission through the second polarizer. However, if stress is present, the polarization vector will be altered, and some light will be transmitted. This method is in fact used to detect strain in glasses and other isotropic optical materials.

Birefringence can also be induced in an isotropic material by applying an electric field to break the symmetry. This effect was first discovered by Kerr in the nineteenth century, and is therefore known as the **Kerr effect**. Kerr discovered that an isotropic medium behaves as a uniaxial crystal when an electric field is applied in the direction transverse to the light direction. The optic axis is parallel to the field, and the induced birefringence is given by:

Since an electromagnetic wave consists of an oscillating transverse electric field, a high-intensity light beam can produce self-induced birefringence through the Kerr effect. This is an example of a a nonlinear optical effect, and will be discussed further in Chapter 11.

$$\Delta n = \lambda K \mathcal{E}^2 , \qquad (2.51)$$

where λ is the vacuum wavelength, K is the Kerr constant, and $\mathcal{E}$ in the field strength. Since the the birefringence is proportional to the square of the field, the Kerr effect is alternatively known as the **quadratic electro-optic effect**. Table 2.2 lists some representative values of the Kerr constant.

The quadratic field dependence of the Kerr effect can be understood in simple terms as follows: the first power of the field breaks the symmetry, and the second power induces the refractive index change. This contrasts with the **Pockels effect** which is observed in anisotropic crystals, where symmetry breaking is not required. Therefore, a refractive index change proportional to the field automatically produces induced birefringence, giving rise to a **linear electro-optic effect**.

In addition to the different functional dependence on the field strength, there are a number of other important differences between the linear and quadratic electro-optic effects.

Table 2.2 Kerr constants of representative substances. In the case of oxide glasses, the Kerr constant generally increases with the PbO content. Data from Kaye & Laby (1986) and Hoffmann (1995).

Substance	K (m V^{-2})
Nitrobenzene	4.4×10^{-12}
CS_2	3.6×10^{-14}
Water	5.2×10^{-14}
Oxide glass	$0.1 - 3 \times 10^{-14}$
Chalcogenide glass (As_2O_3)	8.7×10^{-14}

(1) The Kerr effect can, in principle, be observed in any medium, but the Pockels effect is only observed in anisotropic crystals that lack inversion symmetry.

(2) The Kerr effect is only observed with transverse fields, but the Pockels effect can also be observed with longitudinal fields.

(3) Since no symmetry breaking is required for the Pockels effect, the fields required to induce a particular value of Δn are smaller than for the Kerr effect, so that the Kerr effect is usually negligible in an anisotropic medium that exhibits the Pockels effect. The Kerr effect is normally only studied in isotropic media such as liquids, gases, and glasses, and requires large fields to observe significant effects. (See Exercise 2.21.)

Further discussion of the linear electro-optic effect and Kerr effect is postponed to Sections 11.3.4 and 11.4.3 in Chapter 11.

2.6 Optical chirality

Objects that do not possess reflection symmetry appear reversed when observed in a mirror. Left- and right-handed gloves are obvious examples, as are springs, corkscrews and helices. Such objects are said to possess **chirality** (i.e. handedness). The fact that an optical medium possesses chirality implies that its response to left- and right-circular light will be different. We can therefore quantify optical chirality by the difference of the refractive indices experienced by left- and right-circular light. A difference in the real part of the complex refractive index gives rise to **optical activity**, while a difference in the imaginary part causes **circular dichroism**.

Optical activity is observed in transparent, chiral materials, and causes the direction of linear polarization to rotate as the light propagates through the medium. The rotation can be in either direction, which leads to a sub-classification of the material as being either **dextro-rotatory** or **laevo-rotatory**, depending on whether the rotation is in a clockwise or anti-clockwise sense when looking along the beam from the source. Since optical activity arises from a difference in the refractive index for circularly polarized light, it is sometimes called **circular birefringence**. This should not be confused with the birefringence discussed in Section 2.5.1, which arises from anisotropy, and is quantified by the difference in the refractive indices for orthogonal *linear* polarizations.

The names 'dextro-rotatory' and 'laevo-rotatory' are derived from the Latin words *dexter* and *laevus* which mean 'right' and 'left' respectively. Dextro-rotatory and laevo-rotatory materials are alternatively said to have positive or negative chirality, respectively.

The rotation angle θ of the polarization in an optically-active medium of thickness d is given by (see Exercise 2.22):

$$\theta = \frac{\pi d}{\lambda}(n_{\mathrm{R}} - n_{\mathrm{L}}), \tag{2.52}$$

where n_{R} (n_{L}) is the refractive index for right- (left-) circular light, and λ is the vacuum wavelength. A dextro-rotatory medium has $n_{\mathrm{R}} < n_{\mathrm{L}}$, and vice versa for a laevo-rotatory one.

Circular dichroism is observed in absorbing chiral materials. The chirality manifests itself as a difference in the imaginary part of the refractive index for left- and right-circular light. The word 'dichroic' means 'two-coloured', and the term dichroism is applied to any optical phenomenon that affects two colours differently. In the case of circular dichroism, the absorption coefficient is sensitive to the direction to the circular polarization, which implies that if the absorption bands lie in the visible spectral region, the colour will be different when viewed with left- or right-circular light.

Chirality can arise either from the crystalline structure, or from the molecules that comprise the medium. We concentrate here on the former effect, of which crystalline quartz (SiO_2) is a good example, with both dextro-rotatory and laevo-rotatory forms existing in nature. Quartz is a uniaxial birefringent crystal, and the optical activity is best observed when the light propagates along the optic axis. In this situation, linearly polarized light experiences a refractive index of n_{o} irrespective of its orientation, which means that it emerges unaffected in a non-chiral

The study of the chirality of molecules is very important in chemistry and biology. Since the chirality arises from the molecules themselves, it can be observed even in isotropic media such as liquids. Sugar solution is a well-known example, so much so that dextrose and laevulose (commonly known as fructose) get their names from their dextrorotatory and laevo-rotatory properties respectively.

Table 2.3 Verdet constant of representative substances. Since V varies with λ, the wavelength of the measurement must be specified. Data from Kaye & Laby (1986).

Substance	Wavelength (nm)	V (radian T^{-1} m^{-1})
Fused silica (SiO_2)	546.1	5.0
Hard crown glass	589.3	5.5
Light flint glass	589.3	9.0
Dense flint glass	589.3	11.2
Rock salt (NaCl)	670.8	7.1
Fluorite (CaF_2)	589.3	7.1
Water	589.3	3.8
CS_2	589.3	12.2

uniaxial crystal such as calcite. However, since the unit cell of quartz is chiral, linearly polarized light emerges rotated even when propagating along the optic axis.

The optical activity of crystalline quartz should be contrasted with the absence of optical activity in fused silica (SiO_2 glass). This shows that the optical activity arises from the crystal structure rather than the molecules. The unit cell of quartz belongs to the trigonal crystal class labelled 32, which has no mirror symmetry, and is therefore chiral. By contrast, calcite has the $\overline{3}2/m$ structure, where the 'm' symbol implies mirror symmetry in the unit cell, and therefore a lack of chirality.

Optical chirality can be induced in non-chiral materials by applying a magnetic field, thereby giving rise to a number of **magneto-optical** phenomena. In the case of transparent materials, the field induces optical activity and the phenomenon is called either the **Faraday effect** or the **magneto-optical Kerr effect**, depending on whether the rotation of the polarization is observed in transmission or reflection respectively. If the medium is absorbing, the field can induce circular dichroism, in which case we have **magnetic circular dichroism**. All of these magneto-optical effects ultimately have their origin in the Zeeman effect. (See Exercise 2.23.)

Figure 2.16 illustrates the Faraday effect. The field is applied along the axis of an optical material and this causes a rotation of linearly polarized light. The rotation angle θ is related to the field strength B by:

$$\theta = VBd, \tag{2.53}$$

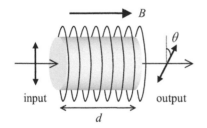

Fig. 2.16 The Faraday effect. A magnetic field B applied along the axis of an optical medium induces optical activity and causes a rotation of linearly polarized light by an angle θ.

where V is the **Verdet coefficient** of the medium and d is its thickness. Table 2.3 lists the values of the Verdet constant for a number of materials. In general, the Verdet coefficients are small in the visible spectral region, and decrease strongly as the detuning from the ultraviolet absorption bands increases. In practice, this means that substantial thicknesses have to be used to obtain significant rotations at the field strengths that are readily available from permanent magnets. (See Exercise 2.24.)

Chapter summary

- The classical model of a solid treats the atoms and molecules as oscillating electric dipoles with characteristic resonant frequencies. The resonances due to the bound electrons occur in the near infrared, visible, and ultraviolet spectral regions (10^{14}–10^{15} Hz), while those associated with vibrations occur in the infrared (10^{12}–10^{13} Hz). Free electrons can be treated in the dipole oscillator model by assuming that the natural resonant frequency is zero.

- The medium absorbs light when the frequency coincides with one of its resonant frequencies. In non-resonant conditions, the medium is transparent, but the velocity of light is reduced by the phase lag due to multiple coherent elastic scattering.

- The absorption coefficient of an individual dipole oscillator has a Lorentzian line shape (cf. eqn 2.21). The spectral width of the absorption line is equal to the damping constant γ. The peak absorption is proportional to $1/\gamma$.

- The refractive index of a dipole oscillator increases as the frequency approaches the resonant frequency, then drops sharply in the absorbing region, and then increases again at higher frequencies. The off-resonant refractive index decreases each time we increase the frequency through an absorption line.

- The relative dielectric constant of a medium with multiple resonant frequencies is given by eqn 2.24. The refractive index and absorption coefficient can be calculated from the real and imaginary parts of ϵ_{r}.

- The dipole oscillator model demonstrates that the absorption and refraction of an optical medium are fundamentally related to each other. This interrelationship is made explicit through the Kramers–Kronig formulae.

- Dispersion in the refractive index originates from the wings of the resonances at transition frequencies. The dispersion is called normal when the refractive index increases with frequency. Group velocity dispersion causes temporal broadening of short pulses.

- Optical anisotropy leads to birefringence. The anisotropy is described through the electric susceptibility tensor or the dielectric constant tensor. Anisotropy can be induced in isotropic media by strain or electric fields, giving rise respectively to the photo-elastic and electro-optic effects.

- Structural chirality causes optical activity and circular dichroism. The chirality can be induced by a magnetic field, which gives rise to magneto-optical effects.

Further reading

The subject matter of this chapter is covered, to a greater or lesser extent, in most electromagnetism and optics textbooks. See, for example: Bleaney and Bleaney (1976), Born and Wolf (1999), Hecht (2001), Smith, King, and Wilkins (2007), or Klein and Furtak (1986).

An excellent collection of optical data on a wide range of solid-state materials can be found in Palik (1985). A detailed discussion on the origin of the dispersion in silica glass and its relationship to ultraviolet and infrared absorption may be found in Smith et al. (2004).

For a fuller description of birefringence and optical activity, see: Hecht (2001), Smith, King, and Wilkins (2007), Born and Wolf (1999) or Klein and Furtak (1986).

Exercises

(2.1) Write down the equations of motion for the frictionless displacements x_1 and x_2 of two masses, m_1 and m_2, connected together by a light spring with a spring constant K_s. Hence show that the angular frequency for small oscillations is equal to $(K_s/\mu)^{1/2}$ where $\mu^{-1} = m_1^{-1} + m_2^{-1}$.

(2.2) A damped oscillator with mass m, natural angular frequency ω_0, and damping constant γ is being driven by a force of amplitude F_0 and angular frequency ω. The equation of motion for the displacement x of the oscillator is:

$$m\frac{\mathrm{d}^2 x}{\mathrm{d}t^2} + m\gamma\frac{\mathrm{d}x}{\mathrm{d}t} + m\omega_0^2 x = F_0\cos\omega t.$$

What is the phase of x relative to the phase of the driving force ?

(2.3) A sapphire crystal doped with titanium absorbs strongly around 500 nm. Calculate the difference in the refractive index of the doped crystal above and below the 500 nm absorption band if the density of absorbing atoms is $1 \times 10^{25}\,\mathrm{m}^{-3}$. The refractive index of undoped sapphire is 1.77.

(2.4) The laser crystal Ni^{2+}:MgF_2 has a broad absorption band in the blue which peaks at 405 nm and has a full width at half maximum of 8.2×10^{13} Hz. The oscillator strength of the transition is 9×10^{-5}. Estimate the maximum absorption coefficient in a crystal with $2\times10^{26}\mathrm{m}^{-3}$ absorbing atoms per unit volume. The refractive index of the crystal is 1.39.

(2.5) Show that the absorption coefficient of a Lorentz oscillator at the line centre does not depend on the value of ω_0.

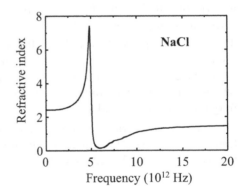

Fig. 2.17 Infrared refractive index of NaCl. After Palik (1985).

(2.6) Figure 2.17 shows the refractive index of NaCl in the infrared spectral region. The data can be modelled approximately by assuming that the resonance feature is caused by the vibrations of the completely ionic Na^+Cl^- molecules. The atomic weights of sodium and chlorine are 23 and 35.5 respectively. Use the data to estimate:

(a) The static dielectric constant of NaCl.

(b) The natural oscillation frequency of the vibrations.

(c) The restoring force for a unit displacement of the oscillator.

(d) The density of NaCl molecules per unit volume.

(e) The damping constant γ for the vibrations.

(f) The peak absorption coefficient.

(2.7) Derive both the relationships between the group

velocity and the phase velocity given in eqn 2.26.

(2.8)* Consider a simple Lorentz oscillator with a single undamped resonance. The dielectric constant will be given by eqn 2.14 with χ and γ both zero. This gives:

$$\epsilon_r(\omega) = 1 + \frac{Ne^2}{\epsilon_0 m_0} \frac{1}{(\omega_0^2 - \omega^2)}.$$

Prove that the group velocity is always less than c.

(2.9)* Consider a dielectric sample placed in a uniform electric field pointing in the z direction as shown in Fig. 2.8. Assume that the atoms are arranged on a cubic lattice and the dipoles are all pointing along the external field direction.
(a) Let us first consider the field generated by the dipoles within the spherical surface. By using the standard formula for the electric field generated by an electric dipole, show that the field at the centre of the sphere is given by

$$\mathcal{E}_{\text{sphere}} = \frac{1}{4\pi\epsilon_0} \sum_j p_j \frac{3z_j^2 - r_j^2}{r_j^5},$$

where the summation runs over all the dipoles within the surface except the one at the centre, and p_j is the dipole moment of the atom at the jth lattice site.
(b) Show that $\mathcal{E}_{\text{sphere}} = 0$ in a homogenous medium where all the values of p_j are identical.
(c) Now consider the uniformly polarized dielectric material outside the spherical hole. Let $\boldsymbol{P}$ be the macroscopic polarization of the medium, which is assumed to be parallel to the external field. Show that the surface charge density on the sphere at an angle θ from the z axis is equal to $-P\cos\theta$. Hence show that the material outside the spherical surface generates a field at the centre of the sphere equal to $-\boldsymbol{P}/3\epsilon_0$.

(2.10) Under what conditions does the Clausius–Mossotti relationship given by eqn 2.35 reduce to the usual relationship between the dielectric constant and the electric susceptibility given in eqn A.4?

(2.11) The relative dielectric constant of N_2 gas at standard temperature and pressure is 1.000588. Calculate χ_a for the N_2 molecule. Show that the electric field strength required to generate a dipole equivalent to displacing the electron by 1 Å (10^{-10} m) is of a similar magnitude to the electric field between a proton and an electron separated by the same distance.

(2.12) A certain material has a single absorption band which runs from photon energy E_1 to E_2. On the assumption that the absorption coefficient has a constant value of α_0 within the band, and is zero elsewhere, use the Kramers–Kronig relationship to show that the refractive index for a photon energy E below E_1 is given by:

$$n(E) = 1 + \frac{c\hbar\alpha_0}{2\pi E} \ln \frac{(E_2 - E)(E_1 + E)}{(E_2 + E)(E_1 - E)}.$$

(2.13) (a) Sellmeier derived the following equation for the wavelength dependence of the refractive index in 1871:

$$n^2 = 1 + \sum_j \frac{A_j \lambda^2}{(\lambda^2 - \lambda_j^2)}.$$

Show that this equation is equivalent to eqn 2.24 in regions of transparency far from any absorption lines. State the values of A_j and λ_j.
(b) Assume that the dispersion is dominated by the closest resonance, so that we only need to include one term (say the one with $j = 1$) in the summation of Sellmeier's equation. Assume that λ_1^2/λ^2 is small, and expand Sellmeier's equation to derive the earlier dispersion formula determined empirically by Cauchy:

$$n = C_1 + \frac{C_2}{\lambda^2} + \frac{C_3}{\lambda^4} + \cdots$$

State the values of C_1, C_2, and C_3 in terms of A_1 and λ_1.

(2.14) The refractive index of crown glass is 1.5553 at 402.6 nm and 1.5352 at 706.5 nm.
(a) Determine the coefficients C_1 and C_2 in Cauchy's formula given in the previous exercise, on the assumption that the term in C_3 is negligible.
(b) Estimate the refractive index for blue light at 450 nm and for red light at 650 nm.
(c) White light strikes a crown glass prism with an apex angle of 60°, as shown in Fig. 2.11. The angle of incidence with the first surface is 45°. Calculate the difference in the angle between the light at 450 nm and 650 nm at the exit surface of the prism.

(2.15) Show that the temporal broadening of a short pulse by a dispersive medium of length L is given approximately by:

$$\Delta\tau = L \left| \frac{\lambda}{c} \frac{d^2 n}{d\lambda^2} \right| \Delta\lambda,$$

*Exercises marked with an asterisk are more difficult.

where λ is the vacuum wavelength and $\Delta\lambda$ is the spectral width of the pulse. Estimate $\Delta\tau$ for a laser pulse with a temporal width of 10 ps in 1 km of optical fibre at 1550 nm, where $|(\lambda/c)\mathrm{d}^2 n/\mathrm{d}\lambda^2| = 17\,\mathrm{ps\,km^{-1}\,nm^{-1}}$.

(2.16) Consider the propagation of a wave with polarization vector components (x, y, z), where $x^2 + y^2 + z^2 = 1$, in a birefringent medium. The dielectric constant experienced by the wave is conveniently described by the index ellipsoid:[1]

$$\frac{x^2}{\epsilon_{11}/\epsilon_0} + \frac{y^2}{\epsilon_{22}/\epsilon_0} + \frac{z^2}{\epsilon_{33}/\epsilon_0} = 1 \,,$$

where the ϵ_{ij} are the components of the dielectric constant tensor defined in eqn 2.49. Use the index ellipsoid to show that the refractive index for the e-ray propagating at an angle θ to the optic axis of a uniaxial crystal as shown in Fig. 2.13(b) is given by:

$$\frac{1}{n(\theta)^2} = \frac{\sin^2\theta}{n_\mathrm{e}^2} + \frac{\cos^2\theta}{n_\mathrm{o}^2} \,,$$

where n_e and n_o are defined in Example 2.3.

(2.17) By using the refractive index data given in Table 2.1, calculate the range of apex angles that will lead to selective total internal reflection of the o-ray in a Glan–Foucault polarizing prism made from calcite.

(2.18) (a) Consider a half-wave plate with linearly polarized light at the input surface, as shown in Fig. 2.15. Show that the wave plate rotates the polarization of the light by an angle 2θ, where θ is the angle between the input polarization and the optic axis.
(b) Show that a quarter-wave plate turns a linearly polarized input beam into circularly-polarized light, and vice versa, if the angle θ is chosen to be $45°$.
(c) Describe the output of the quarter-wave plate when $\theta \neq 45°$.

(2.19) A uniaxial birefringent crystal made from quartz has $n_\mathrm{o} = 1.5443$ and $n_\mathrm{e} = 1.5534$. A retardation plate is made by cutting the crystal so that the optic axis is parallel to the surfaces of the plate, as shown in Fig. 2.15. Calculate the thickness of the crystal that must be chosen so that it behaves as a quarter-wave plate at 500 nm.

(2.20) Look up the crystal structures of the following materials to determine whether they are birefringent or not: (a) NaCl, (b) diamond, (c) graphite

(in the infrared, where it transmits), (d) ZnS (wurtzite structure), (e) ZnS (zinc-blende structure), (f) solid argon at 4 K, (g) sulphur. Specify which, if any, of the birefringent materials are biaxial.

(2.21) (a) A Kerr cell consists of a Kerr medium with contacts attached so that an electric field can be applied. Show that the field strength required to produce a birefringent phase shift equivalent to half a wavelength is given by:

$$\mathcal{E}_{\lambda/2} = 1/\sqrt{2Kd} \,,$$

where K is the Kerr constant, and d is the length of the medium.
(b) Calculate the voltage required to produce a field of $\mathcal{E}_{\lambda/2}$ in a Kerr cell of length 2 cm made from chalcogenide glass with a Kerr constant of $8.7 \times 10^{-14}\,\mathrm{m\,V^{-2}}$, if the lateral width across which the voltage is dropped is 5 mm.

(2.22) (a) A transparent chiral medium of thickness d has a refractive index of n_L and n_R for left- and right-circular light respectively for light of vacuum wavelength λ. By considering linearly-polarized light as a superposition of left- and right-circular light, show that the medium rotates the linear polarization by an angle θ given by

$$\theta = \frac{\pi d}{\lambda}(n_\mathrm{R} - n_\mathrm{L}) \,.$$

(b) The value of $|n_\mathrm{R} - n_\mathrm{L}|$ for crystalline quartz is 7.1×10^{-5} at 589 nm. Calculate the rotatory power of quartz (defined as θ/d) at this wavelength in units of $°/\mathrm{mm}$.

(2.23) Consider an optical medium with a single Lorentzian resonance line at angular frequency ω_0 of width γ. When a magnetic field of strength B is applied, the transition energy shifts by $\pm\mu_\mathrm{B}B$ according to the normal Zeeman effect. For light propagating parallel to the field, the absorption frequency increases to $\omega_0 + \mu_\mathrm{B}B/\hbar$ for σ^+ circular polarization, and decreases to $\omega_0 - \mu_\mathrm{B}B/\hbar$ for σ^- polarization. Magneto-optical phenomena can be explained by considering the difference in the complex refractive index, $(\tilde{n}_+ - \tilde{n}_-)$, for σ^+ and σ^- light.
(a) By considering the real part of $(\tilde{n}_+ - \tilde{n}_-)$, sketch the frequency dependence of the Faraday

[1]The use of the index ellipsoid can be justified by considering the direction of the energy flow through the crystal: see Born and Wolf (1999).

rotation.

(b) By considering the imaginary part of $(\tilde{n}_+ - \tilde{n}_-)$, sketch the frequency dependence of the magnetic circular dichroism.

(2.24) In an optical isolator, the Faraday effect is used to rotate the plane of linearly polarized light by 45°. Given that Verdet coefficient of flint glass is 9.0 radian $T^{-1} m^{-1}$, calculate the thickness of glass required in an isolator with a permanent magnet of strength 0.5 T.

3

Interband absorption

3.1	Interband transitions	62
3.2	The transition rate for direct absorption	64
3.3	Band edge absorption in direct gap semiconductors	68
3.4	Band edge absorption in indirect gap semiconductors	79
3.5	Interband absorption above the band edge	82
3.6	Measurement of absorption spectra	84
3.7	Semiconductor photodetectors	86
Chapter summary		91
Further reading		92
Exercises		92

We noted in Section 1.4.1 that semiconductors and insulators have a fundamental absorption edge in the near-infrared, visible, or ultraviolet spectral regions. The absorption edge is caused by the onset of optical transitions across the fundamental band gap of the material. This naturally leads us to investigate the physical processes that occur when electrons are excited between the bands of a solid by making optical transitions. This process is called **interband absorption**, and is the subject of the present chapter. The opposite process of **interband luminescence**, in which electrons drop from excited state bands by emitting photons, is discussed in Chapter 5.

Interband transitions are observed in all solids. Our objective here is to understand how the absorption spectrum of a given material is related to its band structure, and in particular to the density of states for the transition. We shall postpone the discussion of excitonic effects on the absorption spectra to Chapter 4, and concentrate on crystalline semiconductors, which illustrate the main points clearly. The principles that we shall find can easily be adapted to other materials, as required. This is done, for example, in Section 7.3.2, where we consider the effects of interband transitions on the reflectivity spectra of metals.

The understanding of interband absorption is based on applying the quantum-mechanical treatment of the light–matter interaction to the band states of solids. This presupposes a working knowledge of both quantum mechanics and band theory. A summary of the main results required for the chapter is given in Appendices B and D. The reader is recommended to refer to the quantum-mechanics and solid-state physics texts listed in these appendices if any of the material is unfamiliar.

3.1 Interband transitions

The energy-level diagram of isolated atoms consists of a series of states with discrete energies. Optical transitions between these levels give rise to sharp lines in the absorption and emission spectra. We have to use quantum mechanics to calculate the transition energies and the oscillator strengths. Once we have done this, we can obtain a good understanding of the frequency dependence of the refractive index and absorption coefficient by applying the classical oscillator model described in the previous chapter.

The optical transitions of solids are more complicated to deal with. Some of the properties that apply to the individual atoms carry over, but

new physics arises as a result of the formation of bands with their delocalized states. The classical model has difficulty dealing with continuous
absorption bands rather than discrete lines, and we must develop new
techniques to describe the frequency dependence of the optical properties. We can only expect the classical oscillator model to work with any
accuracy when the frequency is far away from the absorption transitions
between the bands.

Figure 3.1 shows a highly simplified energy diagram of two separated
bands in a solid. The gap in energy between the bands is called the
band gap E_g. Interband optical transitions will be possible between these
bands if the selection rules allow them. During the transition an electron
jumps from the band at lower energy to the one above it by absorbing a
photon. This can only happen if there is an electron in the initial state
in the lower band. Furthermore, the Pauli exclusion principle demands
that the final state in the upper band must be empty. A typical example
of a situation where this applies is the transitions across the fundamental
band gap of a semiconductor or insulator. In this case, a photon excites
an electron from the filled valence band to the empty conduction band.

By applying the law of conservation of energy to the interband transition shown in Fig. 3.1 we can see that:

$$E_f = E_i + \hbar\omega,\qquad(3.1)$$

where E_i is the energy of the electron in the lower band, E_f is the energy
of the final state in the upper band, and $\hbar\omega$ is the photon energy. Since
there is a continuous range of energy states within the upper and lower
bands, the interband transitions will be possible over a continuous range
of frequencies. The range of frequencies is determined by the upper and
lower energy limits of the bands.

It is apparent from Fig. 3.1 that the minimum value of $(E_f - E_i)$ is E_g.
This implies that the absorption shows a threshold behaviour: interband
transitions will not be possible unless $\hbar\omega > E_g$. Interband transitions
therefore give rise to a continuous absorption spectrum from the low
energy threshold at E_g to an upper value set by the extreme limits of
the participating bands. This contrasts with the absorption spectrum of
isolated atoms which consist of discrete lines.

The excitation of the electron leaves the initial state at energy E_i in
the lower band unoccupied. This is equivalent to the creation of a **hole**
in the initial state. The interband absorption process therefore creates
a hole in the initial state and an electron in the final state, and may be
considered as the creation of an **electron-hole pair**.

In the sections that follow, we shall study how the interband absorption rate depends on the band structure of the solid. At this stage we just
make one general distinction based on whether the band gap is **direct**
or **indirect**. This point is illustrated in Fig. 3.2. Figure 3.2(a) shows
the E–k diagram of a solid with a direct band gap, while Fig. 3.2(b)
shows the equivalent diagram for an indirect gap material. The distinction concerns the relative positions of the conduction band minimum
and the valence band maximum in the Brillouin zone. In a direct gap

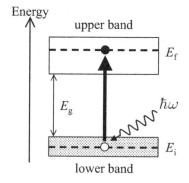

Fig. 3.1 Interband optical absorption
between an initial state of energy E_i in
an occupied lower band and a final state
at energy E_f in an empty upper band.
The energy difference between the two
bands is E_g.

We shall consider the effect of the attractive force between the negative electron and the positive hole in Chapter 4.
At this stage we shall ignore these effects and concentrate on investigating
the general features of interband absorption.

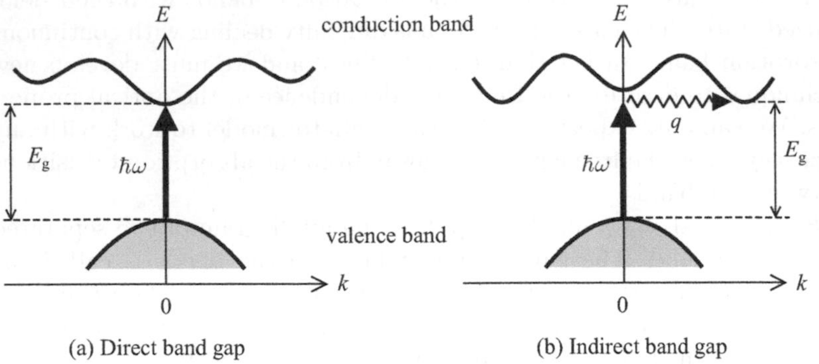

Fig. 3.2 Interband transitions in solids: (a) direct band gap, (b) indirect band gap. The vertical arrow represents the photon absorption process, while the wiggly arrow in part (b) represents the absorption or emission of a phonon.

material, both occur at the zone centre where $k = 0$. In an indirect gap material, however, the conduction band minimum does not occur at $k = 0$, but rather at some other value of k which is usually at the zone edge or close to it.

The distinction between the nature of the band gap has very important consequences for the optical properties. We shall see in Section 3.2 that conservation of momentum implies that the electron wave vector does not change significantly during a photon absorption process. We therefore represent photon absorption processes by vertical lines on E–k diagrams. It is immediately apparent from Fig. 3.2(b) that the electron wave vector must change significantly in jumping from the valence band to the bottom of the conduction band if the band gap is indirect. It is not possible to make this jump by absorption of a photon alone: the transition must involve a phonon to conserve momentum. This contrasts with a direct gap material in which the process may take place without any phonons being involved.

Indirect absorption plays a very significant role in technologically important materials such as silicon. The treatment of indirect absorption is more complicated than direct absorption because of the role of the phonons. We shall therefore begin our discussion of interband transitions by restricting our attention to direct processes. Interband absorption processes in indirect gap materials will be considered in Section 3.4.

3.2   The transition rate for direct absorption

The optical absorption coefficient α is determined by the quantum mechanical transition rate $W_{i \to f}$ for exciting an electron in an initial quantum state ψ_i to a final state ψ_f by absorption of a photon of angular frequency ω. Our task is therefore to calculate $W_{i \to f}$, and hence to derive the frequency dependence of α. As discussed in Appendix B, the transition rate is given by Fermi's golden rule:

$$W_{i \to f} = \frac{2\pi}{\hbar} |M|^2 g(\hbar\omega) \,. \tag{3.2}$$

The transition rate thus depends on two factors:

- the **matrix element** M,
- the **density of states** $g(\hbar\omega)$.

In the discussion below, we consider the matrix element first, and then $g(\hbar\omega)$ afterwards.

The matrix element describes the effect of the external perturbation caused by the light wave on the electrons. It is given by:

$$
\begin{aligned}
M &= \langle \mathrm{f}|H'|\mathrm{i}\rangle \\
&= \int \psi_{\mathrm{f}}^*(\boldsymbol{r})\, H'(\boldsymbol{r})\, \psi_{\mathrm{i}}(\boldsymbol{r})\, \mathrm{d}^3\boldsymbol{r},
\end{aligned} \tag{3.3}
$$

where H' is the perturbation associated with the light wave, and $\boldsymbol{r}$ is the position vector of the electron. We adopt here the semi-classical approach in which we treat the electrons quantum mechanically, but the photons are described by electromagnetic waves.

In classical electromagnetism, the presence of a perturbing electric field $\boldsymbol{\mathcal{E}}$ causes a shift in the energy of a charged particle equal to $-\boldsymbol{p}\cdot\boldsymbol{\mathcal{E}}$, where $\boldsymbol{p}$ is the electric-dipole moment of the particle. The appropriate quantum perturbation to describe the electric-dipole interaction between the light and the electron is therefore:

$$
H' = -\boldsymbol{p}_{\mathrm{e}}\cdot\boldsymbol{\mathcal{E}}_{\mathrm{photon}}, \tag{3.4}
$$

where $\boldsymbol{p}_{\mathrm{e}}$ is the electron dipole moment and is equal to $-e\boldsymbol{r}$. This form for the perturbation is justified more rigorously in Section B.2 of Appendix B.

The light wave is described by plane waves of the form:

$$
\boldsymbol{\mathcal{E}}_{\mathrm{photon}}(\boldsymbol{r}) = \boldsymbol{\mathcal{E}}_0\, \mathrm{e}^{\mathrm{i}\boldsymbol{k}\cdot\boldsymbol{r}}, \tag{3.5}
$$

and the perturbation is thus:

$$
H'(\boldsymbol{r}) = e\boldsymbol{\mathcal{E}}_0\cdot\boldsymbol{r}\, \mathrm{e}^{\mathrm{i}\boldsymbol{k}\cdot\boldsymbol{r}}. \tag{3.6}
$$

The electron states in a crystalline solid are described by Bloch functions. This allows us to write the wave functions as a product of a plane wave and an envelope function that has the periodicity of the crystal lattice. (See eqns 1.30 and D.7.) We therefore write:

$$
\psi_{\mathrm{i}}(\boldsymbol{r}) = \frac{1}{\sqrt{V}}\, u_{\mathrm{i}}(\boldsymbol{r})\, \mathrm{e}^{\mathrm{i}\boldsymbol{k}_{\mathrm{i}}\cdot\boldsymbol{r}} \tag{3.7}
$$

$$
\psi_{\mathrm{f}}(\boldsymbol{r}) = \frac{1}{\sqrt{V}}\, u_{\mathrm{f}}(\boldsymbol{r})\, \mathrm{e}^{\mathrm{i}\boldsymbol{k}_{\mathrm{f}}\cdot\boldsymbol{r}}, \tag{3.8}
$$

where u_{i} and u_{f} are the appropriate envelope functions for the initial and final bands respectively, and V is the normalization volume. $\boldsymbol{k}_{\mathrm{i}}$ and $\boldsymbol{k}_{\mathrm{f}}$ are the wave vectors of the initial and final electron states. On substituting

The bracket symbol $\langle \mathrm{f}|H'|\mathrm{i}\rangle$ is an example of **Dirac notation**. The 'ket' $|\mathrm{i}\rangle$ represents the wave function ψ_{i}, while the 'bra' $\langle \mathrm{f}|$ represents ψ_{f}^*. The closed 'bra-cket' with the perturbation in the middle indicates that we evaluate the expectation value written out explicitly in the second line of eqn 3.3.

Note that we only need to include the spatial dependence of the light wave here. The $\mathrm{e}^{-\mathrm{i}\omega t}$ time dependence of the perturbation has already been included in the derivation of Fermi's golden rule, and is implicitly contained in the conservation of energy statement of eqn 3.1.

the perturbation of eqn 3.6 and the wave functions of eqns 3.7 and 3.8 into eqn 3.3, we obtain:

$$M = \frac{e}{V} \int u_f^*(\boldsymbol{r}) \, e^{-i\boldsymbol{k_f} \cdot \boldsymbol{r}} \left(\boldsymbol{\mathcal{E}}_0 \cdot \boldsymbol{r} \, e^{i\boldsymbol{k} \cdot \boldsymbol{r}} \right) u_i(\boldsymbol{r}) \, e^{i\boldsymbol{k_i} \cdot \boldsymbol{r}} \, d^3\boldsymbol{r} \,, \qquad (3.9)$$

where the limits of the integration are over the whole crystal.

The integral in eqn 3.9 can be simplified by two considerations. First, we invoke conservation of momentum, which demands that the change in crystal momentum of the electron must equal the momentum of the photon, that is:

$$\hbar \boldsymbol{k_f} - \hbar \boldsymbol{k_i} = \hbar \boldsymbol{k} \,. \qquad (3.10)$$

This is equivalent to requiring that the phase factor in eqn 3.9 must be zero. If the phase factor is not zero, the different unit cells within the crystal will be out of phase with each other and the integral will sum to zero. Second, we recall that Bloch's theorem requires that u_i and u_f are periodic functions with the same periodicity as the lattice. This implies that we can separate the integral over the whole crystal into a sum over identical unit cells, because the unit cells are equivalent and, as we have seen above, in phase. We thus obtain:

We are assuming here that the light is linearly polarized, and have arbitrarily chosen the polarization direction as the x axis. In crystals with cubic symmetry, the x, y, and z directions are all equivalent, but this will not be the case for anisotropic materials. At this stage we just consider the general principles, and postpone the consideration of anisotropy to the discussion of quantum wells in Chapter 6. Circularly polarized light is considered in Section 3.3.7.

$$|M| \propto \int_{\text{unit cell}} u_i^*(\boldsymbol{r}) \, x \, u_f(\boldsymbol{r}) \, d^3\boldsymbol{r} \,, \qquad (3.11)$$

where we have defined our axes in such a way that the light is polarized along the x axis. This matrix element represents the electric-dipole moment of the transition. Its evaluation requires knowledge of the envelope functions u_i and u_f. These functions are derived from the atomic orbitals of the constituent atoms, and so each material has to be considered separately.

The conservation of momentum condition embodied in eqn 3.10 can be simplified further by considering the magnitude of the wave vectors of the electrons and photons. The wave vector of the photon is $2\pi/\lambda$, where λ is the wavelength of the light. Optical frequency photons therefore have k values of about $10^7 \, \text{m}^{-1}$. The wave vectors of the electrons, however, are much larger. This is because the electron wave vector is related to the size of the Brillouin zone, which is equal to π/a, where a is the unit cell dimension. Since $a \sim 10^{-10} \, \text{m}$, the photon wave vector is much smaller than the size of a Brillouin zone. Therefore we may neglect the photon momentum in eqn 3.10 in comparison to the electron momentum and write:

$$\boldsymbol{k_f} = \boldsymbol{k_i} \,. \qquad (3.12)$$

A direct optical transition therefore leads to a negligible change in the wave vector of the electron. This is why we represent the absorption processes by vertical arrows in the electron E–k diagrams such as the ones shown in Fig. 3.2.

The factor of $g(\hbar\omega)$ that appears in eqn 3.2 is the **joint density of states** evaluated at the photon energy. As explained in Section 1.5.4, the density of states function describes the distribution of the states within the bands. The *joint* density of states accounts for the fact that both the initial and final electron states lie within continuous bands.

For electrons within a band, the density of states per unit energy range $g(E)$ is obtained from:

$$g(E)\,dE = 2\,g(k)\,dk\,, \tag{3.13}$$

where $g(k)$ is the density of states in momentum space. The extra factor of 2 here compared to eqn 1.32 allows for the fact that there are two electron spin states for each k state. This gives:

$$g(E) = \frac{2g(k)}{dE/dk}\,, \tag{3.14}$$

where dE/dk is the gradient of the E–k dispersion curve in the band diagram. $g(k)$ itself is worked out by calculating the number of k states in the incremental volume between shells in k space of radius k and $k+dk$. This is equal to the number of states per unit volume of k space, namely $1/(2\pi)^3$ (see Exercise 3.1), multiplied by the incremental volume $4\pi k^2 dk$. Hence $g(k)$ is given by the standard formula:

$$\begin{aligned} g(k)dk &= \frac{1}{(2\pi)^3}\, 4\pi k^2 dk \\[2mm] \Rightarrow g(k) &= \frac{k^2}{2\pi^2}. \end{aligned} \tag{3.15}$$

We can then work out $g(E)$ by using eqn 3.14 if we know the relationship between E and k from the band structure of the material. For electrons in a parabolic band with effective mass m^*, $g(E)$ is given by (see Exercise 3.2):

$$g(E) = \frac{1}{2\pi^2}\left(\frac{2m^*}{\hbar^2}\right)^{3/2} E^{1/2}\,. \tag{3.16}$$

This is just the standard formula for free electrons but with the free electron mass m_0 replaced by m^*.

The joint density of states factor is finally obtained by evaluating $g(E)$ at E_i and E_f when they are related to $\hbar\omega$ through the band energies. To proceed further, we therefore need detailed knowledge of the band structure. We shall see how to do this in the case of parabolic bands in Section 3.3.3, and then use the result to derive the frequency dependence of the absorption near the band edge of a direct gap semiconductor in Section 3.3.4. The case of non-parabolic bands is considered in Section 3.5.

One useful general point can be made at this stage. Since the density of atoms in a solid is very large, the density of states within a band is also going to be large. Therefore, the absorption strength for allowed transitions in a solid will, in general, be much larger than that found in dilute media such as gases.

3.3   Band edge absorption in direct gap semiconductors

The basic process for an optical transition across the fundamental band gap of a direct gap semiconductor is shown in Fig. 3.2(a). An electron is excited from the valence band to the conduction band by absorption of a photon. The transition rate is evaluated by working out the matrix element and the density of states, as discussed in the preceding section. These factors are considered separately below.

3.3.1   The atomic physics of the interband transitions

The matrix element to be evaluated is given in eqn 3.11. This allows us to calculate the probability for electric-dipole transitions if we know the atomic character of the envelope wave functions $u_i(\boldsymbol{r})$ and $u_f(\boldsymbol{r})$. The full treatment of this problem employs group theory to determine the character of the bands involved. This approach is beyond the scope of our present discussion, and at this level we just offer a few intuitive arguments.

The semiconductors that we shall be considering all have four valence electrons. This is obvious in the case of the elemental semiconductors such as silicon and germanium, which come from group IV of the periodic table. It is also true, however, for the binary compounds made from elements symmetrically displaced from group IV of the periodic table. The covalent bond in these compounds is made by sharing the electrons in such a way that each atom ends up with four electrons. For example, the bond in the III–V compounds is formed by sharing the five valence electrons from the group V element with the three from the group III element, giving a total of eight electrons for every two atoms. It is energetically favourable to do this because it is then possible to form very stable covalent crystals with a structure similar to diamond. Similar arguments apply to the II–VI semiconductor compounds.

The valence electrons of a four-valent atom are derived from the s and p orbitals. For example, the electronic configuration of germanium is $4s^2 4p^2$. In the crystalline phase the adjacent atoms share the valence electrons with each other in a covalent bond. Figure 3.3 shows schematically the evolution of the s- and p-like atomic states, through the s and p bonding and anti-bonding orbitals of the molecule, to the valence and conduction bands of the crystalline solid. The level ordering shown is appropriate for most III–V and II–VI semiconductors, as well as germanium. The level ordering in silicon is different: see Exercise 3.12.

The evolution of the levels shown in Fig. 3.3 makes it apparent that the top of the valence band has a p-like atomic character, while the bottom of the conduction band is s-like. This is because the four valence electrons occupy the four bonding orbitals, which then evolve into the valence band. The top of the valence band is derived from the p bond-

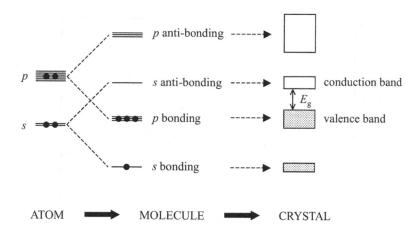

Fig. 3.3 Schematic diagram of the electron levels in a covalent crystal made from four-valent atoms such as germanium or binary compounds such as gallium arsenide. The s and p states of the atoms hybridize to form bonding and anti-bonding molecular orbitals, which then evolve into the conduction and valence bands of the semiconductor.

ing orbitals, while the bottom of the conduction band originates from the s anti-bonding orbitals. Therefore optical transitions from the valence band to the conduction band are from p-like states to s-like states. We know from the selections rules for electric-dipole transitions that $p \rightarrow s$ transitions are allowed. (See Exercise 3.3 and Section B.3 in Appendix B.) Hence we conclude that the transitions between the valence band and the conduction band of a semiconductor with a level ordering such as the one shown in Fig. 3.3 are electric-dipole allowed.

The conclusion of this discussion is that the probability for interband transitions across the band gap in materials like germanium or the III–V compounds is high. Since the density of states factor is also large above the band edge, we therefore expect to observe strong absorption. This is indeed the case, as we shall see below. The discussion of germanium is complicated because it has an indirect band gap. We shall therefore concentrate our attention on the III–V compound semiconductor gallium arsenide. GaAs has a direct band gap, and the level ordering follows the scheme shown in Fig. 3.3. The transitions across the gap are therefore both dipole allowed and direct. This makes GaAs a standard example for considering direct interband transitions. It is also a very important material for opto-electronic applications.

3.3.2 The band structure of a direct gap III–V semiconductor

The band structure of GaAs in the energy range near the fundamental band gap is shown in Fig. 3.4. The energy E of the electrons in the different bands is plotted against the electron wave vector $\mathbf{k}$. GaAs has the zinc-blende structure, which is based on the face-centred cubic (fcc) lattice. The band dispersion is shown for increasing $\mathbf{k}$ along two different directions of the Brillouin zone. The right-hand side of the figure corresponds to moving from the zone centre where $\mathbf{k} = (0,0,0)$ along the (100) direction to the zone edge at $\mathbf{k} = (2\pi/a)(1,0,0)$, a being the length of the cube edge in the fcc lattice. The left-hand side corresponds

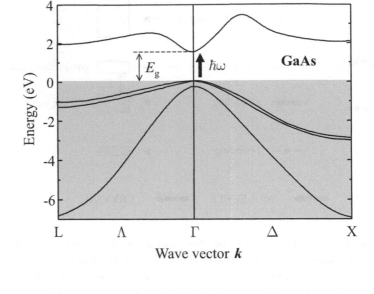

Fig. 3.4 Band structure of GaAs. The dispersion of the bands is shown for two directions of the Brillouin zone: $\Gamma \to X$ and $\Gamma \to L$. The Γ point corresponds to the zone centre with a wave vector of $(0,0,0)$, while the X and L points correspond respectively to the zone edges along the (100) and (111) directions. The valence bands are below the Fermi level and are full of electrons. This is indicated by the shading in the figure. After Chelikowsky and Cohen (1976), © American Physical Society, reprinted with permission.

The high symmetry points of Brillouin zones are given symbolic names. The zone centre where $\boldsymbol{k} = (0,0,0)$ is called the Γ point. The zone edges along the (100) and (111) directions are called the X and L points respectively. In the Brillouin zone of the diamond or zinc-blende lattice, the wave vectors at the X and L points are $\boldsymbol{k} = (2\pi/a)(1,0,0)$ and $(\pi/a)(1,1,1)$ respectively, where a is the length of the cube edge of the face-centred cubic lattice from which the diamond or zinc-blende structure is derived. The $\Gamma \to X$ direction is labelled Δ, while the $\Gamma \to L$ direction is labelled Λ. See Appendix D for further details.

to moving from $\boldsymbol{k} = 0$ along the body-diagonal direction until reaching the zone edge at $\boldsymbol{k} = (\pi/a)(1,1,1)$.

The figure is divided into a shaded region and an unshaded region. The shading represents the occupancy of the levels in the bands: bands that fall in the shaded region are below the Fermi level and are full of electrons. The three bands in the shaded region therefore correspond to valence band states. The single band above the shaded region is empty of electrons and is therefore the conduction band. The three bands in the valence band correspond to the three p bonding orbitals shown in Fig. 3.3, while the single conduction band corresponds to the s anti-bonding state. This correspondence between the bands and the molecular orbitals is strictly valid only at the Γ point at the Brillouin zone centre. The atomic character (or more accurately, the symmetry) of the bands actually changes as k increases, and is only well defined at high symmetry points in the Brillouin zone such as Γ, X, or L.

In this section we are interested in the transitions that take place across the band gap for small k values close to the Γ point. This means that the correspondence to Fig. 3.3 will be justified in our discussion here. We can therefore assume that the transitions are dipole allowed, and concentrate on working out the density of states for the transition. To do this, it is helpful to make use of the simplified four-band model shown in Fig. 3.5. This model band diagram is typical of direct gap III–V semiconductors near $k = 0$. There is a single s-like conduction band and three p-like valence bands. All four bands have parabolic dispersions. The positive curvature of the conduction band on the E–k diagram indicates that it corresponds to an electron (e) band, while the negative curvature of the valence bands correspond to hole states. Two of the hole bands are degenerate at $k = 0$. These are known as the heavy (hh) and light hole (lh) bands, the heavy hole band being the one with the smaller curvature. The third band is split off to lower energy by the spin–orbit coupling, and is known as the split-off (so) hole band. The energy difference between the maximum of the valence band and the minimum of the conduction band is the band gap E_g, while the spin–orbit splitting between the hole bands at $k = 0$ is usually given the symbol Δ.

The schematic diagram of Fig. 3.5 should be compared with the detailed band structure of GaAs shown previously in Fig. 3.4. The maxima of the valence band occur at the Γ point of the Brillouin zone, while the conduction band has a 'camel back' structure, with minima at the Γ point, the L point, and near the X point. We can neglect the subsidiary minima at the L point and near the X point here because momentum conservation does not allow direct transitions to these states from the top of the valence band. The bands near the zone centre are all approximately parabolic, and so the simplified picture in Fig. 3.5 is valid near $k = 0$.

The three valence band states all have p-like atomic character, and so it is possible to have electric-dipole transitions from each of the bands to the conduction band. Two such transitions are indicated on Fig. 3.5. As noted earlier, these absorption processes are represented by vertical arrows on the E–k diagram. This means that the k vector of the electron and hole created by the transition are the same (cf. eqn 3.12). The transition labelled 1 involves the excitation of an electron from the heavy-hole band to the electron band. Transition 2 is the corresponding process originating in the light-hole band. Direct transitions are also possible from the split-off band to the conduction band, but these are not shown in the figure for clarity.

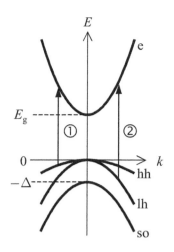

Fig. 3.5 Band structure of a direct gap III–V semiconductor such as GaAs near $k = 0$. $E = 0$ corresponds to the top of the valence band, while $E = E_g$ corresponds to the bottom of the conduction band. Four bands are shown: the heavy-hole (hh) band, the light-hole (lh) band, the split-off hole (so) band, and the electron (e) band. Two optical transitions are indicated. Transition 1 is a heavy-hole transition, while transition 2 is a light-hole transition. Transitions can also take place between the split-off hole band and the conduction band, but these are not shown for the sake of clarity. This four-band model was originally developed for InSb in Kane (1957).

3.3.3 The joint density of states

The frequency dependence of the absorption coefficient can now be calculated if we know the joint density of states factor given in eqn 3.14. This can be evaluated analytically for the simplified band structure shown in Fig. 3.5. The dispersion of the bands is determined by their respective effective masses, namely m_e^* for the electrons, m_{hh}^* for the heavy holes, m_{lh}^* for the light holes, and m_{so}^* for the split-off holes. This allows us to write the following E–k relationships for the electron, heavy-hole, light-hole, and split-off hole bands respectively:

$$E_e(k) = E_g + \frac{\hbar^2 k^2}{2m_e^*} \tag{3.17}$$

$$E_{hh}(k) = -\frac{\hbar^2 k^2}{2m_{hh}^*} \tag{3.18}$$

$$E_{lh}(k) = -\frac{\hbar^2 k^2}{2m_{lh}^*} \tag{3.19}$$

$$E_{so}(k) = -\Delta - \frac{\hbar^2 k^2}{2m_{so}^*}. \tag{3.20}$$

It is evident from Fig. 3.5, that conservation of energy during a heavy-hole or light-hole transition requires that:

$$\hbar\omega = E_g + \frac{\hbar^2 k^2}{2m_e^*} + \frac{\hbar^2 k^2}{2m_h^*}, \tag{3.21}$$

where $m_h^* = m_{hh}^*$ or m_{lh}^* for the heavy- or light-hole transition respectively. We define the reduced electron-hole mass μ according to:

$$\frac{1}{\mu} = \frac{1}{m_e^*} + \frac{1}{m_h^*}. \tag{3.22}$$

This allows us to rewrite eqn 3.21 in the simpler form:

$$\hbar\omega = E_g + \frac{\hbar^2 k^2}{2\mu}. \tag{3.23}$$

We are interested in evaluating $g(E)$ with $E = \hbar\omega$. The joint electron-hole density of states can be worked out by substituting eqn 3.23 into eqns 3.14 and 3.15. This gives:

For $\hbar\omega < E_g$, $g(\hbar\omega) = 0$.

For $\hbar\omega \geq E_g$, $g(\hbar\omega) = \frac{1}{2\pi^2} \left(\frac{2\mu}{\hbar^2} \right)^{3/2} (\hbar\omega - E_g)^{1/2}$. $\qquad$ (3.24)

We therefore see that the density of states factor rises as $(\hbar\omega - E_g)^{1/2}$ for photon energies greater than the band gap.

3.3.4   The frequency dependence of the band edge absorption

Now that we have discussed the matrix element and the density of states, we can put it all together and deduce the frequency dependence of the absorption coefficient α. Fermi's golden rule given in eqn 3.2 tells us that the absorption rate for a dipole allowed interband transition is proportional to the joint density of states given by eqn 3.24. We therefore expect the following behaviour for $\alpha(\hbar\omega)$:

For $\hbar\omega < E_g$, $\alpha(\hbar\omega) = 0$.

For $\hbar\omega \geq E_g$, $\alpha(\hbar\omega) \propto (\hbar\omega - E_g)^{1/2}$. $\qquad$ (3.25)

There is no absorption if $\hbar\omega < E_g$, and the absorption increases as $(\hbar\omega - E_g)^{1/2}$ for photon energies greater than the band gap. We also expect that transitions with larger reduced masses will give rise to stronger absorption due to the $\mu^{3/2}$ factor in eqn 3.24.

The predictions of eqn 3.25 can be compared to experimental data. Figure 3.6 shows results for the absorption coefficient of the direct gap III–V semiconductor indium arsenide at room temperature. The graph plots α^2 against the photon energy in the spectral region close to the band gap. The straight-line relationship between α^2 and $(\hbar\omega - E_g)$ indicates that the model we have developed is a good one. The band gap

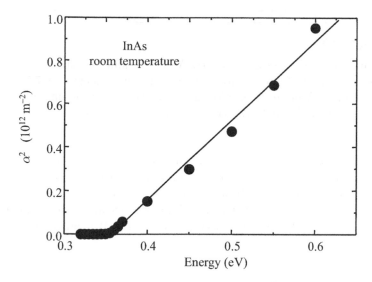

Fig. 3.6 Square of the optical absorption coefficient α versus photon energy for the direct gap III–V semiconductor InAs at room temperature. The band gap can be deduced to be $0.35\,\mathrm{eV}$ by extrapolating the absorption to zero. Data from Palik (1985).

can be read from the data as the point at which the absorption goes to zero. This gives a value of $0.35\,\mathrm{eV}$, which is in good agreement with values deduced from electrical measurements. Note that the values of the absorption coefficient are very large. This is a consequence of the very large density of states in the solid phase.

In many III–V semiconductors, including GaAs itself, it is found that the frequency dependence predicted by eqn 3.25 is only approximately obeyed. There are a number of reasons for this.

- We have neglected the Coulomb attraction between the electron and hole, which can enhance the absorption rate and cause exciton formation. These effects become stronger as the band gap gets larger and the temperature is lowered. This is why we have presented room temperature data for a semiconductor with a smallish band gap in Fig. 3.6. Excitonic effects are very significant in materials like GaAs even at room temperature. This point will be discussed further in Chapter 4, and is clearly apparent in the absorption data for GaAs shown in Fig. 4.3.

- The semiconductor may contain impurity or defect states with energies within the band gap, and these may allow absorption for photon energies less than the band gap. This point is discussed in Section 7.4.2.

- The parabolic band approximations embodied in the dispersion relations of eqns 3.17–3.20 are only valid near $k = 0$. As the photon energy increases above the band gap, the joint density of states will no longer obey the frequency dependence given in eqn 3.24. In these cases we must use the full band structure to evaluate the density of states, as discussed in Section 3.5 below.

Example 3.1

Indium phosphide is a direct gap III–V semiconductor with a band gap of 1.35 eV at room temperature. The absorption coefficient at 775 nm is $3.5 \times 10^6 \ \mathrm{m}^{-1}$. A platelet sample 1 μm thick is made with anti-reflection coated surfaces. Estimate the transmission of the sample at 620 nm.

Solution

The sample is anti-reflection coated, and so we do not need to consider multiple reflections. We therefore calculate the transmission from eqn 1.8 with $R = 0$. The wavelength of 775 nm corresponds to a photon energy of 1.60 eV, which is greater than E_g. Similarly, 620 nm corresponds to a photon energy of 2.00 eV, which is also above E_g. We can therefore use eqn 3.25 and write:

$$\frac{\alpha(620\,\mathrm{nm})}{\alpha(775\,\mathrm{nm})} = \frac{(2.00 - E_g)^{1/2}}{(1.60 - E_g)^{1/2}} = 1.6\,,$$

where we have used $E_g = 1.35$ eV. This implies that $\alpha(620\,\mathrm{nm}) = 5.6 \times 10^6 \ \mathrm{m}^{-1} \equiv 5.6 \ \mu\mathrm{m}^{-1}$, and hence that $\alpha l = 5.6$. We thus obtain the final result:

$$T(620\,\mathrm{nm}) = \exp(-\alpha l) = \exp(-5.6) = 0.37\%\,.$$

The value of T calculated in this example is only an estimate because we have ignored the excitonic effects and we have assumed that the parabolic band approximation is valid, even though we are quite a long way above E_g. The experimental value of $\alpha(620\,\mathrm{nm})$ is actually about 15% larger than the value calculated here.

3.3.5   The Franz–Keldysh effect

The modification of the band edge absorption by the application of an external electric field $\mathcal{E}$ was studied independently by W. Franz and L.V. Keldysh in 1958. They showed that there are two main effects:

- The absorption coefficient for photon energies less than E_g is no longer zero, as stated in eqn 3.25, but now decreases exponentially with $(E_g - \hbar\omega)$. The frequency dependence of α is given by:

$$\alpha(\hbar\omega) \propto \exp\left(-\frac{4\sqrt{2m_e^*}}{3|e|\hbar\mathcal{E}}(E_g - \hbar\omega)^{3/2}\right). \qquad (3.26)$$

  This implies that the band edge shifts to lower energy as the field is increased. (See Exercise 3.14.)

- The absorption coefficient for $\hbar\omega > E_g$ is modulated by an oscillatory function. The oscillations in $\alpha(\hbar\omega)$ are called Franz–Keldysh oscillations.

These two effects are collectively known as the **Franz–Keldysh effect**. They are typically observed when the semiconductor is incorporated as a thin i-region at the junction of a p–n diode. This allows controllable fields to be applied by varying the bias on the device, as explained in Appendix E.

It can be seen from the Kramers–Kronig relationship given in eqn 2.36 that a change in the absorption coefficient will produce changes in the refractive index at frequencies below the band gap. Thus the application of the electric field modulates both the absorption and the refractive index of the material. This modulation of the optical constants by the electric field is an example of an **electro-optic effect**. The changes may be either linear or quadratic in the field, as discussed in Section 2.5.2. In Chapter 11 we shall explain how these effects can be described in terms of nonlinear optical susceptibility tensors.

The changes in the real and imaginary parts of the refractive index produced by the electric field imply that the reflectivity will also be changed through eqn 1.29. This is the basis of the technique of **electro-reflectance**, in which the modulation of the reflectivity in response to an AC electric field is measured as a function of the photon energy. The electro-reflectance technique is widely used to determine important band structure parameters.

3.3.6   Band edge absorption in a magnetic field

It is well known in classical physics that the application of a strong magnetic field with flux density B causes electrons to perform circular motion around the field at the **cyclotron frequency** $\omega_{\rm c}$ given by (see Exercise 3.15):

$$\omega_{\rm c} = \frac{eB}{m_0}. \tag{3.27}$$

In classical physics, the radius of the orbit and the energy can have any values, but in quantum physics, they are both quantized. The quantized energies are given by:

$$E_n = (n + \tfrac{1}{2})\hbar\omega_{\rm c}, \tag{3.28}$$

where $n = 0, 1, 2, \dots$.. These quantized energy levels are called **Landau levels**.

Consider a semiconductor in the presence of a strong magnetic field along the z direction. The motion of the electrons in the conduction band and holes in the valence band will be quantized in the x, y plane, but their motion will still be free in the z direction. Their energies within the bands will thus be given by:

$$E^n(k_z) = (n + \tfrac{1}{2})\frac{e\hbar B}{m^*} + \frac{\hbar^2 k_z^2}{2m^*}, \tag{3.29}$$

where m^* is the appropriate effective mass. The first term gives the energy of the quantized motion in the x, y plane, while the second describes the free motion along the z direction. In absolute terms relative to $E = 0$ at the top of the valence band, the electron and hole energies are given by:

$$
\begin{aligned}
E_n^{\rm e}(k_z) &= E_{\rm g} + (n + \tfrac{1}{2})\frac{e\hbar B}{m_{\rm e}^*} + \frac{\hbar^2 k_z^2}{2m_{\rm e}^*}, \\[2mm]
E_n^{\rm h}(k_z) &= -(n + \tfrac{1}{2})\frac{e\hbar B}{m_{\rm h}^*} - \frac{\hbar^2 k_z^2}{2m_{\rm h}^*}.
\end{aligned}
\tag{3.30}
$$

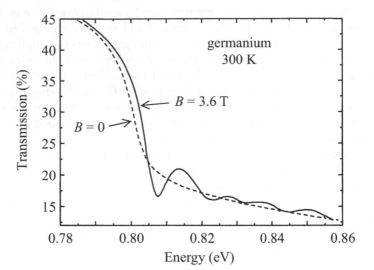

Fig. 3.7 Transmission spectrum of germanium for $B = 0$ and $B = 3.6$ T at 300 K. After Zwerdling et al. (1957), © American Physical Society, reprinted with permission.

These are equivalent to eqns 3.17–3.19, which are valid at $B = 0$.

If the sample is illuminated when the field is applied, an interband transition can take place in which an electron is created in the conduction band and a hole is created in the valence band. It can be shown that the Landau level number n does not change during the interband transition. (See Exercise 3.15.) This selection rule implies that the electron and hole must have the same value of n. Furthermore, the k_z value of both particles must be the same because the photon has negligible momentum. Therefore, the transition energy will be given by:

$$
\begin{aligned}
\hbar\omega &= E_n^e(k_z) - E_n^h(k_z) \\
&= E_g + (n + \tfrac{1}{2})\frac{e\hbar B}{\mu} + \frac{\hbar^2 k_z^2}{2\mu},
\end{aligned}
\tag{3.31}
$$

where μ is the reduced mass given in eqn 3.22. Equation 3.31 should be compared to eqn 3.23 which applies when $B = 0$. The term in k_z is unchanged, but the x and y components of $\boldsymbol{k}$ are now quantized by the magnetic field.

The frequency dependence of the absorption coefficient which follows from eqn 3.31 is considered in detail in Exercise 3.16. In brief, we expect very high absorption at any photon energy that can satisfy eqn 3.31 with $k_z = 0$. This gives rise to a series of equally spaced peaks in the absorption spectrum with energies given by

$$
\hbar\omega = E_g + (n + \tfrac{1}{2})\frac{e\hbar B}{\mu}; \qquad n = 0, 1, 2, \ldots .
\tag{3.32}
$$

One immediate consequence of this result is that we expect the absorption edge to shift to higher energy by $\hbar eB/2\mu$ in the magnetic field.

Figure 3.7 shows the room temperature transmission spectrum of germanium at $B = 0$ and $B = 3.6$ T. We see that at $B = 3.6$ T the absorption edge is indeed shifted to higher energy and there is a regularly spaced series of dips in the transmission, as predicted by eqn 3.32. The spectral width of the dips is determined mainly by line broadening due to scattering. The electron effective mass can be determined from the energies of the minima in the transmission: see Exercise 3.16.

3.3.7   Spin injection

In the absence of a magnetic field, the electrons in the conduction band are equally likely to be in the spin up ($m_s = +1/2$) and the spin down ($m_s = -1/2$) states. This means that there is normally no net spin in the electron gas. However, it is possible to create a net electron spin by absorption of circularly polarized light. This technique is called **optical spin injection**, or **optical orientation**, and is important in the field of **spintronics**, which aims to exploit the electron spin to generate new functionality in electronic devices.

Optical spin injection is possible, in general terms, because circularly polarized photons carry angular momentum components of $\pm\hbar$ along the direction of propagation. The sign of the angular momentum component depends on the sense of rotation, with $+\hbar$ for σ^+ photons and $-\hbar$ for σ^-. This means that the absorption of a circularly polarized light beam imparts angular momentum to the semiconductor as a whole, which can end up as a net spin of the electron gas.

Let us consider a direct gap III–V semiconductor in the four-band model with the band structure shown in Fig. 3.5. We concentrate on transitions at the fundamental band edge (i.e. at $k = 0$), where the heavy-hole and light-hole bands are degenerate. We have seen in Section 3.3.1 that the conduction band is derived from s-like atomic states with orbital angular momentum quantum number $L = 0$, while the valence band is derived from p-like states with $L = 1$. The electrons and holes have spin angular momentum with quantum number $S = 1/2$. In the conduction band we therefore have a single $J = 1/2$ level, whereas in the valence band we have two values of J, namely $J = 3/2$ and $J = 1/2$. At $k = 0$, these two valence band J levels are split by the spin–orbit energy Δ, as shown in Fig. 3.5.

In order to understand how spin injection works, we first need to consider the M_J states of the levels at $k = 0$. Figure 3.8 shows the detailed sub-level structure of the conduction and valence bands at $k = 0$. The conduction band consists of the degenerate $M_J = \pm 1/2$ sub-levels arising from the electron spin. The valence band is more complicated. The four sub-levels of the $J = 3/2$ level correspond to the degenerate heavy- and light-hole bands, with $M_J = \pm 3/2$ for the heavy-hole band and $M_J = \pm 1/2$ for the light-hole band. The $M_J = \pm 1/2$ sub-levels of the $J = 1/2$ level of the valence band correspond to the split-off hole band.

If the semiconductor is illuminated with circularly polarized light, the selection rules only permit specific transitions to occur. Light with positive circular polarization (i.e. σ^+ polarization) induces transitions with $\Delta M_J = +1$. (See Exercise 3.3.) Similarly, negatively circularly

Positive and negative circular light (i.e. σ^+ and σ^-) are defined by the sense of rotation relative to the *source*. This makes σ^+ and σ^- equivalent, respectively, to left and right circular, since the handedness of circular light is defined relative to the *observer*.

The rules for the addition of quantum mechanical angular momenta are summarized in Appendix C. See especially eqn C.7, and the subsequent discussion of the possible values of J.

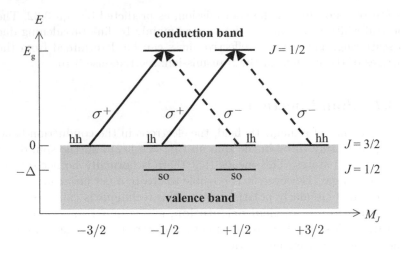

Fig. 3.8 Detailed sub-level structure of a semiconductor with the four-band model of Fig. 3.5 at $k = 0$. Circularly polarized transitions for a photon with energy E_g are shown from the degenerate heavy- and light-hole bands to the conduction band. It is assumed that there is no magnetic field applied.

Note that the removal of an electron with $J_z^{\text{electron}} = M_J \hbar$ from the valence band creates a hole with $J_z^{\text{hole}} = -M_J \hbar$. This is because the valence band has $\sum J_z = 0$ when it is fully occupied, so that $J_z^{\text{hole}} = -J_z^{\text{electron}}$. Thus a σ^+ heavy-hole transition creates an electron with $M_J = -1/2$ and a hole with $M_J = +3/2$. The electron-hole pair therefore has:

$$M_J^{\text{eh}} = M_J^{\text{electron}} + M_J^{\text{hole}} = +1 \,,$$

as required by conservation of angular momentum in the optical transition. Similarly, a σ^+ light-hole transition creates an electron with $M_J = +1/2$ and a hole with $M_J = +1/2$, again giving $M_J^{\text{eh}} = +1$.

In atomic physics, the spin–orbit interaction is proportional to $\boldsymbol{L} \cdot \boldsymbol{S}$, where $\boldsymbol{L}$ and $\boldsymbol{S}$ are the orbital and spin angular momenta respectively.

polarized light (i.e. σ^- polarization) induces transitions with $\Delta M_J = -1$. If the photon energy is just above the band gap energy E_g, then four transitions are possible, as illustrated in Fig. 3.8. In σ^+ light, we have transitions from the $M_J = -3/2$ heavy-hole sub-level to the $M_J = -1/2$ electron sub-level, and from the $M_J = -1/2$ light-hole sub-level to the $M_J = +1/2$ electron sub-level. The signs of the M_J states are reversed for σ^- polarization. It is shown in Exercise 3.9 that the square of the matrix element for heavy-hole transitions is three times larger than that for the light-hole transitions. Therefore, σ^+ light generates three times as many electrons with $M_J = -1/2$ as those with $M_J = +1/2$. The electron spin polarization is defined by:

$$\Pi = \frac{N(+1/2) - N(-1/2)}{N(+1/2) + N(-1/2)} \,, \tag{3.33}$$

where $N(+1/2)$ and $N(-1/2)$ are the number of electrons with spin $+1/2$ and $-1/2$ respectively. Hence we find $\Pi = -50\%$ for σ^+ excitation. Similarly, σ^- photons with energy just above E_g can generate an electron spin polarization with $\Pi = +50\%$. We therefore conclude that the use of circularly polarized light can produce a spin polarization of 50% in a bulk III–V semiconductor. In order to generate 100% electron spin polarization by optical excitation at zero magnetic field, it is necessary to use quantum well structures. (See Section 6.4.5.)

The arguments developed for the electrons should apply in an analogous way to the holes, because the σ^+ and σ^- transitions create holes with well-defined M_J values. However, the electrons, with $L = 0$, experience no spin–orbit interaction, whereas the holes, with $L = 1$, experience the strong spin–orbit interaction that is responsible for the splitting of the $J = 1/2$ and $J = 3/2$ states. This spin–orbit coupling randomizes the hole spin in a very short time, so that it is normally assumed that the hole-spin polarization is negligible. By contrast, the electron-spin polarization can persist for a significant amount of time until it is destroyed by a relatively slow spin-flip process.

It should be pointed out that the electron spin injection picture discussed above only applies to semiconductors that have the cubic zincblende crystal structure, such as InSb and GaAs. In wide gap semiconductors such as GaN or ZnO, the picture breaks down for two reasons. Firstly, the spin–orbit interaction is very small, and secondly, the crystals tends to adopt the hexagonal wurtzite structure. This means that crystal-field interactions have to be considered in addition to the spin–orbit coupling, and the end result is that the Kane four-band model that leads to the band structure shown in Fig. 3.8 does not apply to wide-gap semiconductors that have the wurtzite structure.

The band gap of a semiconductor generally decreases on descending the periodic table. For example:

$$E_g^{GaN} > E_g^{GaP} > E_g^{GaAs} > E_g^{GaSb}.$$

Since the spin–orbit coupling increases with the atomic number Z, this means that spin–orbit effects are more important in narrow gap semiconductors than in wide gap ones.

3.4 Band edge absorption in indirect gap semiconductors

In the previous two sections, we have been concentrating on direct interband transitions. As it happens, several of the most important semiconductors have indirect band gaps, most notably silicon and germanium. Indirect gap semiconductors have their conduction band minimum away from the Brillouin zone centre, as shown schematically in Fig. 3.2(b). Transitions at the band edge must therefore involve a large change in the electron wave vector. Optical frequency photons only have a very small k vector, and it is not possible to make this transition by absorption of a photon alone: the transition must involve a phonon to conserve momentum.

Consider an indirect transition that excites an electron in the valence band in state $(E_i, \mathbf{k}_i)$ to a state $(E_f, \mathbf{k}_f)$ in the conduction band. The photon energy is $\hbar\omega$, while the phonon involved has energy $\hbar\Omega$ and wave vector $\mathbf{q}$. Conservation of energy demands that:

$$E_f = E_i + \hbar\omega \pm \hbar\Omega, \tag{3.34}$$

while conservation of momentum requires that:

$$\hbar\mathbf{k}_f = \hbar\mathbf{k}_i \pm \hbar\mathbf{q}. \tag{3.35}$$

The $\pm$ factors allow for the possibility of phonon absorption or emission, with the $+$ sign corresponding to absorption, and the $-$ sign to emission. We have neglected the photon's momentum in eqn 3.35. This approximation was justified previously in connection with eqn 3.12.

Before considering the shape of the band edge absorption spectrum, we can first make a general point. Indirect transitions involve both photons and phonons. In quantum-mechanical terms, this is a second-order process: a photon must be destroyed, and a phonon must be either created or destroyed. This contrasts with direct transitions which are first-order processes because no phonons are involved. The transition rate for indirect absorption is therefore much smaller than for direct absorption.

The smaller transition rate for indirect processes is clearly shown by the data given in Fig. 3.9, which compares the band edge absorption

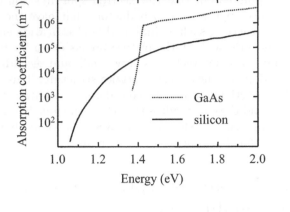

Fig. 3.9 Comparison of the absorption coefficients of GaAs and silicon near their band edges. GaAs has a direct band gap at 1.42 eV, while silicon has an indirect gap at 1.12 eV. Note that the vertical axis is logarithmic. Data from Palik (1985).

The derivation of eqn 3.36 may be found, for example, in Yu & Cardona (1996) or Hamaguchi (2001).

of silicon and GaAs. Silicon has an indirect band gap at 1.12 eV, while GaAs has a direct gap at 1.42 eV. We see that the absorption rises much faster with frequency in the direct gap material, and soon exceeds the indirect material even though its band gap is larger. The absorption of GaAs is roughly an order of magnitude larger than that of silicon for energies greater than ~ 1.43 eV.

The derivation of the quantum-mechanical transition rate for an indirect gap semiconductor is beyond the scope of this book. The results of such a calculation give the following result:

$$\alpha^{\text{indirect}}(\hbar\omega) \propto (\hbar\omega - E_{\text{g}} \mp \hbar\Omega)^2 . \tag{3.36}$$

This shows that we expect the absorption to have a threshold close to E_{g}, but not exactly at E_{g}. The difference is $\mp\hbar\Omega$, depending on whether the phonon is absorbed or emitted. Note that the frequency dependence is different to that for direct gap semiconductors given in eqn 3.25. This provides a convenient way to determine whether the band gap is direct or not. Furthermore, the involvement of the phonons gives other tell-tale signs that the band gap is indirect, as we shall discuss below.

Indirect absorption has been thoroughly studied in materials like germanium. The band structure of germanium is shown in Fig. 3.10. The overall shape of the band dispersion is fairly similar to that of GaAs given in Fig. 3.4. This is hardly surprising, given that gallium and arsenic lie on either side of germanium in the periodic table, so that GaAs and Ge are approximately isoelectronic materials. There is, however, one very important qualitative difference: the lowest conduction band minimum of germanium occurs at the L point, where $\boldsymbol{k} = (\pi/a)(1, 1, 1)$, and not at $k = 0$. This makes germanium an indirect semiconductor with a band gap of 0.66 eV.

Figure 3.11 shows the results of absorption measurements on germanium near its band edge. Figure 3.11(a) focuses on the indirect absorption near the band gap, while Fig. 3.11(b) concentrates on the direct absorption at higher energy. Consider first the indirect absorption edge shown in Fig. 3.11(a). In general we expect contributions both from phonon emission and absorption. Phonon emission is possible at all tem-

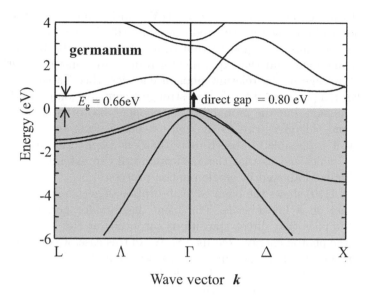

Fig. 3.10 Band structure of germanium. After Cohen and Chelikowsky (1988), © Springer-Verlag, reprinted with permission.

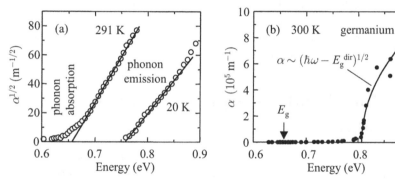

Fig. 3.11 Experimental data for the absorption of germanium near its band edge. (a) Plot of $\alpha^{1/2}$ against photon energy at 291 K and 20 K. (b) Plot of α against photon energy at 300 K. After MacFarlane & Roberts (1955) and Dash & Newman (1955), © American Physical Society, reprinted with permission.

peratures, but phonon absorption is only possible if phonons are thermally excited. The number of phonons of angular frequency Ω excited at temperature T is proportional to the Bose–Einstein formula:

$$f_{\text{BE}}(\hbar\Omega) = \frac{1}{\exp(\hbar\Omega/k_{\text{B}}T) - 1}. \tag{3.37}$$

The variation of the phonon populations implied by eqn 3.37 leads to a characteristic temperature dependence of the indirect absorption edge. As we decrease T, the contributions due to phonon absorption gradually freeze out, and at the lowest temperatures, the absorption edge is determined entirely by phonon emission. This contrasts with direct gap materials, in which the absorption edge merely shifts with the band gap as the temperature is varied.

The absorption spectra shown in Fig. 3.11(a) clearly show the behaviour discussed above. At 20 K only phonon emission is possible, and eqn 3.36 predicts that a graph of $\sqrt{\alpha}$ against photon energy should give a straight line extrapolating back to $(E_{\text{g}} + \hbar\Omega)$. This is clearly observed in the data, from which a value of $E_{\text{g}} + \hbar\Omega \approx 0.76$ eV is deduced. The

The Bose–Einstein formula is normally written:

$$f_{\text{BE}}(E) = \frac{1}{\exp[(E - \mu)/k_{\text{B}}T] - 1},$$

where μ is the chemical potential. When applying this to phonons, the chemical potential is zero because the number of particles does not have to be conserved.

Table 3.1 Phonon energies for germanium at the L point where $\boldsymbol{q} = (\pi/a)(1,1,1)$, a being the unit cell size. Data from Madelung (1996).

Mode	$\hbar\Omega$ (eV)
Longitudinal acoustic (LA)	0.027
Transverse acoustic (TA)	0.008
Longitudinal optic (LO)	0.030
Transverse optic (TO)	0.035

accepted band gap of Ge at 20 K is 0.74 eV, which implies that the average energy of the phonons emitted is around 0.02 eV. The wave vector of the phonon must be equal to that of an electron at the L point of the Brillouin zone, and the experimental results are consistent with a weighted average of the relevant phonon energies. (See Table 3.1.) At the higher temperature of 291 K, the strongest contribution is still from phonon emission, but the probability for phonon absorption is now significant, and this gives rise to the tail extending down to ~ 0.60 eV. Note that it is possible to absorb more than one phonon in an indirect transition, which means that the absorption tail can extend below the energy threshold allowed by single phonon processes.

Figure 3.11(b) shows the band edge absorption of germanium at room temperature on a linear scale. The band diagram for germanium in Fig. 3.10 implies that direct transitions can occur at the Γ point (i.e. $k = 0$) if the photon energy exceeds 0.80 eV. In this case, we expect that the absorption should follow eqn 3.25 instead of eqn 3.36. This is indeed observed in the data, with $\alpha \propto (\hbar\omega - E_g^{\text{dir}})^{1/2}$, where $E_g^{\text{dir}} = 0.805$ eV is the direct band gap. Note that the direct absorption completely dominates once we have crossed the threshold at E_g^{dir}. The indirect absorption below 0.80 eV is much weaker, and is insignificant when plotted on the same scale as the direct absorption. This highlights the second-order nature of indirect transitions.

3.5   Interband absorption above the band edge

Up to this point, we have been concentrating on the absorption near the band edge. As we shall see in Chapter 5, the reason for doing this is that the optical properties at the band edge determine the emission spectra. This does not mean that the rest of the absorption spectrum is uninteresting: it is just more complicated to deal with because the parabolic band approximation does not apply. However, as we shall see below, much useful information about the full band structure can be obtained from analysis of the overall spectrum.

It is not possible to give explicit formulae for the full frequency dependence of the absorption spectrum as we did for the band edge absorption in eqns 3.25 and 3.36. Instead, we have to work out dE/dk in eqn 3.14 from the full band structure. In this section we shall illustrate how this is done for the case of silicon. The principles described here can be applied to other materials if the band structure is known.

Figure 3.12 shows the interband absorption spectrum of silicon up to 10 eV. Two features at about 3.5 eV and 4.3 eV are readily identified in the data. These two energies are labelled E_1 and E_2 and are related to aspects of the band structure, as discussed further below. The absorption coefficient in the spectral region around E_1 and E_2 is extremely large, with values of α in excess of $10^8 \, \text{m}^{-1}$. This should be compared to values of 10^2–$10^6 \, \text{m}^{-1}$ in the spectral region immediately above the

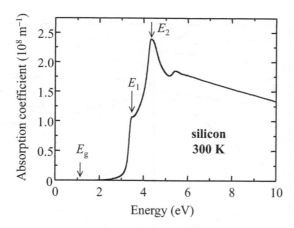

Fig. 3.12 Interband absorption spectrum of silicon at 300 K up to 10 eV. The energies E_1 and E_2 correspond to critical points where the conduction and valence bands are parallel to each other. This can be seen more clearly in the band structure diagram given in Fig. 3.13. Data from Palik (1985).

band gap E_g at 1.1 eV. (See Fig. 3.9.) Indeed, the band edge absorption is completely negligible on the scale of the data shown in Fig. 3.12. This is a consequence of two factors. Firstly, the band edge absorption is weak because it is indirect, and secondly, the density of states at the band edge is comparatively small. The measured absorption spectrum is actually dominated by direct absorption at photon energies where the density of states is very high.

Figure 3.13 shows the band structure of silicon along the (100) and (111) directions. The band gap E_g is indirect and has a value of 1.1 eV, with the conduction band minimum located near the X point of the Brillouin zone. Direct transitions can take place between any state in the valence band and the conduction band states directly above it, if the transitions are dipole-allowed. The minimum direct separation between the conduction and valence bands occurs near the L point, where the transition energy is 3.5 eV. The energy of these transitions is labelled E_1, and corresponds to the sharp increase in the absorption at 3.5 eV observed in the data shown in Fig. 3.12. The separation of the conduction and valence bands near the X point is also significant. This energy is labelled E_2 and corresponds to the absorption maximum at 4.3 eV.

The transitions near the L and X points are particularly important because of the 'camel's back' shape of the conduction band, which means that the conduction band ends up having a negative curvature near these points of the Brillouin zone. The curvature is more or less the same as that in the valence band, so that the conduction and valence bands are approximately parallel to each other. This means that we can have direct transitions with the same photon energy for many different values of k. The joint density of states factor is therefore very high at E_1 and E_2, and we expect the absorption to be correspondingly high. This is indeed observed in the experimental data: the absorption rises sharply at E_1 and reaches a peak at E_2. The absolute values of the absorption coefficient are extremely large: over $10^8 \, \mathrm{m}^{-1}$ as we have already noted.

In a region of the Brillouin zone where the bands are parallel, the photon energy E for direct transitions does not depend on k. This implies

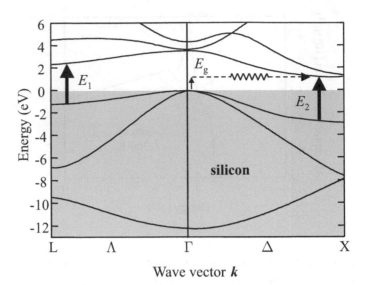

Fig. 3.13 Band structure of silicon. The band gap E_g is indirect and occurs at 1.1 eV. The conduction and valence bands are approximately parallel along the (111) and (100) directions near the zone edges at the L and X points. The separation of the bands in these region are labelled E_1 (3.5 eV) and E_2 (4.3 eV) respectively. The absorption at these energies is very high due to the van Hove singularities in the joint density of states. After Cohen and Chelikowsky (1988), © Springer-Verlag, reprinted with permission.

that dE/dk is zero, and hence that the joint density of states $g(E)$ diverges (cf. eqn 3.14). The energies at which dE/dk vanishes are called **critical points**, and the corresponding divergences in the density of states are called **van Hove singularities**. In practice, the bands are only approximately parallel over a portion of the Brillouin zone, and so $g(\hbar\omega)$ just becomes very large, rather than diverging completely.

The discussion of the absorption coefficient of silicon given here can be adapted to other materials if their band structure is known. The absorption strength will be proportional to the joint density of states, which will be particularly high if the conduction and valence bands are parallel to each other. An example of how this is done in the case of metals is discussed in Section 7.3.2 of Chapter 7.

3.6   Measurement of absorption spectra

The easiest way to measure the absorption coefficient of a material is to make a transmission measurement on a thin platelet sample. If the absorption is strong enough to dampen out any interference effects, then the absorption coefficient can be deduced from eqn 1.8 if the thickness and surface reflectivities are known. However, since the absorption coefficient can vary by several orders magnitude according to the wavelength, this can be more difficult than it sounds, and it is usually necessary to combine several techniques to determine α accurately over a wide range of photon energies.

Figure 3.14 illustrates the basic principles of transmission and reflection measurements. Light from a white-light source is filtered by a monochromator and is incident on the sample. The transmitted and reflected light is recorded by detectors as the photon energy is changed by scanning the monochromator. The detector collecting the reflected

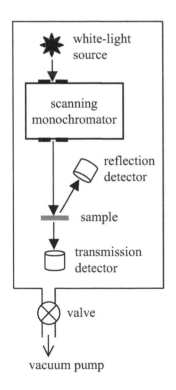

Fig. 3.14 Schematic diagram of the experimental arrangement required to determine the absorption coefficient over a wide spectral range by making reflectivity and transmissivity measurements.

Table 3.2 Experimental considerations for reflection and transmission measurements in different spectral regions. Note that the transition points from the infrared to near infrared and from the ultraviolet to vacuum ultraviolet spectral regions are not uniquely defined, and the values listed here are only approximate.

Spectral region	Wavelength range	Source	Detector
Infrared	> 1600 nm	Black body	Cooled semiconductor
Near infrared	700 – 1600 nm	Black body	Semiconductor
Visible	400 – 700 nm	Black body	Photomultiplier or silicon
Ultraviolet	200 – 400 nm	Xenon lamp	Photomultiplier or silicon
Vacuum ultraviolet	< 200 nm	Specialist UV source	Photomultiplier

light should be positioned so that the angle between the incoming and reflected light is as small as possible, so that the experiment effectively measures the reflection at normal incidence. The transmission coefficient is determined by comparing the signal on the transmission detector for two identical scans, one with the sample present, and the other without it. The reflectivity is determined by comparing the signal on the reflection detector to that obtained from a calibrated mirror. Aluminium is typically chosen as the mirror material, owing to its strong reflectivity up to 15 eV. (See Fig. 7.2.)

The choice of source and detector for a particular experiment depends on the spectral region in which the measurements are being made. (See Table 3.2.) A black-body emitter such as a tungsten bulb can be used as the source for measurements in the visible or infrared spectral regions, but at higher frequencies xenon arc lamps and other specialized ultraviolet sources must be used. Photomultiplier tubes can be used as the detector for the visible and ultraviolet regions, and silicon detectors can be used up to about 1000 nm. In the near infrared spectral region beyond the detection limit of silicon, it is possible to use high-efficiency InGaAs or germanium detectors developed for fibre optics telecommunications at 1550 nm. At longer wavelengths, narrow gap semiconductor detectors are selected according to the criteria discussed in Section 3.7.1.

For measurements in the infrared or vacuum ultraviolet spectral regions, the apparatus must be enclosed within a vacuum chamber to prevent absorption by the air molecules. Specialist optical components must be also used, because silica-based glasses no longer transmit. In some experiments, reflective optics only are used to avoid problems relating to absorption in the windows and lenses. Transmission measurements in the infrared spectral region beyond $\sim 5\,\mu$m become increasingly difficult due to the strong background black-body emission from objects in the laboratory at ambient temperatures. For this reason, a different technique called Fourier transform spectroscopy is commonly used for the long wavelength infrared spectral region.

Figure 3.15 shows a modern experimental arrangement designed for fast transmission measurements in the detection range of silicon, namely ~ 200–1000 nm. The light from a low intensity white-light source is passed through the sample, and the spectrum of the transmitted light

The absorption spectrum of high purity fused silica is given in Fig. 2.7(b). The transmission range is from about 180 nm to 3500 nm. Most other common types of glass include additives that reduce the transmission range.

is recorded with a spectrograph and a silicon diode array detector. The transmission coefficient is determined by calculating the ratio of the light on the detector with and without the sample present. The absorption coefficient is then calculated from the transmission using eqn 1.8, after measuring the reflectivity in a separate experiment. By placing the sample in a helium cryostat, the absorption coefficient can be measured as a function of temperature down to 2 K.

The measurement of the absorption coefficient of a material like silicon over a wide range of photon energies such as that presented in Fig. 3.12 is very difficult by transmission experiments alone. The absorption coefficient varies from about $10^3 \, \text{m}^{-1}$ at the indirect band edge to $> 10^8 \, \text{m}^{-1}$ at the critical points. In an ideal transmission experiment, the thickness of the sample should be of order α^{-1}, so that the absorption produces a measurable change of the transmission without making the sample almost totally opaque. This means that a series of plate thicknesses is required to cover the interesting spectral regions. Since this would require impractical thicknesses of order 10 nm for photon energies near and above the critical points, an alternative method based on reflection measurements is normally used.

In a reflection measurement, the absorption coefficient is calculated from the imaginary part of the complex refractive index by using eqn 1.19. κ itself is deduced from the measured reflectivity spectra $R(\hbar\omega)$ through eqn 1.29. This might seem impossible at first sight, because R depends on both n and κ. However, we know from Section 2.3 that n and κ are not completely independent variables and must be related to each other through the Kramers–Kronig relationships. Hence by self-consistent fitting of the reflectivity spectra using the Kramers–Kronig formulæ given in eqns 2.36 and 2.37, we can determine both n and κ from $R(\hbar\omega)$, and hence deduce α from κ.

In recent years a more refined version of the reflectivity technique called **ellipsometry** has been developed. In this technique the sample is illuminated at an oblique angle by linearly polarized light with the polarization vector outside the s- or p-plane. The reflected light becomes elliptically polarized because of the different reflectivities for s- and p-polarizations, and a careful analysis allows values of the real and imaginary parts of the refractive index to be determined. The accuracy of these values can depend critically on the cleanliness of the surface, because the light only penetrates a very small distance into the material when the absorption coefficient is very high.

For oblique incidence on a plane surface, light polarized parallel or perpendicular to the plane of incidence and reflection is called p-polarized or s-polarized, respectively. The difference in the reflection coefficients for these two polarizations is governed by Fresnel's equations. See, for example, Hecht (2001).

3.7 Semiconductor photodetectors

The strong absorption found in semiconductors is the basis of semiconductor **photodetectors**. Light with photon energy greater than the band gap is absorbed in the semiconductor, and this creates free electrons in the conduction band and free holes in the valence band. The presence of the light can therefore be detected either by measuring a

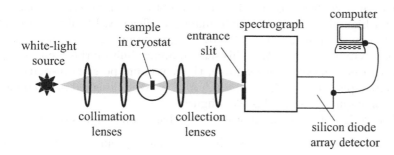

Fig. 3.15 Schematic diagram of a modern experimental arrangement to measure absorption spectra in the wavelength range 200–1000 nm using a silicon diode array detector.

change in the resistance of the sample or by measuring an electrical current in an external circuit. In this section we consider the operating principles of two different types of detector, and then discuss the use of semiconductor detectors in solar cells.

3.7.1   Photodiodes

Figure 3.16 shows a schematic diagram of a photodiode detector. The detector consists of a p–n junction with a thin intrinsic (undoped) layer sandwiched in the depletion region, forming a p–i–n structure. The band alignments and electrostatics of this type of structure are discussed in Appendix E. The diode is operated in reverse bias. This ensures that there is only a very small current in the circuit when no light is present, and applies a very strong DC electric field across the i-region. Photons absorbed in the i-region generate electron-hole pairs, that are rapidly swept towards the contacts by the field, and hence into the external circuit. The current generated in this way is called the **photocurrent**.

Consider a photodiode of active length l illuminated by a light beam of optical power P and angular frequency ω. The flux of photons per unit time on the detector is $P/\hbar\omega$. From the definition of the absorption coefficient given in eqn 1.4, we can deduce that the fraction of light absorbed in a length l is equal to $(1 - e^{-\alpha l})$, where α is the absorption coefficient at frequency ω. Each absorbed photon generates one electron-hole pair, and we define the **quantum efficiency** η as the fraction of these charge carriers that flow into the external circuit. The magnitude of the photocurrent I_{pc} is thus given by:

$$I_{\mathrm{pc}} = e\eta \frac{P}{\hbar\omega}(1 - e^{-\alpha l}). \tag{3.38}$$

We have assumed here that the top surface of the detector has been anti-reflection coated to prevent the wasteful reflection of incident photons. We have also assumed that the absorption in any layers above the active region is negligible.

The **responsivity** of the device is the ratio of the photocurrent I_{pc} to the optical power P, and is given by:

$$\mathrm{Responsivity} = \frac{I_{\mathrm{pc}}}{P} = \frac{\eta e}{\hbar\omega}(1 - e^{-\alpha l}) \quad \text{amps per watt.} \tag{3.39}$$

Many basic detectors just use p–n structures without the i-region. The light is absorbed in the depletion region at the junction, where there are no free carriers. The p–i–n structure is preferable because of the faster response times that can be achieved. It is, however, more complicated to make.

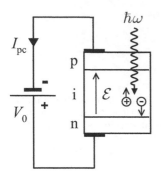

Fig. 3.16 Schematic diagram of a p–i–n photodiode. The diode is operated in reverse bias with a positive voltage V_0 applied to the n-region. This generates a strong DC electric field $\mathcal{E}$ across the i-region. Absorption of photons in the i-region creates free electrons and holes that are driven to the n-region and p-regions respectively by the field. The carriers that reach the doped regions flow into the external circuit, thereby generating the photocurrent I_{pc}.

Equation 3.39 shows us that in order to obtain a large responsivity we need a high absorption and high quantum efficiency. Ideally, we would like to have both η and $(1 - e^{-\alpha l})$ to be equal to unity, in which case the responsivity is simply $e/\hbar\omega$. This sets an upper limit on the responsivity that can be achieved. For example, the maximum possible responsivity for a 2 eV photon ($\lambda = 620$ nm) is 0.5 A W^{-1}. Well-designed photodiodes can come quite close to this ideal figure.

The design of practical photodiodes is based on several criteria.

- The choice of the semiconductor is made to optimize the responsivity while ensuring a fast response and low noise. The most fundamental criterion is that the band gap must be smaller than the photon energy. Having satisfied this criterion, we want E_g to be as large as possible to minimize the noisy dark current that arises from the thermal excitation of electrons and holes across the gap. At the same time, we want a material in which the electron and hole mobilities are high so that the photogenerated carriers can be swept quickly across the device and give a fast response time.

- Materials with direct band gaps are better than those with indirect gaps because the absorption is higher. With typical values of α over 10^6 m^{-1} for direct absorption, the thickness of the active layer needs only to be ~ 1 µm to achieve very strong absorption. In an indirect gap semiconductor, greater thicknesses are required, which increases the constraints on the purity of the material. Furthermore, the direct gap materials can give faster response times because the thinner i-region reduces the transit time of the device.

- The top contact should be designed to transmit as much of the light into the i-region as possible. This means that the top contact should be made very thin. A better solution is to use different semiconductors for the p–n junction and i-region, such that the band gap of the top contact is larger than the energy of the photons to be detected. This is possible with modern epitaxial semiconductor growth technology.

All these physical considerations have to be weighed against the manufacturing costs.

Table 3.3 gives a list of several common types of semiconductor photodetectors. Silicon is extensively used at visible and near infrared wavelengths, despite the fact that the absorption is indirect. This choice is determined by the advanced technology of the silicon industry. Germanium detectors can be used out to 1.9 µm, but for more demanding applications in the wavelength range 1–1.6 µm, the III–V alloy semiconductor InGaAs is becoming increasingly important. This is because it has a direct gap and also has a higher electron mobility than Ge, which means that fast, efficient detectors can be made for the telecommunications wavelengths of 1.3 µm and 1.5 µm.

At wavelengths beyond 1.9 µm, narrow gap semiconductors such as InAs or InSb have to be used. These long wavelength detectors invariably require cryogenic cooling to suppress the thermal dark currents and achieve good signal to noise ratios. The II–VI alloy semiconductor HgCdTe is frequently used for wavelengths beyond 5 μm. It has a band gap which can be varied according to the composition, and detectors with peak sensitivities in the range 5–14 μm are available. HgCdTe detectors are therefore able to cover several technologically important infrared wavelengths, especially 10.6 μm, which corresponds to one of the infrared windows in the atmosphere and also to the emission lines of the CO_2 laser. In Section 6.7 of Chapter 6 we shall describe an alternative detector for 10.6 μm which has recently been developed using GaAs quantum wells. These quantum well detectors operate on a different principle to the interband detectors described here.

Table 3.3 Common semiconductor photodetectors. E_g is the band gap, T the operating temperature, and λ_{max} the maximum wavelength that can be detected. The band gap of alloy semiconductors such as InGaAs and HgCdTe can be varied by altering the composition. The compositions listed here correspond to typical values used in detectors.

Semiconductor	E_g (eV)	T (K)	λ_{max} (μm)
Si	1.1	300	1.1
$In_{0.53}Ga_{0.47}As$	0.75	300	1.65
Ge	0.66	300	1.9
Ge	0.73	77	1.7
InAs	0.42	77	3.0
InSb	0.23	77	5.2
$Hg_{0.8}Cd_{0.2}Te$	0.09	77	14

Example 3.2

Estimate the responsivity of a 10 μm thick anti-reflection coated silicon photodiode at 800 nm. Calculate the photocurrent generated when the diode is illuminated with a 1 mW beam from a semiconductor laser operating at this wavelength.

Solution

The responsivity is given by eqn 3.39. We can read a value of $\alpha \approx 1 \times 10^5 \, \text{m}^{-1} \equiv 0.1 \, \mu\text{m}^{-1}$ for silicon at 800 nm (1.55 eV) from Fig. 3.9. The device is anti-reflection coated, and we therefore assume that no optical power is lost at the front surface. A well-designed photodiode will have negligible absorption in the top contact and quantum efficiency $\eta \approx 1$ at the operating wavelength. We therefore obtain:

$$\text{Responsivity} = \frac{e}{\hbar\omega}(1 - e^{-0.1 \times 10}) = 0.41 \text{ amps per watt}.$$

The photocurrent is given by the product of the responsivity and the optical power. The photocurrent will therefore be 0.41 mA.

3.7.2 Photoconductive devices

An alternative way to make a semiconductor photodetector is to use the photoconductive effect. This relies on the change of the conductivity of the material when illuminated by light. The conductivity is proportional to the density of free electrons and holes. The conductivity therefore increases due to the generation of free carriers after absorption of photons by interband transitions.

The devices consist of a sample with contacts at the ends so that a constant DC current can flow through the semiconductor between the contacts. The resistance between the contacts decreases upon illumination. This alters the voltage dropped across the device, and hence provides the detection mechanism. Photoconductive detectors are simpler to make than photodiodes, but tend to have slow response times.

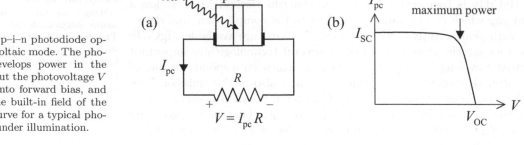

Fig. 3.17 (a) A p–i–n photodiode operating in photovoltaic mode. The photocurrent I_{pc} develops power in the load resistor R, but the photovoltage V puts the device into forward bias, and hence opposes the built-in field of the diode. (b) I–V curve for a typical photovoltaic device under illumination.

Solar cells can be used as renewable energy sources, which makes the development of low-cost, high-efficiency photovoltaic devices a very important research field.

3.7.3   Photovoltaic devices

Semiconductor photodiodes can also be operated in photovoltaic mode. In this mode of operation, the device does not have an external power supply, but instead generates a photovoltage when irradiated by light. The voltage can generate electrical power in an external load, thus converting optical energy into electrical energy. This is the basis of operation of **solar cells**, which generate electrical power from sunlight.

The operating principle of a photovoltaic device relies on the relationship between the photocurrent and the applied bias in a photodiode. The photocurrent is sensitive to the bias because it affects the electric field $\mathcal{E}$ across the depletion region. As explained in Appendix E, the field strength can be quite large even when the external bias is zero. This is because of the alignment of the Fermi levels in the p- and n-regions, which produces a voltage drop across the depletion region called the built-in voltage V_{bi}. The magnitude of V_{bi} is approximately equal to E_g/e. A forward bias approaching V_{bi} must therefore be applied before $\mathcal{E}$ drops to zero. The diode will produce a photocurrent on illumination provided that there is a field to sweep out the electrons and holes. Thus photocurrents can be produced at zero bias and even in forward bias, as long as the forward bias voltage is less than V_{bi}.

Let us suppose that we replace the battery in Fig. 3.16 with an electrical load of resistance R, as shown in Fig. 3.17(a). The voltage on the diode in the dark is zero. If the diode is illuminated, a photocurrent will be generated because the field due to the built-in voltage sweeps the carriers out of the i-region. This photocurrent flows through the load and the device therefore converts optical power to electrical power. The direction of the photocurrent is such that the photovoltage $V \equiv I_{pc}R$ puts the diode into forward bias. This limits the maximum power that can be generated, since I_{pc} drops as V increases towards V_{bi}, as shown schematically in Fig. 3.17(b). The two key parameters identified in Fig. 3.17(b) are the **open-circuit voltage** V_{OC} and the **short-circuit current** I_{SC}. V_{OC} is the voltage generated when the load resistance is very high so that no current flows, while I_{SC} is the current generated when the load resistance is very low, so that no voltage is generated. The power output is equal to $I_{pc}V$, and the maximum efficiency point usually occurs just below V_{OC}, as indicated in Fig. 3.17(b).

Solar radiation has a broad spectrum, and this puts conflicting demands on the optimization of the efficiency of solar cells.

(1) I_{SC} is proportional to the number of photons absorbed, which in turn is proportional to the number of photons in the solar spectrum with $\hbar\omega > E_g$. This favours solar cells with *small* band gaps, so that the largest part of the solar spectrum can be captured.

(2) Since $V_{bi} \sim E_g/e$, the open-circuit voltage increases with E_g, which favours devices with *large* band gaps.

An effective way to beat this trade-off is to develop 'multi-junction' solar cells incorporating two or more materials with varying band gaps within the active region. The high energy solar photons are captured by the large gap material at the front of the device, while the lower energy ones are transmitted through to the smaller gap material underneath. In this way a larger part of the solar energy spectrum can be harnessed with high efficiency.

The maximum power conversion efficiency that can be achieved from a silicon solar cell is in the range 10–25%, and the maximum voltage generated is about 0.6 V. Larger efficiencies ($\sim 40\%$) have been obtained from multi-junction devices. Multi-junction technology is very expensive, and its use is currently restricted to the most demanding applications such as space science.

Chapter summary

- Interband transitions occur when electrons jump to an excited state band by absorption of photons. The absorption process may be considered as the creation of an electron-hole pair.

- Interband absorption is only possible if the photon energy exceeds the band gap energy E_g. The absorption spectrum therefore shows a threshold at E_g.

- The absorption rate for direct transitions is proportional to the product of the joint density of states and the square of the electric-dipole matrix element.

- The photon wave vector is negligible compared to that of the electron, and so the electron wave vector is unchanged in a direct transition. Direct transitions are represented by vertical arrows on E–k band diagrams.

- The frequency dependence of the absorption edge of a direct gap semiconductor near E_g is given by eqn 3.25. At higher frequencies, the absorption coefficient is determined by the detailed frequency dependence of the joint density of states. The absorption is particularly high at critical points.

- The application of an external electric field results in non-zero absorption below the band gap through the Franz–Keldysh effect. The application of a magnetic field causes the absorption edge to shift to higher energy.

- Electron spin polarization can be created in a semiconductor with the zinc-blende structure by excitation with circularly polarized light.

- Interband transitions in indirect gap materials involve the absorption or emission of a phonon to conserve momentum in the process. Indirect absorption is much weaker than direct absorption since it is a second-order process.

- The frequency dependence of the absorption edge in an indirect gap material is given by eqn 3.36. This is different to that observed in direct gap semiconductors, and provides a way for determining the nature of the band gap experimentally.

- The absorption of light by interband transitions can be used to make photodetectors and photovoltaic devices. The photons with energies greater than the band gap generate a current in a photodetector and a voltage in a photovoltaic device. A photovoltaic device generating power from solar radiation is called a solar cell.

Further reading

The electronic states of solids are covered in the companion book of this series by Singleton (2001). They are also covered in all general solid-state physics texts, for example Burns (1985), Ibach & Luth (2003), or Kittel (2005), and in more detail by Harrison (1999).

Detailed information on the interband absorption of semiconductors may be found in Klingshirn (1995), Pankove (1971), Seeger (1997), or Yu & Cardona (1996). Introductory treatments of the application of group theory to interband transitions can be found in Klingshirn (1995) or Yu & Cardona (1996).

The Franz–Keldysh effect and the use of modulation spectroscopy to determine band structure parameters are described in Aspnes (1980), Hamaguchi (2001), Seeger (1997), and Yu & Cardona (1996), while Seeger (1997) gives a good discussion of the effect of magnetic fields on the band edge absorption. A detailed account of the accurate determination of the optical parameters of semiconductors by ellipsometry may be found in Aspnes and Studna (1983).

The physics of semiconductor photodetectors is described in more detail in Bhattacharya (1997), Chuang (1995), Sze (1985), Wilson and Hawkes (1998), or Yariv (1997). Sze (1985) gives a good discussion of solar cells.

Exercises

(3.1) Apply Born–von Karman periodic boundary conditions (i.e. $e^{ikx} = e^{ik(x+L)}$ etc., where L is a macroscopic length) to show that the density of states per unit volume in k space is $1/(2\pi)^3$.

(3.2) Show that the density of states for an electron with $E(k) = \hbar^2 k^2/2m^*$ is given by eqn 3.16.

(3.3) The wave function of an atomic state with principal quantum number n, orbital quantum number l and magnetic quantum number m may be written in the form:

$$\psi_{nlm}(r, \theta, \phi) = R_{nl}(r) Y_{lm}(\theta, \phi),$$

where $R_{nl}(r)$ is the radial wave function, $Y_{lm}(\theta, \phi)$ is the spherical harmonic function, and (r, θ, ϕ) are spherical polar coordinates. The spherical harmonic function itself may be written

$$Y_{lm}(\theta, \phi) = C P_l^m(\cos\theta) e^{im\phi},$$

where $P_l^m(\cos\theta)$ is a polynomial function in $\cos\theta$, and C is a normalization constant. The parity of the spherical harmonic functions is equal to $(-1)^l$.
(a) Explain what is meant by the 'parity' of an atomic wave function.

(b) The matrix element for an electric-dipole transition between states with initial and final wave functions ψ_i and ψ_f respectively is given by

$$M = \left| \int_{r=0}^{\infty} \int_{\theta=0}^{\pi} \int_{\phi=0}^{2\pi} \psi_f^* \, H' \, \psi_i \, r^2 \sin\theta \, \mathrm{d}r\mathrm{d}\theta\mathrm{d}\phi \right| \, ,$$

where $H' = -e\mathbf{r}$. By considering the parity of the wave functions, prove that $M = 0$ unless l changes by an odd number during the transition.[1]

(c) By writing the components of $\mathbf{r}$ in spherical polar coordinates, prove that $\Delta m = 0$ if the light is polarized along the z direction, and $\Delta m = \pm 1$ for light polarized in the x or y direction.

(d) By writing circularly polarized light in the form given in eqn A.40, show that σ^+ and σ^- light induce transitions with $\Delta m = +1$ and $\Delta m = -1$ respectively.

(3.4) Draw a schematic diagram of an experimental arrangement that could be used to obtain the absorption data shown in Fig. 3.6.

Table 3.4 Absorption coefficient α of GaP tabulated against photon energy E at 300 K. Data from Palik (1985).

E (eV)	α (m^{-1})	E (eV)	α (m^{-1})
2.2	3.12×10^1	2.7	7.39×10^5
2.3	7.79×10^3	2.8	3.35×10^6
2.4	2.72×10^4	2.9	5.38×10^6
2.5	6.43×10^4	3.0	6.81×10^6
2.6	1.44×10^5	3.1	8.64×10^6

(3.5) Explain how you would use optical absorption measurements to determine whether a semiconductor has a direct or indirect band gap.

(3.6) Table 3.4 gives absorption data for gallium phosphide at 300 K. What can you deduce about the band structure of GaP from this data?

(3.7) Use the data given in Fig. 3.11 to estimate the absorption coefficient of germanium at 1200 nm.

(3.8) The band parameters of the four-band model shown in Fig. 3.5 are given for GaAs in Table D.2. (a) Calculate the k vector of the electron excited from the heavy-hole band to the conduction band in GaAs when a photon of energy 1.6 eV is absorbed at 300 K. What is the corresponding value

for the light-hole transition ?
(b) Calculate the wave vector of the photon inside the crystal. Does this confirm the validity of the approximation given in eqn 3.12 ? The refractive index of GaAs at 1.6 eV is 3.7.
(c) Calculate the ratio of the joint density of states for the heavy- and light-hole transitions.
(d) What is the wavelength at which transitions from the split-off hole band become possible?

(3.9) Consider an electric-dipole transition with $\Delta J = -1$, as appropriate for transitions from the heavy-hole and light-hole bands to the conduction band at $k = 0$. The matrix elements for σ^-, linear, and σ^+ light are given respectively by:[2]

$$|\langle J-1, M_J-1|\sigma^-|J, M_J\rangle|^2 =$$
$$\tfrac{1}{2}(J+M_J)(J+M_J-1)C \, ,$$
$$|\langle J-1, M_J|z|J, M_J\rangle|^2 =$$
$$(J^2 - M_J)^2 C \, ,$$
$$|\langle J-1, M_J+1|\sigma^+|J, M_J\rangle|^2 =$$
$$\tfrac{1}{2}(J-M_J)(J-M_J-1)C \, ,$$

where C is the same for all three transitions. Use these results to show that, for circularly polarized light, the square of the matrix element for heavy-hole transitions in a semiconductor with the band structure shown in Fig. 3.8 is three times as strong as for light-hole transitions.

(3.10) Explain why the electron spin polarization generated with linearly polarized light is equal to zero.

(3.11) Discuss the variation of the electron spin polarization generated by absorption of circularly polarized photons as the photon energy is increased above the band gap energy.

(3.12)* In silicon the s-like anti-bonding orbital lies at a higher energy than the p-like anti-bonding orbitals, which contrasts with the ordering of the levels for Ge or GaAs shown in Fig. 3.3. This leads to major qualitative differences between the conduction band states of silicon and germanium at the Γ point, as can be seen by comparing Figs 3.10 and 3.13.
(a) Deduce the value of the direct band gap of silicon at the Γ point from the band structure diagram given in Fig. 3.13.

---

[1]By considering the properties of the function $P_l^m(\cos\theta)$, it is possible to prove that the selection rule on Δl is stricter than just being an odd number: Δl must in fact be equal to ± 1.
[2]See, for example, Woodgate (1980), Table 8.1, or Corney (1977), Table 5.1.
*Exercises marked with an asterisk are more challenging

(b) Explain qualitatively how the transitions at energies E_1 and E_2 can be dipole allowed.

(3.13) Where would you expect to measure the optical absorption edge in germanium at $4\,\mathrm{K}$? The indirect band gap is $0.74\,\mathrm{eV}$ at this temperature.

(3.14) Estimate the electric field strength at which the band edge of GaAs is red shifted by $0.01\,\mathrm{eV}$. The electron effective mass is $0.067m_0$.

(3.15)* Show that a classical particle of mass m and charge e performs circular orbits around a magnetic field with an angular frequency of eB/m, where B is the field strength. Show also that the selection rule for the Landau level number n during an interband transition is $\Delta n = 0$.

(3.16)* (a) Show that the density of states of a particle which is free to move in one dimension only is proportional to $E^{-1/2}$, where E is the energy of the particle.

(b) Draw a sketch of the frequency dependence of the optical absorption edge of a one-dimensional direct gap semiconductor.

(c) Explain why a bulk semiconductor in a strong magnetic field can be considered as a one-dimensional system. Hence explain the shape of the optical transmission spectrum of germanium at $300\,\mathrm{K}$ at $3.6\,\mathrm{T}$ given in Fig. 3.7.

(d) Use the data in Fig. 3.7 to deduce values for the band gap and the electron effective mass of Ge on the assumption that $m_{\mathrm{h}}^* \gg m_{\mathrm{e}}^*$. Comment on the values you obtain.

(3.17) The absorption coefficient of germanium is $4.6 \times 10^4\,\mathrm{m}^{-1}$ at $1.55\,\mu\mathrm{m}$ and $7.5 \times 10^5\,\mathrm{m}^{-1}$ at $1.30\,\mu\mathrm{m}$. Calculate the maximum responsivities of a germanium photodiode with a $10\,\mu\mathrm{m}$ thick absorbing layer at these two wavelengths.

(3.18) (a) The capacitance of a reverse biased p–i–n photodiode can be calculated by treating the device as a parallel plate capacitor. Justify this approximation.

(b) Calculate the capacitance of a silicon p–i–n photodiode with an area of $1\,\mathrm{mm}^2$ and an i-region thickness of $10\,\mu\mathrm{m}$. The static dielectric constant of silicon is 11.9.

(c) Estimate the time taken for the photogenerated electrons and holes to drift across the i-region when the reverse bias on the photodiode is $10\,\mathrm{V}$. Assume that the built-in voltage is $1.1\,\mathrm{V}$, and that the electron and hole mobilities of Si at room temperature are $0.15\,\mathrm{m}^2\,\mathrm{V}^{-1}\,\mathrm{s}^{-1}$ and $0.045\,\mathrm{m}^2\,\mathrm{V}^{-1}\,\mathrm{s}^{-1}$ respectively.

(d) At what voltage would the electron transit time be equal to the RC time constant of the diode when connected to a $50\,\Omega$ load?

Excitons

4

In the previous chapter we discussed the absorption of photons by interband transitions. We saw that this process creates an electron in the conduction band and a hole in the valence band, but we neglected the effects of the mutual Coulomb attraction between them. As we shall see in this chapter, the Coulomb interaction can give rise to the formation of new excitations of the crystal called excitons. These excitons have interesting optical properties and are important for opto-electronic applications.

We shall encounter excitons in several different contexts throughout this book. In this chapter we concentrate mainly on their effects on the absorption edge of bulk semiconductors. In Chapter 6 we shall see how the excitonic effects can be enhanced in quantum-confined structures, and then in Chapter 8 we shall see how excitonic effects have a strong influence on the optical properties of molecular materials. Finally, in Chapter 11 we shall briefly study how the presence of excitons can give rise to useful nonlinear optical properties.

4.1	The concept of excitons	95
4.2	Free excitons	96
4.3	Free excitons in external fields	101
4.4	Free excitons at high densities	104
4.5	Frenkel excitons	107
Chapter summary		109
Further reading		110
Exercises		110

4.1 The concept of excitons

The absorption of a photon by an interband transition in a semiconductor or insulator creates an electron in the conduction band and a hole in the valence band. The oppositely charged particles are created at the same point in space and can attract each other through their mutual Coulomb interaction. This attractive interaction increases the probability of the formation of an electron-hole pair, and therefore increases the optical transition rate. Moreover, if the right conditions are satisfied, a bound electron-hole pair can be formed. This neutral bound pair is called an **exciton**. In the simplest picture, the exciton may be conceived as a small hydrogenic system similar to a positronium atom with the electron and hole in a stable orbit around each other.

Excitons are observed in many crystalline materials. There are two basic types:

- **Wannier–Mott excitons**, also called **free excitons**;
- **Frenkel excitons**, also called **tightly bound excitons**.

The Wannier–Mott excitons are mainly observed in semiconductors, while the Frenkel excitons are found in insulators and molecular crystals.

The two types of exciton are illustrated schematically in Fig. 4.1. The diagrams show an electron and hole orbiting around each other

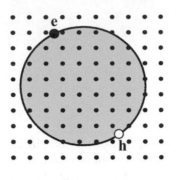

(a) Free exciton

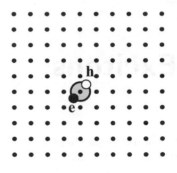

(b) Tightly bound exciton

Fig. 4.1 Schematic diagram of: (a) a free exciton, and (b) a tightly bound exciton. The free excitons illustrated in (a) are also called Wannier–Mott excitons, while the tightly bound excitons illustrated in (b) are also called Frenkel excitons.

within a crystal. The Wannier–Mott type excitons have a large radius that encompasses many atoms, and they are delocalized states that can move freely throughout the crystal: hence the alternative name of 'free' excitons. Frenkel excitons, by contrast, have a much smaller radius which is comparable to the size of the unit cell. This makes them localized states which are tightly bound to specific atoms or molecules; hence their alternative name of 'tightly bound' excitons. Tightly bound excitons are much less mobile than free excitons, and they have to move through the crystal by hopping from one atom site to another.

Stable excitons will only be formed if the attractive potential is sufficient to protect the exciton against collisions with phonons. Since the maximum energy of a thermally excited phonon at temperature T is $\sim k_{\mathrm{B}}T$, where k_{B} is Boltzmann's constant, this condition will be satisfied if the exciton binding energy is greater than $k_{\mathrm{B}}T$. Wannier–Mott excitons have small binding energies due to their large radius, with typical values of around $0.01\,\mathrm{eV}$. Since $k_{\mathrm{B}}T \sim 0.025\,\mathrm{eV}$ at room temperature, the excitons are only observed clearly at cryogenic temperatures in many materials. Frenkel excitons, on the other hand, have larger binding energies of the order 0.1–$1\,\mathrm{eV}$, which makes them stable at room temperature.

In the sections that follow, we first describe the basic properties of free excitons, and then study how they are affected by external electric and magnetic fields. We then discuss the interactions between excitons, which are the basis for the nonlinear optical properties of excitons discussed in Chapter 11. We close the chapter with a brief discussion of the optical properties of Frenkel excitons.

4.2   Free excitons

4.2.1   Binding energy and radius

In a free exciton, the average separation of the electrons and holes is much greater than the atomic spacing, as shown in Fig. 4.1(a). This is effectively the definition of a Wannier exciton, and it specifies more

accurately what is meant by saying that the free exciton is a weakly
bound electron-hole pair. Since the electron-hole separation is so large,
it is a good approximation to average over the detailed structure of the
atoms in between the electron and hole and consider the particles to
be moving in a uniform dielectric material. We can then model the free
exciton as a hydrogenic system similar to positronium.

We know from atomic physics that the motion of hydrogenic atoms
splits into the centre of mass motion and the relative motion. (See Exercise 4.1.) The centre of mass motion describes the kinetic energy of
the atom as a whole, while the relative motion determines the internal
structure. The energies of the bound states can be determined by finding
the eigenvalues of the Schrödinger equation for the relative motion, or
alternatively by using approximation techniques such as the variational
method. (See Exercises 4.2–4.4). The main results are, however, well
explained by using the Bohr model (see Exercise 4.5), and this is the
procedure we adopt here.

In applying the Bohr model to the exciton, we must take account of
the fact that the electron and hole are moving through a medium with
a high dielectric constant ϵ_r. We must also remember that the reduced
mass μ will be given by eqn 3.22, instead of the value of $0.9995m_0$ that
applies to the electron–proton system in a hydrogen atom. With these
two qualifications, we can then just use the standard results of the Bohr
model. The bound states are characterized by the principal quantum
number n, and the energy of the nth level relative to the ionization limit
is given by:

$$E(n) = -\frac{\mu}{m_0}\frac{1}{\epsilon_r^2}\frac{R_H}{n^2} = -\frac{R_X}{n^2}, \qquad (4.1)$$

where R_H is the Rydberg energy of the hydrogen atom (13.6 eV). The
quantity $R_X = (\mu/m_0\epsilon_r^2)R_H$ introduced here is the exciton Rydberg
energy. The radius of the electron-hole orbit is given by

$$r_n = \frac{m_0}{\mu}\epsilon_r\, n^2 a_H = n^2 a_X \qquad (4.2)$$

where a_H is the Bohr radius of the hydrogen atom (5.29×10^{-11}m) and
$a_X = (m_0\epsilon_r/\mu)a_H$ is the exciton Bohr radius. Equations 4.1 and 4.2
show that the ground state with $n = 1$ has the largest binding energy
and smallest radius. The excited states with $n > 1$ are less strongly
bound and have a larger radius.

Table 4.1 lists the exciton Rydberg energy and Bohr radius for a
number of direct gap III–V and II–VI semiconductors. A general pattern
is easily noticed in the data, namely that R_X tends to increase and a_X
to decrease as E_g increases. This is explained by the fact that ϵ_r tends
to decrease and μ to increase as the band gap increases. From eqns 4.1
and 4.2, we see that this causes an increase in the exciton binding energy
and a decrease in the radius. In insulators with band gaps greater than
about 5 eV, a_X becomes comparable to the unit cell size, and the Wannier
model is no longer valid. At the other extreme, R_X is so small in narrow
gap semiconductors such as InSb that it is difficult to observe any free

Table 4.1 Calculated Rydberg
energy and Bohr radius of the free
excitons in several direct gap III–V
and II–VI compound semiconductors. The bracketed figures for InSb
indicate that there has been no
experimental confirmation of the
values.
E_g: band gap,
R_X: exciton Rydberg energy from
eqn 4.1,
a_X: exciton Bohr radius from
eqn 4.2.

Crystal	E_g (eV)	R_X (meV)	a_X (nm)
GaN	3.5	23	3.1
ZnSe	2.8	20	4.5
CdS	2.6	28	2.7
ZnTe	2.4	13	5.5
CdSe	1.8	15	5.4
CdTe	1.6	12	6.7
GaAs	1.5	4.2	13
InP	1.4	4.8	12
GaSb	0.8	2.0	23
InSb	0.2	(0.4)	(100)

exciton effects at all. Hence, free exciton behaviour is best observed in semiconductors with medium-sized band gaps in the range $\sim 1\text{–}3\,\mathrm{eV}$.

Example 4.1

It is not immediately obvious what is the correct dielectric constant or hole effective mass to use for a III–V semiconductor such as GaAs. This is because ϵ_r varies with frequency (see Section 10.2), and the heavy- and light-hole bands are degenerate at $k = 0$. (See Fig. 3.5.) As a rule of thumb, we use the value of ϵ_r for the photon energy that corresponds to R_X, and a weighted average of the heavy- and light-hole masses for m_h^*. In this example, R_X comes out to be $4.2\,\mathrm{meV}$, which is in the far-infrared spectral region. We therefore use the static dielectric constant ϵ_st for ϵ_r.

(i) Calculate the exciton Rydberg energy and Bohr radius for GaAs, which has $\epsilon_\mathrm{r} = 12.8$, $m_\mathrm{e}^* = 0.067m_0$ and $m_\mathrm{h}^* = 0.2m_0$.

(ii) GaAs has a cubic crystal structure with a unit cell size of $0.56\,\mathrm{nm}$. Estimate the number of unit cells contained within the orbit of the $n = 1$ exciton. Hence justify the validity of assuming that the medium can be treated as a uniform dielectric in deriving eqns 4.1 and 4.2.

(iii) Estimate the highest temperature at which it will be possible to observe stable excitons in GaAs.

Solution

(i) We first need to calculate the reduced electron-hole mass μ, which is given by eqn 3.22. With $m_\mathrm{e}^* = 0.067m_0$, and $m_\mathrm{h}^* = 0.2m_0$, we find

$$\mu = \left(\frac{1}{0.067m_0} + \frac{1}{0.2m_0}\right)^{-1} = 0.05m_0\,.$$

We then insert this value of μ and $\epsilon_\mathrm{r} = 12.8$ into eqns 4.1 and 4.2 to obtain:

$$R_\mathrm{X} = \frac{0.05}{12.8^2} \times 13.6\,\mathrm{eV} = 4.2\,\mathrm{meV}\,,$$

and

$$a_\mathrm{X} = \frac{12.8}{0.05} \times 0.0529\,\mathrm{nm} = 13\,\mathrm{nm}\,.$$

(ii) We see from eqn 4.2 that the radius of the $n = 1$ exciton is equal to a_X. The volume occupied by this exciton is $\frac{4}{3}\pi a_\mathrm{X}^3$ which is equal to $9.2 \times 10^{-24}\,\mathrm{m}^3$. The volume of the cubic unit cell is equal to $(0.56\,\mathrm{nm})^3 = 1.8 \times 10^{-28}\,\mathrm{m}^3$. Hence the exciton volume can contain 5×10^4 unit cells. Since this is a large number, the approximation of averaging the atomic structure to a uniform dielectric is justified.

(iii) The $n = 1$ exciton has the largest binding energy with a value of $4.2\,\mathrm{meV}$. This is equal to $k_\mathrm{B}T$ at $49\,\mathrm{K}$. Therefore, we would not expect the excitons to be stable above $\sim 50\,\mathrm{K}$.

4.2.2   Exciton absorption

Free excitons are typically observed in direct gap semiconductors such as GaAs. They are created during direct optical transitions between the valence and conduction bands. As discussed in Section 3.2 this creates an electron-hole pair in which the electron and hole have the same k vector.

Excitons can only be formed if the electron and hole group velocities v_e and v_h are the same. This is a necessary condition for the electrons and holes to be able to move together as a bound pair. The group velocity of an electron in a band is given by (see eqn D.4):

$$v_g = \frac{1}{\hbar}\frac{\partial E}{\partial \boldsymbol{k}} . \tag{4.3}$$

This implies that the condition $v_e = v_h$ can only be satisfied if the gradients of the conduction and valence bands are the same at the point of the Brillouin zone where the transition occurs. All bands have zero gradient at the zone centre. Hence we can form excitons during a direct transition at $\boldsymbol{k} = 0$. In a direct gap semiconductor, these transitions correspond to photon energies of E_g. (See eqn 3.23.) Therefore we expect to observe strong excitonic effects in the spectral region close to the fundamental band gap.

The energy of the exciton created in a direct transition at $\boldsymbol{k} = 0$ is equal to the energy required to create the electron-hole pair, namely E_g, less the binding energy due to the Coulomb interaction, which is given by eqn 4.1. Hence the energy of the exciton will be given by:

$$E_n = E_g - \frac{R_X}{n^2} . \tag{4.4}$$

Whenever the photon energy is equal to E_n, excitons can be formed. The probability for the formation of excitons is expected to be high, because it is energetically favourable for the exciton states to be formed compared to free electron-hole pairs. Therefore we expect to observe strong optical absorption lines at energies equal to E_n. These will appear in the optical spectra at energies just below the fundamental band gap.

The band edge absorption spectrum expected when excitonic effects are included is illustrated schematically in Fig. 4.2. The electron-hole Coulomb interaction causes a series of excitonic absorption lines to appear just below the band gap, and enhances the absorption coefficient just above the band gap. The second point is a consequence of the fact that the Coulomb attraction decreases the size of the free electron and hole wave functions and hence increases their overlap, thereby leading to an increased transition probability.

Free excitons can only be observed in the absorption spectrum of very pure samples. This is because impurities release unpaired free electrons and holes that can screen the Coulomb interaction in the exciton and thereby strongly reduce the binding forces. For this reason, excitonic effects are not usually observed in doped semiconductors or metals, since they contain a very high density of free carriers. Charged impurities also generate electric fields, which tend to ionize the excitons, as discussed in Section 4.3.1.

Free excitons can also be observed near the fundamental band gap of indirect semiconductors such as silicon and germanium. These indirect excitons are more difficult to conceptualize because the electrons and holes have different $\boldsymbol{k}$ vectors. The condition $v_e = v_h$ is satisfied

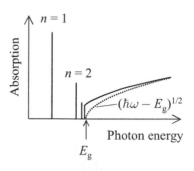

Fig. 4.2 Band edge absorption spectrum for a direct gap semiconductor with excitonic effects included. The dotted line shows the expected absorption when the excitonic effects are ignored.

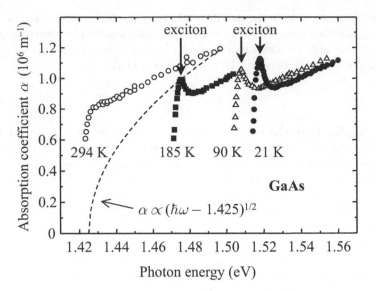

Fig. 4.3 Excitonic absorption of GaAs between 21 K and 294 K. The dashed line is an attempt to fit the absorption edge using eqn 3.25 with a value of E_g equal to 1.425 eV, which is appropriate for GaAs at 294 K. After Sturge (1962), © American Physical Society, reprinted with permission.

because the electron at the conduction band minimum still has $v_e = 0$, even though it has a large $\boldsymbol{k}$ vector. Experimental results give the binding energies of the free excitons in silicon and germanium as 14 meV and 4 meV respectively. These values are slightly higher than the general trends for direct gap semiconductors shown in Table 4.1. This is because of the larger electron mass at the zone edges compared to the Γ point. It is difficult to observe indirect excitons in absorption because of the reduced probability for indirect transitions. They can, however, be clearly observed in emission experiments, as will be discussed briefly in Section 4.4.

4.2.3   Experimental data for free excitons in GaAs

Figure 4.3 gives experimental data for the excitonic absorption of undoped GaAs between 21 K and 294 K. As expected, the data show strong absorption lines at photon energies just below the fundamental band gap of GaAs. At 21 K a peak is observed just below the direct absorption edge. This corresponds to the $n = 1$ exciton. The line is too broad to permit observation of any of the excited states. As the temperature is increased, the band gap shifts to lower energy and the exciton line weakens. At room temperature where $k_B T \gg R_X$, the exciton line has completely gone.

The spectrum at 185 K shows a weak exciton line at the band edge even though $k_B T$ is almost four times greater than R_X. This indicates that the criterion for exciton stability used in Example 4.1(iii), namely $k_B T < R_X$, is too stringent. The main mechanism that causes dissociation of excitons is collisions with longitudinal optic (LO) phonons. As the probability of such collisions increases, the lifetime of the excitons shortens. This leads to a corresponding broadening of the exciton line in the absorption spectrum. In GaAs, the relevant LO phonon has an

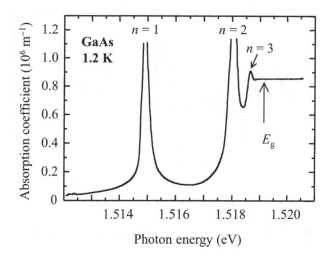

Fig. 4.4 Excitonic absorption of ultra pure GaAs at 1.2 K. After Fehrenbach et al. (1985), © Excerpta Medica Inc., reprinted with permission.

energy of 35 meV, which has a thermal occupation of 11% at 185 K. (See eqn 3.37.) There are therefore still relatively few LO phonons in the crystal at this temperature, and the exciton line is just resolved.

The dashed line in Fig. 4.3 shows the frequency dependence of the absorption edge expected if excitonic effects are ignored. This line is obtained from eqn 3.25 with a value of 1.425 eV for E_g, which is appropriate for GaAs at 294 K. We see that the fit to the data is not good. This tells us that the Coulomb interaction between the electron and hole still enhances the absorption rate considerably, even though there are no clear exciton lines observed in the spectrum.

Figure 4.4 shows more recent data for the excitonic absorption of ultra pure GaAs at 1.2 K. The data clearly show the hydrogen-like energy spectrum of the exciton in the vicinity of the band gap. The exciton lines are more clearly resolved in this data set than in Fig. 4.3 because the temperature is lower and the sample purity is superior. As discussed above, the presence of impurities leads to screening of the Coulomb interaction by free carriers, while lower temperatures reduce the thermal broadening of the absorption lines.

Three exciton states can be clearly identified in the absorption spectrum shown in Fig. 4.4. The energies of the $n = 1$, $n = 2$ and $n = 3$ excitons are 1.5149 eV, 1.5180 eV, and 1.5187 eV respectively. These energies fit eqn 4.4 very well with E_g=1.5191 eV and R_X = 4.2 meV. This value of E_g agrees well with other measurements, while the experimental figure of 4.2 meV for R_X is in excellent agreement with the value calculated in Example 4.1.

In Chapter 6 we shall discuss how the excitonic effects in materials such as GaAs can be enhanced in quantum-confined structures. This has made it possible to observe very strong free exciton absorption lines in GaAs quantum wells even at room temperature.

4.3   Free excitons in external fields

Free excitons are bound together by the electrostatic attraction between the negative electron and the positive hole. External electric and magnetic fields perturb the system through the forces exerted on the charged

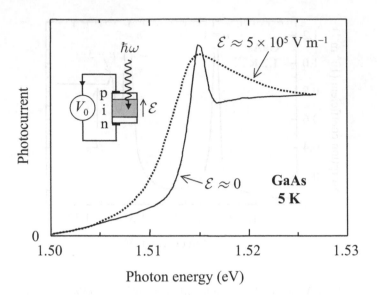

Fig. 4.5 Field ionization of the free excitons in GaAs at 5 K. The data was taken on a GaAs p–i–n diode with an i-region thickness of 1.0 μm. The solid line corresponds to 'flat band' conditions (forward bias = +1.44 V, $\mathcal{E} \approx$ 0), while the dashed line is for a forward bias of +1.00 V, where $\mathcal{E} \approx 5 \times 10^5$ V m^{-1}. No exciton lines are resolved at zero bias. Unpublished data from G. von Plessen and A.M. Tomlinson.

particles. The effects of these perturbations are discussed here, using the excitons in GaAs as an example.

4.3.1   Electric fields

When a DC electric field $\mathcal{E}$ is applied to an exciton, the oppositely charged electrons and holes are pushed away from each other. It is shown in Exercise 4.10 that the order of magnitude of the electric field between the electron and hole in the ground state exciton is equal to $2R_X/ea_X$. If $\mathcal{E}$ exceeds this value, the exciton will break apart. This effect is known as **field ionization**.

Electric fields are applied to excitons by incorporating the semiconductor as the i-region in a p–i–n diode structure, as discussed in Appendix E. The field strength across the i-region when a bias voltage V_0 is applied is given by eqn E.3 as:

$$\mathcal{E} = \frac{|V_{bi} - V_0|}{l_i} \,, \tag{4.5}$$

where V_{bi} is the built-in voltage of the diode and l_i is the intrinsic region thickness. The sign convention is such that positive V_0 corresponds to forward bias.

In a typical GaAs p–i–n diode, the i-region thickness is about 1 μm, and V_{bi} is about 1.5 V. Equation 4.5 then tells us that $\mathcal{E}$ is 1.5×10^6 V m^{-1} at zero bias. At the same time we see from Table 4.1 that in GaAs $2R_X/ea_X$ is of order 6×10^5 V m^{-1}, which is substantially less than the field strength at $V_0 = 0$. We would therefore expect the excitons to be ionized even before we apply bias to the diode.

Figure 4.5 shows experimental data for the field ionization of free excitons in a GaAs p–i–n diode with $l_i = 1.0$ μm at 5 K. In this experiment, the diode is illuminated with light, and the photocurrent generated at a given voltage and wavelength is recorded. The solid line is the photocurrent recorded in 'flat band conditions' ($V_0 = +1.44\,\text{V}$, $\mathcal{E} \approx 0$), while the dashed line is for $V_0 = +1.00\,\text{V}$, where $\mathcal{E} \approx 5 \times 10^5$ V m^{-1}. In the flat band case we observe a well-resolved exciton line at 1.515 eV. However, once we reduce the bias by only a very small amount, we rapidly approach the ionization field, and the exciton broadens significantly. At zero bias (not shown), we are well above the ionization field, and no exciton lines are resolved in the spectrum.

From the discussion above, it is clear that excitonic effects do not play a large part in the physics of bulk semiconductor diodes. The excitons will only be observed over a small range of forward bias voltages just less than V_{bi}. Therefore, the physics of bulk semiconductors in electric fields is dominated more by the effect of the field on the band states, namely the Franz–Keldysh effect discussed in Section 3.3.5. As we shall see in Chapter 6, this is not the case for the enhanced free excitons in GaAs quantum wells. These show very interesting electric field effects even at room temperature.

The wavelength dependence of the photocurrent follows the absorption spectrum. We can see this from eqn 3.38, which shows that the photocurrent is proportional to $(1 - e^{-\alpha l})$. If αl is small, the photocurrent is directly proportional to α. If αl is not small, the photocurrent will still show peaks at wavelengths where α is a maximum.

4.3.2   Magnetic fields

The application of a magnetic field perturbs the free excitons by applying magnetic forces to the electron and hole. The strength of the perturbation is set by the exciton cyclotron energy $\hbar\omega_c$, which is given by

$$\hbar\omega_c = \hbar\frac{eB}{\mu}, \tag{4.6}$$

where B is the magnetic flux density. This is similar to the formula for individual electrons given in eqn 3.27, except that the reduced electron-hole effective mass μ appears instead of the individual electron mass.

The behaviour can be divided into the weak and strong field limits, with the transition point set by the ratio of the exciton Rydberg energy to the cyclotron energy. If $R_X \gg \hbar\omega_c$, we are in the weak field regime, whereas $R_X \ll \hbar\omega_c$ corresponds to the strong field regime. In GaAs, the transition between the two limits occurs around 2 T for the $n = 1$ exciton: see Exercise 4.12.

In the weak field limit we treat the magnetic field as a perturbation on the excitons. The ground state of a hydrogen atom has no net magnetic moment because it is spherically symmetric. Thus the interaction between the $n = 1$ exciton and the magnetic field will be described by diamagnetic effects. The diamagnetic energy shift is given by (see Exercise 4.13):

$$\delta E = +\frac{e^2}{12\mu}r_n^2 B^2. \tag{4.7}$$

The shift is positive because Lenz's law tells us that the field induces a magnetic moment that opposes the applied field. This induced dipole then interacts with the field to give an energy shift proportional to $+B^2$.

In the strong field limit, the interaction of the electrons and holes with the field is stronger than their mutual Coulomb interaction. We therefore

A more detailed discussion of the effects of magnetic fields on excitons may be found in Klingshirn (1995).

consider the Landau energy of the individual electrons and holes first, as in Section 3.3.6. We then add on the Coulomb interaction as a small perturbation. The details of this analysis are beyond the scope of this book. The end result is that the excitonic effects cause a small shift in the energies of the optical transitions between the Landau levels.

4.4   Free excitons at high densities

Wannier excitons behave as if they are hydrogen-like atoms moving freely through the crystal. The atoms in a gas of hydrogen are agitated by thermal motion and interact with each other whenever they get close together. The simplest type of interaction is the tendency to form the H_2 molecule, but other phenomena such as Bose–Einstein condensation are also possible. Excitons show a similar variety of phenomena such as the tendency to form molecules or condense to a liquid phase. The type of behaviour observed in any one material depends very much on the conditions that apply and the details of the interactions between the excitons.

(a) Low density
Separation ≫ diameter

(b) High density
Separation ≈ diameter

Fig. 4.6 Distribution of the free excitons within a crystal. (a) Low densities: the excitons are randomly distributed throughout the excitation volume and the interexciton separation is large. (b) High densities: the wave functions overlap when the exciton-exciton separation becomes comparable to the exciton diameter.

We first consider an experiment in which we take a powerful laser and tune it to one of the exciton absorption lines. The laser creates excitons in the sample, with a density that is proportional to the laser power. At low powers, the density of the excitons is small, and the separation between the excitons is large, as sketched in Fig. 4.6(a). The exciton-exciton interactions are negligible in these conditions. As the power is increased, the density of excitons increases. Eventually, the density will be high enough that the exciton wave functions begin to overlap, as sketched in Fig. 4.6(b). At this point, we expect that the exciton-exciton interactions will become very significant.

We can see from Fig. 4.6(b) that exciton wave function overlap occurs when the exciton-exciton distance is equal to the exciton diameter. The density at which this occurs is called the **Mott density** N_{Mott}. It is given approximately by the inverse volume of the exciton:

$$N_{\text{Mott}} \approx \frac{1}{\frac{4}{3}\pi r_n^3} \, . \qquad (4.8)$$

From Table 4.1 and eqn 4.2, we find that the Mott density for the $n = 1$ excitons in GaAs is $1.1 \times 10^{23} \, \text{m}^{-3}$. This density is easily achievable with a focused laser beam.

When the exciton density approaches N_{Mott}, a number of effects can occur. In GaAs the collisions between the excitons cause the exciton gas to dissociate into an electron-hole plasma, i.e. a neutral gas containing equal numbers of electrons and holes. This causes exciton broadening with a reduction in the absorption strength. Figure 4.7 shows the absorption coefficient at the $n = 1$ exciton in GaAs at $1.2 \, \text{K}$ at three different excitation powers. The weakening and broadening of the exciton line as the carrier density is increased is clearly observed in the data. The density at which these effects occur agrees well with the value

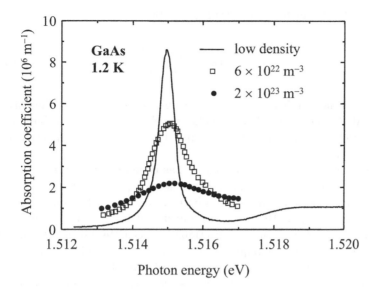

Fig. 4.7 Absorption coefficient of GaAs in the spectral region close to the band edge at 1.2 K at three different excitation powers. The carrier densities generated for the two higher power levels are indicated. After Fehrenbach et al. (1985), © Excerpta Medica Inc., reprinted with permission.

of 1.1×10^{23} m^{-3} given by eqn 4.8. The change of the exciton absorption with increasing power is an example of a nonlinear optical effect: the absorption coefficient depends on the intensity of the light. We shall return to discuss applications of these nonlinear effects in Section 11.4.7 of Chapter 11.

Another effect that can be observed at high exciton densities in other materials is the formation of exciton molecules called **biexcitons**. This is the equivalent process to the formation of an H$_2$ molecule from two isolated hydrogen atoms. Biexcitons have been observed in a number of wide gap semiconductors, including CdS, ZnSe, ZnO, and especially copper chloride. CuCl has a band gap at 3.40 eV, and the ground-state exciton is observed at 3.20 eV, implying that $R_X = 0.2$ eV. At high densities, a new feature that is attributed to biexcitons is observed in the absorption spectrum at 3.18 eV. The energy difference between the two features tells us that the binding energy of the biexciton is 0.02 eV.

In silicon and germanium at high densities, yet another effect occurs. At low densities the excitons may be considered to be in a gaseous phase. As the density increases, the excitons condense to form a liquid. The liquid phase manifests itself in the formation of **electron-hole droplets**, which are observed in the recombination radiation of the excitons at high densities. The droplet appears as a broad feature at lower energy than the free excitons.

The final high-density excitonic effect that we consider here is **Bose–Einstein condensation**. At high temperatures, the particles in a non-interacting boson gas are distributed between the possible energy levels of the system according to Bose–Einstein statistics. As the temperature is lowered, the distribution undergoes a radical range, and a macroscopic number of particles accumulates in the ground state. The critical tem-

Attempts to observe biexcitons in bulk GaAs have been complicated by the nonlinear saturation effects described above. However, biexcitons can readily be observed in GaAs quantum-confined structures such as quantum wells and quantum dots.

perature T_c at which this occurs is given by:

$$N = 2.612 \left(\frac{m k_B T_c}{2 \pi \hbar^2} \right)^{3/2} , \qquad (4.9)$$

where N is the number of particles per unit volume and m is the particle mass. At T_c the thermal de Broglie wavelength is comparable to the interparticle separation, and quantum effects are to be expected. (See Exercise 4.16.)

Bose–Einstein condensation (BEC) has been observed in many boson systems. One of the best-studied examples is liquid helium. In this case, N is fixed, and eqn 4.9 predicts a phase transition as the liquid is cooled through T_c at 2.2 K. However, the physics of BEC in liquid helium is complicated by the strong interactions between the atoms. To achieve pure BEC behaviour, we require that the interactions between the bosons are negligible. This suggests that we need highly dilute gaseous systems such that the interparticle separation is very large. However, from eqn 4.9 we see that the transition temperature for such a dilute system would be very low. It has been an outstanding recent achievement of atomic physics to succeed in observing BEC in extremely dilute gases of atoms at temperatures below 1 µK.

Excitons consist of two spin 1/2 particles, and so their total spin is either 0 or 1. This means that they are bosons, and therefore that BEC should be possible. However, the study of excitonic BEC has a long, chequered history, with many claimed observations that have subsequently been disputed. Part of the reason for the controversy is that it is actually very difficult to prove definitively that condensation has occurred.

We briefly mention here three of the more promising candidate systems for excitonic BEC. Details of the experiments and results may be found by referring to the articles and books listed for Further Reading.

- Copper oxide (Cu_2O) and copper chloride (CuCl). These wide gap semiconductors have particularly strong excitonic effects. In the case of Cu_2O, it is the excitons with zero spin that have given the most encouraging results, while for CuCl it the biexcitons that are of particular interest.

- Coupled GaAs quantum wells. Excitonic effects are enhanced in quantum wells (see Section 6.4.4), and the use of coupled wells leads to long recombination lifetimes. This gives sufficient time for the excitons to form a cold gas, which increases the probability for BEC to occur.

- CdTe quantum wells in a microcavity. The use of the microcavity leads to the formation of a coupled exciton–photon quasiparticle called an 'exciton polariton'. The results for these excitonic polaritons are probably the most convincing to date.

4.5   Frenkel excitons

The free exciton model that leads to eqns 4.1 and 4.2 breaks down when the predicted radius becomes comparable to the interatomic spacing. This occurs in large band gap materials with small dielectric constants and large effective masses. In these materials we observe Frenkel excitons rather than Wannier excitons.

Frenkel excitons are localized on the atom site at which they are created, as shown in Fig. 4.1(b). The excitons may therefore be considered as excited states of the individual atoms or molecules on which they are localized, and they can propagate through the crystal by hopping from atom site to site. They have very small radii and correspondingly large binding energies, with typical values ranging from about 0.1 eV to several eV. This means that Frenkel excitons are usually stable at room temperature.

The theoretical treatment of Frenkel excitons requires techniques more akin to atomic or molecular physics than solid-state physics. There is no simple model similar to the one that led to eqns 4.1 and 4.2 for free excitons. The calculation of the exciton energies usually follows a tight binding approach, in order to emphasize the correspondence to the atomic or molecular states from which the excitons are derived. The calculation is further complicated by the fact that the coupling between the excitons and the crystal lattice is usually very strong. This leads to 'self-trapping' effects, in which the exciton produces a local distortion of the lattice, which then causes further localization of the exciton wave functions.

> The self-trapping of electrons or holes is caused by the electron–phonon coupling. These polaronic effects will be discussed in Section 10.4.

Frenkel excitons have been observed in many inorganic and organic materials. The properties of some of the more widely studied crystals are described briefly below.

4.5.1   Rare gas crystals

The rare gases from group VIII of the periodic table, namely neon, argon, krypton, and xenon, crystallize at cryogenic temperatures. The band gap ranges from 21.6 eV in neon to 9.3 eV in xenon. Neon in fact has the largest band gap of any crystal known in nature. The excitonic absorption of these materials has been thoroughly studied, and the results are summarized in Table 4.2. The excitonic transitions all occur in the vacuum ultraviolet spectral range, and the binding energies are very large.

It has been found experimentally that there is a close correspondence between the $n = 1$ exciton energies in the crystals and the optical transitions of the isolated atoms. For example, the energy of the $n = 1$ exciton in xenon crystals coincides almost exactly with the lowest energy absorption line of xenon atoms in the gaseous phase, namely the $5p^6 \rightarrow 5p^5 6s$ transition. This underlines the point made earlier that the localized nature of the Frenkel excitons makes them equivalent to excited states of the individual atoms or molecules. This correspondence gets weaker for

Table 4.2 Properties of Frenkel excitons in rare gas crystals. All energies are given in eV. Data from Song and Williams (1993).
T_m: melting temperature in K,
E_g: band gap,
E_1: energy of the $n = 1$ exciton,
E_b: binding energy of the $n = 1$ exciton.

Crystal	T_m	E_g	E_1	E_b
Ne	25	21.6	17.5	4.1
Ar	84	14.2	12.1	2.1
Kr	116	11.7	10.2	1.5
Xe	161	9.3	8.3	1.0

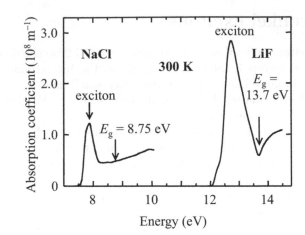

Fig. 4.8 Absorption spectra of NaCl and LiF at room temperature. Data from Palik (1985).

the excitons with larger values of n. As the radius increases with n, the excitons become more and more delocalized, and it eventually becomes valid to use the Wannier model.

4.5.2   Alkali halides

Frenkel excitons are readily observable in the optical spectra of alkali halide crystals. These have large direct band gaps in the ultraviolet spectral region ranging from 5.9 eV in NaI to 13.7 eV in LiF. LiF has the widest band gap of any practical optical material: only argon and neon crystals have larger band gaps, but neither of these are solids at room temperature.

Table 4.3 Properties of Frenkel excitons in selected alkali halide crystals. All energies are given in eV. Data from Song and Williams (1993).
E_g: band gap,
E_1: energy of the $n = 1$ exciton line,
E_b binding energy of the $n = 1$ exciton.

Crystal	E_g	E_1	E_b
KI	6.3	5.9	0.4
KBr	7.4	6.7	0.7
KCl	8.7	7.8	0.9
KF	10.8	9.9	0.9
NaI	5.9	5.6	0.3
NaBr	7.1	6.7	0.4
NaCl	8.7	7.9	0.8
NaF	11.5	10.7	0.8
CsF	9.8	9.3	0.5
RbF	10.3	9.5	0.8
LiF	13.7	12.8	0.9

Table 4.3 lists the band gap of selected alkali halide crystals, together with the energy and binding energy of the $n = 1$ exciton. The data show that E_g tends to increase with decreasing anion and cation size. The exciton binding energy follows a similar general trend. Detailed spectroscopy has established that the excitons are localized at the negative (halogen) ions.

Figure 4.8 shows the absorption spectrum of two representative alkali halide crystals at room temperature, namely NaCl and LiF. Both spectra show a strong excitonic absorption line below the band gap. The binding energies are 0.8 eV and 0.9 eV respectively. These values are well above $k_B T$ at room temperature, which explains why the excitons are observed so strongly. The fine structure of the excitons due to the excited states can be observed by cooling the crystals. Note that the absorption coefficient at the exciton lines is extremely large, with values over 10^8 m^{-1} in both materials.

4.5.3   Molecular crystals

Frenkel excitons can be observed in many molecular crystals and organic thin film structures. In most cases, there is a very strong correspondence between the optical transitions of the isolated molecules and the excitons

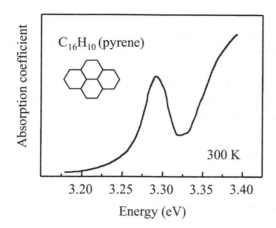

Absorption coefficient

$C_{16}H_{10}$ (pyrene)

300 K

Energy (eV)

3.20 3.25 3.30 3.35 3.40

Fig. 4.9 Absorption spectrum of pyrene ($C_{16}H_{10}$) single crystals at room temperature. After Matsui and Nishimura (1980), reprinted with permission.

observed in the solid state. This is a consequence of the fact that the molecular crystals are held together by relatively weak van der Waals forces, so that the molecular levels are only weakly perturbed when condensing to the solid state.

Figure 4.9 shows the fundamental absorption edge of pyrene crystals at room temperature. The pyrene molecule has a composition of $C_{16}H_{10}$ and is an example of an aromatic hydrocarbon, that is, a carbon–hydrogen compound based on benzene rings. The four-ring structure of pyrene is given in the inset. The absorption spectrum shows a clear excitonic peak at 3.29 eV. Other aromatic hydrocarbons such as anthracene ($C_{14}H_{10}$) also show very strong excitonic effects, but the optical spectra are more complicated because of the strong coupling to the vibrational modes of the molecule. These effects will be discussed in more detail in Section 8.3.1 in Chapter 8. The pyrene spectrum is relatively simple because the four-ring structure makes the molecule very rigid and reduces the effects of the vibrational coupling.

Frenkel excitons are also very important in conjugated polymers, such as polydiacetylene (PDA). Single crystals of PDA can be grown, but the optical properties are often studied by using amorphous films coated onto glass substrates. The strong excitonic effects in conjugated polymers have acquired considerable technological significance in recent years, following the development of organic light-emitting diodes for use in display technology. The optical properties of organic semiconductors such as PDA will be discussed in more detail in Sections 8.3.2 and 8.4 of Chapter 8.

Pyrene, anthracene, and the other aromatic hydrocarbons are examples of *conjugated* molecules. The π electrons of the benzene rings form large delocalized molecular orbitals with optical transitions in the blue/UV spectral range. Polydiacetylene is another example of a conjugated molecule. In Section 1.4.4 we mentioned that the most interesting molecular materials, from the point of view of their optical properties, are those with conjugated bonds. This point will be developed further in Chapter 8.

Chapter summary

- Excitons are electron-hole pairs bound together in stable orbits by the mutual Coulomb attraction between them.

- There are two types of excitons. Wannier (free) excitons have a large radius and move freely throughout the crystal. Frenkel (tightly bound) excitons are localized on individual atoms sites.
- The properties of free excitons can be calculated by treating them as hydrogen-like atoms. The binding energies and radii are given by eqns 4.1 and 4.2 respectively.
- Free excitons are observed in semiconductors at photon energies just below E_g. They have fairly small binding energies, and are observed most clearly at low temperatures. They are easily ionized by electric fields.
- Free excitons can interact with each other, and they show a rich variety of phenomena at high densities due to the exciton-exciton interactions.
- Frenkel excitons have very small radii and large binding energies. They are easily observed at room temperature in insulator crystals and molecular materials. There is a strong correspondence between the excitons observed in the solid state and the excited states of the individual atoms or molecules of which the solid is composed.

Further reading

Supplementary reading on excitons may be found in most of the standard solid-state texts such as Burns (1985) or Kittel (1996). More detailed information on free excitons in semiconductors may be found in Klingshirn (1995), Pankove (1971), Seeger (1997), or Yu and Cardona (1996).

Dexter and Knox (1965) is a classic text on excitons, while Rashba and Sturge (1982) is a more recent authoritative reference work. Reynolds and Collins (1981) give a good overview of excitonic physics, while Song and Williams (1993) give a thorough discussion of the properties of Frenkel excitons.

An overview of high-density exciton effects may be found in Klingshirn (1995). The general phenomenon of Bose–Einstein condensation is discussed in most texts on statistical mechanics, for example, Mandl (1988). Griffin et al. (1995) give a review of measurements of BEC in a wide variety of systems, while Moskalenko and Snoke (2000) give a more detailed account specifically for excitons and biexcitons. Details of recent experimental work on BEC in coupled quantum wells and microcavities may be found in Butov (2007), Kasprzak et al. (2006), and Kavokin et al. (2007).

Exercises

(4.1) Write down the Schrödinger equation for the hydrogen atom. By defining the centre of mass and relative coordinates for the electron and proton, show that the Hamiltonian of the system can be split into two parts, one describing the free motion of the whole atom and the other describing the internal energy of the atom due to the Coulomb energy and orbital motion.

(4.2) The Hamiltonian for the relative motion of an electron-hole pair in a semiconductor is given by:

$$\hat{H} = -\frac{\hbar^2}{2\mu}\nabla^2 - \frac{e^2}{4\pi\epsilon_0\epsilon_r r}.$$

(a) Explain the origin of the two terms in the Hamiltonian.

(b) Show that the wave function $\Psi(r, \theta, \phi) = C \exp(-r/a_0)$ is a solution of the Schrödinger equation

$$\hat{H}\Psi = E\Psi,$$

and find the values of E and a_0. Find also the value of the normalization constant C.

(4.3) Find the radius at which the radial probability density of the hydrogenic wave function given in the previous question reaches its maximum value. Compare this to the expectation value $\langle r \rangle$ defined by

$$\langle r \rangle = \int_{r=0}^{\infty} \int_{\theta=0}^{\pi} \int_{\phi=0}^{2\pi} \Psi^* r \, \Psi \, r^2 \sin\theta \, dr d\theta d\phi.$$

(4.4)* In the variational method, we make an enlightened guess of the wave function, and then vary its parameters to minimize the expectation value of the energy. The variational principle says that the wave function that gives the minimum energy is the best approximation to the true wave function, and that the corresponding expectation value of the energy is the best approximation of the true energy.[1]

(a) Explain why the following function is a sensible guess for the wave function of the ground state of the exciton system:

$$\Psi(r, \theta, \phi) = \left(\frac{1}{\xi} \right)^{3/2} \frac{1}{\sqrt{\pi}} \exp\left(-\frac{r}{\xi} \right).$$

(b) Calculate the expectation value for the energy of an exciton with wave function Ψ:

$$\langle E \rangle = \iiint \Psi^* \hat{H} \Psi \, r^2 \sin\theta \, dr \, d\theta \, d\phi,$$

where $\hat{H}$ is the Hamiltonian given in Exercise 4.2.

(c) Find the value of ξ that minimizes $\langle E \rangle$, and compute $\langle E \rangle$ for this value of ξ.

(d) Compare the minimal values of E and ξ obtained in part (c) to those obtained in Exercise 4.2, and comment on your answer.

(4.5) (a) State the assumptions of the Bohr model of the hydrogen atom.

(b) Use the Bohr model to show that the energy and radius of a hydrogenic atom with reduced mass μ in a medium with a relative dielectric constant ϵ_r are given by eqns 4.1 and 4.2 respectively.

(c) How does $E(n)$ compare to the exact solution of the Schrödinger equation considered in Exercise 4.2?

(d) How does r_n relate to the conclusions of Exercise 4.3?

(4.6) Calculate the binding energy and radius of the $n = 1$ and $n = 2$ free excitons in zinc sulphide (ZnS) which has $m_e^* = 0.28m_0$, $m_h^* = 0.5m_0$ and $\epsilon_r = 7.9$. Would you expect these excitons to be stable at room temperature?

(4.7) Calculate the difference in the wavelengths of the $n = 1$ and $n = 2$ excitons in InP, which has $E_g = 1.424\,\text{eV}$, $m_e^* = 0.077m_0$, $m_h^* = 0.2m_0$, and $\epsilon_r = 12.4$.

(4.8) At $4\,\text{K}$ the $n = 1$ exciton in GaAs has a peak absorption coefficient of $3 \times 10^6\,\text{m}^{-1}$ at $1.5149\,\text{eV}$, with a full width at half maximum equal to $0.6\,\text{meV}$. By applying the bound oscillator model discussed in Chapter 2 to the exciton, determine the magnitude and energy of the local maximum in the refractive index just below the exciton absorption line. The non-resonant refractive index of GaAs at energies below the band gap is 3.5.

(4.9) Excitons can absorb photons by making transitions to excited states in exactly the same way that hydrogen atoms do. Calculate the wavelength of the photon required to promote an exciton in GaAs ($\mu = 0.05m_0$, $\epsilon_r = 12.8$) from the $n = 1$ to the $n = 2$ state.

(4.10) Use the Bohr model to show that the magnitude of the electric field between the electron and hole in the ground state of a free exciton is equal to $2R_X/ea_X$.

(4.11) Direct excitons may be formed in germanium at low temperatures using photon energies close to the direct band gap at $0.898\,\text{eV}$. Calculate the binding energy and radius of the ground state exciton, taking $m_e^* = 0.038m_0$, $m_h^* = 0.1m_0$, and $\epsilon_r = 16$. Calculate the voltage at which the field across the excitons will be equal to the ionization field in a germanium p–i–n diode, which has $V_{bi} = 0.74\,\text{V}$ and an i-region thickness of $2\,\mu\text{m}$.

---

*Exercises marked with an asterisk are more difficult.

[1] This exercise illustrates the use of the variational method to obtain approximate solutions for the wave function and energy of the ground state. These can of course be found by brute force solution of the Schrödinger equation, but the variational method is more intuitive, and can be easily adapted to other problems where no analytic solutions are possible.

(4.12) Show that the magnetic field strength at which the exciton cyclotron energy is equal to the exciton Rydberg energy is given by:

$$B = \frac{\mu^2}{\epsilon_r^2 m_0 \hbar} \left(\frac{R_H}{e} \right) .$$

Evaluate this field strength for GaAs with $\mu = 0.05 m_0$ and $\epsilon_r = 12.8$.

(4.13)* Verify by using eqn A.14 that a vector potential of the form $\boldsymbol{A} = (B/2)(-y, x, 0)$ produces a constant magnetic flux density of magnitude B in the z direction. By following an analysis similar to the one that leads to eqn B.19 in Appendix B, show that the diamagnetic energy shift of an electron in an atom with a wave function ψ is given by:

$$\Delta E = \frac{e^2 B^2}{8 m_0} \langle \psi | (x^2 + y^2) | \psi \rangle .$$

Hence derive eqn 4.7.

(4.14) Calculate the diamagnetic energy shift of the $n = 1$ exciton of GaAs in a magnetic field of $1.0\,\mathrm{T}$. What is the shift in the wavelength of the exciton caused by applying the field? Take $\mu = 0.05 m_0$,

and the energy of the exciton at $B = 0$ to be $1.515\,\mathrm{eV}$.

(4.15) Estimate the Mott densities for the $n = 1$ and $n = 2$ excitons in gallium nitride (GaN), which has $m_e^* = 0.2 m_0$, $m_h^* = 1.2 m_0$ and $\epsilon_r = 10$.

(4.16) Show that the de Broglie wavelength λ_{deB} of a particle of mass m with thermal kinetic energy $\frac{3}{2} k_B T$ is given by:

$$\lambda_{\mathrm{deB}} = \frac{h}{(3 m k_B T)^{1/2}} .$$

Calculate the ratio of the interparticle separation to λ_{deB} at the Bose–Einstein condensation temperature.

(4.17) Calculate the Bose–Einstein condensation temperature for excitons in cuprous oxide when the exciton density is $10^{24}\,\mathrm{m}^{-3}$. The electron and hole effective masses are $1.0 m_0$ and $0.7 m_0$ respectively.

(4.18) The values of μ and ϵ_r for sodium iodide (NaI) are $0.18 m_0$ and 2.9 respectively. The unit cell size is $0.65\,\mathrm{nm}$. Would you expect the Wannier model to be valid for the $n = 1$ exciton? What about the $n = 2$ exciton?

Luminescence

5

In Chapter 3 we considered how light can be absorbed in solids by exciting interband transitions. Then in Chapter 4 we considered how the absorption spectrum is modified by the interactions that lead to the formation of excitons. We now consider the reverse process in which electrons in an excited state drop to lower levels by emitting photons. This is the solid-state equivalent to light emission in atoms by spontaneous emission, which is reviewed in Appendix B.

The physical mechanisms responsible for light emission in solids vary considerably from material to material. In this chapter we start by giving a few general principles that apply to all materials, and then focus on the emission of light by interband transitions in bulk semiconductors. This will provide the framework for discussing the light emission processes in quantum-confined structures in Chapter 6, and will also serve as a general introduction to the light emission processes in other types of materials.

5.1	**Light emission in solids**	**113**
5.2	**Interband luminescence**	**115**
5.3	**Photoluminescence**	**118**
5.4	**Electroluminescence**	**126**
Chapter summary		**136**
Further reading		**137**
Exercises		**138**

5.1 Light emission in solids

Atoms emit light by spontaneous emission when electrons in excited states drop down to a lower level by radiative transitions. In solids the radiative emission process is called **luminescence**. Luminescence can occur by a number of mechanisms, but in this book we mainly consider just two:

- **Photoluminescence**: the re-emission of light after absorbing a photon of higher energy.
- **Electroluminescence**: the emission of light caused by running an electrical current through the material.

The physical processes involved in both photoluminescence and electroluminescence are more complicated than those in absorption. This is because the generation of light by luminescence is intimately tied up with the energy relaxation mechanisms in the solid. Furthermore, the shape of the emission spectrum is affected by the thermal distributions of the electrons and holes within their bands. Therefore, we have to consider the emission rates and the thermal spread of the carriers before we can gain a good understanding of the emission efficiency and the luminescence spectrum.

Figure 5.1 gives an overview of the main processes that occur when light is emitted from a solid. The photon is emitted when an electron

The diagram in Fig. 5.1 applies to emission between bands, but the basic idea that the carriers relax to the lowest excited state level before emitting the photon is usually applicable even if the levels are discrete.

inject electrons

relaxation   EXCITED
                STATE

τ_{NR} τ_R ⟶ $\hbar\omega$

                GROUND
                STATE

inject holes

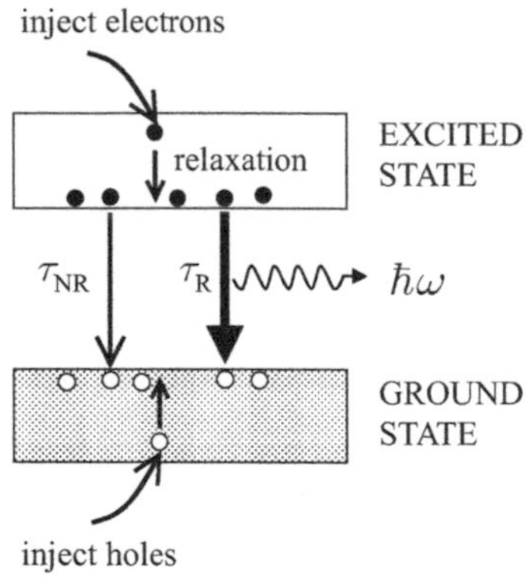

Fig. 5.1 General scheme of luminescence in a solid. Electrons are injected into the excited state band and relax to the lowest available level before dropping down to empty levels in the ground-state band by emitting a photon. These empty levels are generated by the injection of holes. The radiative recombination rate is determined by the radiative lifetime τ_R. Radiative emission has to compete with non-radiative recombination, which has a time constant τ_{NR}. The luminescent efficiency is determined by the ratio of τ_R to τ_{NR}, and is given by eqn 5.5.

in an excited state drops down into an empty state in the ground-state band. For this to be possible, we must first inject electrons, which then relax to the state from where the emission occurs. This could be the bottom of the conduction band, but it might also be a discrete level. The photon cannot be emitted unless the lower level for the transition is empty, because the Pauli principle does not permit us to put two electrons into the same quantum state. The empty lower level is produced by injecting holes into the ground-state band in an entirely analogous way to the injection of the electrons into the excited state.

The spontaneous emission rate for radiative transitions between two levels is determined by the Einstein A coefficient. (See Appendix B.) If the upper level has a population N at time t, the radiative emission rate is given by:

$$\left(\frac{\mathrm{d}N}{\mathrm{d}t}\right)_{\text{radiative}} = -AN. \tag{5.1}$$

This shows that the number of photons emitted in a given time is proportional to both the A coefficient of the transition and also to the population of the upper level. The rate equation can be solved to give:

$$N(t) = N(0)\exp(-At) = N(0)\exp(-t/\tau_R), \tag{5.2}$$

where $\tau_R = A^{-1}$ is the **radiative lifetime** of the transition.

Equation B.11 in Appendix B tells us that the Einstein A coefficient is directly proportional to the B coefficient, which determines the probability for absorption. This means that transitions which have large absorption coefficients also have high emission probabilities and short radiative lifetimes. However, the fact that the absorption and emission probabilities are closely related to each other does not imply that the absorption and emission spectra are the same. This is because of the population factor that enters eqn 5.1. A transition might have a high emission probability, but no light will be emitted unless the upper level is populated.

We can summarize these points by writing the luminescent intensity at frequency ν as:

$$I(h\nu) \propto |M|^2 g(h\nu) \times \text{level occupancy factors}, \tag{5.3}$$

where the occupancy factors give the probabilities that the relevant upper level is occupied and the lower level is empty. The other two terms are the matrix element and the density of states for the transition, which determine the quantum mechanical transition probability by Fermi's golden rule. (See Section B.2 in Appendix B.)

The occupancy factors that enter eqn 5.3 will be discussed in detail in Section 5.3. The main point is that the electrons relax very rapidly to the lowest levels within the excited state band, and then form a thermal distribution that can be calculated by statistical mechanics. In normal circumstances the electrons will relax to within $\sim k_B T$ of the bottom of the excited state band. The holes follow a similar series of relaxation processes. The light is emitted between the electron and hole states that are thermally occupied, and will therefore only be emitted within a narrow energy range from the lowest levels in the excited state band. This contrasts with the absorption spectrum, where photons can be absorbed to any state within the excited state band, no matter how far it is above the bottom of the band.

Radiative emission is not the only mechanism by which the electrons in an excited state can drop down to the ground state. The alternative pathway between the excited state and ground state bands in Fig. 5.1 indicates the possibility of **non-radiative** relaxation. The electron might, for example, lose its excitation energy as heat by emitting phonons, or it may transfer the energy to impurities or defects called 'traps'. If these non-radiative relaxation processes occur on a faster time scale than the radiative transitions, very little light will be emitted.

The luminescent efficiency η_R (sometimes also called the quantum efficiency) can be calculated by writing down the rate equation for the population of the excited state when non-radiative processes are possible:

$$\left(\frac{\mathrm{d}N}{\mathrm{d}t}\right)_{\text{total}} = -\frac{N}{\tau_R} - \frac{N}{\tau_{NR}} = -N\left(\frac{1}{\tau_R} + \frac{1}{\tau_{NR}}\right). \tag{5.4}$$

The two terms on the right-hand side of eqn 5.4 represent the radiative and non-radiative rates respectively. τ_{NR} is the non-radiative lifetime. η_R is given by the ratio of the radiative emission rate to the total de-excitation rate. This is obtained by dividing eqn 5.1 by eqn 5.4 to obtain

$$\eta_R = \frac{AN}{N(1/\tau_R + 1/\tau_{NR})} = \frac{1}{1 + \tau_R/\tau_{NR}}, \tag{5.5}$$

where we have used the fact that $A = \tau_R^{-1}$. If $\tau_R \ll \tau_{NR}$ then η_R approaches unity and the maximum possible amount of light is emitted. On the other hand, if $\tau_R \gg \tau_{NR}$ then η_R is very small and the light emission is very inefficient. Thus efficient luminescence requires that the radiative lifetime should be much shorter than the non-radiative lifetime.

The principles discussed here are very general and apply to a wide range of light emission phenomena in solids. In the rest of this chapter we concentrate on the luminescence generated by interband transitions in a bulk semiconductor. In subsequent chapters we shall consider the light emission processes in quantum-confined structures (Chapter 6), molecular materials (Chapter 8), and luminescent impurities (Chapter 9).

5.2   Interband luminescence

Interband luminescence occurs in a semiconductor when an electron that has been excited into the conduction band drops back to the valence band by the emission of a photon. This simultaneously reduces the number of electrons in the conduction band and holes in the valence band by one. Interband luminescence thus corresponds to the annihilation of an electron-hole pair, and is known as radiative **electron-hole recombination**. This should be contrasted with interband absorption, which is equivalent to the creation of an electron-hole pair.

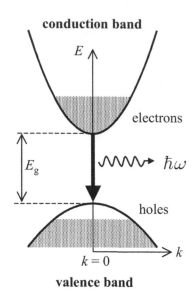

conduction band

valence band

Fig. 5.2 Schematic diagram of the interband luminescence process in a direct gap semiconductor. The shading indicates that the states are occupied by electrons. The filled states at the bottom of the conduction band and the empty states at the top of the valence band are created by injecting electrons and holes into the semiconductor.

We noted in Chapter 3 that there are very important differences between the optical properties of direct and indirect band gap materials. This is particularly true when we come to consider the interband emission processes. We must therefore consider them separately, beginning with direct gap materials.

5.2.1   Direct gap materials

Figure 5.2 shows the band diagram for an interband luminescence process in a direct gap semiconductor. The photons are emitted when electrons at the bottom of the conduction band recombine with holes at the top of the valence band. As discussed in Chapter 3, the optical transitions between the valence and conduction bands of typical direct gap semiconductors are dipole-allowed and have large matrix elements. This implies through eqn B.30 that the radiative lifetime will be short, with typical values in the range 10^{-8}–10^{-9} s. (See Exercise 5.3.) The luminescent efficiency is therefore expected to be high.

The processes by which the electrons and holes are injected into the bands will be discussed in Sections 5.3 and 5.4. We shall also see in Section 5.3.1 that the injected electrons and holes relax very rapidly to the lowest energy states within their respective bands by emitting phonons. This means that the electrons accumulate at the bottom of the conduction band before they recombine, as indicated in Fig. 5.2. By contrast, holes move *upwards* on energy band diagrams when they relax. This is because band diagrams show electron energies, rather than hole energies, so that the hole energy is zero at the top of the valence band and increases as we move further down into the valence band. Holes therefore accumulate at the top of the valence band after relaxation.

Since the momentum of the photon is negligible compared to the momentum of the electron, the electron and hole that recombine must have the same $\boldsymbol{k}$ vector (cf eqn 3.12). Therefore, the transition is represented by a downward vertical arrow on the band diagram, as indicated in Fig. 5.2. The emission takes place near $k = 0$, and corresponds to a photon of energy E_g. No matter how we excite the electrons and holes in the first place, we always obtain luminescence at energies close to the band gap.

Figure 5.3 shows the luminescence and absorption spectra of the direct gap semiconductor gallium nitride at 4 K. The band gap is 3.5 eV at this temperature. The luminescence spectrum consists of a narrow emission line close to the band gap energy, while the absorption shows the usual threshold at E_g with continuous absorption for $h\nu > E_g$.

The data shown in Fig. 5.3 illustrate the point that the emission and absorption spectra are not the same, even though they are determined by the same matrix element and density of states. The band gap corresponds to the threshold for optical absorption, but to the energy of the optical emission. This means that the criteria for choosing the best material to act as an emitter or detector for a particular wavelength are different. When we are designing an emitter, we must choose a material

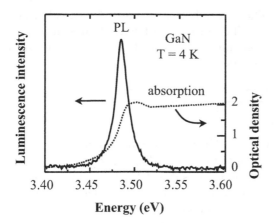

Fig. 5.3 Luminescence spectrum (solid line) and absorption (dotted line) of a GaN epilayer of thickness 0.5 μm at 4 K. The photoluminescence (PL) was excited by absorption of 4.9 eV photons from a frequency doubled copper vapour laser. Unpublished data from K.S. Kyhm and R.A. Taylor.

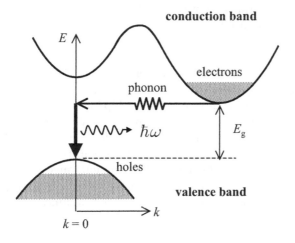

Fig. 5.4 Schematic diagram of the interband luminescence process in an indirect gap material. The transition must involve the absorption or emission of a phonon to conserve momentum.

that has a band gap corresponding to the desired wavelength. Detectors, on the other hand, will work at any wavelength provided that the photon energy exceeds E_g.

5.2.2 Indirect gap materials

Figure 5.4 illustrates the processes that occur during interband emission in an indirect gap material. This is the reverse of the indirect absorption process shown in Fig. 3.2(b). In an indirect gap material, the conduction band minimum and valence band maximum are at different points in the Brillouin zone. Conservation of momentum requires that a phonon must either be emitted or absorbed when the photon is emitted.

The requirement of emitting both a phonon and a photon during the transition makes it a second-order process, with a relatively small transition probability. The radiative lifetime is therefore much longer than for direct transitions. We can see from eqn 5.5 that this makes the luminescent efficiency small, because of the competition with non-radiative recombination. For this reason, indirect gap materials are generally bad light emitters. They are only used when there is no alternative direct gap material available. Two of the most important semiconductors, namely silicon and germanium, have indirect band gaps and are therefore not used as light emitters.

Example 5.1

The band gap of the III–V semiconductor alloy $Al_xGa_{1-x}As$ at $k = 0$ varies with composition according to $E_g(x) = (1.420 + 1.087x + 0.438x^2)$ eV. The band gap is direct for $x \leq 0.43$, and indirect for larger values of x. Light emitters for specific wavelengths can be made by appropriate choice of the composition.
(a) Calculate the composition of the alloy in a device emitting at 800 nm.
(b) Calculate the range of wavelengths than can usefully be obtained from an AlGaAs emitter.

Solution
(a) The photons at 800 nm have an energy of 1.55 eV. The device will emit at the band gap wavelength, so we must choose x such that $E_g(x) = 1.55$ eV. On substituting into the relationship for $E_g(x)$, we find $x = 0.11$.
(b) The long wavelength limit is set by the smallest band gap that can be obtained in the alloy, namely 1.420 eV for $x = 0$. The short wavelength limit is set by the largest direct band gap that can be obtained, namely 1.97 eV for $x = 0.43$. The useful emission range is therefore 1.42–1.97 eV, or 630–870 nm. Alloy compositions with $x > 0.43$ are not useful because indirect gap materials have very low luminescent efficiencies.

5.3   Photoluminescence

In this section we consider the re-emission of light by interband luminescence after a direct gap semiconductor has been excited by a photon with energy greater than E_g. As noted at the start of Section 5.1, this process is called photoluminescence.

5.3.1   Excitation and relaxation

The band diagram corresponding to the photoluminescence process in a direct gap material is given in Fig. 5.5(a). This is a more detailed version of the diagram already given in Fig. 5.2. Photons are absorbed from an excitation source such as a laser or lamp, and this injects electrons into the conduction band and holes into the valence band. This will be possible if the frequency ν_L of the source is chosen so that $h\nu_L$ is greater than E_g.

It is apparent from Fig. 5.5(a) that the electrons are initially created in states high up in the conduction band. The electrons do not remain in

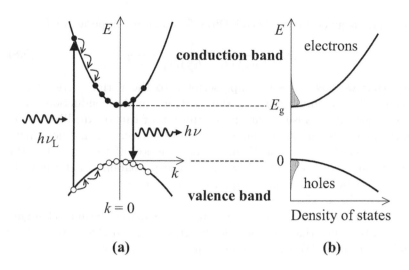

(a)

(b)

Fig. 5.5 (a) Schematic diagram of the processes occurring during photoluminescence in a direct gap semiconductor after excitation at frequency ν_L. The electrons and holes rapidly relax to the bottom of their bands by phonon emission before recombining by emitting a photon. (b) Density of states and level occupancies for the electrons and holes after optical excitation. The distribution functions shown by the shading apply to the classical limit where Boltzmann statistics are valid. Note that the distribution functions and density of states are not on the same scale: the level occupancies are always small in the Boltzmann limit.

these initial states for very long, because they can lose their energy very rapidly by emitting phonons. This process is indicated by the cascade of transitions within the conduction band shown in Fig. 5.5(a). Each step corresponds to the emission of a phonon with the correct energy and momentum to satisfy the conservation laws. The electron–phonon coupling in most solids is very strong and these scattering events take place on time scales as short as $\sim 100\,\text{fs}$ (i.e. $\sim 10^{-13}\,\text{s}$). This is much faster than the radiative lifetimes which are in the nanosecond range, and the electrons are therefore able to relax to the bottom of the conduction band long before they have had time to emit photons. The same conditions apply to the relaxation of the holes in the valence band.

After the electrons and holes have relaxed as far as they can by phonon emission, they must wait at the bottom of the bands until they can emit a photon or recombine non-radiatively. This leaves time to form thermal distributions, as sketched in Fig. 5.5(b). The shading indicates the occupancy of the available states. These occupancy factors can be calculated by applying statistical physics to the electron and hole distributions.

The distributions of the optically excited electrons and holes in their bands can be calculated by Fermi–Dirac statistics. The total number density N_e of electrons is determined by the power of the illumination source (see Exercises 5.6 and 5.7), and must satisfy the following equation:

$$N_e = \int_{E_g}^{\infty} g_c(E) f_e(E) \, dE \, , \tag{5.6}$$

where $g_c(E)$ is the density of states in the conduction band and $f_e(E)$ is the Fermi–Dirac distribution for the electrons. $g_c(E)$ is given by eqn 3.16 with m^* replaced by m_e^*:

$$g_c(E) = \frac{1}{2\pi^2} \left(\frac{2m_e^*}{\hbar^2} \right)^{3/2} (E - E_g)^{1/2} \, , \tag{5.7}$$

When we apply statistical mechanics to the carriers generated by optical excitation, it is important to realize that we are dealing with a non-equilibrium situation: there are more electrons and holes present than there would normally be just from the thermal excitation of electrons across the band gap. The system is therefore in a state of 'quasi-equilibrium'. This means that the electrons and holes form thermal distributions but with separate Fermi energies. This should be contrasted with full thermal equilibrium in which the electrons and holes share the same Fermi energy. Full thermal equilibrium can only be restored by turning off the excitation source, or by waiting for the excess electrons and holes created by a pulsed light source to recombine.

and $f_e(E)$ is given by the Fermi–Dirac formula at temperature T:

$$f_e(E) = \frac{1}{\exp\left(E - E_F^c\right)/k_B T + 1}.\tag{5.8}$$

Note that we have added a superscript c to the Fermi energy E_F to indicate that it only applies to the electrons in the conduction band. This is needed because we are in a situation of **quasi-equilibrium** in which there is no unique Fermi energy, and the electrons and holes have different Fermi levels. The Fermi–Dirac function of the holes has the same form as eqn 5.8, and $f_h(E)$ gives the probability that the state is occupied by a hole. This is equal to the probability that the state is *unoccupied* by an electron.

The Fermi integrals can be put in a more transparent form by changing the variables such that we start the electron energy at the bottom of the conduction band. We then combine eqns 5.6–5.8 to obtain

$$N_e = \int_0^\infty \frac{1}{2\pi^2} \left(\frac{2m_e^*}{\hbar^2}\right)^{3/2} E^{1/2} \left[\exp\left(\frac{E - E_F^c}{k_B T}\right) + 1\right]^{-1} dE,\tag{5.9}$$

where E_F^c is now measured relative to the bottom of the conduction band. In the same way, we can write

$$N_h = \int_0^\infty \frac{1}{2\pi^2} \left(\frac{2m_h^*}{\hbar^2}\right)^{3/2} E^{1/2} \left[\exp\left(\frac{E - E_F^v}{k_B T}\right) + 1\right]^{-1} dE,\tag{5.10}$$

for the holes, where $E = 0$ corresponds to the top of the valence band and the energy is measured downwards. The Fermi energy for the holes E_F^v is also measured downwards from the top of the valence band. Note that N_e must equal N_h here because the photo-excitation process creates equal numbers of electrons and holes.

Equations 5.9 and 5.10 can be used to determine the electron and hole Fermi energies for a given carrier density. Once these are known, the occupancy factors required to calculate the emission spectrum using eqn 5.3 can be computed. Unfortunately, the general solution of eqns 5.9 and 5.10 requires numerical methods. However, the equations simplify in two important limits. These are discussed separately below.

5.3.2 Low carrier densities

At low carrier densities, the electron and hole distributions will be described by classical statistics. The distributions shown in Fig. 5.5(b) are drawn for this limit. In this situation the occupancy of the levels is small and we can ignore the +1 in the denominator of eqn 5.8. The occupancies are then just given by Boltzmann statistics:

$$f(E) \propto \exp\left(-\frac{E}{k_B T}\right).\tag{5.11}$$

Equation 5.11 will be valid for the electrons if E_F^c is large and negative. Exercise 5.9 explores this limit. It is reasonably obvious that it will be valid at low carrier densities and high temperatures.

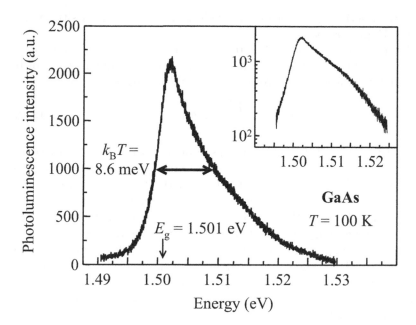

Fig. 5.6 Photoluminescence spectrum of GaAs at 100 K. The excitation source was a helium–neon laser operating at 632.8 nm. The inset gives a semi-logarithmic plot of the same data. Unpublished data from A.D. Ashmore and M. Hopkinson.

The frequency dependence of the emission spectrum in the classical limit can be calculated if we assume that the matrix element in eqn 5.3 is independent of frequency. We can then evaluate all the factors in eqn 5.3 and obtain:

$$I(h\nu) \propto (h\nu - E_g)^{1/2} \exp\left(-\frac{h\nu - E_g}{k_B T}\right). \tag{5.12}$$

The $(h\nu - E_g)^{1/2}$ factor arises from the joint density of states for the interband transition (cf. eqn 3.24). The final factor arises from the Boltzmann statistics of the electrons and holes: see Exercise 5.8. The luminescence spectrum described by eqn 5.12 rises sharply at E_g and then falls off exponentially with a decay constant of $k_B T$ due to the Boltzmann factor. We thus expect a sharply peaked spectrum of width $\sim k_B T$ starting at E_g.

Figure 5.6 shows the photoluminescence spectrum of GaAs at 100 K. The spectrum was obtained by using 1.96 eV photons from a helium–neon laser as the excitation source. The spectrum shows a sharp rise at E_g due to the $(h\nu - E_g)^{1/2}$ factor in eqn 5.12, and then falls off exponentially due to the Boltzmann factor. The full width at half maximum of the emission line is very close to $k_B T$, as expected. The fact that the high energy decay is exponential is clearly shown by the semi-logarithmic plot of the same data given in the inset. The slope of the decay is consistent with the carrier temperature of 100 K.

At very low temperatures, the emission spectrum from a direct gap semiconductor begins to depart from the form predicted by eqn 5.12, even for very low carrier densities. This is caused by the formation of excitons, and the possibility of radiative recombination involving impurities.

5.3.3   Degeneracy

At high carrier densities, the classical limit is no longer be valid. The Fermi energies are positive, and it is essential to use Fermi–Dirac statis-

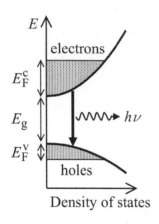

Fig. 5.7 Occupancy of the conduction and valence band states in the degenerate limit at $T = 0$. The electrons and holes have separate Fermi energies E_F^c and E_F^v respectively which are determined by the number of carriers injected into the bands. The conduction and valence bands are filled up to their respective Fermi levels, as shown by the shading.

tics to describe the electron and hole distributions. This situation is called **degeneracy**.

In the extreme limit of $T = 0$, all the states up to the Fermi energy are filled and all states above it are empty. The Fermi energies can be calculated explicitly (see Exercise 5.10) and are given by:

$$E_F^{c,v} = \frac{\hbar^2}{2m_{e,h}^*}(3\pi^2 N_{e,h})^{2/3} . \tag{5.13}$$

The distribution of the carriers in this limit is shown in Fig. 5.7. Electron–hole recombination can occur between any states in which there is an electron in the upper level and a hole in the lower level. Recombination is thus possible for a range of photon energies between E_g and $(E_g + E_F^c + E_F^v)$. We therefore expect to observe a broad emission spectrum starting at E_g up to a sharp cut-off at $(E_g + E_F^c + E_F^v)$.

At finite temperatures the carriers will still be degenerate provided that $E_F^{c,v} \gg k_B T$, where $E_F^{c,v}$ is calculated using eqn 5.13. As T increases, the Fermi–Dirac functions smear out around the Fermi energies, and we expect to observe that the cut-off at $(E_g + E_F^c + E_F^v)$ will be broadened over an energy range $\sim k_B T$.

Figure 5.8 shows the emission spectrum of the III–V alloy semiconductor $Ga_{0.47}In_{0.53}As$ in the degenerate limit. $Ga_{0.47}In_{0.53}As$ has a direct gap of 0.81 eV at the lattice temperature T_L of 10 K. The spectra were obtained using the techniques of time-resolved photoluminescence spectroscopy described in Section 5.3.5 below. The figure shows the emission spectrum recorded at two different times after the sample has been excited with an ultrashort (< 8 ps) pulse from a dye laser operating at 610 nm. Each pulse has an energy of 6 nJ and is able to excite an initial carrier density of 2×10^{24} m^{-3}.

The spectrum taken 24 ps after the pulse arrives rises sharply at E_g, and then shows a flat plateau up to ~ 0.90 eV. The spectrum then gradually falls off to zero at higher energies. The flat plateau is a signature of the degenerate carriers, while the high energy tail is an indication that the effective carrier temperature is higher than T_L due to the 'hot carrier' effect discussed in the next paragraph. In this case, the effective carrier temperature is 180 K. At 250 ps the carrier density is lower because a significant number of the electrons and holes have recombined, and the carriers have also cooled to a temperature of 55 K. At still longer times, the spectrum continues to narrow as the carrier density decreases and the carriers cool further towards the lattice temperature of 10 K. Eventually, the carrier density falls to the point where classical statistics are appropriate, and the emission only occurs at energies close to E_g. The analysis of this data is explored in more detail in Exercise 5.14.

Effective temperatures higher than T_L are possible in time-resolved photoluminescence experiments because the carriers are not in full thermal equilibrium with the lattice. The carriers are 'hot' in the same sense that boiling water that has just been poured into a cold cup is hot: the temperatures are different initially, but gradually converge as heat flows from the water to the cup. In the case we are considering here, the

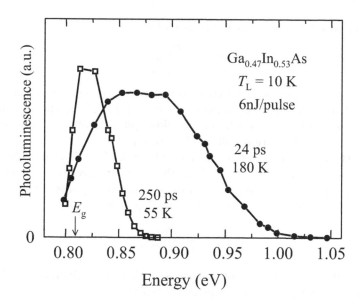

Fig. 5.8 Time-resolved photoluminescence spectra of the direct gap III–V alloy semiconductor $Ga_{0.47}In_{0.53}As$ at a lattice temperature T_L of 10 K. The sample was excited with laser pulses at 610 nm with an energy of 6 nJ and a duration of 8 ps. This generated an initial carrier density of $2 \times 10^{24} \, m^{-3}$. Spectra are shown for time delays of 24 ps (filled circles) and 250 ps (open squares). The effective carrier temperature at the two time delays is indicated. After Kash and Shah (1984), © American Institute of Physics, reprinted with permission.

electrons and holes are created high up the bands. This gives them a large amount of kinetic energy, which implies that their initial effective temperature is very high, since the temperature is just a measure of the distribution of the carriers among the energy levels of the system. The temperature decreases rapidly as energy flows from the carriers to the lattice by phonon emission. The cooling towards T_L is therefore determined by the electron–phonon interactions in the material.

5.3.4   Optical orientation

Optical orientation is the phenomenon by which angular momentum is imparted to electrons by interaction with photons. In Section 3.3.7 we studied how a net electron spin polarization can be created by excitation with circularly polarized light, and we now wish to understand how this affects the polarization of the light that is emitted.

The luminescence polarization is defined by:

$$P = \frac{I^+ - I^-}{I^+ + I^-}, \tag{5.14}$$

where I^+ and I^- are the intensities of the σ^+ and σ^- circular polarizations respectively. We consider a zinc-blende semiconductor excited by circularly polarized light, as shown in Fig. 3.8. As explained in Section 3.3.7, this creates an initial electron spin polarization of 50%, and negligible hole spin. The selection rules shown in Fig. 3.8 apply in both directions, and this leads to an expected luminescence polarization of 25%. (See Exercise 5.15.)

The actual polarization observed experimentally is smaller than 25% because the electron spin can change during the lifetime of the carriers. If the carrier lifetime and spin relaxation time are τ and τ_S respectively, then the measured polarization is given by:

$$P = \frac{P_0}{1 + \tau/\tau_S} , \tag{5.15}$$

where P_0 is the polarization expected if there is no spin relaxation. This shows that if the spin relaxation is fast (i.e. $\tau_S \ll \tau$), the measured polarization will be small, whereas for slow spin relaxation (i.e. $\tau_S \gg \tau$), the measured polarization will be close to P_0.

The **Hanle effect** provides an elegant method to determine both τ and τ_S in a single experiment. The Hanle effect describes the loss of optical polarization caused by the precession of the spin in a transverse magnetic field B. The measured polarization is given by:

$$P(B) = \frac{P(0)}{1 + (\Omega T_S)^2} , \tag{5.16}$$

where $P(0)$ is the polarization measured at $B = 0$, as given by eqn 5.15. Ω is the Larmor precession frequency given by

$$\Omega = \frac{g_e \mu_B B}{\hbar} , \tag{5.17}$$

where g_e is the electron g-factor, and

$$\frac{1}{T_S} = \frac{1}{\tau} + \frac{1}{\tau_S} . \tag{5.18}$$

τ and τ_S are determined by measuring $P(0)$ and the field at which the polarization drops to half its value at $B = 0$. (See Exercise 5.16.)

There are a number of different mechanisms that can cause the spin of an electron to relax in a semiconductor, the most important of these being:

- the **Elliott–Yafet** (EY) mechanism. This arises from the spin–orbit interaction, which mixes the spin up and down wave functions, thereby allowing momentum scattering events to randomize the spin.

- the **Dyakonov–Perel** (DP) mechanism. This also arises from the spin–orbit interaction. In crystals that lack inversion symmetry, the spin degeneracy of the electrons is lifted for $|\mathbf{k}| > 0$. The splitting is equivalent to an effective magnetic field, the axis of which fluctuates during a momentum scattering event, thereby depolarizing the spin by a series of random fractional rotations.

- the **Bir–Aronov–Pikus** (BAP) mechanism. This mechanism is important when a population of unpolarized holes is present. Electron spin flips can then occur by exchange interactions with holes.

By studying the way the polarization changes with the temperature and doping levels, it is possible to determine which of these mechanisms is dominant in any particular sample. The EY mechanism, for example, is expected to be particularly important in narrow gap semiconductors,

The Hanle effect was originally studied in atomic physics, and refers to the depolarization of resonance fluorescence by external magnetic fields. The derivation of eqn 5.16 may be found, for example, in Meier and Zakharchenya (1984).

since these have strong spin–orbit interactions, while the BAP mechanism is likely to be important in p-type materials. The DP mechanism should occur in all zinc-blende samples, and its effectiveness depends inversely (i.e. counter-intuitively) on the electron scattering rate. Further details about electron spin relaxation mechanisms may be found in the works cited for Further Reading.

5.3.5 Photoluminescence spectroscopy

Photoluminescence spectroscopy is mainly used as a diagnostic and development tool in semiconductor research. The usual goal is to develop electroluminescent devices such as light-emitting diodes and lasers. This is usually only achieved after the emission mechanisms have been studied in detail by photoluminescence spectroscopy.

Photoluminescence spectra can be recorded with an experimental arrangement such as the one shown in Fig. 5.9. The sample is mounted in a variable temperature cryostat and is illuminated with a laser or bright lamp with photon energy greater than E_g. If a liquid helium cryostat is used, sample temperatures from 2 K upwards are easily obtained. The luminescence is emitted at lower frequencies and in all directions. A portion is collected with a lens and focused onto the entrance slit of a spectrometer. The spectrum is recorded by scanning the spectrometer and measuring the intensity at each wavelength with a sensitive detector such as a photomultiplier tube. Alternatively, the whole spectrum is recorded at once by using an array of detectors such as a charge-coupled device (CCD).

A number of useful variations of the basic photoluminescence technique have been developed over the years. In **photoluminescence excitation spectroscopy** (PLE), the sample is excited with a tunable laser, and the intensity of the luminescence at the peak of the emission is measured as the laser wavelength is tuned. Since the shape of the emission spectrum is independent of the way the carriers are excited, the signal strength is simply proportional to the carrier density, which in turn is determined by the absorption coefficient. (See Exercise 5.6.) Hence the signal is proportional to the absorption coefficient at the laser wavelength. This might seem to be a very complicated way to measure the absorption, but it is actually very useful. Many semiconductor samples are grown as thin layers on top of a thick substrate which is opaque at the wavelengths of interest. This makes it impossible to perform direct transmission measurements, and the use of the PLE technique allows the absorption spectrum to be measured in conditions where it would not be possible otherwise.

In **time-resolved photoluminescence spectroscopy** the sample is excited with a very short light pulse and the emission spectrum is recorded as a function of time after the pulse arrives. The spectra are obtained using the arrangement shown in Fig. 5.9 but with an ultra-fast pulse laser as the excitation source. Lasers emitting pulses shorter than 1 ps are now readily available, and the time resolution is usually

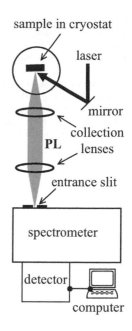

Fig. 5.9 Experimental arrangement used for the observation of photoluminescence (PL) spectra. The sample is excited with a laser or lamp with photon energy greater than the band gap. The spectrum is obtained by recording the emission as a function of wavelength using a computer-controlled spectrometer and detector. In photoluminescence excitation spectroscopy (PLE), the detection wavelength is fixed and the excitation wavelength is scanned. In time-resolved photoluminescence spectroscopy, a pulsed laser is used, and the emission at each wavelength is recorded on a fast detector as a function of time after the pulse has arrived.

Fig. 5.10 (a) Layer structure and (b) circuit diagram for a typical electroluminescent device. The thin active region at the junction of the p- and n-layers is not shown, and the dimensions are not drawn to scale. The thickness of the epitaxial layers will be only ~ 1 μm, whereas the substrate might be ~ 500 μm thick. The lateral dimensions of the device might be several millimetres.

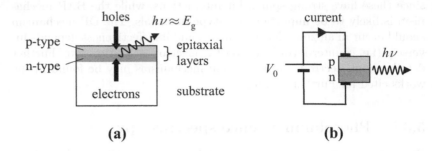

limited by the response time of the detector. Time resolutions down to ~ 100 ps can be obtained with fast photon-counting photomultiplier tubes or avalanche photodiodes, while resolutions down to 1 ps or better are possible with 'streak camera' or 'up-conversion' techniques. The time dependence of the emission spectrum gives direct information about the carrier relaxation and recombination mechanisms, and allows the radiative lifetimes to be measured. Figure 5.8 gives an example of the data that can be obtained by using this technique.

5.4 Electroluminescence

Electroluminescence is the process by which luminescence is generated while an electrical current flows through an opto-electronic device. There are two main types of device:

- **light-emitting diodes** (LEDs)
- **laser diodes**.

We shall look at both types of device here, concentrating on inorganic semiconductors, and postponing the discussion of molecular LEDs to Section 8.4. We conclude by briefly considering the related technique of cathodoluminescence.

5.4.1 General principles of electroluminescent devices

Figure 5.10 shows the layer structure and circuit diagram for a typical electroluminescent device. The device consists of several **epitaxial** layers grown on top of a thick crystal **substrate**. The epitaxial layers consist of a p–n diode with a thin **active region** at the junction. The diode is operated in forward bias with a current flowing from the p-layer through to the n-layer underneath. The luminescence is generated in the active region by the recombination of electrons that flow in from the n-type layer with holes that flow in from the p-type side.

The microscopic mechanisms that determine the emission spectrum are exactly the same as the ones discussed in the context of photoluminescence in Sections 5.3.1–5.3.3. The only difference is that the carriers are injected electrically rather than optically. At room temperature we therefore expect a single emission line of width $\sim k_\mathrm{B} T$ at the band gap energy E_g. Hence E_g determines the emission wavelength.

We pointed out in Section 5.2 that the radiative efficiency of indirect gap materials is low. Modern commercial electroluminescent devices are therefore made from direct gap compounds. Any direct gap semiconductor can, in principle, be used for the active region, but in practice only a few materials are commonly employed. The main factors that determine the choice of the material are:

(1) the size of the band gap;

(2) constraints relating to lattice matching;

(3) the ease of p-type doping.

The first point is obvious: the band gap determines the emission wavelength. The second and third points are practical ones relating to the way the devices are made. These are discussed further below.

The term **lattice matching** relates to the relative size of the lattice parameters of the epitaxial layers and the substrate. The thin epitaxial layers are grown on top of a substrate crystal, as shown in Fig. 5.10(a). This is done for practical reasons. It is hard to grow large crystals with sufficient purity to emit light efficiently. We therefore grow thin ultrapure layers on top of a substrate of poorer optical quality by various techniques of crystal **epitaxy**. The crystal growth conditions constrain the epitaxial layers to form with the same unit cell size as the substrate crystal. This means that the epitaxial layers will be highly strained unless they have the same lattice constant as the substrate, that is, that we have 'lattice matching' between the epitaxial layers and the substrate. If this condition is not satisfied, crystal dislocations and other defects are likely to form in the epitaxial layers, leading to a severe degradation of the optical quality.

Figure 5.11 plots the band gap of a number of III–V materials used in electroluminescent devices against their lattice constant. The lattice constants of the commonly used substrate crystal are indicated at the top of the figure. The materials separate into two distinct groups. On the right we have the arsenic and phosphorous compounds which crystallize with the cubic zinc-blende structure, while on the left we have the nitride compounds which have the hexagonal wurtzite structure. We shall discuss the cubic materials first, and then consider the nitrides afterwards.

For many years, the opto-electronics industry was mainly based on GaAs and its alloys. GaAs emits in the infrared at 870 nm, and by mixing it with AlAs to form $Al_x Ga_{1-x} As$, light emitters for the range 630–870 nm can be produced. (See Example 5.1.) AlGaAs can easily be grown lattice-matched to GaAs substrates because of the convenient coincidence that the lattice constants of GaAs and AlAs are almost identical. AlGaAs emitters operating at 850 nm are widely used in local area fibre-optic networks and infrared free-space data links, while devices with higher Al content are used in red LEDs.

In the past, some indirect gap materials have been used for lack of practical direct gap alternatives. For example, gallium phosphide was used for yellow and green LEDs, and silicon carbide for blue ones. The active regions of these devices were often doped to promote recombination via impurities and hence increase the luminescent quantum efficiency. The advent of efficient direct gap nitride LEDs in 1995 has made these indirect gap devices obsolete.

Epitaxy is the name given to any crystal growth technique involving the formation of thin high-quality layers on top of a thicker substrate crystal. The substrate acts as a support for the epitaxial layers, and also serves as a heat sink where needed. There are a number of techniques commonly used. Medium-quality crystals are grown by liquid-phase epitaxy (LPE), but the highest quality materials are grown by metal–organic vapour-phase epitaxy (MOVPE)—also called metal–organic chemical vapour deposition (MOCVD)—and molecular beam epitaxy (MBE). These techniques are crucial to the successful growth of the high quality quantum well structures described in the next chapter.

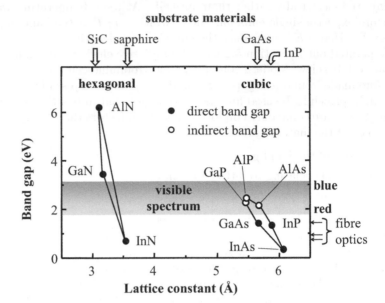

Fig. 5.11 Band gap of selected III–V semiconductors as a function of the their lattice constant. The materials included in the diagram are the ones commonly used for making LEDs and laser diodes. The lattice constants of readily available substrate crystals are indicated along the top axis. The nitride materials on the left grow with the hexagonal wurtzite structure, whereas the phosphides and arsenides on the right have the cubic zinc-blende structure. Data from Madelung (1996).

Table 5.1 Band gap energy E_g and emission wavelength λ_g for several compositions of the direct band gap quaternary III–V alloy $Ga_xIn_{1-x}As_yP_{1-y}$. The compositions indicated all satisfy the lattice-matching condition for InP substrates, namely $x \approx 0.47y$. Data from Madelung (1996).

x	y	E_g (eV)	λ_g (μm)
0	0	1.35	0.92
0.27	0.58	0.95	1.30
0.40	0.85	0.80	1.55
0.47	1	0.75	1.65

The value of the band gap of InN has been the subject of some controversy. Older texts (including the first edition of this book) quote values around 2 eV, but recent results indicate that the gap is much smaller.

AlGaAs is an example of a 'ternary' alloy which contains three elements. 'Quaternary' alloys such as $(Al_yGa_{1-y})_xIn_{1-x}P$ can also be formed. All of these arsenic and phosphorous alloys suffer from the problem that they become indirect as the band gap gets larger. This limits their usefulness to the red and near-infrared spectral range.

Applications in the fibre optics industry require light-emitting devices that operate around 1.3 μm and 1.55 μm. These are the wavelengths at which silica fibres have the lowest dispersion and loss respectively. Emitters for these wavelengths tend to be made from the quaternary alloy $Ga_xIn_{1-x}As_yP_{1-y}$. Lattice matching to InP substrates can be achieved if $x \approx 0.47y$. This allows a whole range of direct gap compounds to be made with emission wavelengths varying from 0.92 μm to 1.65 μm. See Table 5.1.

Until fairly recently, it was very difficult to make efficient electroluminescent devices for the green and blue spectral regions from III–V compounds. This is because of the problem that has already been mentioned, namely that the arsenic and phosphorous compounds become indirect as the band gap gets larger. However, in 1995 Shuji Nakamura at Nichia Chemical Industries in Japan made an important breakthrough and reported the successful development of LEDs based on gallium nitride compounds. GaN has a direct band gap of 3.5 eV at 4 K (see Fig. 5.3) and 3.4 eV at room temperature. By alloying it with InN, the emission wavelength can be varied from the ultraviolet to the red spectral regions. This enables the entire visible spectrum to be covered by using nitrides for the blue and green colours, and AlGaInP alloys for the reds.

It is interesting to consider why it took so long to develop the nitride devices. It was well known that the nitrides would in principle make good blue/green emitters, but no commercial devices were available. The reason for this relates to the third point on our list of factors affecting the choice of electroluminescent materials, namely the difficulty of p-type doping. This is a problem that has also dogged other wide band gap materials. For example, direct gap II–VI compounds like ZnSe and CdSe should also, in principle, make good LEDs for the blue/green/yellow spectral regions, but they have never found widespread commercial application due to the doping problem.

P-type doping is difficult in wide band gap semiconductors because they have very deep acceptor levels. The energies of the acceptors are given by eqn 7.29 with m_e^* replaced by m_h^*. The high value of m_h^* and the relatively small value of ϵ_r in wide gap materials increases the acceptor energies, and hence reduces the number of holes that are thermally excited into the valence band at room temperature. This last point follows from the Boltzmann factor (eqn 5.11) with E equal to the acceptor binding energy, which is significantly larger than $k_B T$. The low hole density gives the layers a high resistivity, which causes ohmic heating when the current flows, and hence device failure. Nakamura's breakthrough came after discovering new techniques to activate the holes in p-type GaN by annealing the layers in nitrogen at 700 °C.

There is another point that is surprising about the development of nitride LEDs. Lattice-matching considerations suggest that the devices should ideally be grown on silicon carbide substrates, or, better still, GaN itself. (See Fig. 5.11.) However, both of these materials are expensive, and the commercial devices tend to be grown on cheaper sapphire substrates. Conventional wisdom would suggest that the radiative efficiency should be low due to the large defect density arising from the lattice mismatch. However, the radiative efficiency can in fact be very high. One factor that has made this possible is the growth of a thick 'buffer' layer immediately above the substrate, which has the effect of reducing the number of crystal dislocations in the active region. Another factor is the relatively poor diffusion coefficients of the electrons and holes in GaN, combined with the high radiative probability. The electrons and holes then tend to recombine radiatively before they have time to diffuse to a defect and recombine non-radiatively.

In the next chapter we shall describe how the use of quantum well layers has led to further developments in the field of electroluminescent materials. In fact, many commercial devices—especially laser diodes, but also many LEDs—now routinely use quantum wells in the active region. Another important recent breakthrough has been the combination of nitride light-emitting diodes with phosphor technology to make efficient white light sources. These devices, which form the basis of the solid-state lighting industry, will be considered in Section 9.5.

5.4.2   Light-emitting diodes

The operating principle of a light-emitting diode (LED) can be understood with reference to the band diagram shown in Fig. 5.12. The p- and n-regions are both very heavily doped to produce degenerate distri-

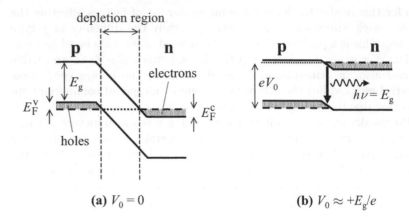

depletion region

p   **n**

E_g

electrons

E_F^v E_F^c

holes

(a) $V_0 = 0$

p   **n**

eV_0

$h\nu = E_g$

(b) $V_0 \approx +E_g/e$

Fig. 5.12 Band diagram of a light-emitting diode at (a) zero bias, and (b) forward bias $V_0 \approx E_g/e$. The device consists of a p–n diode with heavily doped p- and n-regions. The dashed lines indicate the positions of the Fermi levels in the p- and n-regions, which must be aligned when $V_0 = 0$. Light is emitted in (b) when the electrons in the n-region recombine with holes in the p-region at the junction.

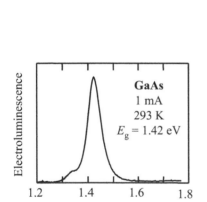

GaAs
1 mA
293 K
$E_g = 1.42$ eV

Energy (eV)

Fig. 5.13 Electroluminescence spectrum of a GaAs LED at room temperature. Unpublished data from A.D. Ashmore.

butions of holes in the p-region and electrons in the n-region. Note that this is a different type of degeneracy to that considered in Section 5.3.3. Degeneracy here means that the carrier density produced by the doping is so large that the Fermi energies in the p- and n-regions are positive with respect to the band edges. There is full thermal equilibrium at $V_0 = 0$, with a unique Fermi energy for the whole device, and the bands therefore align as shown in Fig. 5.12(a). At the junction, a depletion region is formed, with neither electrons nor holes present. No light can be emitted, because there is no point within the device where there is a significant population of both electrons and holes.

The situation is different when a forward bias of $V_0 \sim E_g/e$ is applied to drive a current through the device. In this non-equilibrium condition, the Fermi levels in the p- and n-regions shift relative to each other as shown in Fig. 5.12(b). The depletion region shrinks, allowing the electrons in the n-region to diffuse into the p-region, and vice versa. This creates a region at the junction where both electrons and holes are present. The electrons recombine with the holes, emitting photons at energy E_g by interband luminescence. The electrons and holes that recombine are replenished by the current flowing through the device from the external circuit, which was given previously in Fig. 5.10(b).

Figure 5.13 shows the spectrum of a forward biased GaAs p–i–n diode with a current of 1 mA flowing through the device. The light is generated in the thin i-region at the junction between the p- and n-regions. As mentioned previously, GaAs has a band gap of 1.42 eV at room temperature, which gives emission in the near-infrared around 870 nm. The full width at half maximum of the emission line is 58 meV, which is about twice $k_B T$ at 293 K.

5.4.3 Diode lasers

Semiconductor lasers are more difficult to make than LEDs, but they give superior performance in terms of their output efficiency, spectral linewidth, beam quality, and response speed. They are therefore used for the more demanding applications, leaving the simpler ones for the cheaper LED devices. They are mainly made from GaAs-based materials, and operate in the red and near-infrared spectral regions. However, blue laser diodes have recently become available following the development of efficient nitride based emitters.

The acronym 'laser' stands for 'Light Amplification by Stimulated Emission of Radiation'. As the name suggests, laser operation is based on the quantum-mechanical process of **stimulated emission**. This should be distinguished from the process of spontaneous emission that is responsible for luminescence. (See Section B.1 in Appendix B.) Stimulated emission causes an *increase* in the photon number as the light interacts with the atoms of the medium, which in turn leads to optical amplification. This contrasts with the process of absorption which *reduces* the number of photons, and hence causes attenuation.

Consider the interaction between a light wave of frequency ν and a medium containing atoms with an electronic transition at energy $h\nu$, as illustrated in Fig. B.2. The absorption processes cause beam attenuation, while stimulated emission causes amplification. The transition rates for the two processes are given by eqns B.5 and B.6 respectively. In the normal conditions of thermal equilibrium, the population of the lower level N_1 will be greater than the population of the upper level N_2 by the Boltzmann factor given in eqn B.8. This means that the absorption rate exceeds the stimulated emission rate, and there is net beam attenuation. However, if we were somehow to arrange for N_2 to be larger than N_1, then the reverse would be true. The stimulated emission rate would exceed the absorption rate, and there would be net beam amplification. The non-equilibrium condition with $N_2 > N_1$ is called **population inversion**. It is a necessary condition for laser oscillation to occur.

In Section 5.3.1 we explained how the carrier distributions after injection of electrons and holes only reach quasi-equilibrium rather than full thermal equilibrium. The top of the valence band is empty of electrons, while the bottom of the conduction band is filled with them. We therefore have population inversion at the band gap frequency E_g/h. This gives rise to net optical gain, which can be used to obtain laser operation if an optical cavity is provided.

Figure 5.14 shows a schematic diagram of a laser cavity formed from a gain medium with mirrors at either end. This is the typical arrangement for a semiconductor laser diode, which usually consists of just the semiconductor chip itself. The reflectivities of the semiconductor–air surfaces at the edge of the crystal are typically around 30%. (See Exercise 5.18.) This may be sufficient in itself to obtain lasing, although in what follows we assume that the reflectivities R_1 and R_2 at the two ends are different, and that $R_1 \gg R_2$.

On passing a current through the p–n junction of a laser diode, light at frequency $\nu \approx E_g/h$ is generated by electroluminescence. This light is reflected back and forth within the cavity, and experiences gain due to the population inversion between the conduction and valence bands. At

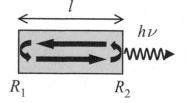

Fig. 5.14 Schematic diagram of a laser cavity formed by reflections from the end surfaces of the gain medium of length l. The reflectivities of the surfaces are taken to be R_1 and R_2 respectively, with $R_1 \gg R_2$.

some particular value of the **injection current** I_{in} called the **threshold current** I_{th}, the laser will begin to oscillate. For current values above I_{th}, the light output power of the laser increases linearly with I_{in}. This is illustrated in Fig. 5.15(a). The output power is coupled out of the cavity by transmission through the mirror with the lower reflectivity, which is called the **output coupler** of the laser.

Once the laser is oscillating, the emission spectrum will be determined by the resonant **longitudinal modes** of the optical cavity. The resonant modes must satisfy the condition that they form standing waves between the mirrors, and hence that there are an integer number of half wavelengths within the cavity. This condition can be written:

$$\text{integer} \times \frac{\lambda'}{2} = l \,, \tag{5.19}$$

where λ' is the wavelength inside the crystal, which is equal to λ/n, λ being the air wavelength and n the refractive index. This means that the frequencies of the longitudinal modes must satisfy:

$$\nu = \text{integer} \times \frac{c}{2nl} \,. \tag{5.20}$$

The laser will oscillate at one or several of these resonant frequencies. Some semiconductor lasers oscillate on just one single longitudinal mode, and have emission linewidths in the MHz range. This is many orders of magnitude smaller than that of the equivalent LED.

The condition for stable oscillation of the laser is that the light intensity in the cavity should not change with time. This implies that the gain in the laser medium must exactly balance any losses suffered by the light during a round trip of the cavity. This condition allows us to work out the value of the gain in the medium when the laser is oscillating.

We assume that there is population inversion inside the medium, and hence that there is optical amplification at the transition frequency ν. We define the incremental gain coefficient γ_ν as:

$$dI = +\gamma_\nu \, dx \times I(x) \,. \tag{5.21}$$

This is exactly the same definition as for the absorption coefficient in eqn 1.3, except that the intensity is now growing with distance rather than diminishing. Integration of eqn 5.21 yields:

$$I(x) = I_0 e^{\gamma_\nu x} \,. \tag{5.22}$$

We follow the light at frequency ν around a round trip of the cavity shown in Fig. 5.14. In stable laser oscillation, the increase of the intensity due to the gain must exactly balance the losses due to the imperfect reflectivity of the end mirrors and any other losses that may be present in the medium. This condition may be written:

$$R_1 R_2 \, e^{2\gamma_\nu l} \, e^{-2\alpha_b l} = 1 \,. \tag{5.23}$$

The factor of 2 in the two exponentials allows for the fact that the light passes through the gain medium twice during a round trip. The

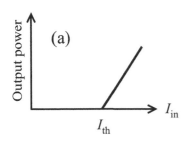

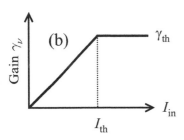

Fig. 5.15 (a) Power output and (b) gain coefficient γ_ν as a function of injection current I_{in} in a semiconductor laser diode. I_{th} is the threshold injection current, and γ_{th} is the threshold gain required for stable laser oscillation. Note that these are idealized curves, and that real devices may depart somewhat from the behaviour shown here.

attenuation coefficient α_b in eqn 5.23 accounts for scattering losses and absorption due to processes other than interband transitions, for example impurity absorption. The oscillation condition in eqn 5.23 can be re-written:

$$\gamma_{th} = \alpha_b - \frac{1}{2l}\ln(R_1 R_2). \qquad (5.24)$$

This defines the threshold gain γ_{th} required to make the laser oscillate. Direct gap semiconductors such as GaAs have very large gain coefficients due to their high density of states and short radiative lifetimes. This makes it possible to overcome the output coupling losses with cavity lengths of order 1 mm or less.

We assume that the gain coefficient increases linearly with the injection current I_{in}, as indicated in Fig. 5.15(b). When $I_{in} = I_{th}$, the gain reaches the value γ_{th} defined by eqn 5.24, at which point the laser begins to oscillate. Once the laser is oscillating, the gain must be clamped at the value of γ_{th}, because otherwise the gain would exceed the losses, and the stability condition set out in eqn 5.23 would not hold. Thus for $I_{in} > I_{th}$, the extra electrons and holes injected into the junction do not produce any more gain, but recombine directly by stimulated emission, and cause the output power to increase, as indicated in Fig. 5.15(a).

The output power P_{out} above threshold can be written:

$$P_{out} = \eta \frac{h\nu}{e}(I_{in} - I_{th}). \qquad (5.25)$$

where η is the quantum efficiency. η defines the fraction of injected electron-hole pairs that generate laser photons. The quantum efficiency determines the **slope efficiency** in watts per amp through

$$\text{slope efficiency} = \frac{P_{out}}{(I_{in} - I_{th})} = \frac{\eta h\nu}{e}. \qquad (5.26)$$

In an ideal laser diode we would have $\eta = 1$ and the slope efficiency would be equal to the theoretical maximum of $h\nu/e$. Many of the best diode lasers come quite close to this ideal limit.

One of the main reasons why η might be less than unity in a real laser diode relates to issues of **optical confinement** and **electrical confinement**. The device will not work efficiently unless we can arrange that the injection current is confined to the same part of the device where the light is confined. This is not necessarily an easy task due to the inherently planar nature of semiconductor lasers. The devices have very small dimensions (e.g. 1 μm) in the vertical (z) direction, and much larger directions (e.g. several hundred microns) in the horizontal x, y plane. The light is generated in the thin active region, and is emitted from the edge of the chip. In such a planar structure, the light tends to spread out in the y, z plane, while the current tends to spread out in the x, y direction. This leads to the possibility that the current and light might not overlap properly in the x, y plane, in which case we would have poor quantum efficiency.

There are many different ways to achieve optical and electrical confinement, and we can understand the basic principles by looking at a

New types of lasers called vertical-cavity surface-emitting lasers have different geometries to the planar lasers discussed here. The light is emitted from the top of the chip, rather than from its sides. See the references given in the Further Reading list for more details.

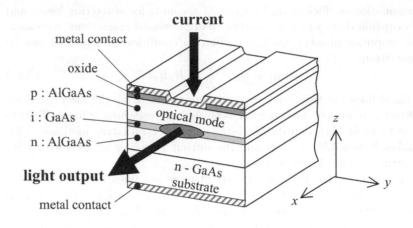

Fig. 5.16 Schematic diagram of an oxide-confined GaAs–AlGaAs heterojunction stripe laser. The current flows in the $-z$ direction, while the light propagates in the $\pm x$ direction. The stripe is defined by the gap in the insulating oxide layers deposited on the top of the device during the fabrication process. The active region is the intrinsic GaAs layer at the junction between the n- and p-type AlGaAs cladding layers.

A heterojunction is one with different materials at the junction, and contrasts with a homojunction, where all the materials are the same. The names are derived from the Greek words *heteros*, meaning 'other', and *homos* meaning 'same'. The heterojunction laser was independently invented by Zhores I. Alferov and Herbert Kroemer in 1963, for which they were awarded the Nobel Prize for physics in 2000.

specific example. Figure 5.16 gives a schematic diagram of an oxide-confined GaAs–AlGaAs heterostructure stripe laser. The 'stripe' is defined by the gap in the insulating oxide layers deposited on the top of the device during the fabrication process. The current flows in the $-z$ direction, between the top and bottom metal contacts. The top contact only connects to the p-region in between the oxide layers, and so the current is confined to the long thin rectangular strip of the x, y plane defined by the fabrication process.

The light, on the other hand, propagates in the $\pm x$ direction. The shape of the laser mode in the y, z plane is determined by **waveguide** effects. This refers to the confinement of a light beam in the direction perpendicular to its propagation instead of the usual divergence due to diffraction. The confinement in the z direction is achieved through the tendency of the light to propagate in the region with the largest refractive index, which can be understood in terms of repeated total internal reflections at the interfaces between the high and low refractive index materials. This vertical confinement is easily achieved in **heterojunction** devices such as the one shown in Fig. 5.16. In the example given, the active region is made of GaAs, which has a higher refractive index than the AlGaAs 'cladding' layers on either side.

The optical confinement in the y direction is more difficult. It is either achieved by index guiding or gain guiding. Index guiding is the same effect as that used to produce the vertical confinement. The lateral patterning of the top of the chip can produce small variations in the effective refractive index in the y direction through strain or other effects. Gain guiding, on the other hand, follows as a consequence of current confinement. The semiconductor layers have very strong absorption at the laser wavelength except in the regions where there is gain due to population inversion. Hence the optical mode will be extremely lossy except in the gain regions defined by the current confinement. This is the case with the example shown in Fig. 5.16.

The reader is referred to the references given in the Further Reading list for more detailed information about the many different types of semiconductor laser that have been made. In the next chapter, we shall explain how the use of quantum wells in the active region has led to superior performance and greater flexibility in the emission wavelength.

5.4.4 Cathodoluminescence

Cathodoluminescence is the phenomenon by which light is emitted from a solid in response to excitation by cathode rays, that is: electron beams (e-beams). Since the light is generated in response to the electron current of the e-beam, it can be considered as a type of electroluminescence, although some texts list it as a separate sub-category of luminescence. Cathodoluminescence is extensively used in cathode ray tubes, and it is also a powerful research tool.

The basic processes that occur when an e-beam strikes a crystal are illustrated in Fig. 5.17. The electrons in the e-beam are called *primary* electrons, and have an energy which is determined by the applied voltage, which might typically be 1–100 kV. Some of the primary electrons are scattered elastically by the atoms (i.e. without significant energy loss) and give rise to high energy *back-scattered* electrons. These back-scattered electrons can be collected and used to from an image of the sample, as happens in an electron microscope. The remaining electrons are scattered inelastically many times as they penetrate the crystal, and their direction gets randomized in the process. The region of the crystal that interacts with the e-beam is called the excitation volume, and the distance the beam travels is called the penetration depth (or electron range R_e). The penetration depth increases with increasing primary electron energy, and typical values of R_e are in the range 1–10 μm. However, R_e can be significantly less than 1 μm for e-beam energies below ~ 10 keV.

The electrons that penetrate the surface transfer their energy to the crystal by exciting electron-hole pairs. The number of electron-hole pairs generated per primary electron is given by:

$$N^{\mathrm{eh}} = (1 - \gamma)\frac{E^{\mathrm{p}}}{E^{\mathrm{i}}}, \qquad (5.27)$$

where γ is the fractional energy loss due to back scattering, E^{p} is the energy of the primary electron, and E^{i} is the ionization energy (i.e., the energy required to form an electron-hole pair.) The electron-hole pairs are produced by a complicated multi-step process involving the re-emission and subsequent inelastic scattering of *secondary* electrons. However, for a wide range of materials it has been found that E^{i} is given by the following simple semi-empirical formula:

$$E^{\mathrm{i}} = 2.8 E_{\mathrm{g}} + E', \qquad (5.28)$$

where E' depends only on the material, and has a magnitude in the range 0–1 eV. A primary electron with an energy of ~ 10 keV can therefore generate thousands of electron-hole pairs in a semiconductor with a band gap of 1–3 eV. These electrons and holes are created high up in their

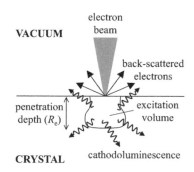

Fig. 5.17 Generation of cathodoluminescence following excitation of a crystal by an electron beam.

bands, and emit photons in all directions with energy $\hbar\omega \gtrsim E_\mathrm{g}$ after having relaxed to the bottom of their bands. It is these photons that comprise the cathodoluminescence signal.

As a research tool, cathodoluminescence is usually studied by collecting and detecting the light emitted by the sample in an electron microscope. It is particularly useful for investigating wide band gap materials and nanostructures. In the former case, photoluminescence experiments may be impractical due to the lack of an excitation source with a suitably high photon energy, leaving cathodoluminescence as the only viable technique for studying the light emission processes. In the latter case, the ability to focus the e-beam to a very small spot allows the selective excitation of structures with submicron dimensions. This spatial selectivity is limited by the spreading of the electrons within the excitation volume, but resolutions of $\sim 100\,\mathrm{nm}$ or less can be achieved with low energy beams (e.g. 5 keV).

In commercial applications, cathodoluminescence is widely used in cathode ray tubes. In these devices an electron beam is scanned across a screen coated with a light-emitting material called a phosphor. In monochrome displays such as those found in oscilloscopes, a singe beam and a single phosphor is used. However, in colour displays such as those used in some computer monitors, three separate e-beams must be used, together with three different phosphors: one for each primary colour, namely red, green and blue. Each pixel of the screen consists of red, green and blue sub-pixels, and by addressing these with separate e-beams, the full range of colours can be obtained. Further details of the physics of phosphors may be found in Section 9.5.

Chapter summary

- Luminescence is the generic name for light emission by spontaneous emission in solids. Photoluminescence is the re-emission of light following absorption of higher-energy photons. Electroluminescence is the luminescence generated by electrical excitation.

- The emission rate is proportional to the matrix element for the transition, the density of states, and the occupancy factors of the upper and lower levels.

- Transitions with high absorption coefficients have short radiative lifetimes. Efficient luminescence is only obtained when the radiative lifetime is shorter than the non-radiative lifetime.

- Interband luminescence occurs when an electron in the conduction band drops to the valence band with the emission of a photon. The process is equivalent to the recombination of an electron-hole pair. The transition is represented by a downward vertical arrow on the band diagram.

- The interband luminescence spectrum is usually independent of the way the material is excited. The emission wavelength corresponds to the fundamental band gap of the material.

- Direct gap materials have short radiative lifetimes ($\sim 1\,\mathrm{ns}$) and are strong emitters. Indirect gap materials have much longer radiative lifetimes and are generally very inefficient emitters.

- The carriers generated by photoexcitation rapidly relax to the bottom of their bands before recombining, and come to a state of quasi-equilibrium with separate Fermi energies for the electrons and holes. The luminescence spectrum can be calculated from the thermal distributions of the carriers.

- The depolarization of the luminescence following excitation by circularly polarized light gives information about electron spin relaxation processes.

- Light-emitting diodes consist of p–n diodes with the light-emitting material in the active region at the junction between the p- and n-layers. Light is emitted when the diode is forward biased. LEDs are usually made from direct gap semiconductors.

- The injection of electrons and holes into the conduction and valence bands can produce population inversion at the band gap frequency. This can support laser operation if the gain due to stimulated emission balances the round trip losses in the optical cavity.

- Semiconductor lasers are usually planar structures with the light emitted from the edge of the chip. The cavity is formed between the end mirrors at the air–semiconductor interfaces.

- The light emitted following excitation by cathode rays is called cathodoluminescence.

Further reading

A good introductory overview of luminescent processes in solids may be found in Elliott and Gibson (1974). Interband luminescence in semiconductors is discussed in Pankove (1971) and Yu and Cardona (1996). More detailed discussions may be found in Landsberg (1991) or Voos et al. (1980).

The definitive work on optical orientation experiments performed up to 1984 is Meier and Zakharchenya (1984). More recent work is reviewed in Awschalom et al. (2002), Dyakonov (2008), and Kusrayev & Landwehr (2008). An authoritative discussion of time-resolved luminescence spectroscopy may be found in Shah (1999).

The physics of electroluminescence is discussed in most opto-electronics texts, for example Bhattacharya (1997), Chuang (1995), Sze (1981), Sze (1985), or Wilson and Hawkes (1998). A thorough account of the principles of light-emitting diodes may be found in Schubert (2006). The development of nitride light emitters is discussed in Nakamura et al. (2000), while detailed information about semiconductor laser diodes may be found in Silfvast (2004), Svelto (1998), or Yariv (1997). The physics of cathodoluminescence is covered in depth in Yacobi and Holt (1990) or Gustafsson et al. (1998).

Exercises

(5.1) Explain why it is difficult to make light-emitting devices out of indirect gap materials.

(5.2) When a direct gap semiconductor is excited by absorption of photons with energy greater than the band gap, it is generally found that the luminescence spectrum is independent of the excitation frequency. Explain this phenomenon.

(5.3)* The wave functions for atomic hydrogen may be written in the form:

$$\Psi_{nlm}(r, \theta, \phi) = R_{nl}(r)\, Y_{l,m}(\theta, \phi)\,.$$

The radial wave functions for the $1s$ and $2p$ states are given by:

$$R_{10}(r) = \frac{2}{a_H^{3/2}} e^{-r/a_H}\,,$$

and

$$R_{21}(r) = \frac{r}{\sqrt{24} a_H^{5/2}} e^{-r/2a_H}\,,$$

where a_H is the Bohr radius of hydrogen. The spherical harmonic functions of the same states are given by

$$Y_{0,0}(\theta, \phi) = \frac{1}{\sqrt{4\pi}}\,,$$

$$Y_{1,0}(\theta, \phi) = \sqrt{\frac{3}{4\pi}} \cos\theta\,,$$

and

$$Y_{1,\pm 1}(\theta, \phi) = \mp\sqrt{\frac{3}{8\pi}} e^{\pm i\phi} \sin\theta\,.$$

Use eqn B.31 in Appendix B to calculate the Einstein A coefficient for the $2p \rightarrow 1s$ transition. Hence calculate the radiative lifetime of the $2p$ state.

(5.4) The radiative lifetime τ_R of the laser transition in titanium doped sapphire is $3.9\,\mu s$. The lifetime τ of the excited state is measured to be $3.1\,\mu s$ at $300\,K$ and $2.2\,\mu s$ at $350\,K$. Explain why τ is different from τ_R, and suggest a reason why τ decreases with increasing temperature. Calculate the radiative efficiencies at the two temperatures.

(5.5) A semiconductor crystal is found to emit efficiently at $540\,nm$ when excited with the $488\,nm$ line from an argon ion laser. Use the data in Table D.3 to make a guess at what the crystal is.

(5.6) A continuous wave laser beam is incident on a material which has an absorption coefficient of α at the laser frequency ν.
(a) Show that electron-hole pairs are generated at a rate equal to $I\alpha/h\nu$ per unit volume per unit time, where I is the intensity in the material.
(b) By considering the balance between carrier generation and recombination in steady state conditions, show that the carrier density N within the illuminated volume is equal to $I\alpha\tau/h\nu$, where τ is the recombination lifetime of the electrons and holes.
(c) Calculate N when a laser beam of power $1\,mW$ is focused to a circular spot of radius $50\,\mu m$ on an anti-reflection coated sample with an excited state lifetime of $1\,ns$. Take the absorption coefficient to be $2 \times 10^6\,m^{-1}$ at the laser wavelength of $514\,nm$.

(5.7) A very short laser pulse at $780\,nm$ is incident on a thick crystal which has an absorption coefficient of $1.5 \times 10^6\,m^{-1}$ at this wavelength. The pulse has an energy of $10\,nJ$ and is focused to a circular spot of radius $100\,\mu m$.
(a) Calculate the initial carrier density at the front of the sample.
(b) If the radiative and non-radiative lifetimes of the sample are $1\,ns$ and $8\,ns$ respectively, calculate the time taken for the carrier density to drop to 50% of the initial value.
(c) Calculate the total number of luminescent photons generated by each laser pulse.

(5.8) Explain why the emission probability for an interband transition is proportional to the product of the electron and hole occupancy factors f_e and f_h respectively. In the classical limit where Boltzmann statistics apply, show that the product $f_e f_h$ is proportional to $\exp\left(-(h\nu - E_g)/k_B T\right)$.

(5.9) In the classical limit, show that the number of electrons in the conduction band of a semiconductor is given by

$$N_e = \frac{e^{E_F^c/k_B T}}{2\pi^2} \left(\frac{2m_e^* k_B T}{\hbar^2}\right)^{3/2} \int_0^\infty x^{1/2} e^{-x}\,dx\,.$$

Given that $\int_0^\infty x^{1/2} e^{-x}\,dx = \sqrt{\pi}/2$, evaluate E_F^c at $300\,K$ for GaAs ($m_e^* = 0.067 m_0$) when (a) $N_e = 1 \times 10^{20}\,m^{-3}$ and (b) $N_e = 1 \times 10^{24}\,m^{-3}$. Discuss whether the approximations used to derive this equation are justified in the two cases.

*Exercises marked with an asterisk are more difficult.

(5.10) Show that at $T = 0$ the Fermi integrals given in eqns 5.9 and 5.10 simplify to:

$$N_{e,h} = \int_0^{E_F^{c,v}} \frac{1}{2\pi^2} \left(\frac{2m_{e,h}^*}{\hbar^2} \right)^{3/2} E^{1/2} \, dE \, .$$

Evaluate the integral to derive eqn 5.13.

(5.11) A laser excites a semiconductor which has $m_e^* = 0.1m_0$ and $m_h^* = 0.5m_0$. Calculate the electron and hole Fermi energies for carrier densities of (a) $1 \times 10^{21} \, m^{-3}$, and (b) $1 \times 10^{24} \, m^{-3}$, on the assumption that the distributions are degenerate. Write down a condition on the temperature for the degeneracy conditions to apply in each case and comment on the answers you obtain.

(5.12) Show that in the degenerate limit during photoluminescence, the k vectors corresponding to the conduction and valence band Fermi energies are the same, even though the Fermi energies are different.

(5.13)* The photoluminescence spectrum of CdTe, which has a direct band gap at 1.61 eV and a refractive index of 2.7, is measured using the apparatus shown in Fig. 5.9. An argon ion laser of power 1 mW and photon energy 2.41 eV is focused to a small spot on the sample. The luminescence is collimated with a lens of diameter 25 mm and focal length 100 mm.
(a) Calculate the solid angle subtended by the lens at the sample.
(b) Estimate the fraction of the photoluminescence collected by the lens. Assume that the luminescence is emitted uniformly in all directions inside the crystal, and is then both reflected and refracted at the front surface.
(c) Calculate the total luminescent power emitted by the atoms in terms of the radiative quantum efficiency η_R of the sample.
(d) Hence estimate the luminescent power collected by the collimation lens in terms of η_R.

(5.14)* Figure 5.8 shows the emission spectrum from the direct gap semiconductor $Ga_{0.47}In_{0.53}As$ at two time delays after $2 \times 10^{24} \, m^{-3}$ carriers have been excited using an ultrashort laser pulse.
(a) Calculate the electron Fermi energy for the initial carrier density if $T = 0$. ($m_e^* = 0.041m_0$.)
(b) Calculate the hole Fermi energy in the same conditions, on the assumption that the densities of states from the light- and heavy-hole bands can just be added together. ($m_{hh}^* = 0.47m_0$ and $m_{lh}^* = 0.05m_0$.)
(c) The effective carrier temperature for the 24 ps

spectrum is 180 K. Are the carriers degenerate?
(d) Explain the shape of the 24 ps spectrum, given that the band gap of $Ga_{0.47}In_{0.53}As$ is 0.81 eV.
(e) Use the data at 250 ps to obtain a rough estimate of the carrier density at this time delay. Estimate the average lifetime of the carriers.

(5.15) Consider a zinc-blende III–V semiconductor with an initial electron spin polarization of 50% and zero hole polarization. By considering the relative populations of the electron spin levels, and the relative weights of the possible transitions, show that luminescence is expected to have a circular polarization of 25%.

(5.16) Optical orientation and Hanle effect experiments are performed on a sample with electron g-factor g_e. Derive expressions for the carrier lifetime τ and spin lifetime τ_S in terms of the degree of polarization at zero field (i.e. $P(0)/P_0$) and the Hanle half field $B_{1/2}$ (i.e. the field for which $P(B) = P(0)/2$).

(5.17) GaP has an indirect gap at 2.27 eV and a direct gap at 2.78 eV. The band gap of the alloy semiconductor $GaAs_{1-x}P_x$ varies approximately linearly with composition, and is direct for $x \leq 0.45$. The band gap of GaAs is 1.42 eV.
(a) What is the shortest wavelength that can be produced efficiently by a $GaAs_{1-x}P_x$ light-emitting diode?
(b) Estimate the composition of the alloy in an LED emitting at 670 nm.

(5.18) GaAs has a refractive index of 3.5 at its band gap.
(a) Calculate the reflectivity at the interface between the air and the GaAs crystal.
(b) Calculate the frequency separation of the longitudinal modes of a GaAs laser diode of length 1 mm.
(c) The laser diode of part (b) is coated so that one end of the chip has a reflectivity of 95%. The other end is uncoated. Calculate the threshold gain coefficient for the laser if the scattering and other impurity losses are negligibly small.

(5.19) A laser diode emits at 830 nm when operating at an injection current of 100 mA.
(a) Calculate the maximum possible power that can be emitted by the device.
(b) Calculate the power conversion efficiency, if the actual power output is 50 mW and the operating voltage is 1.9 V.
(c) The threshold current of the laser is 35 mA. What is the slope efficiency and the quantum efficiency?

(5.20) Show that the electron-hole pair density generated when an electron beam with current density J strikes a sample is given by:

$$N = \frac{J\tau}{e\,R_\mathrm{e}}(1 - \gamma)\frac{E^\mathrm{p}}{E^\mathrm{i}}$$

where τ is the carrier lifetime, R_e is the penetration depth, and the other symbols are defined in eqn 5.27.

Quantum confinement

In this chapter we give an overview of the optical properties of quantum-confined semiconductors. These are artificial structures in which the electrons and holes are confined in one or more directions. The structures that we consider generally have sizes in the nanometre range, and may thus be considered as examples of **nanostructures**. We concentrate mainly on quantum wells, in which the confinement is just in one dimension, since these illustrate the main points most clearly. We then give an introduction to the physics of quantum dots, which is a subject that has advanced very rapidly since the first edition of this book. As we shall see, quantum wells and dots have very interesting optical properties that readily lend themselves to applications in opto-electronics. Furthermore, the physical principles that we shall study here for inorganic semiconductors can readily be adapted to other types of quantum-confined systems, such as the carbon nanostructures considered in Section 8.5.

The optical properties of quantum-confined semiconductors are derived from the physics of interband absorption, excitons, and interband luminescence discussed in Chapters 3–5. We presuppose here that these subjects have been fully assimilated, and the new material in this chapter provides a good opportunity to practise and develop the principles that have been learnt previously.

6.1	Quantum-confined structures	141
6.2	Growth and structure of quantum wells	144
6.3	Electronic levels	146
6.4	Quantum well absorption and excitons	152
6.5	The quantum-confined Stark effect	160
6.6	Optical emission	164
6.7	Intersubband transitions	166
6.8	Quantum dots	167
	Chapter summary	174
	Further reading	175
	Exercises	176

6.1 Quantum-confined structures

The optical properties of solids do not usually depend on the size of the crystal. Rubies, for example, have the same red colour irrespective of how big they are. This statement is only true as long as the dimensions of the crystal are large. If we make very small crystals, then the optical properties do in fact depend on the size. A striking example of this is semiconductor-doped glasses. As discussed in Section 6.8.2, these contain very small semiconductor nanocrystals within a colourless glass host, and the colour can be altered just by changing the size of the crystals.

The size dependence of the optical properties in very small crystals is a consequence of the **quantum confinement** effect. The Heisenberg uncertainty principle tells us that if we confine a particle to a region of the x axis of length Δx, then we introduce an uncertainty in its momentum given by:

$$\Delta p_x \sim \frac{\hbar}{\Delta x}.$$ (6.1)

If the particle is otherwise free, and has a mass m, the confinement in

Table 6.1 Number of degrees of freedom tabulated against the dimensionality of the quantum confinement. The final column shows the functional form of the density of states for free electrons.

Structure	Quantum confinement	Number of free dimensions	Electron density of states
Bulk	none	3	$E^{1/2}$
Quantum well/superlattice	1-D	2	E^0
Quantum wire	2-D	1	$E^{-1/2}$
Quantum dot/box	3-D	0	discrete

the x direction gives it an additional kinetic energy of magnitude

$$E_{\text{confinement}} = \frac{(\Delta p_x)^2}{2m} \sim \frac{\hbar^2}{2m(\Delta x)^2} \,. \tag{6.2}$$

This confinement energy will be significant if it is comparable to or greater than the kinetic energy of the particle due to its thermal motion in the x direction. This condition may be written:

The principle of equipartition of energy tells us that we have a thermal energy of $k_{\text{B}}T/2$ for each degree of freedom of the motion.

$$E_{\text{confinement}} \sim \frac{\hbar^2}{2m(\Delta x)^2} > \frac{1}{2}k_{\text{B}}T \,, \tag{6.3}$$

and tells us that quantum size effects will be important if

$$\Delta x \lesssim \sqrt{\frac{\hbar^2}{mk_{\text{B}}T}} \,. \tag{6.4}$$

This is equivalent to saying that Δx must be comparable to or smaller than the de Broglie wavelength $\lambda_{\text{deB}} \equiv h/p_x$ for the thermal motion.

The criterion given in eqn 6.4 gives us an idea of how small the structure must be if we are to observe quantum confinement effects. At room temperature, we find that we must have $\Delta x \lesssim 5\,\text{nm}$ for an electron in a typical semiconductor with $m_{\text{e}}^* = 0.1m_0$. Thus a 'thin' semiconductor layer of thickness $1\,\mu\text{m}$ is not thin by the standards of the electrons. It is in fact a bulk crystal that would not exhibit any quantum size effects except at extremely low temperatures. (See Exercise 6.1.) To observe quantum size effects we need thinner layers.

Quantum-confined structures are generally classified by their dimensionality. Table 6.1 summarizes the three basic types of quantum-confined structures that can be produced, for which the following nomenclature is usually adopted:

Certain types of quantum well structures are called superlattices (see Section 6.2), while quantum dots are sometimes called quantum boxes.

- **quantum wells**: 1-D confinement;
- **quantum wires**: 2-D confinement;
- **quantum dots**: 3-D confinement.

Table 6.1 also lists the number of degrees of freedom associated with the type of quantum confinement. The electrons and holes in bulk semiconductors are free to move within their respective bands in all three directions, giving them three degrees of freedom, and hence three-dimensional

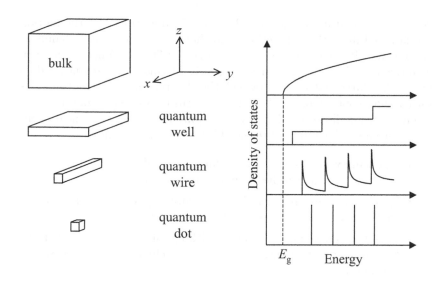

Fig. 6.1 Schematic representation of quantum wells, wires, and dots. The generic shape of the density of states function for electrons in the conduction band of a semiconductor with band gap E_g is shown for each type of structure.

(3-D) physics. The electrons and holes in a quantum well, by contrast, are confined in one direction, and therefore only have two degrees of freedom. This means that they effectively behave as 2-D materials. Similarly, quantum wire structures have 1-D physics, while quantum dots have '0-D' physics. This last point means that the motion of the electrons and holes is quantized in all three dimensions, so that they are completely localized in the quantum dot. Quantum wells, wires, and dots are thus all examples of **low-dimensional structures**.

The quantization of the motion of the electrons and holes has two main consequences:

(1) The energy of a particle at rest is increased by the quantum confinement energy.

(2) The functional form of the density of states is changed.

These two points will be discussed at length throughout the chapter, and at this stage it is useful just to make some general comparisons, as illustrated schematically in Fig. 6.1.

The conduction band electrons in a bulk semiconductor can have any energy above the band gap energy E_g and the density of states is proportional to $(E - E_g)^{1/2}$. This is a consequence of the free motion in all three dimensions. The density of states for a quantum well is determined by the 2-D free motion and the shift of the energy due to the quantum confinement. As shown in Exercise 6.3, the density of states is independent of the energy, and so we have a sequence of steps in the density of states for each quantized level. Note that the band edge is effectively shifted to higher energy by the quantized energy for the quantum-confined motion in the third direction.

The argument can be repeated for 1-D quantum wire and 0-D quantum dot systems. In the case of quantum wires, the density of states has an $E^{-1/2}$ dependence (see Exercise 6.4) which leads to peaks at each

See, for example, eqn 3.16, which shows that the density of states for free electrons varies as $E^{1/2}$ in three dimensions. For an electron in the conduction band of a semiconductor, the energy must be measured relative to the bottom of the conduction band. Note that the $(E - E_g)^{1/2}$ dependence only applies within the parabolic band approximation.

new quantized state as shown in Fig. 6.1. In quantum dots the motion is quantized in all three directions and there are no continuous bands at all. The density of states consists of a series of Dirac δ-functions at each quantized level, as illustrated in Fig. 6.1. In this sense, quantum dots behave like 'artificial atoms' in which the electrons have discrete energies rather than continuous bands as is the norm in solid-state physics.

The very small crystal dimensions required to observe quantum confinement effects have to be produced by special techniques.

- Quantum wells are made by techniques of advanced epitaxial crystal growth. This will be explained in Section 6.2.
- Quantum wires are made by lithographic patterning of quantum well structures, or by epitaxial growth on patterned substrates.
- Quantum dots can be made by lithographic patterning of quantum wells or by spontaneous growth techniques, as discussed in Section 6.8.

In the sections that follow, we mainly concentrate on quantum well structures. This is because they illustrate the physical principles very well, and are already widely used in many commercial opto-electronic devices. We also briefly consider the optical properties of quantum dots. We do not mention quantum wires further here, due to the difficulties associated with making them, although we shall return to 1-D materials when we consider carbon nanotubes in Section 8.5.3.

6.2   Growth and structure of quantum wells

Semiconductor quantum wells are examples of **heterostructure** crystals. Heterostructures are artificial crystals that contain layers of different materials grown on top of a thicker substrate crystal. The structures are made by the specialized epitaxial crystal growth techniques introduced previously in Section 5.4.1. The two most important ones are **molecular beam epitaxy** (MBE) and **metal–organic chemical vapour deposition** (MOCVD), which is also called **metal–organic vapour-phase epitaxy** (MOVPE) by some authors. The layer thicknesses of the crystals grown by these techniques can be controlled with atomic precision. This makes it easy to achieve the thin layer thicknesses required to observe quantum confinement of the electrons in a semiconductor at room temperature.

Figure 6.2(a) shows a schematic diagram of the simplest type of quantum well that can be grown. In this particular case, a GaAs/AlGaAs structure grown on a GaAs substrate is shown. The structure consists of a GaAs layer of thickness d sandwiched between much thicker layers of the alloy semiconductor AlGaAs. d is chosen so that the motion of the electrons in the GaAs layer is quantized according to the criterion given in eqn 6.4. We set up axes so that the z axis corresponds to the crystal

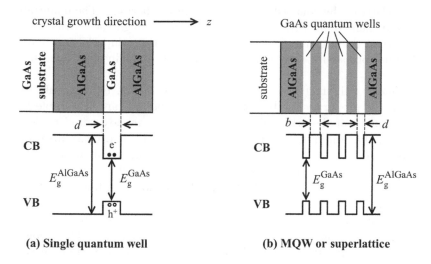

(a) Single quantum well

(b) MQW or superlattice

Fig. 6.2 (a) Schematic diagram of a single GaAs/AlGaAs quantum well. The quantum well is formed in the thin GaAs layer sandwiched between AlGaAs layers which have a larger band gap. The lower half of the figure shows the spatial variation of the conduction band (CB) and the valence band (VB). (b) Schematic diagram of a GaAs/AlGaAs multiple quantum well (MQW) or superlattice structure. The distinction between an MQW and a superlattice depends on the thickness b of the barrier separating the quantum wells.

growth direction, while the x and y axes lie in the plane of the layers. We thus have quantized motion in the z direction, and free motion in the x, y plane.

The bottom half of Fig. 6.2(a) shows the spatial variation of the conduction and valence bands that corresponds to the change of the composition along the z direction. The band gap of AlGaAs is larger than that of GaAs, and the bands line up so that the lowest conduction and valence band states of the GaAs lie within the gap of the AlGaAs. This means that electrons in the GaAs layer are trapped by potential barriers at each side due to the discontinuity in the conduction band. Similarly, holes are trapped by the discontinuity in the valence band. These barriers quantize the states in the z direction, but the motion in the x, y plane is still free. We thus effectively have a 2-D system in which the electrons and holes are quantized in one direction and free in the other two.

Epitaxial techniques are very versatile, and they allow the growth of a great variety of quantum well structures. Figure 6.2(b) shows one such variant derived from the single well structure shown in Fig. 6.2(a). The crystal consists of a series of repeated GaAs quantum wells of width d separated from each other by AlGaAs layers of thickness b. This type of structure is either called a **multiple quantum well** (MQW) or a **superlattice**, depending on the parameters of the system. The distinction depends mainly on the value of b.

MQWs have large b values, so that the individual quantum wells are isolated from each other, and the properties of the system are essentially the same as those of single quantum wells. They are often used in optical applications to give a usable optical density. It would be very difficult to measure the optical absorption of a single 10 nm thick quantum well, simply because there is so little material to absorb the light. By growing many identical quantum wells, the absorption will increase to a measurable value.

Superlattices, by contrast, have much thinner barriers. The quantum wells are then coupled together by tunnelling through the barrier, and new extended states are formed in the z direction. Superlattices have additional properties over and above those of the individual quantum wells.

Quantum well structures of the type shown in Fig. 6.2 can only be made if the properties of the constituent compounds are favourable to the formation of the artificial crystals. We have already noted in Section 5.4.1 that the unit cell size of GaAs and AlAs (and hence also the $Al_xGa_{1-x}As$ alloy) are almost identical: see Fig. 5.11. This means that both the GaAs and AlGaAs layers in the quantum well structure are lattice matched to the GaAs substrate, enabling dislocation-free crystals to be grown.

In recent years it has been realized that it is also possible to make quantum wells from materials with different unit cell sizes. This allows much more flexibility in the combinations of materials that can be used. The mismatch in the lattice constants introduces strain into the structure, but high-quality crystals can still be grown provided the total layer thickness is kept below a critical value. We briefly mention the application of these non-lattice-matched quantum wells in light-emitting diodes and laser diodes in Section 6.6.

6.3   Electronic levels

The wave functions and energies of the quantized states in the conduction and valence bands of a quantum well can be calculated by using Schrödinger's equation and the effective mass approximation. Fortunately, we do not have to solve the Schrödinger equation in three dimensions because the problem separates naturally between the free motion in the x, y plane and the quantized motion in the z direction. In this section, we first explain how this separation of variables works, and then go on to discuss the quantized states in the z direction in two different approximations. We treat the electron and holes separately here, postponing till Section 6.4 the discussion of the effects of electron-hole Coulomb interaction that leads to the formation of excitons.

6.3.1   Separation of the variables

The electrons and holes in a quantum well layer are free to move in the x, y plane but are confined in the z direction. This allows us to write the wave functions in the form:

$$\Psi(x, y, z) = \psi(x, y)\,\varphi(z)\,, \tag{6.5}$$

and then solve separately for $\psi(x,y)$ and $\varphi(z)$. The states of the system are described by two parameters: a wave vector $\boldsymbol{k}$ to specify the free motion in the x, y plane, and a quantum number n to indicate the energy level for the z direction. The total energy is then obtained by adding together the separate energies for the z and x, y motion, according to:

$$E^{\text{total}}(n, \boldsymbol{k}) = E_n + E(\boldsymbol{k}), \qquad (6.6)$$

where E_n is the quantized energy of the nth level.

We can deal with the x, y plane motion very quickly. Since the motion is free, the electron and hole wave functions are described by plane waves of the form:

$$\psi_k(x, y) = \frac{1}{\sqrt{A}} \, e^{i\boldsymbol{k} \cdot \boldsymbol{r}}, \qquad (6.7)$$

where $\boldsymbol{k}$ is the wave vector of the particle, and A is the normalization area. Note that $\boldsymbol{k}$ and $\boldsymbol{r}$ only span the two-dimensional x, y plane here. The energy corresponding to this motion is just the kinetic energy determined by the effective mass:

$$E(\boldsymbol{k}) = \frac{\hbar^2 \boldsymbol{k}^2}{2m^*}. \qquad (6.8)$$

The total energy for an electron or hole in the nth quantum level is therefore given by:

$$E^{\text{total}}(n, \boldsymbol{k}) = E_n + \frac{\hbar^2 \boldsymbol{k}^2}{2m^*}. \qquad (6.9)$$

6.3.2   Infinite potential wells

The calculation of the wave functions and energies for the quantized states in the z direction is determined by the spatial dependence of the conduction and valence bands. We begin by considering the simplest case in which we assume that the confining barriers are infinitely high. This allows us to model the states by those of a 1-D potential well with infinite barriers, as shown in Fig. 6.3.

We consider a quantum well of thickness d and define position and energy coordinates such that the potential is zero for $-d/2 < z < +d/2$ (i.e. inside the well) and ∞ elsewhere, as indicated in Fig. 6.3. The choice of $z = 0$ at the centre of the well is convenient, since it corresponds to the symmetry axis of the potential. The Schrödinger equation within the well is:

$$-\frac{\hbar^2}{2m^*} \frac{d^2\varphi(z)}{dz^2} = E\varphi(z). \qquad (6.10)$$

Since the barriers are infinitely high, there is no probability that the particle can tunnel out of the well. The solutions of eqn 6.10 are therefore subject to the boundary condition that $\varphi = 0$ at the interfaces.

It can be checked by substitution that the normalized wave functions that satisfy eqn 6.10 and the boundary conditions are of the form:

$$\varphi_n(z) = \sqrt{\frac{2}{d}} \sin\left(k_n z + \frac{n\pi}{2}\right), \qquad (6.11)$$

where n is an integer that gives the quantum number of the state, and

$$k_n = \frac{n\pi}{d}. \qquad (6.12)$$

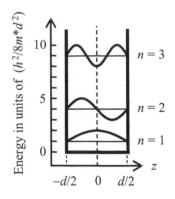

Fig. 6.3 The infinite one-dimensional potential well. The first three energy levels and corresponding wave functions are shown.

In a bulk semiconductor like GaAs, the heavy and light hole states are degenerate at $k = 0$. This is a consequence of the high symmetry of the cubic lattice. The lifting of this degeneracy is caused by the difference in the effective masses, but it can also be seen as a consequence of the lower symmetry of the quantum well. The bulk crystals are isotropic, but quantum wells are not: the z direction is physically distinguishable from the other two. As explained in Section 1.5.1, we therefore expect certain degeneracies to be lifted, in the same way that a magnetic field splits the heavy- and light-hole states of bulk GaAs via their different magnetic energies.

This form of wave function describes a standing wave inside the well with nodes at the interfaces. The energy that corresponds to the nth level is given by

$$E_n = \frac{\hbar^2 k_n^2}{2m^*} = \frac{\hbar^2}{2m^*}\left(\frac{n\pi}{d}\right)^2. \tag{6.13}$$

The wave functions of the first three levels are shown in Fig. 6.3. Equation 6.13 describes an infinite ladder of levels with quantization energy increasing in proportional to n^2 in units of $(\hbar^2\pi^2/2m^*d^2)$. The ground state is the $n = 1$ level, and the levels of higher n are the excited states of the system.

The energies of the first two levels for an electron with $m^* = 0.1m_0$ in a 10 nm quantum well are 38 meV and 150 meV respectively. These values should be compared to the thermal energy k_BT, which is 25 meV at room temperature. It is clear that the quantization energy is greater than the thermal energy at room temperature, and thus that the quantum description of the motion is appropriate. The comparison of the quantization energies to the thermal energy gives a criterion to decide whether a particular quantum well will in fact exhibit quantum effects at a particular temperature. This criterion can be compared to the one based on the Heisenberg uncertainty principle given in eqn 6.4. It is easy to show that the two criteria predict a crossover from classical to quantum behaviour at roughly the same value of d. (See Exercise 6.2.)

Although real semiconductor quantum wells have finite barriers, the infinite barrier model is a good starting point for a discussion of their properties. The accuracy of the model will be best for states with small quantization energies in material combinations that give rise to high barriers at the interfaces. A few useful general points emerge from the analysis:

(1) The energy of the levels is inversely proportional to the effective mass and the square of the well width. This means that low mass particles in narrow quantum wells have the highest energies.

(2) Since the energy depends on the effective mass, the electrons, heavy holes, and light holes will all have different quantization energies. In the valence band, the heavy holes will have the lowest energy, and are dominant in most situations because they form the ground-state level.

(3) The wave functions can be identified by their number of nodes, i.e. the number of zero crossings within the well. It is evident from Fig. 6.3 that the nth level has $(n-1)$ nodes.

(4) The states are also labelled by their **parity** with respect to inversion about the centre of the well, that is, whether $\varphi(-z) = +\varphi(z)$ (even parity) or $\varphi(-z) = -\varphi(z)$ (odd parity). States of odd n have even parity, and vice versa.

These points also apply to more realistic models of quantum wells in which the barriers at the interfaces are only of finite height. As we shall see below, the infinite well model overestimates the quantization energy.

In real quantum wells with finite barriers, the particles are able to tunnel into the barriers to some extent, and this allows the wave functions to spread out further, thereby reducing the confinement energy.

6.3.3 Finite potential wells

Figure 6.4 shows the band diagram of a more realistic quantum well which has a finite potential barrier of height V_0 at each interface. There are now only a finite number of bound states with energy $E < V_0$. These bound states are labelled by a quantum number n, and it can be shown that there is always at least one, no matter how small V_0 is: see Exercise 6.5.

The Schrödinger equation within the quantum well is the same as before (eqn 6.10). We therefore have sine and cosine solutions of the form:

$$\varphi_{\mathrm{w}}(z) = C\sin(kz) \tag{6.14}$$

and

$$\varphi_{\mathrm{w}}(z) = C\cos(kz)\,, \tag{6.15}$$

where

$$\frac{\hbar^2 k^2}{2m_{\mathrm{w}}^*} = E\,. \tag{6.16}$$

Note that we have added a subscript 'w' to the effective mass to clarify that it is the value for the semiconductor used in the quantum well layer. By comparison with eqn 6.11 and also Fig. 6.3, we see that the bound states with odd values of n have cosine solutions, while those with even n have sine solutions with a node at $z = 0$. The wave functions given in eqns 6.14 and 6.15 are valid for $-d/2 \le z \le +d/2$.

We now consider the extension of the wave functions into the barrier regions. This occurs because the finite potential discontinuity allows the electrons and holes to tunnel into the barriers. We therefore no longer have nodes at the interfaces. The Schrödinger equation in the barrier regions is given by:

$$-\frac{\hbar^2}{2m_{\mathrm{b}}^*}\frac{\mathrm{d}^2\varphi(z)}{\mathrm{d}z^2} + V_0\varphi(z) = E\varphi(z), \tag{6.17}$$

where m_{b}^* is the effective mass of the barrier material. In general, m_{b}^* and m_{w}^* will not be the same because the materials that constitute the quantum well and barrier regions have different band structures. The solutions of eqn 6.17 are exponentials of the form:

$$\varphi_{\mathrm{b}}(z) = C'\,\mathrm{e}^{\pm\kappa z} \tag{6.18}$$

where κ satisfies:

$$\frac{\hbar^2\kappa^2}{2m_{\mathrm{b}}^*} = V_0 - E\,. \tag{6.19}$$

For bound states we require that the solutions decay in the barrier, and so we choose $\varphi(z) = C'\exp(-\kappa z)$ for $z \ge +d/2$ and $\varphi(z) = C'\exp(+\kappa z)$ for $z \le -d/2$.

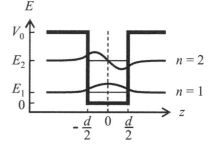

Fig. 6.4 First two bound states of a finite potential well of depth V_0 and width d.

The wave functions and energies of the bound states can be found by applying the appropriate boundary conditions at the interfaces. These tell us that both the wave function $\varphi(z)$ and the particle flux $(1/m^*)\mathrm{d}\varphi/\mathrm{d}z$ must be continuous at $\pm d/2$. Hence we must have

$$\varphi_{\mathrm{w}}(\pm d/2) = \varphi_{\mathrm{b}}(\pm d/2)\,, \tag{6.20}$$

and

$$\frac{1}{m_{\mathrm{w}}^*}\left(\frac{\mathrm{d}\varphi_{\mathrm{w}}}{\mathrm{d}z}\right)_{z=\pm d/2} = \frac{1}{m_{\mathrm{b}}^*}\left(\frac{\mathrm{d}\varphi_{\mathrm{b}}}{\mathrm{d}z}\right)_{z=\pm d/2}. \tag{6.21}$$

The wave functions must have symmetry about $z = 0$, so we just concentrate on $z = +d/2$. We consider the solutions with cosine solutions in the quantum wells first. The wave function continuity requires that

$$C\cos(kd/2) = C'\exp(-\kappa d/2)\,, \tag{6.22}$$

while flux continuity requires that

$$-C\frac{k}{m_{\mathrm{w}}^*}\sin(kd/2) = -C'\frac{\kappa}{m_{\mathrm{b}}^*}\exp(-\kappa d/2)\,. \tag{6.23}$$

On dividing eqn 6.23 by eqn 6.22, we find:

$$\tan(kd/2) = \frac{m_{\mathrm{w}}^*\kappa}{m_{\mathrm{b}}^*k}\,. \tag{6.24}$$

On following a similar procedure for the solutions with sine solutions in the quantum well, we find:

$$\tan(kd/2) = -\frac{m_{\mathrm{b}}^*k}{m_{\mathrm{w}}^*\kappa}\,. \tag{6.25}$$

On substituting the values of k and κ from eqns 6.16 and 6.19, we can now solve for the energy E of the bound states, and hence find the wave functions. Unfortunately, there is no analytic solution for E. The equations have to be solved numerically or graphically. An example of how this is done is given in Example 6.1.

It is useful to make a few general observations about the solutions, as we did for the case of the infinite well in the previous section.

(1) The spreading of the wave functions into the barrier by tunnelling increases k and hence reduces the quantum confinement energy compared to a well with infinite barriers.

(2) The decay constant can be found by substituting E_n for E in eqn 6.19. This means that the levels near the top of the well with E_n close to V_0 tunnel more into the barrier regions because they have a smaller decay constant.

The possibility of tunnelling into the barriers gives rise to a whole series of electronic and opto-electronic quantum well tunnelling devices.

(3) The eigenstates can be identified by the number of nodes, just as for infinite wells. The nth bound state has $(n-1)$ nodes. The potential energy has even symmetry about $z = 0$, and so the eigenstates have well-defined parities.

Figure 6.4 sketches the wave functions of a typical finite well with two bound states. The similarity between these wave functions and the first two states of the infinite well shown in Fig. 6.3 is apparent. The main difference is that the wave functions of the finite well spread out more by tunnelling into the barrier, whereas the wave functions of the infinite well stop abruptly at the interfaces.

It is useful to compare directly the predictions of the finite and infinite well models. Table 6.2 tabulates the energies of the bound states of a 10 nm GaAs quantum well with $Al_{0.3}Ga_{0.7}As$ barriers for the two models. In all cases, the infinite well model overestimates the quantization energy. The discrepancy gets worse for the higher levels. Note that the quantization energies of the heavy holes are smaller than those of the electrons because of their heavier effective mass. Note also that the separation of the first two electron levels is more than three times the thermal energy at room temperature, where $k_B T \sim 25$ meV. This confirms that we expect to observe 2-D physics for the electrons at 300 K. The quantum confinement of the heavy holes is less good, but is still acceptable since $E_2 - E_1$ is comparable to $k_B T$ at 300 K. Although the infinite well model overestimates the confinement energies, it is a useful starting point for the discussion of the physics because of its simplicity.

Table 6.2 Bound states of a 10 nm GaAs/Al$_{0.3}$Ga$_{0.7}$As quantum well calculated using the finite and infinite well models. The states are labelled by the particle type (e for electron, hh for heavy hole and lh for light hole) and by the quantum number n. All energies are in meV.

State	Finite well	Infinite well
e1	32	57
e2	120	227
e3	247	510
hh1	7	11
hh2	30	44
hh3	66	100
hh4	112	177
lh1	21	40
lh2	78	160

Example 6.1

Calculate the energy of the first electron bound state in a GaAs/AlGaAs quantum well with $d = 10$ nm and $V_0 = 0.3$ eV. Take $m_w^* = 0.067 m_0$ and $m_b^* = 0.092 m_0$. Compare this value to the one calculated for an infinite quantum well.

Solution

The first bound state has a maximum at $z = 0$, and so we look for the solutions with cosine wave function in the well region. By making the substitution $x = kd/2$, we can use eqns 6.16 and 6.19 to recast eqn 6.24 in the form:

$$x \tan x = \left(\frac{m_w^*}{m_b^*} \right)^{1/2} \sqrt{\xi - x^2}\,, \qquad (6.26)$$

where

$$\xi = \frac{m_w^* d^2 V_0}{2\hbar^2}\,, \qquad (6.27)$$

and

$$E = \frac{2\hbar^2 x^2}{m_w^* d^2}\,. \qquad (6.28)$$

Our task is thus to solve eqn 6.26, with

$$\left(\frac{m_w^*}{m_b^*} \right)^{1/2} = \left(\frac{0.067}{0.092} \right)^{1/2} = 0.85\,,$$

and

$$\xi = \frac{0.067 m_0 \times (10^{-8})^2 \times 0.30 \text{ eV}}{2\hbar^2} = 13.2\,.$$

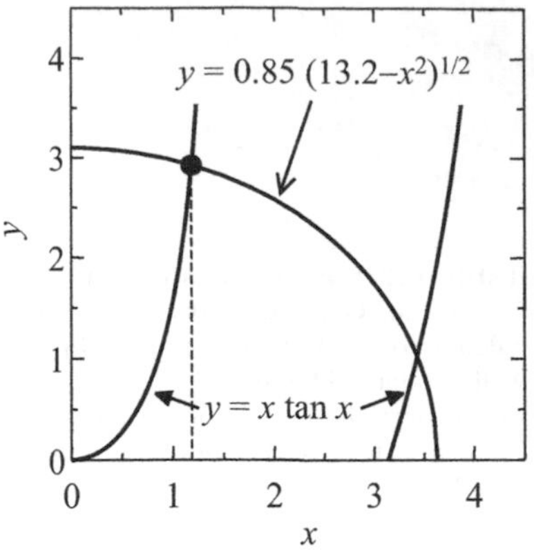

We are using Dirac notation here. See
the marginal comment on p. 65.

Figure 6.5 plots the functions $y = x \tan x$ and $y = 0.85\sqrt{13.2 - x^2}$ on the same scales, and shows that the first value of x where the two functions are the same is $x = 1.18$. Hence from eqn 6.28 we find the required bound state energy:

$$E = \frac{2\hbar^2(1.18)^2}{0.067m_0 \times (10^{-8})^2} = 31.5\,\text{meV}.$$

This value of E can be compared to that given by eqn 6.13 for an infinite well:

$$E_1 = \frac{\hbar^2\pi^2(1)^2}{2 \times 0.067m_0 \times (10^{-8})^2} = 57\,\text{meV}.$$

The infinite well model thus overestimates the energy of the bound state by a factor of 1.8.

Fig. 6.5 Graphical solution of eqn 6.26 for the parameters given in Example 6.1.

6.4 Quantum well absorption and excitons

In Sections 3.2 and 3.3 of Chapter 3 we used Fermi's golden rule to calculate the absorption spectrum of a bulk semiconductor. Then in Section 4.2 of Chapter 4 we studied how the spectrum is altered by excitonic effects. We now follow a similar approach for quantum wells, beginning with the selection rules and density of states for the optical transitions, and then moving on to consider the excitonic effects.

6.4.1 Selection rules

We consider a quantum well irradiated by light of angular frequency ω propagating in the z direction, as shown in Fig. 6.6. The photons are absorbed by exciting electrons from an initial state $|i\rangle$ at energy E_i in the valence band to a final state $|f\rangle$ at energy E_f in the conduction band. Conservation of energy requires that $E_f = (E_i + \hbar\omega)$.

Fermi's golden rule tells us that the absorption rate is determined by the density of states and the square of the electric-dipole matrix element. (See Section B.2 in Appendix B.) The transition rate can be calculated by combining eqns 3.2, 3.3 and 3.6 to obtain:

$$W_{i \to f} = \frac{2\pi}{\hbar}\,|\,\langle f|-e\boldsymbol{r}\boldsymbol{\cdot}\boldsymbol{\mathcal{E}}|i\rangle\,|^2\,g(\hbar\omega), \tag{6.29}$$

where $\boldsymbol{r}$ is the position vector of the electron, $\boldsymbol{\mathcal{E}}$ is the electric field amplitude of the light wave, and $g(\hbar\omega)$ is the density of states. We have simplified the form of the electric-dipole perturbation here by setting the $e^{\pm i\mathbf{k} \cdot \mathbf{r}}$ factor of the light in eqn 3.6 equal to unity. As discussed in connection with eqn 3.12, this approximation is justified because the photon wave vector is negligible in comparison to that of the electron.

We first consider the matrix element for the transition. This will allow us to work out important selection rules. With photons incident in the

z direction as shown in Fig. 6.6, the polarization vector of the light is in the x, y plane. We therefore have to evaluate matrix elements of the form:

$$M = \langle \mathrm{f}|x|\mathrm{i}\rangle = \int \Psi_\mathrm{f}^*(\boldsymbol{r})\, x\, \Psi_\mathrm{i}(\boldsymbol{r})\, \mathrm{d}^3\boldsymbol{r}\,. \qquad (6.30)$$

When we considered matrix elements of this type in Section 3.2, it made no difference whether we evaluated $\langle \mathrm{f}|x|\mathrm{i}\rangle$ or $\langle \mathrm{f}|y|\mathrm{i}\rangle$ or $\langle \mathrm{f}|z|\mathrm{i}\rangle$. This was a consequence of the isotropy of the cubic semiconductors that we were considering. In the case of the quantum well, however, the x and y directions are equivalent, but the z direction is physically different. Therefore, for quantum wells we have:

$$\langle \mathrm{f}|x|\mathrm{i}\rangle \;=\; \langle \mathrm{f}|y|\mathrm{i}\rangle \;\neq\; \langle \mathrm{f}|z|\mathrm{i}\rangle\,. \qquad (6.31)$$

In this section we concentrate on x, y polarized light, which is the usual experimental arrangement.

We discuss the effect of having the light polarized in the z direction in Section 6.7.

We are interested in evaluating eqn 6.30 for transitions between bound quantum well states in the valence and conduction bands. Figure 6.7 illustrates the type of transition we are considering. The figure specifically shows a transition from an $n = 1$ hole level to an $n = 1$ electron level, and from an $n = 2$ hole level to an $n = 2$ electron level.

We consider a general transition from the nth hole state to the n'th electron state. In analogy to the Bloch functions of eqns 3.7 and 3.8, we can use eqns 6.5 and 6.7 to write the initial and final quantum well wave functions in the form:

$$\Psi_\mathrm{i} \equiv |\mathrm{i}\rangle \;=\; \frac{1}{\sqrt{A}}\, u_\mathrm{v}(\boldsymbol{r})\, \varphi_{\mathrm{h}n}(z)\, \mathrm{e}^{\mathrm{i}\boldsymbol{k}_{xy}\cdot\boldsymbol{r}_{xy}} \qquad (6.32)$$

$$\Psi_\mathrm{f} \equiv |\mathrm{f}\rangle \;=\; \frac{1}{\sqrt{A}}\, u_\mathrm{c}(\boldsymbol{r})\, \varphi_{\mathrm{e}n'}(z)\, \mathrm{e}^{\mathrm{i}\boldsymbol{k}'_{xy}\cdot\boldsymbol{r}_{xy}}\,. \qquad (6.33)$$

The three factors in these wave functions denote the envelope function for the valence or conduction band as appropriate, the bound states of the quantum well in the z direction, and the plane waves for the free motion in the x, y plane. We have written explicit subscripts to show that the plane waves only span the 2-D x, y coordinates. A is the normalization area in the x, y plane.

The momentum of the photon is very small in comparison to that of the electrons, and so conservation of momentum in the transition requires that $\boldsymbol{k}_{xy} = \boldsymbol{k}'_{xy}$. This is the 2-D equivalent of eqn 3.12 for 3-D bulk semiconductors. Therefore, on substituting eqns 6.32 and 6.33 into eqn 6.30, we see that the matrix element breaks into two factors:

$$M = M_\mathrm{cv}\, M_{nn'} \qquad (6.34)$$

where M_cv is the valence–conduction band dipole moment:

$$M_\mathrm{cv} = \langle u_\mathrm{c}|x|u_\mathrm{v}\rangle \;=\; \int u_\mathrm{c}^*(\boldsymbol{r})\, x\, u_\mathrm{v}(\boldsymbol{r})\, \mathrm{d}^3\boldsymbol{r}, \qquad (6.35)$$

and $M_{nn'}$ is the electron-hole overlap given by:

$$M_{nn'} = \langle \mathrm{e}n'|\mathrm{h}n\rangle \;=\; \int_{-\infty}^{+\infty} \varphi_{\mathrm{e}n'}^*(z)\, \varphi_{\mathrm{h}n}(z)\, \mathrm{d}z\,. \qquad (6.36)$$

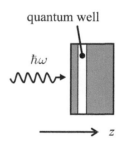

Fig. 6.6 Photons incident on a quantum well with light propagating in the z direction.

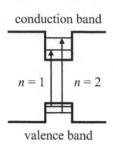

conduction band

$n = 1$ $n = 2$

valence band

Fig. 6.7 Interband optical transitions in a quantum well. The $n = 1$ and $n = 2$ transitions are indicated.

It will usually be the case that the constituent material of the quantum well (e.g. GaAs) has strongly allowed electric-dipole transitions between the conduction and valence bands. We considered this point in Section 3.3.1. Therefore, we can assume that M_{cv} is large and non-zero. Hence the matrix element for the optical transitions is proportional to the overlap of the electron and hole states given by eqn 6.36. This allows us to work out some straightforward selection rules on $\Delta n = (n' - n)$.

Consider first an infinite quantum well with wave functions of the form given by eqn 6.11. The overlap factor is:

$$M_{nn'} = \frac{2}{d} \int_{-d/2}^{+d/2} \sin\left(k_n z + \frac{n\pi}{2}\right) \sin\left(k_{n'} z + \frac{n'\pi}{2}\right) \, \mathrm{d}z. \qquad (6.37)$$

This is unity if $n = n'$ and zero otherwise. (See Exercise 6.7.) Hence we obtain the following selection rule for an infinite quantum well:

$$\Delta n = 0. \qquad (6.38)$$

This is why we only showed $\Delta n = 0$ transitions in Fig. 6.7.

In finite quantum wells the electron and hole wave functions with differing quantum numbers are not necessarily orthogonal to each other because of their different decay constants in the barrier regions. This means that there are small departures from the selection rule of eqn 6.38. However these $\Delta n \neq 0$ transitions are usually weak, and are strictly forbidden if Δn is an odd number, because the overlap of states with opposite parities is zero. (See Exercise 6.7.)

6.4.2   Two-dimensional absorption

The shape of the absorption spectrum in a quantum well can be understood by applying the selection rules we have just derived. If we increase the photon energy from zero, no transitions will be possible until we cross the threshold for exciting electrons from the ground state of the valence band (the $n = 1$ heavy-hole level) to the lowest conduction band state (the $n' = 1$ electron level). This is a $\Delta n = 0$ transition and is therefore allowed. This threshold occurs at a photon energy given by

$$\hbar\omega = E_g + E_{hh1} + E_{e1}, \qquad (6.39)$$

where E_g is the band gap of the quantum well material. This immediately gives us a very important result. The optical absorption edge of the quantum well has been shifted by $(E_{hh1} + E_{e1})$ compared to the bulk semiconductor. Since the confinement energies can be varied by choice of the well width, this gives us a way to tune the frequency of the absorption edge.

The right-hand side of Fig. 6.8 shows the E–k_{xy} diagram for the transition between the $n = 1$ levels. The bands have parabolic dispersions according to eqn 6.9. Conservation of momentum and the negligible $\boldsymbol{k}$ vector of the photon imply that the electron and hole states have the

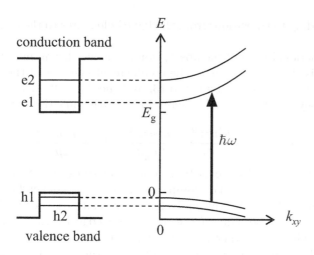

conduction band

e2

e1

E_g

$\hbar\omega$

h1

h2

valence band

k_{xy}

Fig. 6.8 The $n = 1$ interband optical transition in a quantum well at finite k_{xy}.

same k_{xy} values. The energy of the transition shown by the vertical arrow is given by:

$$
\begin{aligned}
\hbar\omega &= E_g + \left(E_{\mathrm{hh1}} + \frac{\hbar^2 k_{xy}^2}{2m_{\mathrm{hh}}^*} \right) + \left(E_{\mathrm{e1}} + \frac{\hbar^2 k_{xy}^2}{2m_{\mathrm{e}}^*} \right) \\
&= E_g + E_{\mathrm{hh1}} + E_{\mathrm{e1}} + \frac{\hbar^2 k_{xy}^2}{2\mu} \,, \qquad (6.40)
\end{aligned}
$$

where μ is the electron–hole reduced effective mass defined in eqn 3.22. This makes it clear that the transitions with $\hbar\omega = (E_g + E_{\mathrm{hh1}} + E_{\mathrm{e1}})$ occur at $k_{xy} = 0$.

Equation 6.40 can be compared directly to eqn 3.23 for the bulk semiconductor. We have already noted the shift of the absorption threshold from E_g to $(E_g + E_{\mathrm{hh1}} + E_{\mathrm{e1}})$. The other crucial difference is that the wave vector in eqn 6.40 for the quantum well spans only the 2-D x, y coordinates, instead of the full 3-D x, y, z space. This has a very important consequence for the joint density of states factor that enters the transition rate in eqn 6.29. The 3-D bulk semiconductor had a parabolic density of states given by eqn 3.16, which led to the absorption edge given in eqn 3.25. By contrast, the joint density of states for a 2-D material is independent of energy and is given by (see Exercise 6.3):

$$
g_{2\mathrm{D}}(E) = \frac{\mu}{\pi\hbar^2} \,. \qquad (6.41)
$$

This means that the absorption coefficient will have a step-like structure, being zero up to the threshold energy given in eqn 6.39, and then having a constant non-zero value for larger photon energies.

The argument above can be repeated for the other allowed optical transitions in the quantum well. The next $\Delta n = 0$ transition for the heavy-hole states occurs at an energy of $(E_g + E_{\mathrm{hh2}} + E_{\mathrm{e2}})$ which corresponds to exciting an electron from the $n = 2$ heavy-hole state to the $n' = 2$ electron level. Once the photon energy crosses this threshold, the absorption coefficient will show a new step. There will also be other steps

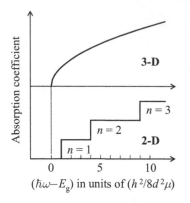

Fig. 6.9 The absorption coefficient for an infinite quantum well of width d compared to the equivalent bulk semiconductor. μ is the electron-hole reduced mass. Excitonic effects are ignored.

corresponding to transitions from the light-hole states to the conduction band.

The functional form of the absorption coefficient for an infinite quantum well is shown in Fig. 6.9. The confinement energies of the electron and hole states are given by eqn 6.13, and the $\Delta n = 0$ selection rule is strictly obeyed. The threshold energy for the nth transition is thus given by:

$$\hbar\omega = E_{\mathrm{g}} + \frac{\hbar^2 n^2 \pi^2}{2m_{\mathrm{e}}^* d^2} + \frac{\hbar^2 n^2 \pi^2}{2m_{\mathrm{h}}^* d^2} = E_{\mathrm{g}} + \frac{\hbar^2 n^2 \pi^2}{2\mu d^2} . \tag{6.42}$$

The spectrum therefore consists of a series of steps with threshold energies given by eqn 6.42. For comparison, the energy-dependence of the absorption coefficient for the equivalent bulk semiconductor is also plotted in Fig. 6.9. The shift of the absorption edge by the confinement energy is evident, together with the change of shape from the parabola of the bulk semiconductor to the step-like structure for the quantum well caused by the change in the density of states on going from 3-D to 2-D.

Example 6.2

Estimate the difference in the wavelength of the absorption edge of a 20 nm GaAs quantum well and bulk GaAs at 300 K.

Solution
We see from eqn 6.39 that the absorption edge of a quantum well occurs at $E_{\mathrm{g}} + E_{\mathrm{hh1}} + E_{\mathrm{e1}}$. We can estimate the confinement energies by using the infinite potential well model. By using eqn 6.13 and the effective mass data for GaAs given in Table D.2, we find that $E_{\mathrm{hh1}} = 2\,\mathrm{meV}$ and $E_{\mathrm{e1}} = 14\,\mathrm{meV}$. These energies are small compared to typical quantum well barrier heights, and so the infinite well approximation should be reasonably accurate. The band edge therefore shifts from 1.424 eV to $(1.424 + 0.002 + 0.014) = 1.440\,\mathrm{eV}$. This corresponds to a blue shift of 10 nm.

6.4.3   Experimental data

Figure 6.10 shows the absorption spectrum of a high-quality GaAs MQW structure containing 40 quantum wells of width 7.6 nm. The barriers were made of AlAs, and the sample temperature was 6 K. It is clear that the predicted step-like behaviour shown in Fig. 6.9 is well reproduced in the data, although the experimental spectrum is complicated by excitonic effects, which give rise to the strong peaks in the absorption at the edge of each step. These excitonic effects will be discussed further in Section 6.4.4, and we concentrate for now on the gross features of the absorption spectrum.

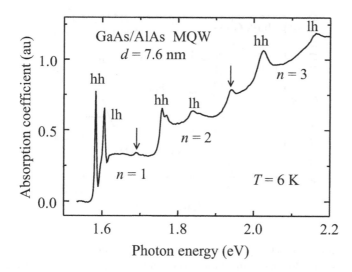

Fig. 6.10 Absorption coefficient of a 40 period GaAs/AlAs MQW structure with 7.6 nm quantum wells at 6 K in arbitrary units. After Fox (1996), © Taylor & Francis Ltd, reprinted with permission.

The most pronounced steps in the spectrum are due to the $\Delta n = 0$ transitions. The first of these occurs for the $n = 1$ heavy-hole transition at 1.59 eV. This is closely followed by the step due to the $n = 1$ light-hole transition at 1.61 eV. This should be compared with the low temperature band edge absorption spectra of bulk GaAs shown in Figs 4.3 and 4.4. We see that the band edge has been shifted in the quantum well by 0.07 eV.

The steps at the band edge are followed by a flat spectrum up to 1.74 eV in which the absorption is practically independent of energy. At 1.77 eV there is a further step in the spectrum due to the onset of the $n = 2$ heavy-hole transition. This is followed by the step for the $n = 2$ light-hole transition at 1.85 eV. Further steps due to the $n = 3$ heavy- and light-hole transitions are observed at 2.03 eV and 2.16 eV respectively.

The two weak peaks identified by arrows are caused by parity-conserving $\Delta n \neq 0$ transitions. The one at 1.69 eV is the hh3 → e1 transition, while that at 1.94 eV is the hh1 → e3 transition.

6.4.4   Excitons in quantum wells

We now return to consider the excitonic effects that give rise to the sharp peaks that are very prominent in the experimental data shown in Fig. 6.10. As discussed in Chapter 4, excitons are bound electron-hole pairs held together by their mutual Coulomb attraction. Since the optical transition can be considered as the creation of an electron-hole pair, the Coulomb attraction increases the absorption rate because it enhances the probability of forming the electron-hole pair. Hence we observe peaks at the resonant energies for exciton formation. These peaks occur at the sum of the single particle energies less the binding energy of the bound pair. Detailed analysis of the data shown in Fig. 6.10 reveals that the binding energies of the quantum well excitons are about 10 meV. This is substantially higher than the value of 4.2 meV in bulk GaAs. (See Section 4.2.)

The enhancement of the excitonic binding energy in the quantum well

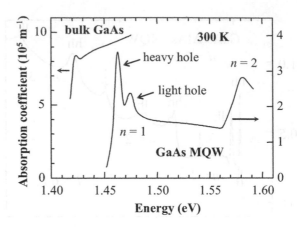

Fig. 6.11 Absorption spectrum of a GaAs/$Al_{0.28}Ga_{0.72}As$ MQW at room temperature. The structure contained 77 GaAs quantum wells of width 10 nm. The absorption spectrum of bulk GaAs at the same temperature is shown for comparison. After Miller et al. (1982), © American Institute of Physics, reprinted with permission.

The reason why the exciton binding energy is not enhanced by a factor of four is that a real quantum well is not a perfect 2-D system. The quantum well has a finite width, and the wave functions extend into the barriers due to tunnelling.

A striking difference between the quantum well and bulk absorption shown in Fig. 6.11 is the lifting of the degeneracy of the heavy- and light-hole states. The bulk sample shows a single excitonic shoulder at the band edge, but the quantum well shows two separate peaks. As discussed in point (2) of Section 6.3.2, this follows from the different effective masses of the heavy and light holes, and it highlights the lower symmetry of the quantum well sample.

is a consequence of the quantum confinement of the electrons and holes. This forces the electrons and holes to be closer together than they would be in a bulk semiconductor, and hence increases the attractive potential. It is possible to show that the binding energy of the ground-state exciton in an ideal 2-D system is enhanced by a factor of four compared to the bulk material. (See Exercise 6.10.) This should be compared with the factor of ~ 2.5 deduced from the experimental data. Although we do not observe perfect 2-D enhancement of the binding energy, the increase is still substantial. The enhancement of the excitonic effects in quantum wells is very useful for device applications, as we discuss further in the next section.

One of the most useful consequences of the enhancement of the exciton binding energy in quantum wells is that the excitons are still stable at room temperature. This contrasts with bulk GaAs, which only shows strong excitonic effects at low temperatures. This can be clearly seen in the data shown in Fig. 6.11, which compares the absorption coefficient of a GaAs MQW structure with 10 nm quantum wells to that of bulk GaAs at room temperature. The bulk sample merely shows a weak shoulder at the band edge, but the MQW shows strong peaks for both the heavy-hole and the light-hole excitons. The more or less flat absorption coefficient expected for quantum wells above these peaks is also evident.

6.4.5   Spin injection in quantum wells

In Section 3.3.7 we studied how the excitation of a bulk semiconductor with circularly polarized light can lead to the generation of spin-polarized electrons. In the case of a III–V semiconductor with the zinc-blende structure such as GaAs, we found that the maximum spin polarization that can be generated is 50%. We now wish to reconsider this situation for the case of a quantum well. As we shall see, the lifting of the degeneracy of the heavy- and light-hole bands now makes it possible to generate fully spin-polarized electrons.

Figure 6.12 shows the optical transitions that can occur in a quantum well such as GaAs/AlGaAs in the presence of circularly polarized light.

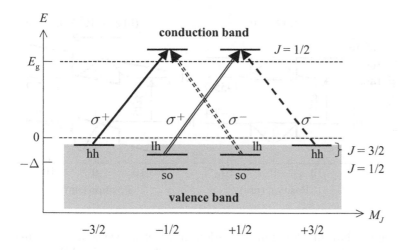

Fig. 6.12 Selection rules for circularly polarized light in a quantum well with the zinc-blende band structure. This diagram should be compared carefully to Fig. 3.8, which applies to bulk semiconductors. Note especially that the heavy- and light-hole transitions are no longer degenerate.

As discussed in Section 3.3.7, the valence-band states are derived from atomic p states split by the spin–orbit interaction, with the heavy- and light-hole levels corresponding respectively to the $M_J = \pm 3/2$ and $M_J = \pm 1/2$ sub-levels of the $J = 3/2$ level. The conduction band has s-like atomic character, with $M_J = \pm 1/2$ corresponding to spin up and down. The $\sigma^\pm$ transitions have selections rules of $\Delta M_J = \pm 1$ in absorption, which means that the transitions between the different sub-levels have the well defined circular polarizations indicated in Fig. 6.12.

Figure 6.12 should be compared carefully to the equivalent one for bulk GaAs given in Fig. 3.8. The key difference is that the energies of the quantum well states are all shifted by the quantum confinement. The heavy- and light-hole levels are therefore no longer degenerate, and the energies of the hh1 $\rightarrow$ e1 and lh1 $\rightarrow$ e1 transitions are split by the difference in the hh1 and lh1 confinement energies. Hence by using circularly polarized light with energy in the range:

$$E_{\mathrm{g}} + E_{\mathrm{e1}} + E_{\mathrm{hh1}} \leq \hbar\omega < E_{\mathrm{g}} + E_{\mathrm{e1}} + E_{\mathrm{lh1}}, \qquad (6.43)$$

it is possible to excite electrons with 100% spin polarization. The direction of the electron spin is determined by the polarization of the light, with σ^- or σ^+ light corresponding respectively to spin-up or spin-down electrons.

The splitting of the heavy- and light-hole states caused by the quantum confinement has another important consequence. In a bulk semiconductor, the degeneracy of the heavy- and light-hole bands at $k = 0$ means that the hole states have a mixed character, and therefore that the hole spin relaxation is very fast. This is no longer the case in a quantum well, where the $M_J = \pm 3/2$ heavy-hole states are pure eigenstates, and circularly polarized photons with energy in the range set by eqn 6.43 create 100% spin-polarized heavy holes as well as electrons. When considering the spin relaxation processes in quantum wells, we therefore have to think about the holes as well as the electrons.

Fig. 6.13 Electron and heavy-hole wave functions for the first quantized levels of a 10 nm GaAs/Al$_{0.3}$Ga$_{0.7}$As quantum well at (a) zero field, and (b) $\mathcal{E}_z = 10^7$ V m^{-1}. The energies are defined relative to the top of the valence band at the centre of the well (i.e. at $z = 0$), and the band gap of GaAs is taken to be 1425 meV. In both diagrams the normalized probability densities (i.e. $\varphi^*\varphi$) are plotted, and the numbers adjacent to the e1 and hh1 states give the level energies. The field direction in (b) is from left to right.

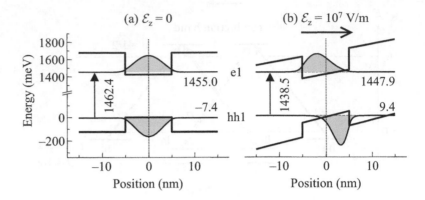

The mechanisms for spin relaxation in quantum wells are the same as those in bulk semiconductors, namely the EY, DP, and BAP processes. (See Section 5.3.4 for an explanation of these acronyms.) However, their relative importance can be affected by the quantum confinement. For example, the enhancement of excitonic effects in quantum wells and the fact that the heavy- and light-hole states are no longer degenerate means that we usually have to consider the spin relaxation of excitons rather than of individual electrons or holes. As a consequence, the BAP mechanism, which relies on electron-hole exchange, is more important in quantum wells than in the bulk. The reader is referred to the Further Reading list for a detailed discussion of the different spin relaxation processes that can occur.

6.5   The quantum-confined Stark effect

The quantum-confined Stark effect describes the response of a quantum-confined system to an external electric field. This makes it a type of electro-optic effect. As we discuss here, the response can be either linear or quadratic.

In Section 4.3.1 of Chapter 4 we considered the effects of a DC electric field on the excitons in bulk GaAs. We found that relatively small electric fields can ionize the excitons by pushing the electrons and holes in opposite directions. The situation in quantum wells is different if the field is applied along the z direction. The field still pushes the electrons and holes in opposite directions, but the barriers prevent the exciton from breaking apart. Hence the excitons are stable up to very high field strengths. These quantum-confined excitons interact with the field and shift to lower energy. In atomic physics, the shift of the energy levels in an electric field is called the Stark effect. The shift of the quantum-confined energy levels in a quantum well is therefore called the **quantum-confined Stark effect**.

When an electric field $\mathcal{E}_z$ is applied to a semiconductor along the z axis, the potential energy of the electron is given by:

$$\Delta E = -p_z \mathcal{E}_z, \tag{6.44}$$

where p_z is the z component of the electron dipole. The electron is negatively charged, and so we can write:

$$p_z = -ez \,, \tag{6.45}$$

where e is the modulus of the electron charge and z is the position along the z axis. The potential energy of the electron is then given by

$$\Delta E_{\mathrm{e}} = +ez\mathcal{E}_z \,. \tag{6.46}$$

The application of the field therefore causes the potential energy of the electron to change linearly as a function of distance along the z axis.

Figure 6.13 illustrates the quantum-confined Stark effect for the first electron and heavy-hole states of a GaAs/Al$_{0.3}$Ga$_{0.7}$As quantum well with a thickness of 10 nm. Figure 6.13(a) shows the probability densities for the e1 and hh1 wave functions at zero field, while Fig. 6.13(b) shows the equivalent quantities at $\mathcal{E}_z = 10^7$ V m^{-1}. In part (b) the linear increase of the valence and conduction band energy as a function of z is caused by adding a potential of the form given by eqn 6.46 to the potential well shown in part (a). The energies of the e1 and hh1 levels are indicated in both parts, with the zero of energy defined as the top of the valence band at the centre of the well (i.e. at $z = 0$).

A number of important points are immediately obvious from comparison of Figs. 6.13(a) and (b).

(1) For the electrons we have a confinement energy of $(E_{\mathrm{e1}} - E_{\mathrm{g}}^{\mathrm{GaAs}}) = (1455.0 - 1425) = 30.0$ meV at $\mathcal{E}_z = 0$ and $(1447.9 - 1425) = 22.9$ meV at $\mathcal{E}_z = 10^7$ V m^{-1}, while for the holes the equivalent energies are 7.4 meV and -9.4 meV respectively. The field thus causes a decrease of the energies of the confined states.

(2) The Stark shift of the levels causes the energy difference between the e1 and hh1 levels (i.e. $E_{\mathrm{e1}} - E_{\mathrm{hh1}}$) to decrease from 1462.4 meV at zero field to 1438.7 meV at $\mathcal{E}_z = 10^7$ V m^{-1}. We thus expect a *red* shift of the transition energy.

(3) At zero field the wave functions are symmetric about the centre of the well, but the application of the field breaks the symmetry and causes the electron and hole probability densities to shift in opposite directions.

(4) The electron-hole overlap (see eqn 6.36) at zero field is nearly perfect with $|\langle \varphi_{\mathrm{e1}} | \varphi_{\mathrm{hh1}} \rangle|^2 = 0.99$. The skewing of the electron and hole wave functions in opposite directions at $\mathcal{E}_z = 10^7$ V m^{-1} reduces the overlap to 0.38.

From these points it is apparent that the quantum-confined Stark effect is expected to cause a red shift of the lowest energy transition and a reduction in its oscillator strength.

The wave functions and level energies shown in Fig. 6.13 were calculated on a computer by numerical methods, which is the only option available for quantum wells with finite barriers. For infinite potential

Hole energies are measured *downwards* on electron band diagrams, and the confinement energy is therefore defined as $-[E_{\mathrm{hh1}} - E^{\mathrm{v}}(z = 0)]$. The fact that this comes out to be negative at $\mathcal{E}_z = 10^7$ V m^{-1} is not significant. The hole level actually lies 40.6 meV *above* the bottom of the well, which occurs at $z = +5$ nm.

The e1–hh1 overlap is less than 100% at $\mathcal{E}_z = 0$ because of the difference in the effective masses and barrier heights, which both affect the barrier penetration due to tunnelling. Perfect overlap is only expected for infinite barriers: see Exercise 6.7.

An exact solution exists for a particle confined in an infinite potential well with an electric field applied. See, for example, Miller (2008). Since this solution makes use of Airy functions, the mathematics is somewhat complicated, and is not discussed further here.

wells, however, analytic solutions exist. Perturbation theory gives the shift to the $n = 1$ level in a small electric field as (see Exercise 6.14):

$$\Delta E = -24 \left(\frac{2}{3\pi}\right)^6 \frac{e^2 \mathcal{E}_z^2 m^* d^4}{\hbar^2},\tag{6.47}$$

where d is the well width, and $\mathcal{E}_z$ is the component of the field in the z direction. This result is analogous to the quadratic Stark effect in atomic hydrogen: the levels shift to lower energy in proportion to $-\mathcal{E}_z^2$.

The quadratic red shift of the levels can be understood as follows. The expectation value of the energy shift caused by the field is given by eqn 6.44 as:

$$\langle \Delta E \rangle = -\langle p_z \rangle \mathcal{E}_z,\tag{6.48}$$

where $\langle p_z \rangle$ is the expectation value of the electron dipole. It follows from eqn 6.45 that $\langle p_z \rangle$ is given by:

$$\langle p_z \rangle = -e\langle z_e \rangle,\tag{6.49}$$

where $\langle z_e \rangle$ is the expectation value of the electron's position along the z axis:

$$\langle z_e \rangle = \int_{-\infty}^{+\infty} \varphi_e(z)^* \, z \, \varphi_e(z) \, dz.\tag{6.50}$$

At zero field, the $n = 1$ electron wave function is symmetric about the centre of the well. This implies that both $\langle z_e \rangle$ and $\langle p_z \rangle$ are zero when no field is applied. With a finite field applied in the positive z direction, the electrons are pushed towards negative z, and $\langle z_e \rangle$ acquires a negative value. This shifting of the average electron position in the opposite direction to the field is clearly shown by the e1 wave function in Fig. 6.13(b). The shift of the electron creates a positive dipole which has a magnitude that is proportional to $\mathcal{E}_z$ at small fields. Hence $\langle p_z \rangle \propto +\mathcal{E}_z$ and therefore $\langle \Delta E \rangle \propto -\mathcal{E}_z^2$. The same argument can be applied to the holes.

We have seen in Section 6.4.4 that the absorption edge of a quantum well is actually dominated by excitonic effects. The shift of the exciton transition energy will be given by the sum of the shifts of the electron and hole levels, less any reduction in the exciton binding energy caused by the field. The latter effect is relatively small, and so the shift in the transition energy is well approximated by the sum of the shifts of the electron and hole levels:

$$\begin{aligned}
\Delta(\hbar\omega) &= \langle \Delta E_e \rangle + \langle \Delta E_h \rangle,\\
&= -\left(\langle p_z^e \rangle + \langle p_z^h \rangle\right)\mathcal{E}_z,\\
&= -\left(-e\langle z_e \rangle + e\langle z_h \rangle\right)\mathcal{E}_z,\\
&= -e\left(\langle z_h \rangle - \langle z_e \rangle\right)\mathcal{E}_z.
\end{aligned}\tag{6.51}$$

This shows that it is the displacement of the electron relative to the hole that causes the red shift of the transition. Furthermore, eqn 6.47 makes it clear that it is the hole that contributes the most to the energy shift, due to its larger effective mass. As discussed above, at small fields we would expect the wave function displacements to be proportional to $\mathcal{E}_z$, giving rise to a quadratic Stark shift. However, this cannot continue indefinitely, because the displacement of the electron relative to the hole is limited by the width of the well. Hence the electron-hole dipole saturates at a value of order $+ed$ at large fields, and the energy shift becomes linear in $\mathcal{E}_z$. There is some analogy here with the linear Stark effect of atomic physics, although the explanation is qualitatively different. In fact, Fig. 6.13(b) is close to the linear saturation limit, since a large field was deliberately chosen to exaggerate the electron-hole displacement.

The quantum-confined Stark effect can be observed by growing the quantum wells in the i-region of a p–i–n diode, as shown in Fig. 6.14. By operating the diode in reverse bias like a photodiode, strong DC electric fields can be applied in the growth direction. The magnitude of the field is given by eqn E.3 in Appendix E:

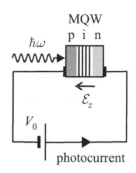

$$\mathcal{E}_z = \frac{|V_{bi} - V_0|}{l_i}, \tag{6.52}$$

where V_{bi} is the built-in voltage of the diode, V_0 is the applied bias, and l_i is the i-region thickness. V_0 is negative in reverse bias, and thus the applied voltage augments the field due to the built-in voltage. As discussed in Section 4.3.1, the photocurrent generated in the device follows the frequency dependence of the absorption.

Fig. 6.14 Experimental set-up required to observe the quantum-confined Stark effect. The quantum wells are grown in the i-region of a p–i–n diode, and the device is operated in reverse bias. This produces a strong electric field $\mathcal{E}_z$ across the quantum wells. The photocurrent generated when light is incident is determined by the absorption of the MQW layer.

Figure 6.15 shows the photocurrent spectra of a GaAs/Al$_{0.3}$Ga$_{0.7}$As MQW p–i–n diode at bias voltages of 0 V and -10 V. The well width was 9.0 nm, and the temperature was 300 K. The i-region thickness was 1 μm, and V_{bi} was 1.5 V. From eqn 6.52, we see that the voltages correspond to field strengths of 1.5×10^6 V m^{-1} and 1.15×10^7 V m^{-1} respectively. The $n = 1$ heavy- and light-hole exciton lines are clearly resolved at both field strengths, despite the fact that the temperature is 300 K and that $\mathcal{E}_z$ considerably exceeds the exciton ionization field of bulk GaAs, which, as discussed in Section 4.3.1, is of order 6×10^5 V m^{-1}.

The spectrum at -10 V shows a clear red-shift for both the heavy- and light-hole excitons. As expected, the shift is larger for the heavy hole excitons due to their larger mass. The decrease in the exciton absorption at the higher field strength is caused by the reduction in the electron-hole overlap integral, as discussed in point (4) above. Two parity-forbidden transitions are clearly identified in Fig. 6.15, namely the hh2 → e1 and hh1 → e2 lines.

The parity forbidden transitions with Δn equal to an odd number become allowed when the field is applied due to the lowering of the symmetry of the system. (See Exercise 6.17.)

The ability to control the shape of the absorption spectrum by applying a voltage opens the possibility for making different types of electro-optic devices. The absorption at 1.44 eV (864 nm) in the quantum wells studied in Fig. 6.15 can be switched on and off by applying the bias. This allows us to make a voltage-tunable photodetector using the same arrangement as that shown in Fig. 6.14. The same device can also function as an intensity modulator by introducing a voltage-dependent loss on a 864 nm beam transmitted through the quantum wells. Moreover, the change of the absorption must also produce a change in the refractive index via the Kramers–Kronig relationships (see Section 2.3), and

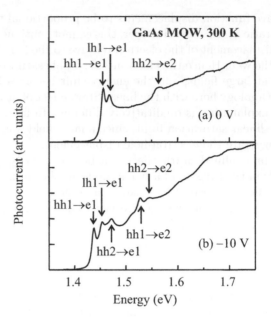

Fig. 6.15 Photocurrent spectra for a GaAs/Al$_{0.3}$Ga$_{0.7}$As MQW p–i–n diode with a 1 μm thick i-region at room temperature. (a) $V_0 = 0$, (b) $V_0 = -10$ V. The quantum well thickness was 9.0 nm. The transitions are labelled by the electron and hole states that participate. After Fox (1996), © Taylor & Francis Ltd, reprinted with permission.

so we can also use the device as a phase modulator.

6.6   Optical emission

The use of quantum well structures in electroluminescent devices is their main commercial application at present. By inserting quantum wells into the active region, a greater range of emission wavelengths can be obtained, together with an increase in the efficiency of the device.

The general principles of light emission in semiconductors were discussed in Chapter 5. The light is generated when electrons in the conduction band recombine with holes in the valence band. We saw in Sections 5.2–5.4 that the luminescence spectrum generally consists of a peak at the band gap energy with a width determined by the carrier density and the temperature.

The physical processes responsible for light emission in quantum wells are essentially the same as those in bulk semiconductors. The electrons and holes injected electrically or optically rapidly relax to the bottom of their bands before emitting photons by radiative recombination. In a quantum well, the lowest levels available to the electrons and hole correspond to the $n = 1$ confined states. Hence the low-intensity luminescence spectrum consists of a peak of spectral width $\sim k_{\mathrm{B}}T$ at energy

The shape of the emission spectrum is only slightly affected by reducing the dimensionality of the system from 3-D to 2-D. The low-intensity emission spectrum of a bulk semiconductor is given by eqn 5.12. In a quantum well, the $(h\nu - E_{\mathrm{g}})^{1/2}$ factor from the 3-D density of states will be replaced by the unit step function derived from the 2-D density of states. In both 3-D and 2-D the net result is that we get a peak of width $\sim k_{\mathrm{B}}T$ starting at the threshold energy for absorption.

$$h\nu = E_{\mathrm{g}} + E_{\mathrm{hh1}} + E_{\mathrm{e1}} \,. \tag{6.53}$$

This shows that the emission peak is shifted by the quantum confinement of the electrons and holes to higher energy compared to the bulk semiconductor.

Figure 6.16 shows the photoluminescence spectrum of a 2.5 nm ZnCdSe

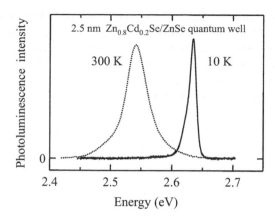

Fig. 6.16 Emission spectrum of a 2.5 nm $Zn_{0.8}Cd_{0.2}Se/ZnSe$ quantum well at 10 K and at room temperature. The spectra have been normalized so that their peak heights are the same. Unpublished data from C.J. Stevens and R.A. Taylor.

quantum well with a cadmium concentration of 20%. $Zn_{0.8}Cd_{0.2}Se$ is a II–VI alloy semiconductor with a direct band gap of 2.55 eV at 10 K, which corresponds to the blue/green spectral region. The barriers of the quantum wells are made from ZnSe, which has a band gap of 2.82 eV. The spectrum at 10 K peaks at 2.64 eV (470 nm) and has a full width at half maximum of 16 meV. The emission energy is about 0.1 eV larger than the band gap of the bulk material, and the linewidth is limited by the inevitable fluctuations in the well width that occur during the epitaxial growth. (See Exercise 6.18.) At room temperature the peak emission energy has shifted to 2.55 eV (486 nm) and the spectrum has broadened so that the linewidth is about 48 meV ($\sim 2k_BT$).

The shift to lower energy between 10 K and 300 K is caused by reduction of the ZnCdSe band gap with temperature. The emission energy is still about 0.1 eV above the band gap of the bulk $Zn_{0.8}Cd_{0.2}Se$ at 300 K.

As mentioned at the start of this section, the use of quantum wells in light-emitting devices is one of the main motivations for their development. Quantum wells offer three main advantages over the equivalent bulk materials:

- The shift of the luminescence peak by the confinement energy ($E_{e1} + E_{hh1}$) allows the wavelength of light-emitting devices to be tuned by choice of the well width.

- The increased overlap between the electron and hole wave functions in the quantum well means that the emission probability is higher. This shortens the radiative lifetime, and the radiative recombination wins out over competing non-radiative decay mechanisms. The radiative efficiency is therefore higher in the quantum wells, which makes it easier to make bright light-emitting devices.

- The total thickness of the quantum wells in an electroluminescent device is very small (~ 10 nm). This is well below the critical thickness for dislocation formation in non-latticed-matched epitaxial layers. This allows the use of non lattice-matched combinations of materials, and hence gives even greater flexibility in emission wavelengths that can be obtained.

The blue and green emitting GaN-based structures grown on sapphire substrates mentioned in Section 5.4.1 are typical examples of a non-lattice-matched opto-electronic device. The commercial devices actually incorporate quantum wells in the active region. Other important examples include the 980 nm $Ga_xIn_{1-x}As$ quantum well lasers grown on GaAs substrates for use with optical fibre amplifiers: see Fig. 9.14 in Section 9.4.

Electroluminescent devices can easily be made from quantum wells by incorporating them in the active region at the junction of a p–n diode, as discussed in Section 5.4 for bulk materials. The devices are operated in forward bias, and the light is emitted when the electrons and holes injected by the current recombine at the junction. GaAs quantum wells emitting around 800 nm are widely used as the lasers in compact disc players and printers. GaAs-based alloys are used to shift the wavelength into the red spectral region or into the infrared to match the optimal wavelengths for optical fibre systems at 1.3 μm and 1.55 μm. (See Exercise 6.19.)

Example 6.3

Estimate the emission wavelength of a 15 nm GaAs quantum well laser at 300 K.

Solution

The emission wavelength is given by eqn 6.53. We estimate the confinement energies from eqn 6.13. Using the effective mass data given in Table D.2, we find $E_{hh1} = 3$ meV and $E_{e1} = 25$ meV. The emission energy is therefore $1.424 + 0.003 + 0.025 = 1.452$ eV, which corresponds to a wavelength of 854 nm.

6.7   Intersubband transitions

An **intersubband transition** is one in which we excite electrons and holes between the levels (or 'subbands') within the conduction or valence band. This contrasts with the interband transitions that we have been considering up till now in which the electrons move from the valence band to the conduction band and vice versa. Figure 6.17 illustrates a typical intersubband absorption transition in which an electron in the $n = 1$ level of a quantum well is excited to the $n = 2$ level by absorption of a photon.

A quick glance at Table 6.2 tells us that intersubband transitions occur at much lower photon energies than interband transitions. For example, the energy spacing between the $n = 1$ and $n = 2$ electron levels in a 10 nm GaAs/AlGaAs quantum well is of order 0.1 eV. This corresponds to an infrared wavelength of about 12 μm. We can therefore use intersubband transitions to make detectors and emitters for the infrared spectral region by using GaAs quantum wells. This offers potential advantages over narrow gap semiconductors both in terms of performance and ease of fabrication.

Intersubband transitions are excited by light polarized along the z direction. The matrix element for such a transition from the nth to the n'th subband is $\langle n|z|n' \rangle$, and the selection rule on $\Delta n = (n - n')$ is that Δn must be an odd number. (See Exercise 6.20.)

The requirement that the polarization must be along the z direction creates some technical difficulties. If the light is incident normal to the

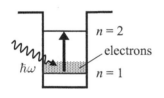

Fig. 6.17 The e1 → e2 intersubband transition in an n-doped quantum well.

surface as shown in Fig. 6.6, there is no polarization component along the z direction. To create a z component, the light must be incident at an angle. However, it is only possible to couple about 10% of the light to the intersubband transitions in this way, due to the high refractive index of semiconductors like GaAs. (See Exercise 6.21.) A better solution is to incorporate a metallic grating on the top of the sample. This can produce a substantial z component even for light incident normal to the surface.

Intersubband detectors have been under development for applications in infrared imaging since the late 1980s. They are made with n-doped quantum wells so that there is a large population of electrons in the $n = 1$ level of the conduction band to absorb the light. In 1994 an intersubband laser called the 'quantum cascade laser' was reported. These intersubband lasers normally work at wavelengths in the infrared spectral region (i.e $\sim 5 - 15\,\mu\text{m}$), and are required for many important applications such as atmospheric sensing. The lower wavelength limit is determined by the difference in the conduction band energies of the component semiconductors, while the upper wavelength limit is ultimately limited by the operating temperature, since the energy level spacing must satisfy $E_2 - E_1 \gtrsim k_\text{B}T$ to prevent thermal occupation of the upper level. In 2002 an important breakthrough was achieved when the maximum wavelength for a quantum cascade laser was extended into the terahertz frequency range, that is $\nu \lesssim 10\,\text{THz} \equiv \lambda \gtrsim 30\,\mu\text{m}$. Terahertz frequency sources are important for imaging and short-range wireless network applications.

Experimental results show that the condition $T \lesssim (E_2 - E_1)/k_\text{B}$ is actually too strict, and that quantum cascade lasers can operate at temperatures above this limit because they are non-equilibrium systems. Nevertheless, the highest operating temperature achieved at longer wavelengths does indeed scale with the laser frequency.

6.8 Quantum dots

We mentioned in Section 6.1 that other types of quantum-confined semiconductor structures can be made in addition to quantum wells. With reference to Table 6.1 and Fig. 6.1, we see that if we confine the electrons in all three directions, we have a quantum dot structure. With dimensions in the nanometre range, these quantum dots typically contain 10^4–10^6 atoms. It has been discovered that quantum dots can form spontaneously in some materials during epitaxial growth and colloidal synthesis. This makes it relatively easy to fabricate them, and accounts for the tremendous expansion in quantum dot research in recent years. Examples of these spontaneous dots will be given below, after first considering a few general principles of the physics of quantum dots.

6.8.1 Quantum dots as artificial atoms

In a quantum dot, the motion of the electrons and holes is confined in all three directions, so that it is appropriate to refer to a dot as as a zero-dimensional structure. The electron and holes states are completely quantized, and the energy spectrum consists of a series of discrete lines, as illustrated schematically in Fig. 6.1. We have seen that the confinement of the electrons and holes in quantum wells gives rise to a host

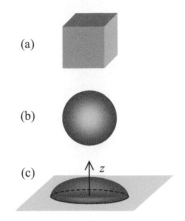

(a)

(b)

(c)

Fig. 6.18 Three different quantum dot shapes: (a) rectangular box, (b) sphere, (c) inverted lens.

of interesting optical effects and improvements in the performance of opto-electronic devices. It is thus to be expected that the increased degree of confinement in quantum dots should give further benefits. For example, by confining the carriers in all three dimensions, we increase the electron-hole overlap and thus increase the radiative quantum efficiency. Furthermore, the discrete nature of the density of states reduces the thermal spread of the carriers within their bands.

The task of calculating the level energies in a quantum dot is quite complicated, since many different shapes are possible. We consider below the shapes shown schematically in Fig. 6.18, namely the rectangular box, sphere, and inverted lens. These examples will serve to illustrate three important general points about quantum dots:

(1) The energy spectrum and density of states are generically similar to those of an atom, with quantized states at discrete energies. Quantum dots are therefore sometimes described as 'artificial atoms'.

(2) The confinement energy scales as $1/d^2$ where d is the size of the dot. This is consistent with the general argument about quantum confinement that leads to eqn 6.3.

(3) The ground state is unique, but the excited states are degenerate. The ground state can therefore accommodate just two electrons (spin-up and spin-down) while the exited states can accommodate more.

Rectangular box dots

The simplest shape to consider is that of a rectangular box with dimensions (d_x, d_y, d_z), as shown in Fig. 6.18(a). In this case the problem separates, with the motion for each of the three axes given by a quantum well Schrödinger equation. If we assume that there are infinite potential barriers at the edges of the box, then the energy levels are given by (cf. eqn 6.13):

$$E(n_x, n_y, n_z) = \frac{\pi^2 \hbar^2}{2m^*} \left(\frac{n_x^2}{d_x^2} + \frac{n_y^2}{d_y^2} + \frac{n_z^2}{d_z^2} \right), \tag{6.54}$$

where the integer quantum numbers n_x, n_y and n_z specify the quantized levels in each direction. The energy spectrum of a cubic dot with $d_x = d_y = d_z$ is considered in Exercise 6.22. The ground state is unique, but the first and second excited states are both triply degenerate.

Spherical dots

Spherical dots can be modelled by assuming that the electron experiences a potential of $-V_0$ inside the dot and zero elsewhere. The spherical symmetry of the problem makes it convenient to work in spherical polar co-ordinates (r, θ, ϕ). The Schrödinger equation for a dot of radius R_0 is then given by:

$$\left[-\frac{\hbar^2}{2m^*}\boldsymbol{\nabla}^2 + V(r)\right]\Psi(\boldsymbol{r}) = E\Psi(\boldsymbol{r})\,, \qquad (6.55)$$

where $V(r) = -V_0$ for $r \leq R_0$, and $V(r) = 0$ for $r > R_0$. Since the potential only depends on r, the wave function can be separated into radial and angular components with:

$$\Psi(\boldsymbol{r}) = R_{nl}(r)Y_{lm}(\theta,\phi)\,. \qquad (6.56)$$

The angular wave functions $Y_{lm}(\theta,\phi)$ are the spherical harmonic functions that appear in all 'central field' problems (i.e. those in which the force points along the radial direction from the origin, so that $V(\boldsymbol{r})$ is a function of r only). The energy is found by solving the radial Schrödinger equation:

$$\left[-\frac{\hbar^2}{2m^*}\frac{1}{r^2}\frac{\mathrm{d}}{\mathrm{d}r}\left(r^2\frac{\mathrm{d}}{\mathrm{d}r}\right) + \frac{\hbar^2 l(l+1)}{2m^* r^2} + V(r)\right]R_{nl}(r) = E_{nl}\,R_{nl}(r)\,, \qquad (6.57)$$

where n is an integer. The energy can in general be written in the form:

$$E_{nl} = \frac{\hbar^2}{2m^*}\frac{C_{nl}^2 \pi^2}{R_0^2}\,. \qquad (6.58)$$

The values of C_{nl} for the first five bound states of an infinite well with $V_0 = \infty$ are listed in Table 6.3. It is a straightforward exercise to show that C_{nl} is an integer when $l = 0$. (See Exercise 6.23.)

The notation used to designate angular momentum states is summarized in Section C.2. The subscripts l and m refer to the orbital and magnetic quantum numbers respectively. In atomic physics, the properties of the Coulomb potential in the hydrogen atom require that $l \leq (n-1)$, where n is the principal quantum number. Since we do not have a Coulomb potential here, this restriction does not apply.

Inverted lens dots

Quantum dots with the inverted lens shape shown in Fig. 6.18(c) are assumed to have cylindrical symmetry about the z (vertical) axis. This makes is convenient to work in cylindrical polar co-ordinates (r, ϕ, z), and to separate the lateral and vertical motions. It is further assumed that the size of the dots is much smaller in the z direction than in the lateral direction, and that the lateral potential approximates to a shallow harmonic oscillator with $V(r,\phi) \propto r^2$ for small r. With these approximations, we may write the wave function in the form $\Psi(\boldsymbol{r}) = \psi(r,\phi)\varphi(z)$, and obtain separate Schrödinger equations for the lateral and vertical motions:

$$\left[-\frac{\hbar^2}{2m^*}\boldsymbol{\nabla}_{xy}^2 + \frac{1}{2}m^*\omega_0^2 r^2\right]\psi(r,\phi) = E^{xy}\psi(r,\phi)\,, \qquad (6.59)$$

$$\left[-\frac{\hbar^2}{2m^*}\frac{\mathrm{d}^2}{\mathrm{d}z^2} + V(z)\right]\varphi(z) = E^z\varphi(z)\,, \qquad (6.60)$$

with total energy given by:

$$E = E^{xy} + E^z\,. \qquad (6.61)$$

Consider first the vertical (z) motion. Equation 6.60 describes a potential well, which will have bound-state solutions with quantized energies.

Table 6.3 First five bound states of a spherical potential well with an infinite barrier. g denotes the degeneracy including the spin. After Bimberg et al. (1999).

Level	n	l	g	C_{nl}
Ground	1	0	2	1
1st excited	1	1	6	1.43
2nd excited	1	2	10	1.83
3rd excited	2	0	2	2
4th excited	1	3	14	2.22

Since the dimensions are very small, it is likely that there will only be one bound state, or if there are more, that the energy difference to the second level will be much larger than the energy separation of the levels associated with the lateral motion. It is therefore reasonable to assume that we only need to consider one bound state for the vertical motion with energy E_1^z.

Now consider the lateral motion. The Schrödinger equation of eqn 6.59 describes a two-dimensional harmonic oscillator. By reverting to Cartesian co-ordinates, it is easy to show that the energy is given by:

$$E^{xy} = (n+1)\hbar\omega_0 \,, \tag{6.62}$$

where $n = 0, 1, 2 \cdots$. (See Exercise 6.25.) The total energy is therefore given by:

$$E_n = E_1^z + (n+1)\hbar\omega_0 \,. \tag{6.63}$$

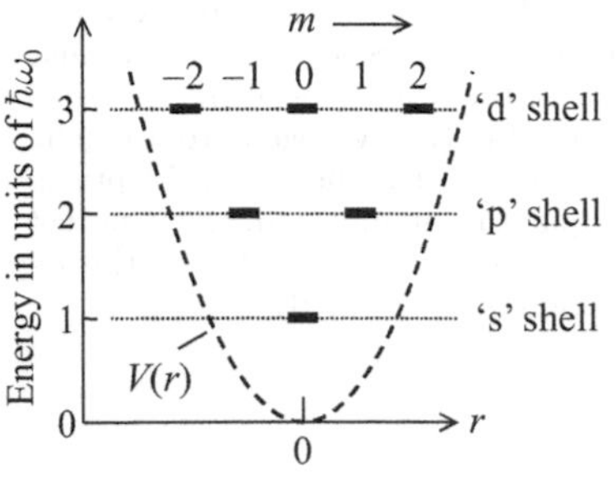

The nth level has n degenerate sub-levels which are identified by the quantum number m. This quantifies the angular momentum about the z axis, and can take integer values from $-n$ to $+n$ in steps of two. The allowed values of m for the first three levels are shown in Fig. 6.19. In analogy with atomic physics, the levels are often labelled as 's', 'p', 'd', etc., shells, although the analogy is not exact, since the notation originates from the analysis of 3-D central fields.

The arguments above can be generalized to dots with more complicated shapes but which still satisfy the criterion that the vertical dimensions are significantly smaller than the lateral ones, and that the lateral confining potential is relatively weak. In this case it will normally be the case that we only need to consider the first bound state for the vertical motion, and the first few excited states arise from the lateral motion. When the cylindrical symmetry about the z axis is lost (e.g. if the dot has an elliptical shape in the x–y plane), we would no longer expect the sub-levels of the nth level to be degenerate.

Fig. 6.19 First three energy levels of a two-dimensional harmonic oscillator. The allowed values of m are indicated, together with the functional form of the potential.

6.8.2   Colloidal quantum dots

One of the ways that quantum dots can be made is by techniques of colloidal synthesis. The resulting dots, which are commonly made from II–VI or III–V semiconductors, are spherical in shape, and can be studied in solution or in the solid state. In the latter case, the dots are either deposited as thin films on transparent substrates or are doped into a glass.

Figure 6.20 shows the absorption spectra of two different types of II–VI colloidal dots. Figure 6.20(a) shows the data for CdSe dots deposited on sapphire substrates at 10 K, while Fig. 6.20(b) shows that for CdTe dots in solution at room temperature. In both cases, the spectra for several different sizes of dots are shown. The diameters quoted represent the average values. In practice there is always a spread of sizes, and this causes broadening of the optical spectra. In fact, a large part of the art of making colloidal quantum dots is precisely in obtaining good size

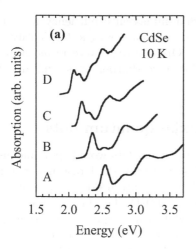
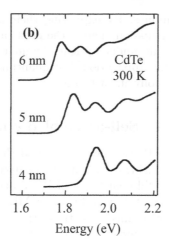

Fig. 6.20 (a) Absorption spectra of close-packed CdSe colloidal quantum dots deposited on sapphire substrates at 10 K. The dot diameters were: (A) 3.03 nm, (B) 3.94 nm, (C) 4.80 nm, and (D) 6.21 nm. (b) Close-up of the band edge of CdTe colloidal quantum dots in solution at room temperature. After Kagan et al. (1996), © American Physical Society, and De Giorgi et al. (2005), © Elsevier, reprinted with permission.

control. For example, the standard deviation of the diameters of the dots shown in Fig. 6.20(a) was < 5%.

In both sets of spectra in Fig. 6.20, there is a clear size-dependent shift of the absorption edge relative to the band gap. ($E_g = 1.85$ eV for CdSe at 10 K and 1.50 eV for CdTe at room temperature.) Excitonic peaks associated with the confined electron and hole states are clearly resolved at both temperatures. The absorption edge occurs at the energy of the exciton from the first confined state of the electrons and holes. If we make the assumption that $C_{10} \approx 1$ for both the electrons and holes, then eqn 6.58 implies that the band edge should occur at:

$$\hbar\omega \approx E_g + \frac{2\pi^2\hbar^2}{\mu d^2} - E_X , \qquad (6.64)$$

where μ is the electron-hole reduced mass, d is the dot diameter, and E_X is the exciton binding energy. The shift of the absorption edge in proportion to $1/d^2$ is a very clear demonstration of the quantum size effects that we have been discussing throughout this chapter. Note that the magnitude of the quantum size effect is very large, with a shift of over 0.6 eV for the CdSe sample with $d = 3.03$ nm.

One application of colloidal dots is in semiconductor-doped glasses. As mentioned in Section 1.4.5, II–VI semiconductors such as CdS, CdSe, ZnS, and ZnSe can be introduced into a glass during the melt process, and colloidal quantum dots can be formed in the right conditions. The size of the dots depends on the way the glass is produced, and good uniformity can be achieved with careful preparation. The size-dependent shift of the absorption shown in Fig. 6.20 provides a way to control the colour of filters made from semiconductor-doped glass. In this way it is possible to make colour-glass filters spanning most of the visible spectral region just by altering the size of the dots.

Another important application of colloidal quantum dots is in fluorescence imaging in chemistry and biology. The dots are prepared in solution and are attached to the molecules under study by synthetic techniques. By exploiting their high luminescence efficiency and tunable emission energy, the dots can act as 'marker tags' to identify the molecule under optical illumination. Organic dyes were formerly used for this application, but quantum dots give sharper lines, and are also more convenient to use.

6.8.3   Self-assembled epitaxial quantum dots

In the 1990s it was discovered that quantum dots would spontaneously form during epitaxial crystal growth in the Stranski–Krastanow regime. Many different types of III–V and II–VI quantum dots have now been made in this way by using MBE or MOCVD. The fact that the dots can be formed directly during epitaxial growth means that they can easily be incorporated into laser diode structures, and the 0-D physics causes a reduction in both the magnitude of the threshold current and its sensitivity to the temperature.

The most common example of Stranski–Krastanow dots is InAs in GaAs. InAs is a narrow gap semiconductor with a unit cell size that is 7% larger than that of GaAs. The dots are formed when thin layers of InAs molecules are deposited on GaAs during MBE growth. The InAs molecules try to adopt the lattice constant of the GaAs, and this leads to the formation of a highly strained layer called the 'wetting layer' on the surface of the crystal. The energy required to strain the layer is so large that it ceases to be favourable to form a uniform layer when the thickness exceeds only a few atoms. Instead, the InAs molecules coalesce into clusters, leading to the formation of InAs quantum dots on top of the wetting layer. By depositing layers of GaAs on top of the dots, the electrons and holes are then confined in both the vertical and lateral directions.

Figure 6.21 shows transmission electron microscope (TEM) images of InAs quantum dots grown by the Stranski–Krastanow technique. Part (a) shows a plan view, while part (b) shows a side view looking down the wafer edge at higher resolution. The TEM images show that the lateral and vertical dimensions of the dot are both in the nanometre range, leading to strong confinement in all directions. It is apparent from Figs 6.21(a) and (b) that the shape of the dots approximates to the inverted lens modelled in Section 6.8.1. The uniformity of the size distribution is apparent in Fig. 6.21(a): most of the dots are roughly of the same size, but some variation is clearly visible.

The appropriateness of describing quantum dots as 'artificial atoms' becomes most clear when techniques are used to isolate the optical spectra of individual dots. Figure 6.22 illustrates typical spectra than can be observed in this way. The lower spectrum in Fig. 6.22(a) shows the photoluminescence spectrum obtained when an InAs/GaAs quantum dot wafer is excited by a laser with a large spot size. Several million dots fall within the illuminated area, and this results in a broad spectrum that reflects the size distribution of the dots. Note that the quantum confinement increases the emission energy by almost 1 eV from the band gap

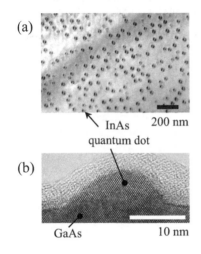

(a)

InAs   200 nm
quantum dot

(b)

GaAs   10 nm

Fig. 6.21 (a) Plan view of an uncapped layer of InAs quantum dots formed during Stranski–Krastanow growth on a GaAs crystal. (b) Side image of one of the InAs dots looking down the edge of the wafer. The mottled pattern above the dot originates from the adhesive used to hold the sample in position. Both images were taken with a transmission electron microscope. After M. Hopkinson (unpublished), and Fry et al. (2000), © American Physical Society, reprinted with permission.

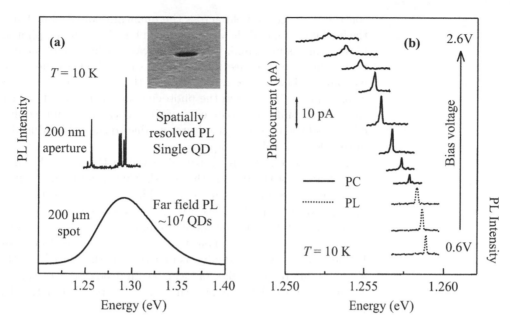

Fig. 6.22 (a) Photoluminescence (PL) spectra of InAs/GaAs self-organized quantum dots at 10 K. The spectrum obtained for a large area laser spot is compared to that obtained through a nano-aperture similar to the one shown in the inset. (b) Voltage dependence of the PL and photocurrent (PC) spectra of a single InGaAs/GaAs quantum dot incorporated into a Schottky diode at 10 K. The spectra were observed through a nano-aperture of diameter 800 nm, and are given for bias voltages from 0.6 V to 2.6 in steps of 0.2 V. After J.J. Finley & M.S. Skolnick (unpublished), and Oulton et al. (2002), © American Physical Society, reprinted with permission.

of bulk InAs (0.4 eV). The upper spectrum in Fig. 6.22(a) illustrates the results obtained when exciting the dots through a nano-aperture of diameter 200 nm. This size of the aperture is chosen so that the light from only a few dots, at most, is collected. The spectrum breaks up into a series of sharp emission lines from the ground- and excited-state excitons of individual quantum dots. If there is more than one dot under the aperture, the change of the dot energy due to inevitable size and shape variations allows the lines of individual dots to be identified by spectral selection.

The width of the exciton line from a single quantum dot is ultimately limited by the radiative lifetime. With typical values of τ_R being around 1 ns, linewidths as small as a few μeV have been observed. Since a single quantum dot can only emit one photon at a time on a particular transition, it can act as a **single-photon source**, with many potential applications in quantum optics.

Figure 6.22(b) shows the voltage dependence of the excitonic spectra from a single InGaAs/GaAs quantum dot embedded within a Schottky diode at 10 K. A combination of photoluminescence (PL) and photocurrent (PC) techniques were used, with a tunable Ti:sapphire as the excitation source. For the PL spectra, the excitation energy was set at ∼ 1.35 eV (i.e. well above the exciton energy), while for the PC measurements, the laser was tuned through the exciton line. The bias voltage

Some of the lines observed in the spectrum of the single dot might arise from charged excitons. For example, the dot might trap a free electron from the conduction band, and then a negatively charged exciton containing two electrons and one hole can be formed when an electron-hole pair is excited by a photon. The energies of these charged excitons are slightly shifted from those of the neutral ones by Coulomb interactions.

controls the electric field across the quantum dot. At low bias voltages, the field is small, and the excitons recombine by emitting photons. As the bias voltage is increased, the field increases, which leads to an increased probability that the electrons and holes tunnel out of the dot before recombining. This causes a reduction of the luminescence efficiency, with a concomitant increase in the photocurrent efficiency. At the same time, the exciton shifts to lower energy due to the quantum-confined Stark effect, and the line strength decreases due to the field-induced reduction in the electron-hole overlap.

All of these effects are clearly observed in Fig. 6.22(b). The most obvious feature of the data is the red shift with increasing voltage due to the quantum-confined Stark effect. The PL efficiency is high at 0.6 V but then drops rapidly above 1.0 V. At the same time, the PC efficiency rises above 1.0 V, reaching its maximum value (assumed to be unity) at 1.8 V. The line intensity then decreases again as the electron-hole overlap decreases, and broadening is observed as the lifetime of the exciton shortens due to tunnelling of electrons and holes out of the dot. These results are typical for many other types of III–V and II–VI quantum dots.

Chapter summary

- Quantum confinement occurs when the dimensions of the structure are small enough that the confinement energy is greater than the thermal energy at that temperature.

- A structure with confinement in one dimension is called a quantum well. Structures with confinement in two or three dimension are called quantum wires and quantum dots respectively.

- Semiconductor quantum wells are made by epitaxial growth of very thin layers. The quantum confinement arises from the potential barriers at the interfaces between different semiconductors due to their different band gaps. The electrons and holes exhibit two-dimensional physics.

- A multiple quantum well is a crystal containing many quantum wells that are separated from each other. A superlattice is a similar structure but with thinner barriers, so that adjacent wells are coupled together by tunnelling through the barriers.

- The energies of the confined states in a quantum well can be calculated by modelling the system as a one-dimensional potential well with a depth determined by the difference in the band gaps of the constituent semiconductors.

- The quantum confinement shifts the absorption edge of a quantum well to higher energy compared to the bulk semiconductor. The absorption spectrum is mainly determined by the 2-D density of states of the quantum well and consists of a series of steps. The splitting of the heavy- and light-hole bands allows the generation of 100% spin polarized electrons by absorption of circularly polarized light.

- Excitonic effects are enhanced in quantum wells. Exciton absorption peaks are readily observed at room temperature in the absorption spectra.

- The quantum-confined Stark effect is the shift of the quantum well levels induced by an electric field, which causes a red shift in the band edge and exciton energies. The effect can be used to make optical modulators.

- The emission energy for luminescence is larger than in a bulk semiconductor due to the quantum confinement of the electrons and holes. Quantum wells make bright light-emitting devices, and the emission wavelength can be tuned by choice of the quantum well parameters.

- Intersubband transitions occur when electrons are excited between the subbands of a quantum well by absorption of a photon. The transitions occur at infrared wavelengths.

- Quantum dots exhibit zero-dimensional physics, with fully quantized energies and a discrete density of states. The dots can be made by colloidal synthesis or by self-organization during epitaxial growth in the Stranski–Krastanow regime. The absorption and emission spectra show broad lines due to variations in the dot sizes, but very sharp lines can be observed by isolating the spectra of individual quantum dots by nano-aperture techniques.

Further reading

The seminal paper on semiconductor quantum wells is Esaki and Tsu (1970). Complementary introductory reading to the treatment given here may be found in Burns (1985) or Singleton (2001). The subject is treated at a more thorough level in Yu and Cardona (1996), or in a number of specific quantum well texts, such as Bastard (1990), Jaros (1989), Kelly (1995), Singh (1993), or Weisbuch and Vinter (1991).

The spin dynamics of electrons, holes, and excitons in quantum wells is discussed in the collection of articles in Awschalom et al. (2002) and Dyakonov (2008). A specific review of the subject may be found in Viña (1999).

The Stark effect in hydrogen is covered in most quantum mechanics texts, for example Gasiorowicz (1996) or Schiff (1969), and also in atomic physics texts, such as Woodgate (1980). A thorough discussion of the effects of electric fields on particles confined in potential wells may be found in Miller (2008), while Chuang (1995) covers the physics of the quantum-confined Stark effect in depth.

Blood (1999) gives a review of the use of quantum wells in visible-emitting diode lasers. The physics of intersubband transitions is discussed in Helm (2000), Liu and Capasso (2000a), and Liu and Capasso (2000b). The terahertz quantum cascade laser is reviewed in Williams (2007), and a collection of articles describing the significance of terahretz technology may be found in Davies et al. (2004).

The physics of quantum dots is described by Bimberg et al. (1999), Harrison (2005), and Woggon (1997). Murray et al. (2000) give a detailed description of the synthe-

sis and characterization of colloidal quantum dots, while a discussion of the spectra of single quantum dots and their application in quantum optics may be found in Michler (2003).

Exercises

(6.1) Estimate the temperature at which quantum size effects would be important for a semiconductor layer of thickness 1 μm if the effective mass of the electrons is $0.1m_0$.

(6.2) A particle of mass m^* is confined to move in a quantum well of width d with infinite barriers. Show that the energy separation of the first two levels is equal to $k_BT/2$ when d is equal to $d = (3h^2/4m^*k_BT)^{1/2}$. Evaluate d for electrons of effective mass m_0 and $0.1m_0$ at 300 K. Hence show that this value of d is smaller than the value of Δx given in eqn 6.4 by a factor of $\sqrt{3\pi^2}$.

(6.3) Consider a gas of spin 1/2 particles of mass m moving in a two-dimensional layer. Apply Born–von Karman periodic boundary conditions (i.e. $e^{ikx} = e^{ik(x+L)}$, etc, where L is a macroscopic length) to show that the density of states in k-space is $1/(2\pi)^2$. By considering the incremental area enclosed by two circles in k-space differing in radius by dk, show that the number of states with k vectors between k and $k + dk$ is given by $g(k)$d$k = (k/2\pi)$dk. Hence show that if the energy dispersion is given by $E(k) = \hbar^2k^2/2m$, the density of states in energy space is given by

$$g_{2D}(E)\,\mathrm{d}E = \frac{m}{\pi\hbar^2}\,\mathrm{d}E\,.$$

(6.4) Repeat Exercise 6.3 for a gas of free spin 1/2 particles of mass m moving in a one-dimensional wire to show that the density of states in energy space is given by:

$$g_{1D}(E)\,\mathrm{d}E = \sqrt{m/2\hbar^2\pi^2}E^{-1/2}\,\mathrm{d}E\,.$$

(6.5) Explain, with reference to eqn 6.26, why a finite quantum well will always have at least one bound state, no matter how small V_0 is.

(6.6) Calculate the energy of the first heavy-hole bound state of a GaAs/AlGaAs quantum well with $d =$

10 nm. Take $V_0 = 0.15$ eV, $m_w^* = 0.34m_0$ and $m_b^* = 0.5m_0$. How does this energy compare to that of an equivalent well with infinite barriers?

(6.7) Consider the electron-hole overlap integral $M_{nn'}$ for a quantum well given by:

$$M_{nn'} = \int_{-\infty}^{+\infty} \varphi_{en'}^*(z)\,\varphi_{hn}(z)\,\mathrm{d}z.$$

(a) Show that M'_{nn} is unity if $n = n'$ and zero otherwise in a quantum well with infinite barriers.
(b) Show that $M_{nn'}$ is zero if $(n - n')$ is an odd number in a quantum well with finite barriers.

(6.8) Draw a sketch of the energy dependence of the absorption spectrum of a 5 nm GaAs quantum well at 300 K between 1.4 eV and 2.0 eV. Assume that the confining barriers are infinite and ignore excitonic effects. See Table D.2 for band structure data on GaAs.

(6.9) Discuss how the spectrum in Exercise 6.8 would change if (a) the barrier height were finite, and (b) excitonic effects were included.

(6.10)* The variational technique introduced in Exercise 4.4 can be used to calculate the energy and radius of a 2-D exciton.[1] The Hamiltonian for the relative motion of an electron-hole pair in a 2-D material is given in polar coordinates by

$$\hat{H} = -\frac{\hbar^2}{2\mu}\left(\frac{1}{r}\frac{\partial}{\partial r}\left(r\frac{\partial}{\partial r}\right) + \frac{1}{r^2}\frac{\partial^2}{\partial\phi^2}\right) - \frac{e^2}{4\pi\epsilon_0\epsilon_r r}\,,$$

where $r^2 = (x^2 + y^2)$. As for the 3-D exciton considered in Exercise 4.4, we guess a trial wave function with a 1s-like radial dependence:

$$\Psi(r,\phi) = \left(\frac{2}{\pi\xi^2}\right)^{1/2}\exp\left(-\frac{r}{\xi}\right),$$

where ξ is the variational parameter.
(a) Verify that the trial wave function is properly

*Exercises marked with an asterisk are more difficult.

[1]The 2-D exciton problem can also be solved exactly using Laguerre polynomials, but the variational approach is easier.

normalized.

(b) The variational energy $\langle E \rangle_{var}$ is given by:

$$\langle E \rangle_{var} = \int_{r=0}^{\infty} \int_{\phi=0}^{2\pi} \Psi^* \hat{H} \Psi \, r \, dr d\phi.$$

Show that $\langle E \rangle_{var}$ is given by:

$$\langle E \rangle_{var} = \frac{\hbar^2}{2\mu\xi^2} - \frac{e^2}{2\pi\epsilon_r\epsilon_0\xi}.$$

(c) Vary $\langle E \rangle_{var}$ with respect to ξ to obtain the best estimate for the energy. Show that this is four times larger than that of the equivalent bulk semiconductor.

(d) Show the Bohr radius of the 2-D exciton, namely the value of ξ that minimizes $\langle E \rangle_{var}$, is half that of the equivalent 3-D exciton.

(6.11)* Discuss qualitatively how you would expect the exciton binding energy in a GaAs/Al$_{0.3}$Ga$_{0.7}$As quantum well to vary with the quantum well thickness, given that the binding energy of the excitons in bulk GaAs and Al$_{0.3}$Ga$_{0.7}$As are 4 meV and 6 meV respectively.

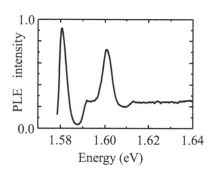

Fig. 6.23 Photoluminescence excitation (PLE) spectrum of a GaAs/AlAs quantum well at 4 K. Unpublished data from R.A. Taylor.

(6.12) Figure 6.23 shows the absorption spectrum of a GaAs/AlAs quantum well at 4 K measured by using the photoluminescence excitation technique. Relevant band structure data for GaAs is given in Table D.2.

(a) Explain the principles of photoluminescence excitation spectroscopy.

(b) Account for the shape of the absorption spectrum.

(c) Estimate the width of the quantum wells by assuming that they behave like a perfect 2-D system with infinite barriers. Would you expect the

true well width to be larger or smaller than the answer you have worked out this way?

(d) Deduce the binding energies of the heavy- and light-hole excitons, and comment on the values you obtain.

(6.13) Discuss the variation of the electron spin polarization created with σ^+ light in a GaAs/AlGaAs quantum well as the photon energy is varied above the fundamental absorption edge.

(6.14)* The Stark shift of the confined levels in a quantum well can be calculated by using second-order perturbation theory. Consider the interaction between the electrons in a quantum well of width d and a DC electric field of strength $\mathcal{E}_z$ applied along the z (growth) axis. Equation 6.46 shows that the perturbation is of the form $H' = e\mathcal{E}_z z$.

(a) Explain why the first-order shift of the energy levels, given by:

$$\Delta E^{(1)} = \int_{-\infty}^{+\infty} \varphi(z)^* H' \varphi(z) \, dz,$$

is zero.

(b) The second-order energy shift of the $n = 1$ level is given by:

$$\Delta E^{(2)} = \sum_{n>1} \frac{|\langle 1|H'|n\rangle|^2}{E_1 - E_n},$$

where

$$\langle 1|H'|n\rangle = \int_{-\infty}^{+\infty} \varphi_1(z)^* H' \varphi_n(z) \, dz.$$

This can be evaluated exactly if we have infinite confining barriers. Within this approximation, show that the Stark shift is given approximately by:

$$\Delta E = -24 \left(\frac{2}{3\pi}\right)^6 \frac{e^2\mathcal{E}_z^2 m^* d^4}{\hbar^2}.$$

(6.15) The magnitude of the $n = 1$ heavy-hole exciton red shift is 10.5 nm at -10 V for the MQW diode studied in Fig. 6.15. The p–i–n sample has an i-region thickness of 1.0 µm and a built-in voltage of 1.5 V.

(a) Compare the magnitude of the Stark shift to that predicted in Fig. 6.13, and account for any difference.

(b) Estimate the magnitude of the red shift at -5 V.

(c) Estimate the average relative displacement of the electron and hole probability densities at -10 V.

Table 6.4 Dependence of the $n = 1$ heavy-hole transition on the electric field strength $\mathcal{E}_z$ for two GaAs quantum well samples. Sample A had a well width of 10 nm, while sample B had a well width of 18 nm. The transition energies are given in eV.

$\mathcal{E}_z$ (Vm^{-1})	A	B
0	1.548	1.524
3×10^6	1.547	1.518
6×10^6	1.543	1.497
9×10^6	1.535	1.470

(6.16) Table 6.4 gives experimental data for the red shift of the $n = 1$ heavy-hole transition due to the quantum-confined Stark effect in two GaAs quantum well samples with well widths of 10 nm and 18 nm respectively. Compare the experimental data with the predictions of Exercise 6.14(b), and account qualitatively for any major discrepancies. Band structure data for GaAs is given in Table D.2.

(6.17) A DC electric field of magnitude $\mathcal{E}_z$ is applied along the growth (z) axis of a quantum well. Use symmetry arguments to explain why transitions between confined electron and holes states with Δn equal to an odd number are forbidden at $\mathcal{E}_z = 0$, but not at finite $\mathcal{E}_z$.

(6.18) By assuming that the confinement energy varies as d^{-2}, estimate the shift in the luminescence emission energy caused by a $\pm 5\%$ change in d for a 2.5 nm ZnCdSe quantum well, given that the total confinement energy for the electrons and holes at $d = 2.5$ nm is 0.1 eV. Compare this value to the measured linewidth of the 10 K data shown in Fig. 6.16, and comment on the answer. The unit cell size of the crystal is 0.28 nm, and the electron and hole effective masses of Zn$_{0.8}$Cd$_{0.2}$Se are $0.15m_0$ and $0.5m_0$ respectively.

(6.19) A Ga$_{0.47}$In$_{0.53}$As quantum well laser is designed to emit at 1.55 μm at room temperature. Estimate the width of the quantum wells within the device. ($E_g = 0.75$ eV, $m_e^* = 0.041m_0$, $m_{hh}^* = 0.47m_0$.)

(6.20)* The matrix element for an intersubband transition between the nth and n'th subbands of a quantum well is given by:

$$\langle n|z|n' \rangle = \int_{-\infty}^{+\infty} \varphi_n^*(z)\, z\, \varphi_{n'}(z)\, \mathrm{d}z.$$

(a) By considering the parity of the states, prove that $\Delta n = (n - n')$ must be an odd number.
(b) Compare the relative strengths of the $1 \rightarrow 2$

and the $1 \rightarrow 4$ transitions in a 20 nm GaAs quantum well with infinite barriers. What is the wavelength of the $1 \rightarrow 2$ transition? ($m_e^* = 0.067m_0$.)

(6.21) Linearly polarized light is incident on a quantum well sample at an angle θ to the normal (z) direction. The polarization direction lies within the plane of incidence. What is the maximum fraction of the power in the beam that can be absorbed by intersubband transitions if the refractive index of the crystal is 3.3?

(6.22) Write down the energies of the first seven quantized energies of a cubic quantum dot with infinite barriers of dimension d. What is the degeneracy of each level?

(6.23) Show that $R(r) = \sin kr/r$ is a solution of the radial equation of a spherical dot for states with $l = 0$. By requiring that $R(R_0) = 0$ in a dot with infinite barriers, show that the confinement energy of states with $l = 0$ is equal to $(\hbar^2/2m^*)(n\pi/R_0)^2$, where n is an integer.

(6.24) Compare the confinement energies of the ground states of cubic and spherical dots of the same volume and with infinite potential barriers in both cases.

(6.25) Consider a two-dimensional harmonic oscillator with a Schrödinger equation given by:

$$\left[-\frac{\hbar^2}{2m_e} \nabla_{2D}^2 + \frac{1}{2} m_e \omega_0^2 r^2 \right] \psi(\boldsymbol{r}) = E\psi(\boldsymbol{r}),$$

with $r^2 = x^2 + y^2$. The form of the operator ∇_{2D}^2 is:

$$\nabla_{2D}^2 = \frac{1}{r}\frac{\partial}{\partial r}\left(r\frac{\partial}{\partial r}\right) + \frac{1}{r^2}\frac{\partial^2}{\partial \phi^2}$$

in polar co-ordinates, and

$$\nabla_{2D}^2 = \frac{\partial^2}{\partial x^2} + \frac{\partial^2}{\partial y^2}$$

in Cartesian co-ordinates.
(a) By working in Cartesian co-ordinates, show that the Schrödinger equation separates into two quantum harmonic oscillator equations for the x and y motion, and hence that:

$$E = (n+1)\hbar\omega_0,$$

where $n = n_x + n_y$, n_x and n_y being the quantum numbers for the x and y axis harmonic oscillators. Hence write down the first five energy levels and their degeneracies.
(b) By working in polar co-ordinates, show that the solutions take the form $\psi(r, \phi) = R(r)e^{im\phi}$,

where m is an integer.

(c)* The wave functions of a 1-D harmonic oscillator are given by:

$$\psi_n(x) = \left(\frac{1}{\sqrt{\pi}\, 2^n n!}\right)^{1/2} H_n(\xi)\, e^{-\xi^2/2},$$

where $\xi = \sqrt{\alpha}\, x$, $\alpha = m_e \omega_0 / \hbar$, and $H_n(\xi)$ are the Hermite polynomials with $H_0(\xi) = 1$, $H_1(\xi) = 2\xi$, $H_2(\xi) = 2 - 4\xi^2$, etc. Similarly, the first six wave functions $\psi_{n,m}(r, \phi)$ of the 2-D harmonic oscillator in polar co-ordinates are of the form:

$$
\begin{aligned}
\psi_{0,0}(r,\phi) &\propto \exp(-\alpha r^2/2) \\
\psi_{1,\pm 1}(r,\phi) &\propto r \exp(-\alpha r^2/2)\, e^{\pm i\phi} \\
\psi_{2,0}(r,\phi) &\propto (\alpha r^2 - 1) \exp(-\alpha r^2/2) \\
\psi_{2,\pm 2}(r,\phi) &\propto r^2 \exp(-\alpha r^2/2)\, e^{\pm i 2\phi}.
\end{aligned}
$$

By comparing the wave functions with the same energies in polar and Cartesian co-ordinates, verify that the allowed values of m for the first three levels are as shown in Fig. 6.19.

7

Free electrons

7.1	Plasma reflectivity	180
7.2	Free carrier conductivity	183
7.3	Metals	185
7.4	Doped semiconductors	191
7.5	Plasmons	198
7.6	Negative refraction	207
	Chapter summary	209
	Further reading	210
	Exercises	211

In this chapter we investigate the optical properties associated with free electrons. As the name suggests, these are systems in which the electrons experience no restoring force from the medium when driven by the electric field of a light wave. The two main solid-state systems that exhibit strong free electron effects are:

- **Metals**. Metals contain large densities of free electrons that originate from the valence electrons of the metal atoms.

- **Doped semiconductors**. n-type semiconductors contain free electrons, while p-type materials contain free holes. The free carrier density is determined by the concentration of impurities used for the doping.

We begin our discussion of the optical properties by using the Drude–Lorentz model introduced in Section 2.1.3 of Chapter 2. This will enable us to explain the main optical property of metals that we mentioned in Section 1.4.3, namely that they reflect strongly in the visible spectral region. We then apply our knowledge of interband transitions from Chapter 3 to obtain a better understanding of the detailed form of the reflectivity spectra of metals such as aluminium and copper. Next we apply the Drude–Lorentz model to doped semiconductors to explain why doping causes infrared absorption. We then consider the collective oscillations of the whole free carrier gas, which will naturally lead us to the notion of plasmons, both bulk and surface. Finally, we conclude with a brief discussion of negative refraction.

7.1 Plasma reflectivity

A neutral gas of charged particles is called a **plasma**. Metals and doped semiconductors can be treated as plasmas because they contain equal numbers of fixed positive ions and free electrons. The free electrons experience no restoring forces when they interact with electromagnetic waves. This contrasts with bound electrons that have natural resonant frequencies in the near-infrared, visible, or ultraviolet spectral regions owing to the restoring forces of the medium.

In this section we derive a formula for the relative permittivity of an electron plasma using the classical oscillator model discussed in Section 2.2 of Chapter 2. As noted in Section 2.1.3, this approach combines the **Drude model** of free electron conductivity with the **Lorentz**

model of dipole oscillators, and is therefore known as the **Drude–Lorentz model**.

We begin by considering the oscillations of a free electron induced by the AC electric field $\mathcal{E}(t)$ of an electromagnetic wave. The equation of motion for the displacement x of the electron is:

$$m_0\frac{\mathrm{d}^2 x}{\mathrm{d}t^2} + m_0\gamma\frac{\mathrm{d}x}{\mathrm{d}t} = -e\mathcal{E}(t) = -e\mathcal{E}_0 e^{-i\omega t},\tag{7.1}$$

where ω is the angular frequency of the light, and $\mathcal{E}_0$ is its amplitude. The first term represents the acceleration of the electron, while the second is the frictional damping force of the medium. The term on the right-hand side is the driving force exerted by the light. Equation 7.1 is the same as the equation of motion for a bound oscillator given in eqn 2.5, except that there is no restoring force term because we are dealing with free electrons.

By substituting $x = x_0 e^{-i\omega t}$ into eqn 7.1, we obtain

$$x = \frac{e\mathcal{E}}{m_0(\omega^2 + i\gamma\omega)}.\tag{7.2}$$

The polarization P of the gas is equal to $-Nex$, where N is the number of electrons per unit volume. By recalling the definitions of the electric displacement D and the relative permittivity ϵ_r (cf. eqns A.2 and A.3), we can write:

$$\begin{aligned}D &= \epsilon_\mathrm{r}\epsilon_0\mathcal{E}\\ &= \epsilon_0\mathcal{E} + P\\ &= \epsilon_0\mathcal{E} - \frac{Ne^2\mathcal{E}}{m_0(\omega^2 + i\gamma\omega)}.\end{aligned}\tag{7.3}$$

Therefore:

$$\epsilon_\mathrm{r}(\omega) = 1 - \frac{Ne^2}{\epsilon_0 m_0}\frac{1}{(\omega^2 + i\gamma\omega)}.\tag{7.4}$$

This equation is identical to eqn 2.14 for the bound oscillator except that the resonant frequency ω_0 is zero and we have not yet considered the effects of background polarizability. Equation 7.4 is frequently written in the more concise form:

$$\epsilon_\mathrm{r}(\omega) = 1 - \frac{\omega_\mathrm{p}^2}{(\omega^2 + i\gamma\omega)},\tag{7.5}$$

where

$$\omega_\mathrm{p} = \left(\frac{Ne^2}{\epsilon_0 m_0}\right)^{1/2}.\tag{7.6}$$

ω_p is known as the **plasma frequency**.

Let us first consider a lightly damped system. In this case, we put $\gamma = 0$ in eqn 7.5 so that

$$\epsilon_\mathrm{r}(\omega) = 1 - \frac{\omega_\mathrm{p}^2}{\omega^2}.\tag{7.7}$$

We have assumed here that the light is polarized along the x direction. The model is not affected by this arbitrary choice provided that the medium is isotropic.

We shall see in Section 7.5 that ω_p corresponds to the natural resonant frequency of the whole free carrier gas. This contrasts with the resonant frequency of the individual electrons, which is of course zero because they are free.

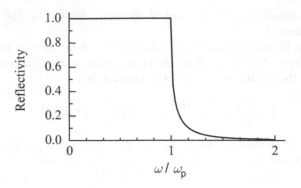

Fig. 7.1 Reflectivity of an undamped free carrier gas as a function of frequency.

The fact that the refractive index is imaginary below ω_p means that the extinction coefficient κ is large, and hence that the medium is highly attenuating. This point will be explained further in Section 7.2. It is a general property of systems with high extinction coefficients that they also have high reflectivities.

The complex refractive index $\tilde{n}$ of the medium is related to the complex dielectric constant by $\tilde{n} = \sqrt{\epsilon_r}$ (cf. eqn 1.22). This means that $\tilde{n}$ is imaginary for $\omega < \omega_p$ and positive for $\omega > \omega_p$, with a value of zero precisely at $\omega = \omega_p$. The reflectivity R can be calculated from eqn 1.29:

$$R = \left| \frac{\tilde{n} - 1}{\tilde{n} + 1} \right|^2 . \tag{7.8}$$

By substituting the frequency dependence of $\tilde{n}$ into this formula, we see that R is unity for $\omega \leq \omega_p$, and then decreases for $\omega > \omega_p$, approaching zero at $\omega = \infty$. This frequency dependence is plotted in Fig. 7.1.

The basic conclusion is that we expect the reflectivity of a gas of free electrons to be 100% for frequencies up to ω_p. This result is very well confirmed by experimental data. In Sections 7.3 and 7.4 below we shall see how the plasma reflectivity effect is observed in both metals and doped semiconductors.

One of the best examples of plasma reflectivity effects is the reflection of radio waves from the upper atmosphere. The atoms in the ionosphere are ionized by the ultraviolet light from the Sun to produce a plasma of ions and free electrons. The plasma frequency is in the MHz range, and so the low-frequency waves used for AM radio transmissions are reflected, but not the higher-frequency waves used for FM radio or television. (See Exercise 7.2.)

Example 7.1

Aluminium is a trivalent metal with $6.0 \times 10^{28} \, \text{m}^{-3}$ atoms per unit volume. Account for the shiny appearance of aluminium.

Solution

Aluminium has three valence electrons per atom, and so the free electron density N is $3 \times (6.0 \times 10^{28}) = 1.8 \times 10^{29} \, \text{m}^{-3}$. We use this value of N in eqn 7.6 to find that $\omega_p = 2.4 \times 10^{16} \, \text{rad/s}$. The free electrons in aluminium will reflect all frequencies below ω_p. Now ω_p corresponds to a wavelength of $2\pi c/\omega_p = 79 \, \text{nm}$, which is in the ultraviolet spectral region, and this means that all visible wavelengths are reflected. Aluminium therefore has a shiny surface that can be used for making mirrors.

7.2   Free carrier conductivity

In deriving eqn 7.7, we neglected the damping of the free carrier oscillations. We can recast the equation of motion in a way that makes the physical significance of the damping term more apparent. To do this we note that $\dot{x}$ is the electron velocity v. Hence we can rewrite eqn 7.1 as:

$$m_0 \frac{dv}{dt} + m_0 \gamma v = -e\mathcal{E} . \tag{7.9}$$

Since $m_0 v$ is the momentum p, we see that:

$$\frac{dp}{dt} = -\frac{p}{\tau} - e\mathcal{E} , \tag{7.10}$$

where we have replaced the damping rate γ by $1/\tau$, where τ is the damping time. This shows that the electron is being accelerated by the field, but loses its momentum in time τ. In other words, τ is the **momentum scattering time**.

In an AC field of the form $\mathcal{E}(t) = \mathcal{E}_0 e^{-i\omega t}$, we look for solutions to the equation of motion with $x = x_0 e^{-i\omega t}$. This implies that $|v| = \dot{x}$ also has a time variation of the form $v = v_0 e^{-i\omega t}$. On substituting this into eqn 7.9, we obtain:

$$v(t) = \frac{-e\tau}{m_0} \frac{1}{1 - i\omega\tau} \mathcal{E}(t). \tag{7.11}$$

The current density j is related to the velocity and field through:

$$j = -Nev = \sigma\mathcal{E} , \tag{7.12}$$

where σ is the electrical conductivity. On combining eqns 7.11 and 7.12, we obtain the **AC conductivity** $\sigma(\omega)$:

$$\sigma(\omega) = \frac{\sigma_0}{1 - i\omega\tau} , \tag{7.13}$$

where

$$\sigma_0 = \frac{Ne^2\tau}{m_0}. \tag{7.14}$$

σ_0 is the conductivity measured with DC electric fields. We can thus deduce the momentum scattering time from the DC conductivity through eqn 7.14. For a typical metal or doped semiconductor, this gives values of τ in the range 10^{-14} to 10^{-13} s at room temperature.

By comparing eqns 7.4 and 7.13, we see that the AC conductivity and the dielectric constant are related to each other through:

$$\epsilon_r(\omega) = 1 + \frac{i\sigma(\omega)}{\epsilon_0\omega} . \tag{7.15}$$

Thus optical measurements of $\epsilon_r(\omega)$ are equivalent to AC conductivity measurements of $\sigma(\omega)$, and the free carrier reflectivity spectrum can be discussed in terms of the conductivity rather than the dielectric constant.

At very low frequencies that satisfy $\omega \ll \tau^{-1}$, we can derive a useful relationship between the conductivity of the free carrier gas and the absorption coefficient for electromagnetic waves. This can be achieved by first splitting $\epsilon_r(\omega)$ into its real and imaginary components in accordance with eqn 1.21, i.e. $\epsilon_r \equiv \epsilon_1 + i\epsilon_2$. Equation 7.5 with $\gamma = \tau^{-1}$ gives:

$$\epsilon_1 = 1 - \frac{\omega_p^2 \tau^2}{1 + \omega^2 \tau^2} \tag{7.16}$$

$$\epsilon_2 = \frac{\omega_p^2 \tau}{\omega(1 + \omega^2 \tau^2)}. \tag{7.17}$$

We then work out n and κ, the real and imaginary parts of the complex refractive index, by using eqns 1.25 and 1.26, and hence deduce the absorption coefficient α from κ. Since $\omega\tau \ll 1$ implies that $\epsilon_2 \gg \epsilon_1$, we can obtain solutions with $n \approx \kappa = (\epsilon_2/2)^{1/2}$. Using eqn 1.19 we therefore obtain:

$$\alpha = \frac{2\omega(\epsilon_2/2)^{1/2}}{c} = \left(\frac{2\omega_p^2 \tau \omega}{c^2}\right)^{1/2}. \tag{7.18}$$

We can put this equation in a more accessible form by noting from eqn 7.14 that $\omega_p^2 \tau = \sigma_0/\epsilon_0$ and from eqn A.28 that $c^2 = 1/\epsilon_0\mu_0$. This gives:

$$\alpha = (2\sigma_0\omega\mu_0)^{1/2}. \tag{7.19}$$

Hence we see that the absorption coefficient is proportional to the square root of the DC conductivity and the frequency.

Equation 7.19 implies that AC electric fields can only penetrate a short distance into a conductor such as a metal. This well-known phenomenon is called the **skin effect**. If the field strength varies as $\exp(-z/\delta)$ with the distance z from the surface, then the power falls off as $\exp(-2z/\delta)$. By comparison with the definition of α in eqn 1.4, we see that:

$$\delta = \frac{2}{\alpha} = \left(\frac{2}{\sigma_0\omega\mu_0}\right)^{1/2}. \tag{7.20}$$

δ is known as the **skin depth**.

The fields that decay exponentially in the conductor are called **evanescent waves**. We have seen in the previous section that we expect the reflectivity of a metal to be very high for frequencies below ω_p. From what we have seen here, it is now apparent that this only applies if the thickness l of the medium is much larger than the skin depth. If l is comparable to, or smaller than, δ, the evanescent fields will not have decayed fully by the back of the medium, and some of the energy will be transmitted. Conservation of energy then demands that the reflectivity R must drop accordingly. This implies that R depends on l when $l \lesssim \delta$, and ultimately drops to zero for a very thin medium.

A treatment of the variation of R with l may be found, for example, in Born and Wolf (1999).

At higher frequencies the relationship given in eqn 7.19 breaks down because the approximation $\omega\tau \ll 1$ is no longer valid. In this case we can derive a different frequency dependence for the attenuation coefficient. This will be discussed when considering the absorption due to free carriers in doped semiconductors in Section 7.4.1.

Example 7.2

The DC electrical conductivity of copper is $6.5 \times 10^7 \, \Omega^{-1}\mathrm{m}^{-1}$ at room temperature. Calculate the skin depth at 50 Hz and 100 MHz.

Solution
The skin depth is given by eqn 7.20. At a frequency of 50 Hz we have $\omega = 2\pi \times 50 = 314 \, \mathrm{rad/s}$. On inserting this value of ω into eqn 7.20 and using $\sigma_0 = 6.5 \times 10^7 \, \Omega^{-1}\mathrm{m}^{-1}$, we obtain $\delta = 8.8 \, \mathrm{mm}$. At 100 MHz, $\omega = 6.28 \times 10^8 \, \mathrm{rad/s}$, and the skin depth δ is only $6.2 \, \mu\mathrm{m}$.

7.3 Metals

The free electron model of metals was proposed by P. Drude in 1900. The model provides a basic explanation for why metals are good conductors of heat and electricity, and is the starting point for more sophisticated theories. As we shall see here, it is also successful in explaining why metals tend to be good reflectors. On the other hand, band theory is needed to explain why some metals (e.g. copper and gold) are coloured.

7.3.1 The Drude model

The Drude free electron model of metals considers the valence electrons of the atoms to be free. When an electric field is applied, the free electrons accelerate and then undergo collisions with the characteristic scattering time τ introduced in eqn 7.10. The electrical conductivity is therefore limited by the scattering, and measurements of σ allow the value of τ to be determined through eqn 7.14.

The free electron density N in the Drude model is equal to the density of metal atoms multiplied by their valency. Table 7.1 lists the Drude free electron densities for a number of common metals. The values of N are in the range 10^{28}–$10^{29} \, \mathrm{m}^{-3}$. These very large free electron densities explain why metals have high electrical and thermal conductivities. The plasma frequencies calculated using eqn 7.6 are also tabulated in Table 7.1, together with the wavelength λ_p that corresponds to ω_p. It is apparent that the very large values of N lead to plasma frequencies in the ultraviolet spectral region.

In the visible spectral region where $\omega/2\pi \sim 10^{15}$ Hz, we are usually in a situation with $\omega \gg \gamma$. This is because $\tau = \gamma^{-1}$ is typically of order 10^{-14} s. Therefore the simplification of eqn 7.5 to eqn 7.7 is a good approximation. With ω_p in the ultraviolet, the visible photons have frequencies below ω_p and thus ϵ_r is negative. As discussed in Section 7.1, this means that the reflectivity is expected to be 100% up to ω_p. This explains the first and most obvious optical property of metals, namely that they tend to be good reflectors at visible frequencies.

Table 7.1 Free electron density and plasma properties of some metals. The figures are for room temperature unless stated otherwise. The electron densities are based on data taken from Wyckoff (1963). The plasma frequency ω_p is calculated from eqn 7.6, and λ_p is the wavelength corresponding to this frequency.

Metal	Valency	N ($10^{28}\,\mathrm{m}^{-3}$)	$\omega_p/2\pi$ ($10^{15}\,\mathrm{Hz}$)	λ_p (nm)
Li (77 K)	1	4.70	1.95	154
Na (5 K)	1	2.65	1.46	205
K (5 K)	1	1.40	1.06	282
Rb (5 K)	1	1.15	0.96	312
Cs (5 K)	1	0.91	0.86	350
Cu	1	8.47	2.61	115
Ag	1	5.86	2.17	138
Au	1	5.90	2.18	138
Be	2	24.7	4.46	67
Mg	2	8.61	2.63	114
Ca	2	4.61	1.93	156
Al	3	18.1	3.82	79

A striking implication of the free carrier model is that the dielectric constant changes from being negative to positive as we go through the plasma frequency. This means that the reflectivity ceases to be 100% above ω_p (see Fig. 7.1) and some of the light can be transmitted through the metal. Thus we expect that all metals will eventually become transmitting if we go far enough into the ultraviolet so that $\omega > \omega_p$. This phenomenon is known as the **ultraviolet transparency of metals**.

In order to observe the ultraviolet transmission threshold at the plasma frequency, it is necessary that there should be no other absorption processes occurring at ω_p. This condition is best satisfied in the alkali metals. Table 7.2 lists the wavelengths of the ultraviolet transmission edges observed in the alkalis. The experimental wavelengths can be compared with those predicted from the calculated plasma frequency tabulated in Table 7.1. The experimental results are in reasonable agreement with the predictions, and show the correct trend on descending the periodic table. The discrepancies can be explained to a large extent by replacing the free electron mass with the electron effective mass derived from the band structure of the metal. (See Exercise 7.4.)

Figure 7.2 shows the measured reflectivity of aluminium as a function of photon energy from the infrared to the ultraviolet spectral region. As noted in Example 7.1, the plasma frequency occurs in the ultraviolet spectral region, and hence the reflectivity is expected to be high for all visible frequencies. The data show that the reflectivity is over 80% for all photon energies up to $\sim 15\,\mathrm{eV}$, and then drops off to zero at higher energies. Thus aluminium shows the characteristic ultraviolet transparency edge predicted by the Drude model. The relatively featureless reflectivity at visible frequencies is exploited in commercial mirrors.

Table 7.2 Ultraviolet transmission threshold wavelength λ_{UV} for the alkali metals. Data from Givens (1958).

Metal	λ_{UV} (nm)
Li	205
Na	210
K	315
Rb	360
Cs	440

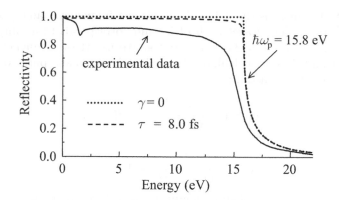

Fig. 7.2 Experimental reflectivity of aluminium as a function of photon energy. The experimental data are compared to predictions of the free electron model with $\hbar\omega_p = 15.8\,\text{eV}$. The dotted curve is calculated with no damping. The dashed line is calculated with $\tau = 8.0 \times 10^{-15}\,\text{s}$, which is the value deduced from DC conductivity. Experimental data from Ehrenreich et al. (1963), © American Physical Society, reprinted with permission.

The plasma frequency for aluminium listed in Table 7.1 corresponds to a photon energy of 15.8 eV. The dotted line in Fig. 7.2 gives the reflectivity predicted from eqn 7.7 with $\hbar\omega_p = 15.8\,\text{eV}$. On comparing the experimental and theoretical results, we see that the model accounts for the general shape of the spectrum, but there are some important details that are not explained.

We shall see in Section 7.5.1 that the plasma frequency can be determined directly by using electron energy-loss spectroscopy.

An improved attempt to model the experimental data can be made by including the damping term in the dielectric constant. Example 7.3 explains how this is done. The reflectivity calculated from eqn 7.5 for the value of τ deduced from the DC conductivity, namely $8.0 \times 10^{-15}\,\text{s}$, is plotted as the dashed line in Fig. 7.2. The main difference between the two calculated curves is that the damping causes the reflectivity to be less than unity below ω_p, and the ultraviolet transmission edge is slightly broadened. However, this is only a relatively small effect because $\omega_p \gg \tau^{-1}$.

The inclusion of damping makes a small improvement in the fit to the data, but there are two important features that are still not explained. Firstly, the reflectivity is significantly lower than predicted, and secondly, there is a dip around 1.5 eV, where we would have expected a featureless curve. Both of these points can be explained by considering the interband absorption rates. These are discussed in the next section.

Example 7.3

The conductivity of aluminium at room temperature is $4.1 \times 10^7\,\Omega^{-1}\text{m}^{-1}$. Calculate the reflectivity at 500 nm according to the Drude–Lorentz model.

Solution

We first work out the damping time τ from the conductivity using eqn 7.14. Taking the value of $N = 1.81 \times 10^{29}\,\text{m}^{-3}$ from Table 7.1, we find:

$$\tau = \frac{m_0\sigma_0}{Ne^2} = 8.0 \times 10^{-15}\,\text{s}.$$

Table 7.1 also gives us the value of the plasma frequency, namely $\omega_\mathrm{p} = 2.4 \times 10^{16}\,\mathrm{rad/s}$. The wavelength of $500\,\mathrm{nm}$ corresponds to an angular frequency $\omega = 2\pi c/\lambda = 3.8 \times 10^{15}\,\mathrm{rad/s}$. We use these frequencies in eqns 7.16 and 7.17 to calculate the real and imaginary parts of the complex dielectric constant:

$$\epsilon_1 = 1 - \frac{\omega_\mathrm{p}^2 \tau^2}{1 + \omega^2 \tau^2} = -39\,,$$

and

$$\epsilon_2 = \frac{\omega_\mathrm{p}^2 \tau}{\omega(1 + \omega^2 \tau^2)} = 1.3\,.$$

We then work out the real and imaginary parts of the complex refractive index using eqns 1.25 and 1.26. This gives:

$$n = \frac{1}{\sqrt{2}}\left(-39 + [(-39)^2 + (1.3)^2]^{1/2}\right)^{1/2} = 0.10\,,$$

and

$$\kappa = \frac{1}{\sqrt{2}}\left(+39 + [(-39)^2 + (1.3)^2]^{1/2}\right)^{1/2} = 6.2\,.$$

We finally obtain the reflectivity from eqn 1.29:

$$R = \frac{(n-1)^2 + \kappa^2}{(n+1)^2 + \kappa^2} = \frac{(-0.9)^2 + (6.2)^2}{(1.1)^2 + (6.2)^2} = 99\%\,.$$

This shows that the inclusion of the damping reduces the reflectivity by only 1% in this case.

7.3.2 Interband transitions in metals

The absorption of light by direct interband transitions was discussed in detail in Chapter 3. Direct transitions involve the promotion of electrons to a higher band by absorption of a photon with the correct energy. The electron does not change its k vector significantly because of the very small momentum of the photon. Thus the transitions appear as vertical arrows on the E–k band diagram of the solid.

Interband absorption is important in metals because the electromagnetic waves penetrate a short distance into the surface, and if there is a significant probability for interband absorption, the reflectivity will be reduced. We consider below the reflectivity spectra of aluminium and copper in order to illustrate the effects of interband absorption, and then make some general comments about other metals such as silver and gold.

Aluminium

The band diagram of aluminium is shown in Fig. 7.3. Aluminium has an electronic configuration of $[\mathrm{Ne}]3\mathrm{s}^2 3\mathrm{p}^1$ with three valence electrons. The crystal structure is face-centred cubic, which has a body-centred cubic

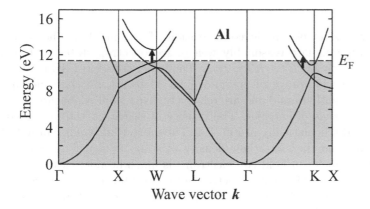

Fig. 7.3 Band diagram of aluminium. The transitions at the W and K points that are responsible for the reflectivity dip at 1.5 eV are labelled. After Segall (1961), © American Physical Society, reprinted with permission.

(bcc) reciprocal lattice, as shown in Fig. D.5 of Appendix D. The first Brillouin zone is completely full, and the valence electrons spread into the second, third and slightly into the fourth zones. The band structure appears quite complex due to the irregular shape of the bcc Brillouin zone. However, the bands are actually very close to the free electron model, with significant departures only in the vicinity of the Brillouin zone boundaries. The bands are filled up to the Fermi energy E_F, which is marked on the diagram. Direct transitions can take place from any of the states below the Fermi level to unoccupied bands directly above them on the E–k diagram.

Fermi's golden rule given in eqn 3.2 tells us that the absorption rate is proportional to the density of states for the transition. The dip in the reflectivity at 1.5 eV which is apparent in Fig. 7.2 is a consequence of the 'parallel band' effect. This occurs when there is a band above the Fermi level that is approximately parallel to another band below E_F. In this case, the interband transitions from a large number of occupied k states below the Fermi level will all occur at the same energy. Hence the density of states at the energy difference between the two parallel bands will be very high, which will result in a particularly strong absorption at this photon energy.

Inspection of the band diagram of aluminium shows that the parallel band effect occurs at both the W and K points of the Brillouin zone. These transitions have been identified on Fig. 7.3. The energy separation of the parallel bands is approximately 1.5 eV in both cases. The enhanced transition rate at this photon energy thus explains the reflectivity dip observed at 1.5 eV in the experimental data. Moreover, we can see from the band diagram that there will be further transitions between bands below the Fermi level to unoccupied bands above E_F at a whole range of photon energies greater than 1.5 eV. The density of states for these transitions will be lower than at 1.5 eV because the bands are not parallel. However, the absorption rate is still significant, and accounts for the reduction of the reflectivity to a value below that predicted by the Drude model in the visible and ultraviolet spectral regions.

We came across a similar example of parallel bands when we discussed the absorption rate at the critical points in the band structure of silicon in Section 3.5.

The positions of the W and K points on the fcc Brillouin zone boundary are shown in Fig. D.5.

Copper

Copper has an electronic configuration of $[\text{Ar}]3d^{10}4s^1$. The outer 4s bands approximate reasonably well to free electron states with a dispersion given by $E = \hbar^2 k^2 / 2m_0$. They therefore form a broad band covering a wide range of energies. The 3d bands, on the other hand, are more tightly bound and are relatively dispersionless, occupying only a narrow range of energies. The density of states for the two bands is illustrated schematically in Fig. 7.4. The narrow 3d bands can hold ten electrons, and therefore their density of states is sharply peaked. The 4s band, which can hold two electrons, is much broader, with a smaller maximum. The 11 valence electrons of copper fill up the 3d band, and half fill the 4s band. The Fermi energy therefore lies within the 4s band above the 3d band. Interband transitions are possible from the filled 3d bands to unoccupied states in the 4s band above E_F, as illustrated in Fig. 7.4. This implies that there will be a well-defined threshold for interband transitions from the 3d bands to the 4s band.

Figure 7.5 shows the actual band structure and density of states of copper. The general features indicated in Fig. 7.4 are clearly shown in the calculated curves. The 4s band is the parabola starting at the Γ point at $-9\,\text{eV}$, while the 3d bands are the five curves bunched together in the energy range $-5 \to -2\,\text{eV}$. The 4s band crosses the 3d bands and then re-emerges as the single band with energy $> -2\,\text{eV}$. It is apparent that the 3d electrons lie in relatively narrow bands with very high densities of states, while the 4s band is much broader with a lower density of states. The Fermi energy lies in the middle of the 4s band above the 3d band. Interband transitions are possible from the 3d bands below E_F to unoccupied levels in the 4s band above E_F. The lowest energy transitions are marked on the band diagram in Fig. 7.5. The transition energy is $2.2\,\text{eV}$ which corresponds to a wavelength of $560\,\text{nm}$.

Figure 7.6 shows the measured reflectivity of copper from the infrared to the ultraviolet spectral region. Based on the plasma frequency given in Table 7.1, we would expect near 100% reflectivity for photon energies below $10.8\,\text{eV}$, which corresponds to an ultraviolet wavelength of $115\,\text{nm}$. However, the experimental reflectivity falls off sharply above $2\,\text{eV}$ owing to the interband absorption edge discussed above. This explains why copper has a reddish colour.

In atomic physics, transitions between d and s states are forbidden for electric-dipole processes. (See Table B.1 in Appendix B.) The matrix element for the 3d→4s transitions is therefore relatively small, but this is compensated by the very high density of states in the solid, which results in strong absorption.

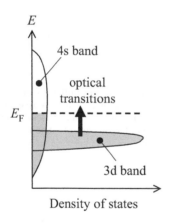

Fig. 7.4 Schematic density of states for the 3d and 4s bands of a transition metal such as copper.

Silver and gold

The arguments used for copper can be applied to the other noble metals. The important parameter is the energy gap between the d bands and the Fermi energy, as shown in Fig. 7.4. In gold the interband absorption threshold occurs at a slightly higher energy than copper, which explains why it has a yellowish colour. In silver, on the other hand, the interband absorption edge is around $4\,\text{eV}$. This frequency is in the ultraviolet, and so the reflectivity remains high throughout the whole visible spectrum. (See Fig. 1.5.) This explains why silver does not have any particular colour, and also why it is so widely used for making mirrors. Gold is also

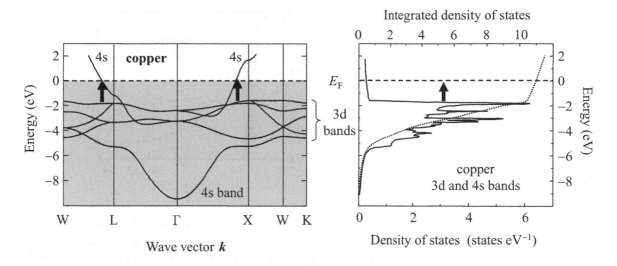

Fig. 7.5 Calculated band structure of copper. The transitions from the 3d bands responsible for the interband transitions around 2 eV are identified. The right-hand side of the figure shows the density of states calculated from the band structure. The strongly peaked features between about −2 eV and −5 eV are due to the 3d bands. The dotted line is the integrated density of states. The Fermi level (defined here as $E = 0$) corresponds to the energy at which the integrated density of states is equal to 11. After Moruzzi et al. (1978).

used for mirrors, but only at infrared wavelengths.

7.4   Doped semiconductors

The controlled doping of semiconductors with impurities is an essential part of solid-state technology. The general principles are discussed in Section D.1 of Appendix D. The introduction of **donor** impurities provides an excess of electrons, while **acceptor** impurities lead to a deficit of electrons, which is equivalent to an excess of holes. Doping that produces excess electrons is called **n-type**, while doping that produces excess holes is called **p-type**.

Experimental measurements on doped semiconductors show that the presence of impurities gives rise to new absorption mechanisms and also to a free carrier plasma reflectivity edge. Our aim here is to explain these effects by applying a suitably modified version of the free carrier model and by considering the quantized levels created by the impurity atoms. In the two subsections that follow, we first consider the free carrier effects, and then move on to discuss the absorption associated with the impurity levels.

In silicon and germanium, which come from group IV of the periodic table, n-type doping is achieved by adding atoms from group V, while p-type doping is achieved by adding atoms from group III. In compound semiconductors such as the III–Vs, the way the doping works is more complicated. If the impurity sits on the group III atom site, then a group II element gives p-type doping, and a group IV element gives n-type. On the other hand, if the impurity sits on the group V atom site, then a group IV impurity gives p-type doping, while a group VI element gives n-type. Group IV doping of a III–V semiconductor can therefore give either n- or p-type doping, depending on how the impurities fit into the crystal.

7.4.1   Free carrier reflectivity and absorption

The free electron model developed in Sections 7.1 and 7.2 can be applied to doped semiconductors if we make two appropriate modifications. Firstly, we must account for the fact that the electrons and holes

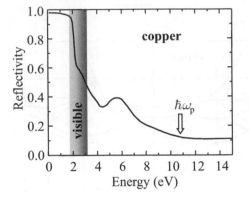

Fig. 7.6 Reflectivity of copper from the infrared to the ultraviolet spectral region. The reflectivity drops sharply above 2 eV due to interband transitions. Data from Lide (1996).

are moving in the conduction or valence band of a semiconductor. This is easily achieved by assuming that the carriers behave as particles with an effective mass m^* rather than the free electron mass m_0. Secondly, we must remember that the semiconductor has a high relative permittivity at the frequencies of interest even before the dopants are added.

The two modifications mentioned above can be handled if we rewrite eqn 7.3 in the following form:

$$
\begin{aligned}
D &= \epsilon_r \epsilon_0 \mathcal{E} \\
&= \epsilon_0 \mathcal{E} + P_{\text{other}} + P_{\text{free carrier}} \\
&= \epsilon_{\text{opt}} \epsilon_0 \mathcal{E} - \frac{N e^2 \mathcal{E}}{m^*(\omega^2 + i\gamma\omega)}.
\end{aligned}
\tag{7.21}
$$

The term P_{other} accounts for the polarizability of the bound electrons before the dopants are added, while the effective mass m^* accounts for the band structure of the semiconductor. The carrier density N that appears in this equation is the density of free electrons or holes generated by the doping process. Note that the sign of the charge cancels, and so the only difference between electrons and holes in this treatment is in the effective mass that is used.

The free carrier effects due to doping are most noticeable in the spectral region 5–30 μm, where we would normally expect the semiconductor to be completely transparent. Hence the value of ϵ_{opt} that we use in eqn 7.21 is the one measured in the transparent spectral region below the interband absorption edge. This value is known from the refractive index of the undoped semiconductor: $\epsilon_{\text{opt}} = n^2$. (See eqn 1.27 with $\kappa = 0$ below the band edge.)

Equation 7.21 tells us that the frequency dependence of the dielectric constant is given by:

$$
\epsilon_r(\omega) = \epsilon_{\text{opt}} - \frac{N e^2}{m^* \epsilon_0} \frac{1}{(\omega^2 + i\gamma\omega)}.
\tag{7.22}
$$

This can be rewritten as:

$$
\epsilon_r(\omega) = \epsilon_{\text{opt}} \left(1 - \frac{\omega_p^2}{(\omega^2 + i\gamma\omega)} \right),
\tag{7.23}
$$

As explained in Section 2.2.2 of Chapter 2, solids have a number of resonant frequencies, each of which can be modelled by dipole oscillators. There are resonant frequencies in the infrared due to the phonons, and others in the near-infrared, visible, or ultraviolet due to the bound electrons. The phonon absorption bands are discussed in detail in Chapter 10, and occur in the range 30–100 μm for a typical III–V semiconductor.

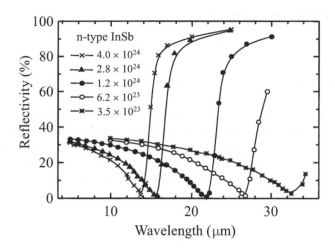

Fig. 7.7 Infrared reflectivity spectra of n-type InSb at room temperature for different values of the free electron density. After Spitzer & Fan (1957), © American Physical Society, reprinted with permission.

where the plasma frequency ω_p is now given by:

$$\omega_p^2 = \frac{Ne^2}{\epsilon_{opt}\epsilon_0 m^*}. \tag{7.24}$$

We have written the dielectric constant in this way to make the link to the Drude model apparent. The difference between the plasma frequency for the semiconductor given in eqn 7.24 and that given in eqn 7.6 is that we have replaced m_0 by m^*, and we have included ϵ_{opt} to account for the background polarizability of the undoped semiconductor.

If we assume that the system is lightly damped, then we can ignore the damping term in eqn 7.23. This then implies that ϵ_r is negative below ω_p and positive at higher frequencies. We thus expect to observe a plasma reflectivity edge at ω_p just as we did in metals. Since the carrier density is much smaller than in metals, the plasma edge occurs at frequencies in the infrared spectral range. This prediction is very well borne out by infrared reflectivity data.

Figure 7.7 shows the measured reflectivity of n-type InSb as a function of the electron density. The fundamental absorption edge at the band gap of InSb occurs at 6 μm, while the phonon absorption band lies around 50 μm. Thus we would expect pure InSb to be transparent in the wavelength range shown and have a featureless reflectivity spectrum. Instead, the data show a well defined reflectivity edge, which shifts to shorter wavelengths as the electron density increases, in accordance with eqn 7.24.

The data shown in Fig. 7.7 demonstrate the phenomenon of the plasma reflectivity edge more clearly than many of the equivalent results obtained for metals. This is because it is not possible to vary the electron density in metals. Moreover, in metals the plasma frequencies are much higher, and the reflectivity edge is frequently obscured by interband transitions.

One very striking feature of the data is the zero in the reflectivity at wavelengths just below the plasma edge. This occurs at a frequency given by (see Exercise 7.8):

$$\omega^2 = \frac{\epsilon_{\text{opt}}}{\epsilon_{\text{opt}} - 1}\, \omega_{\text{p}}^2 . \tag{7.25}$$

By fitting this formula to the data, the effective mass of InSb can be determined. (See Exercise 7.9.)

At frequencies above ω_{p}, the presence of free carriers leads to the absorption of light. This effect is called **free carrier absorption**, and can be observed in the infrared spectral region below the fundamental absorption edge at the band gap, where the semiconductor would normally be transparent. To see how this effect arises, we split the dielectric constant given in eqn 7.23 into its real and imaginary parts. This gives:

$$\epsilon_1 = \epsilon_{\text{opt}} \left(1 - \frac{\omega_{\text{p}}^2 \tau^2}{1 + \omega^2 \tau^2} \right) , \tag{7.26}$$

$$\epsilon_2 = \frac{\epsilon_{\text{opt}} \omega_{\text{p}}^2 \tau}{\omega(1 + \omega^2 \tau^2)} , \tag{7.27}$$

where we have made the usual substitution of τ^{-1} for γ. In a typical semiconductor, with $\tau \sim 10^{-13}$ s at room temperature, it is safe to make the approximation $\omega\tau \gg 1$ at frequencies in the near-infrared. Furthermore, the free carrier term in ϵ_{r} will be small. Therefore we can assume $\epsilon_1 \approx \epsilon_{\text{opt}}$, and that $\epsilon_2 \ll \epsilon_1$. In these conditions we find solutions to eqns 1.25 and 1.26 with $n = \sqrt{\epsilon_{\text{opt}}}$ and $\kappa = \epsilon_2/2n$. This allows us to deduce the absorption coefficient using eqn 1.19. The result is:

The skin effect considered in Section 7.2 may also be considered as a type of free carrier absorption. In the skin effect, however, we are considering the absorption at low frequencies below ω_{p} where the material is highly reflective. We are now considering absorption above ω_{p} where the material should be transparent.

$$\alpha_{\text{free carrier}} = \frac{\epsilon_{\text{opt}} \omega_{\text{p}}^2}{nc\omega^2\tau} = \frac{Ne^2}{m^*\epsilon_0 nc\tau}\frac{1}{\omega^2} . \tag{7.28}$$

This shows that the free carrier absorption is proportional to the carrier density and should vary with frequency as ω^{-2}.

Experimental data on a number of n-doped samples lead to the conclusion that $\alpha_{\text{free carrier}} \propto \omega^{-\beta}$, where β is in the range 2–3. The departure of β from the predicted value of 2 is caused by the failure of our assumption that τ is independent of ω. To see why this is important, we illustrate the physical processes that are occurring during free carrier absorption in Fig. 7.8. The figure shows the conduction band of an n-type semiconductor, which is filled up to the Fermi level determined by the free carrier density. Absorption of a photon excites an electron from an occupied state below the Fermi level to an unoccupied level above E_{F}. The photon only has a very small momentum compared to the electron, and therefore cannot change the electron's momentum significantly. It is obvious from Fig. 7.8 that a scattering event must occur to conserve momentum in the process. Hence the absorption must be proportional to the scattering rate $1/\tau$, in accordance with the prediction of eqn 7.28.

The approximation that τ is independent of ω effectively says that the relaxation time of the electrons does not depend on their initial energy. This is equivalent to the energy independent relaxation time approximation of the Boltzmann equation used in electron transport theory. It is well known that this approximation is only valid in a limited range of conditions. See Ashcroft and Mermin (1976) for further details.

The mechanisms that can contribute to the momentum-conserving process in free carrier absorption include phonon scattering and scattering from the ionized impurities left behind by the release of the free

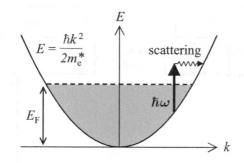

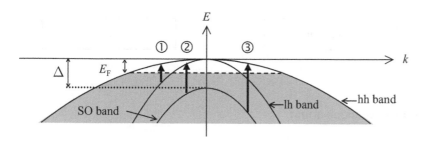

Fig. **7.8** A free carrier transition in a doped semiconductor.

Fig. **7.9** Intervalence band absorption in a p-type semiconductor. E_F is the Fermi energy determined by the doping density. The labelled arrows indicate: (1) transitions from the light-hole (lh) band to the heavy-hole (hh) band; (2) transitions from the split-off (SO) band to the lh band; and (3) transitions from the SO band to the hh band.

electrons from their dopants. It is a sweeping oversimplification to characterize all the possible scattering processes with a single frequency-independent scattering time τ deduced from the DC conductivity. Thus it is hardly surprising that the experimental data do not exactly show an ω^{-2} dependence.

The free carrier reflectivity and absorption of p-type semiconductors can be modelled by a similar treatment to the one developed here for n-type samples. The only change that has to be made is in the effective mass that is used in the calculation. Thus we would expect that all the main results will hold, provided we take account of the fact that the scattering time for holes is not necessarily the same as that for electrons. However, p-type samples also show another effect, which is discussed next.

Figure 7.9 shows the valence band of a p-type III–V semiconductor. This is a larger scale version of the band structure diagram given previously in Fig. 3.5, except that now there are unfilled states near $k = 0$ owing to the p-type doping. Optical transitions can take place in which an electron is promoted from an occupied state below E_F in the light-hole (lh) band to an empty one in the heavy-hole (hh) band above E_F. This is called **intervalence band absorption**. Other intervalence band transitions are possible in which an electron is promoted from the split-off (SO) band to either the lh or hh band. The range of energies over which these transitions occur can be calculated from the effective masses, the doping density and the spin–orbit energy Δ (see Exercise 7.12). The absorption occurs in the infrared, and measurements of the spectrum can give values for Δ and the ratio of the hole effective masses. The absorption can be a strong process because no scattering events are required

Note that intervalence band transitions are forbidden at $k = 0$ since all the hole bands are derived from p-like atomic states. The atomic character of the bands is less well defined for finite k, and this makes the transitions possible away from the centre of the Brillouin zone.

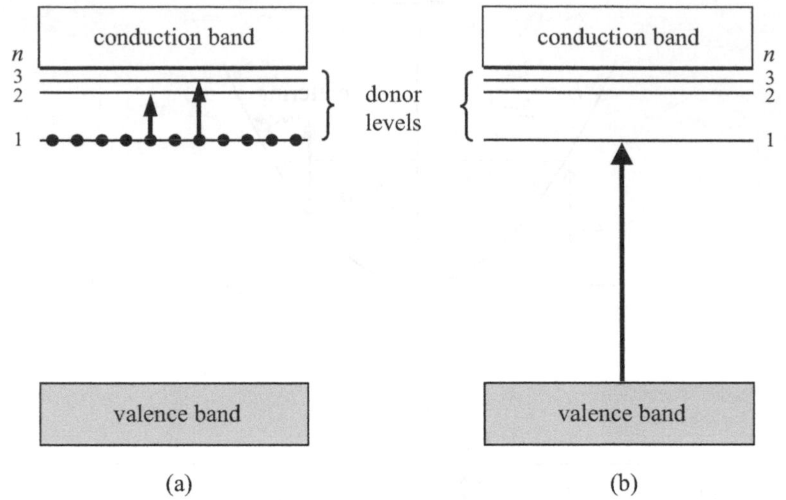

Fig. 7.10 Impurity absorption mechanisms in an n-type semiconductor: (a) transitions between donor levels; (b) transitions from the valence band to empty donor levels. The donor level energy spacings have been exaggerated in this diagram to make the mechanisms clearer.

to conserve momentum.

7.4.2 Impurity absorption

The n-type doping of a semiconductor with donor atoms introduces a series of hydrogenic levels just below the conduction band. These quantized states are called **donor levels**, and are illustrated schematically in Fig. 7.10. The impurity levels give rise to two new absorption mechanisms, in addition to the free carrier effects discussed in the previous section. If the donor states are occupied, it will be possible to absorb photons by exciting electrons between the levels as illustrated in Fig. 7.10(a). On the other hand, if the states are empty, then it will be possible to absorb light by exciting electrons from the valence band to the donor states as illustrated in Fig. 7.10(b).

We consider first the transitions between the donor levels illustrated in Fig. 7.10(a). For such a process to occur, the donor levels must be occupied. This will be the case at low temperatures, when there is insufficient thermal energy to promote the electrons from the donor levels into the conduction band.

The frequencies of the donor-level transitions can be calculated if the energies of the impurity states are known. In the simplest model, we assume that the electron is released into the crystal, and is then attracted back towards the positively charged impurity atom. The electron and the ionized impurity then form a hydrogenic system bound together by their mutual Coulomb attraction. As a first approximation, we can use the Bohr formula, provided we use the effective mass m_e^* instead of the free electron mass m_0, and also include the dielectric constant ϵ_r for the semiconductor. Hence the energy of the donor levels E_n^D will be given by:

$$E_n^D = -\frac{m_e^*}{m_0}\frac{1}{\epsilon_r^2}\frac{R_H}{n^2}, \tag{7.29}$$

Equation 7.29 is very similar to eqn 4.1 for the exciton binding energy except that the electron effective mass appears instead of the reduced electron-hole mass. This is because we are now considering the attraction of an electron to a heavy ion which is bound in the lattice, instead of that between a free electron and a free hole.

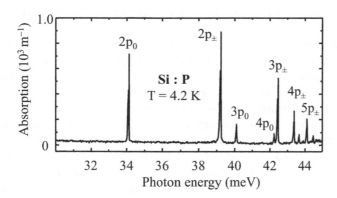

Fig. 7.11 Infrared absorption spectrum of n-type silicon doped with phosphorus at a density of $1.2 \times 10^{20}\,\mathrm{m}^{-3}$. The temperature was 4.2 K. After Jagannath et al. (1981), © American Physical Society, reprinted with permission.

where R_H is the hydrogen Rydberg energy (13.6 eV) and n is an integer.

At low temperatures we can assume that all the electrons from the donors will be in the $n = 1$ ground state impurity level. Optical transitions can then take place in which the electrons are promoted to higher donor levels or into the conduction band by absorption of a photon. Figure 7.10(a) illustrates two possible transitions of this type, in which the electron is promoted to either the $n = 2$ or the $n = 3$ donor level. These transitions give rise to absorption lines analogous to the hydrogen Lyman series with frequencies given by:

$$h\nu = \frac{m_\mathrm{c}^*}{m_0}\frac{R_\mathrm{H}}{\epsilon_\mathrm{r}^2}\left(1 - \frac{1}{n^2}\right), \qquad (7.30)$$

where n is the quantum number of the final impurity level. If we insert typical values into eqn 7.30 we find that the photon energies are in the range 0.01–0.1 eV. This means that the transitions occur in the infrared spectral region.

Figure 7.11 shows the absorption spectrum of n-type silicon at liquid helium temperatures. The sample was doped with phosphorus at a density of $1.2 \times 10^{20}\,\mathrm{m}^{-3}$. The absorption lines correspond to transitions exciting electrons from the $n = 1$ shell to higher shells. In the language of atomic physics, these are $1\mathrm{s} \rightarrow n\mathrm{p}$ transitions. These transitions converge at high n to the donor ionization energy of phosphorus in silicon, which is 45 meV.

The spectrum shown in Fig. 7.11 is actually more complicated than eqn 7.30 would suggest. It consists of two series of transitions, which are labelled as either $n\mathrm{p}_0$ or $n\mathrm{p}_\pm$. The $n\mathrm{p}_0$ series obey eqn 7.30 very well, but the $n\mathrm{p}_\pm$ transitions have a different frequency dependence. This complexity is caused by the anisotropy of the effective mass of silicon. The frequency dependence of the two series can be modelled by assigning different effective Rydberg energies for the '0' and '±' states. (See Exercise 7.13.)

We now consider the absorption mechanism shown in Fig. 7.10(b). These transitions can be observed at temperatures when the donor levels are partly unoccupied owing to the thermal excitation of the electrons into the conduction band. Absorption processes can then occur in which

The absolute value of the absorption coefficient for the impurity transitions in Fig. 7.11 is around $10^3\,\mathrm{m}^{-1}$. This is much smaller than for interband transitions which typically have values in the range 10^6–$10^8\,\mathrm{m}^{-1}$. However, if we were to assume that the absorption strength is simply proportional to the number of atoms that contribute, we would expect the impurity absorption to be weaker than the interband absorption by about a factor of $\sim 10^9$. The measured ratio is much larger because the impurity lines are very sharp, whereas the interband transitions spread out into bands.

electrons are excited from the top of the valence band to the empty donor levels.

The valence band → donor level transitions occur at photon energies just below the band gap E_g, with a threshold given by $E_g - E_1^D$. However, the transitions tend to be broadened into a continuum both by thermal effects and by the fact that transitions can take place from a whole range of states within the valence band. Hence the impurity transitions cause a smearing of the absorption edge compared with the abrupt edge found at the band gap of pure semiconductors. The absorption strength will always be weak compared to the interband and excitonic transitions due to the relatively small number of impurity atoms compared to the density of states within the conduction band. On the other hand, the transitions occur in the spectral region just below the band gap where we would normally expect no absorption at all. Hence these transitions do have an effect on the fundamental absorption edge, and make precise determinations of E_g from the absorption spectra at room temperature more difficult.

We have restricted our attention here to n-type semiconductors for the sake of simplicity. The same effects can of course occur in p-type materials.

In many direct-gap semiconductors it is found experimentally that the absorption decreases exponentially below the band gap. This is called the Urbach tail on account of **Urbach's rule**, which states that the frequency dependence of the absorption for $\hbar\omega < E_g$ is given by:

$$\alpha(\hbar\omega) \propto \exp\left(\frac{\sigma(\hbar\omega - E_g)}{k_B T}\right),$$

where σ is a phenomenological fitting parameter.

7.5 Plasmons

Equation 7.7 tells us that the relative permittivity of a lightly damped gas of free electrons is expected to be zero at ω_p. This suggests that something interesting might happen at this frequency. This is indeed the case, as we discuss here.

7.5.1 Bulk plasmons

A plasma consists of a gas of charged particles in dynamic equilibrium. The particles are in constant motion, and this can create local charge fluctuations. If a fluctuation were to create a small region with an excess charge, the charges in that volume would be repelled away by the surrounding charges. The velocity acquired in this process could cause the excess charges to overshoot their original position, in which case they would then be pushed back in the opposite direction. This process can lead to oscillatory motion called **plasma oscillations**. These plasma oscillations are well known in gas discharge tubes, and they can also occur in the free electron plasmas found in metals and doped semiconductors, which is our interest here.

The frequency of the plasma oscillations can be calculated as follows. Consider a region of a conducting medium of volume V enclosed by a surface S. Conservation of charge requires that the net flow of current into or out of a region must be balanced by the change of the total charge inside the surface. This continuity condition can be written:

$$\oint_S \boldsymbol{j} \cdot \mathbf{d}\boldsymbol{S} = -\frac{\partial}{\partial t} \int_V \rho \, \mathrm{d}V \,, \tag{7.31}$$

where $\boldsymbol{j}$ is the current density, $\mathbf{d}\boldsymbol{S}$ is a surface element, ρ is the local charge density, and $\mathrm{d}V$ is a volume element. On applying the divergence theorem, we find that:

$$\int_V \boldsymbol{\nabla} \cdot \boldsymbol{j} \, \mathrm{d}V = -\int_V \frac{\partial \rho}{\partial t} \, \mathrm{d}V \,, \tag{7.32}$$

and hence (since the volume integrated over is arbitrary):

$$\boldsymbol{\nabla} \cdot \boldsymbol{j} = -\frac{\partial \rho}{\partial t} \,. \tag{7.33}$$

The divergence theorem of mathematics requires that:

$$\oint_S \boldsymbol{j} \cdot \mathbf{d}\boldsymbol{S} = \int_V \boldsymbol{\nabla} \cdot \boldsymbol{j} \, \mathrm{d}V \,,$$

where the volume integral is over the region enclosed by the surface S.

Equation 7.33 is called the **charge continuity equation**.

We now consider the case in which we have a collective motion of the free electrons relative to the fixed lattice of positive ions in a metal or doped semiconductor. The overall charge density is zero, but the movement of the electrons can create local currents. Since the positive charges on the ions are stationary, they do not generate a current, and we can apply the charge continuity equation to the electron current alone, giving

$$\boldsymbol{\nabla} \cdot \boldsymbol{j} = -\frac{\partial \rho_e}{\partial t} \,, \tag{7.34}$$

where ρ_e is the electron charge density. Then, on substituting for ρ_e from Gauss's law (i.e. $\boldsymbol{\nabla} \cdot \boldsymbol{\mathcal{E}} = \rho_e / \epsilon_0$), we find:

$$\boldsymbol{\nabla} \cdot \left(\boldsymbol{j} + \epsilon_0 \frac{\partial \boldsymbol{\mathcal{E}}}{\partial t} \right) = 0 \,. \tag{7.35}$$

Now a vector that has zero divergence can always be written as the curl of another vector, and from Maxwell's fourth equation (eqn A.13), we realize that this vector must be $\boldsymbol{B}/\mu_0$, giving:

$$\boldsymbol{j} + \epsilon_0 \frac{\partial \boldsymbol{\mathcal{E}}}{\partial t} = \frac{1}{\mu_0} \boldsymbol{\nabla} \times \boldsymbol{B} \,. \tag{7.36}$$

On taking the time derivative and substituting from Maxwell's third equation (eqn A.12), we find:

$$\begin{aligned}
\frac{\partial \boldsymbol{j}}{\partial t} + \epsilon_0 \frac{\partial^2 \boldsymbol{\mathcal{E}}}{\partial t^2} &= \frac{1}{\mu_0} \boldsymbol{\nabla} \times \frac{\partial \boldsymbol{B}}{\partial t} \\
&= -\frac{1}{\mu_0} \boldsymbol{\nabla} \times (\boldsymbol{\nabla} \times \boldsymbol{\mathcal{E}}) \,.
\end{aligned} \tag{7.37}$$

The electrons will move in response to the local electric field according to their equation of motion:

$$m\dot{\boldsymbol{v}} = -e\boldsymbol{\mathcal{E}} \,. \tag{7.38}$$

Now the current density is given by $j = -Nev$, which implies that

$$\frac{\partial j}{\partial t} = \frac{Ne^2}{m}\boldsymbol{\mathcal{E}}.$$ (7.39)

On substituting into eqn 7.37 and re-arranging, we find:

$$\frac{\partial^2 \boldsymbol{\mathcal{E}}}{\partial t^2} + \omega_p^2 \boldsymbol{\mathcal{E}} = -c^2 \,\boldsymbol{\nabla} \times (\boldsymbol{\nabla} \times \boldsymbol{\mathcal{E}}),$$ (7.40)

where we have substituted for ω_p from eqn 7.6 and used $c^2 = 1/\mu_0\epsilon_0$ (cf. eqn A.28).

At this stage it is helpful to split the electric field into transverse and longitudinal components:

$$\boldsymbol{\mathcal{E}} = \boldsymbol{\mathcal{E}}_t + \boldsymbol{\mathcal{E}}_l,$$ (7.41)

where $\boldsymbol{\nabla} \cdot \boldsymbol{\mathcal{E}}_t = 0$ and $\boldsymbol{\nabla} \times \boldsymbol{\mathcal{E}}_l = 0$. On substituting into eqn 7.40 and using the vector identity of eqn A.24, this gives:

$$\frac{\partial^2 \boldsymbol{\mathcal{E}}_t}{\partial t^2} + \omega_p^2 \boldsymbol{\mathcal{E}}_t - c^2 \nabla^2 \boldsymbol{\mathcal{E}}_t = -\left(\frac{\partial^2 \boldsymbol{\mathcal{E}}_l}{\partial t^2} + \omega_p^2 \boldsymbol{\mathcal{E}}_l\right).$$ (7.42)

It becomes apparent that both sides of eqn 7.42 must be equal to zero by taking the divergence and curl of the equation.

This implies that we have two independent equations of motion for the transverse and longitudinal components:

$$\frac{\partial^2 \boldsymbol{\mathcal{E}}_t}{\partial t^2} + \omega_p^2 \boldsymbol{\mathcal{E}}_t - c^2 \nabla^2 \boldsymbol{\mathcal{E}}_t = 0,$$ (7.43)

$$\frac{\partial^2 \boldsymbol{\mathcal{E}}_l}{\partial t^2} + \omega_p^2 \boldsymbol{\mathcal{E}}_l = 0.$$ (7.44)

We look for wave-like solutions in which the field varies with time and position as $\exp i(\boldsymbol{k} \cdot \boldsymbol{r} - \omega t)$. For the transverse solutions we find:

$$c^2 k^2 = \omega^2 - \omega_p^2.$$ (7.45)

This describes the dispersion of conventional transverse electromagnetic waves in the plasma. The dispersion is plotted in Fig. 7.16(a) below. There are no travelling solutions with $\omega < \omega_p$ because the waves are reflected by the plasma.

In the case of the longitudinal modes, we simply have:

$$\omega = \omega_p.$$ (7.46)

In a metal, the frequency of the longitudinal modes does in fact vary slightly with k. The correction term is very small, and arises from the breakdown of some of the approximations used in this derivation. See Exercise 7.18.

This shows that the medium can support longitudinal modes at the plasma frequency, and that the modes are dispersionless: i.e. ω is independent of k. These longitudinal modes correspond to the plasma oscillations that we discussed qualitatively at the start of the section. Figure 7.12(a) shows a schematic diagram of the longitudinal electron displacements within a plasma oscillation and the fields that they generate.

The existence of longitudinal solutions at the plasma frequency is a consequence of the fact that $\epsilon_r = 0$ at ω_p. In a medium with zero average

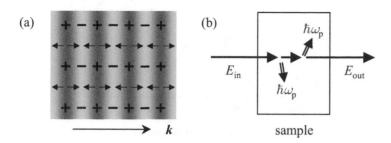

Fig. 7.12 (a) Charge fluctuations in a free carrier plasma oscillation. The lighter regions denote areas with excess electron densities. The small arrows indicate the direction of the electric fields, which are parallel to the direction of propagation of the wave, as indicated by its wave vector k. (b) Excitation of plasmons by inelastic scattering of particles. The case in which two plasmons are excited is shown. For metals, electrons with keV energies are used, but for doped semiconductors, optical frequency photons have sufficient energy.

charge density, Gauss's law (eqn A.10) combined with eqn A.3 tells us that

$$\nabla \cdot D = \nabla \cdot (\epsilon_r \epsilon_0 \mathcal{E}) = 0 \,. \tag{7.47}$$

If $\epsilon_r \neq 0$, we then deduce that $\nabla \cdot \mathcal{E} = 0$. This is the normal situation for transverse electromagnetic waves in which the electric field is perpendicular to the direction of the wave. However, if $\epsilon_r = 0$, we can satisfy eqn 7.47 with waves that have $\nabla \cdot \mathcal{E} \neq 0$, i.e. longitudinal waves. We thus conclude that a dielectric can support longitudinal electric field waves at frequencies that satisfy $\epsilon_r(\omega) = 0$.

We shall come across another example of longitudinal modes at frequencies where $\epsilon_r = 0$ when we consider phonons in Chapter 10. (See Section 10.2.2.)

Equation 7.44 shows us that the longitudinal oscillations of the plasma behave as harmonic oscillators with a natural resonant frequency at ω_p. The derivation is completely classical, and the oscillator can have any energy. However, we know in fact that the energy of harmonic oscillators is quantized. We therefore expect the energy of the plasma oscillations to be quantized in units of $\hbar\omega_p$. The quasi-particles that correspond to these quantized plasma oscillations are called **plasmons**. As shown in eqn 7.46, the frequency of the plasmons is independent of their wave vector.

Since plasmons are associated with longitudinal plasma oscillations, they cannot be excited directly by light, which is a transverse wave. Instead, they have to be observed by techniques of inelastic scattering, in which a beam of particles excites plasmons while passing through the medium, as illustrated in Fig. 7.12(b). The energy E_{in} of the incoming particles must be significantly larger than the plasmon energy. Conservation of energy requires that:

$$E_{out} = E_{in} - n\hbar\omega_p \,, \tag{7.48}$$

where E_{out} is the energy of the outgoing particle, and n is the number of plasmons emitted. The detection of particles with energies given by eqn 7.48 establishes that plasmons have been excited.

In the case of metals, the plasmon energies are several eV, and so electrons with keV energies are typically used. By measuring the energy spectrum of the electrons emerging from a thin sample, the plasma

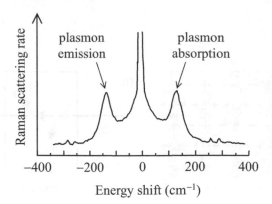

Fig. 7.13 Raman scattering measurements on n-type GaAs at 300 K. The doping density was $1.75 \times 10^{23}\,\text{m}^{-3}$. The data are displayed as a function of the energy shift of the outgoing photons relative to the incoming ones in wave number units. After Mooradian (1972), © Excerpta Medica Inc., reprinted with permission.

Two weak peaks at $\pm 272\,\text{cm}^{-1}$ and $\pm 296\,\text{cm}^{-1}$ are also present in Fig. 7.13. These are caused by optical phonons. (See Section 10.5.2.) The phonon signals are linearly polarized, and have been strongly suppressed in the data by the use of orthogonal polarizers in front of the detector. Note that it is very common to use wave number units in Raman spectroscopy. The wave number $\bar{\nu}$ is equal to the reciprocal of the wavelength: $\bar{\nu} = 1/\lambda$. It is effectively a unit of energy with $1\,\text{cm}^{-1} \equiv 0.124\,\text{meV}$.

frequency can be determined. This technique is called **electron energy-loss spectroscopy**.

Plasmons can also be observed in doped semiconductors. Since the plasma frequencies are much lower, it is possible to use inelastic light scattering techniques (i.e. Raman scattering) to measure the plasmon energies. The general principles of Raman scattering will be discussed in Section 10.5. The basic point is that the energy $\hbar\omega_{\text{out}}$ of the outgoing photon must satisfy:

$$\hbar\omega_{\text{out}} = \hbar\omega_{\text{in}} \pm \hbar\omega_{\text{p}}, \qquad (7.49)$$

where $\hbar\omega_{\text{in}}$ is the energy of the incoming photon. The $+$ sign corresponds to plasmon absorption and the $-$ sign to plasmon emission. Plasmon absorption is possible here, but not in the case of metals, because the plasmon energies are comparable to the thermal energy $k_{\text{B}}T$. This means that there might already be plasmons excited in the sample before the incident photon arrives, and thus there is some probability that a plasmon might be destroyed and its energy transferred to the photon.

Figure 7.13 shows the results of a Raman scattering experiment on n-type GaAs at 300 K. The doping density was $1.75 \times 10^{23}\,\text{m}^{-3}$. The Raman intensity is plotted as a function of the frequency shift of the light in wave number units. The data show two clear peaks shifted by $\pm 130\,\text{cm}^{-1}$ relative to the incoming laser beam due to plasmon emission and absorption. The electron effective mass of GaAs is $0.067m_0$ and ϵ_{opt} is 10.6. Hence from eqn 7.24 we find $\omega_{\text{p}} = 2.8 \times 10^{13}\,\text{rad/s}$, which is equivalent to $150\,\text{cm}^{-1}$. The experimental data are thus in reasonably good agreement with the model.

7.5.2   Surface plasmons

Careful analysis of the electron energy-loss spectra from a metal typically reveals that there are two different types of plasmon within the metal, namely bulk and surface plasmons. We have considered the first type in the previous subsection, and our task now is to explain the second. Interest in these surface plasmons has increased dramatically in recent years, and a new field of research has burgeoned called **plasmonics**.

Surface plasmons are quantized electromagnetic surface waves that are localized at the interface between a plasma and a dielectric material. We are interested here in the case where the plasma is a metal. The dielectric is usually the air, although it might also typically be glass or a semiconductor. The waves propagate along the interface plane, and the electron charge density fluctuations in the metal generate electric field lines as shown in Fig. 7.14. The amplitude of the electric field decays exponentially on either side of the interface.

It is apparent from Fig. 7.14 that the surface plasmons have both transverse and longitudinal electric field components, which contrasts with bulk plasmons which are purely longitudinal. The presence of the transverse component means that surface plasmons can interact directly with photons. This interaction is sufficiently strong that we need to consider the photon and plasmon as a coupled system called a **polariton**. In general, polaritons are coupled electric *polarization*–pho*ton* waves. Several different types of polariton are possible, and the type of polariton that we are considering here is called a **surface plasmon polariton**.

The dispersion of the surface plasmon polaritons can be found by solving Maxwell's equations. We define axes so that the plane $z = 0$ corresponds to the interface, with positive and negative z corresponding to the dielectric and metal respectively, as shown in Fig. 7.15(a). The wave is assumed to be propagating in the x direction. The electric field has components in both the x and z directions, and its amplitude decays exponentially as a function of the distance from the interface, as shown in Fig. 7.15(b). In this geometry, the electric and magnetic fields can be written in the form:

$$
\begin{aligned}
\boldsymbol{\mathcal{E}}^{\mathrm{d}}(x,z,t) &= [\mathcal{E}_x^{\mathrm{d}}, 0, \mathcal{E}_z^{\mathrm{d}}]\, \mathrm{e}^{\mathrm{i}(k_x^{\mathrm{d}}x - \omega t)}\, \mathrm{e}^{-k_z^{\mathrm{d}}z}, \\
\boldsymbol{H}^{\mathrm{d}}(x,z,t) &= [0, H_y^{\mathrm{d}}, 0]\, \mathrm{e}^{\mathrm{i}(k_x^{\mathrm{d}}x - \omega t)}\, \mathrm{e}^{-k_z^{\mathrm{d}}z}, \\
\boldsymbol{\mathcal{E}}^{\mathrm{m}}(x,z,t) &= [\mathcal{E}_x^{\mathrm{m}}, 0, \mathcal{E}_z^{\mathrm{m}}]\, \mathrm{e}^{\mathrm{i}(k_x^{\mathrm{m}}x - \omega t)}\, \mathrm{e}^{+k_z^{\mathrm{m}}z}, \\
\boldsymbol{H}^{\mathrm{m}}(x,z,t) &= [0, H_y^{\mathrm{m}}, 0]\, \mathrm{e}^{\mathrm{i}(k_x^{\mathrm{m}}x - \omega t)}\, \mathrm{e}^{+k_z^{\mathrm{m}}z},
\end{aligned}
\tag{7.50}
$$

where the labels 'd' and 'm' refer to the dielectric and metal respectively. Note that the opposite sign of the decay term in the z direction in the dielectric and the metal ensures that the fields are localized at the interface.

The fields given in eqn 7.50 must satisfy Maxwell's equations and the boundary conditions that apply when there is no net free charge density. Consider first the boundary conditions. The tangential components of $\boldsymbol{\mathcal{E}}$ and $\boldsymbol{H}$, together with the normal component of the electric displacement $\boldsymbol{D}$, must be continuous at the interface. On recalling that $\boldsymbol{D} = \epsilon_{\mathrm{r}}\epsilon_0\boldsymbol{\mathcal{E}}$, we then have that:

$$
\begin{aligned}
\mathcal{E}_x^{\mathrm{d}} &= \mathcal{E}_x^{\mathrm{m}}, \\
H_y^{\mathrm{d}} &= H_y^{\mathrm{m}}, \\
\epsilon_{\mathrm{d}}\mathcal{E}_z^{\mathrm{d}} &= \epsilon_{\mathrm{m}}\mathcal{E}_z^{\mathrm{m}},
\end{aligned}
\tag{7.51}
$$

where ϵ_{d} and ϵ_{m} are the relative permittivities of the dielectric and metal respectively. The requirement that these conditions apply along all the

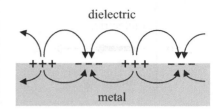

dielectric

metal

Fig. 7.14 Electric fields associated with electron charge density fluctuations at the surface of a metal.

Phonon polaritons are considered in Section 10.3.

In principle, it might also be possible to have waves with electric and magnetic field components along the y and x directions respectively. However, it can be shown that these solutions are not possible. See Maier (2007).

Note that the most interesting case to consider is when $\omega < \omega_{\mathrm{p}}$, where ϵ_{m} is *negative*. $\mathcal{E}_z^{\mathrm{d}}$ and $\mathcal{E}_z^{\mathrm{m}}$ therefore point in opposite directions, as shown in Fig. 7.14.

(a)
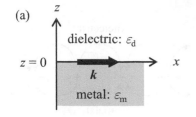

(b)
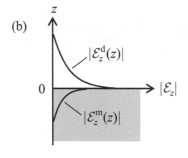

Fig. 7.15 (a) Definition of axes for the interface between a dielectric medium and a metal with relative permittivities of ϵ_d and ϵ_m respectively. The plane $z = 0$ defines the interface, and the polariton propagates in the x direction, as shown by the $\boldsymbol{k}$ vector. (b) Exponential decay of the field amplitudes as a function of the distance from the interface.

surface implies that

$$k_x^d = k_x^m \equiv k_x ,\tag{7.52}$$

where k_x is the common x component of the wave vector on both sides of the interface.

Now consider Maxwell's fourth equation (eqn A.13) with $\boldsymbol{j} = 0$ and $\boldsymbol{D} = \epsilon_r \epsilon_0 \boldsymbol{\mathcal{E}}$. The fact that $\mathcal{E}_y = 0$ and $H_x = 0$ allows us to relate H_y to $\mathcal{E}_x$:

$$-\frac{\partial H_y}{\partial z} = \epsilon_r \epsilon_0 \frac{\partial \mathcal{E}_x}{\partial t} .\tag{7.53}$$

On substituting the fields from eqn 7.50, this gives:

$$\begin{aligned} k_z^d\, H_y^d &= -i\epsilon_d\epsilon_0\omega\, \mathcal{E}_x^d , \\ -k_z^m\, H_y^m &= -i\epsilon_m\epsilon_0\omega\, \mathcal{E}_x^m . \end{aligned}\tag{7.54}$$

By using $\mathcal{E}_x^d = \mathcal{E}_x^m$ and $H_y^d = H_y^m$ from eqn 7.51, we can rearrange this to find:

$$\frac{k_z^d}{\epsilon_d} + \frac{k_z^m}{\epsilon_m} = 0 .\tag{7.55}$$

Note that this is consistent with both k_z^d and k_z^m being positive when ϵ_m is negative, i.e. when $\omega < \omega_p$. Since the overall charge density is zero, the fields must satisfy (cf. eqns A.25 and A.28 with $\mu_r = 1$):

$$\nabla^2 \boldsymbol{\mathcal{E}} = \frac{\epsilon_r}{c^2} \frac{\partial^2 \boldsymbol{\mathcal{E}}}{\partial t^2} .\tag{7.56}$$

On inserting the fields from eqn 7.50 and using eqn 7.52, this gives:

$$\begin{aligned} k_x^2 - (k_z^d)^2 &= \frac{\epsilon_d}{c^2}\, \omega^2 , \\ k_x^2 - (k_z^m)^2 &= \frac{\epsilon_m}{c^2}\, \omega^2 . \end{aligned}\tag{7.57}$$

Then, on using eqn 7.55 to eliminate the k_z components, we finally obtain:

$$k_x = \frac{\omega}{c} \left(\frac{\epsilon_m \epsilon_d}{\epsilon_m + \epsilon_d} \right)^{1/2} = \frac{\omega \sqrt{\epsilon_d}}{c} \left(\frac{\epsilon_m}{\epsilon_m + \epsilon_d} \right)^{1/2} .\tag{7.58}$$

This equation gives the dispersion curve for the surface plasmon polaritons.

Figure 7.16 compares the dispersion of the surface plasmon polaritons to the photon dispersion in the bulk of a metal with a dielectric constant given by eqn 7.7, namely:

$$\epsilon_m = 1 - \frac{\omega_p^2}{\omega^2} .\tag{7.59}$$

The relative permittivity of the dielectric is assumed to be real and independent of frequency.

Consider first the dispersion in the bulk of the metal, which is shown in Fig. 7.16(a). For frequencies below ω_p, the light is reflected, and there are

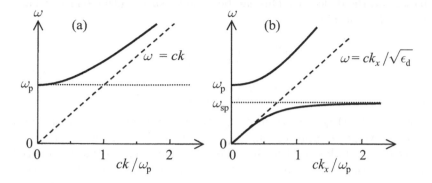

Fig. 7.16 (a) Photon dispersion in the bulk of a metal with a dielectric function given by eqn 7.7. (b) Surface plasmon polariton dispersion for the same metal. The curve is drawn for the case where the dielectric is air (i.e. $\epsilon_d = 1$). The dashed line in (b) shows the dispersion for light in the dielectric.

just evanescent fields in the medium. There are therefore no propagating modes with $\omega < \omega_p$. For $\omega > \omega_p$, the dispersion is given by (see eqn 7.45):

$$\omega = \left(\omega_p^2 + c^2 k^2\right)^{1/2} . \tag{7.60}$$

This asymptotes to $\omega = ck$ at large wave vectors.

The situation for the surface plasmon polaritons is qualitatively different, as shown in Fig. 7.16(b). The 'light line' for the dielectric defined by $\omega = ck_x/\sqrt{\epsilon_d}$ is shown for comparison. Three distinct frequency regions can be identified.

(1) $0 < \omega < \omega_p/\sqrt{1 + \epsilon_d}$. In this frequency region, both ϵ_m and $(\epsilon_m + \epsilon_d)$ are negative, so that k_x is real. For small ω, $|\epsilon_m|$ is large. Therefore the plasmon dispersion curve approaches the light line for small k_x.

(2) $\omega_p/\sqrt{1 + \epsilon_d} < \omega < \omega_p$. In this region ϵ_m is negative, but $(\epsilon_m + \epsilon_d)$ is positive. k_x is therefore imaginary, and there are no propagating modes.

(3) $\omega > \omega_p$. Both ϵ_m and $(\epsilon_m + \epsilon_d)$ are now positive, so that real solutions for k_x are again found. At high frequencies, $\epsilon_m \to 1$, and the dispersion approaches the limit with $\omega = ck_x\sqrt{1 + \epsilon_d}/\sqrt{\epsilon_d}$.

In region (1) at large k_x the group velocity (i.e. $d\omega/dk$) is zero and $\omega \to \omega_{sp}$. This asymptotic frequency limit is called the surface plasmon frequency. Equation 7.58 shows us that $k_x \to \infty$ when $(\epsilon_m + \epsilon_d) \to 0$, and so we can find ω_{sp} by solving $\epsilon_m(\omega) = -\epsilon_d$. For an undamped plasma with $\epsilon_m(\omega)$ given by eqn 7.59, we find:

$$\omega_{sp} = \frac{\omega_p}{\sqrt{1 + \epsilon_d}} . \tag{7.61}$$

In electron energy-loss experiments on metals, it is common to observe peaks corresponding to both bulk and surface plasmons. This gives a convenient method for measuring both ω_p and ω_{sp}.

Note that $\omega_{sp} = \omega_p/\sqrt{2}$ when the dielectric is air.

The behaviour of the surface plasmon polaritons in region (1) where $\omega < \omega_{sp}$ is the most interesting, since this corresponds to propagating modes for frequencies below the plasma frequency. The spatial extent of the fields in the z direction can be found from the values of k_z in the metal and the dielectric. This can be done by substituting eqn 7.58 back into 7.57 to obtain:

$$k_z^{\mathrm{d}} = \frac{\omega}{c} \left(\frac{-\epsilon_{\mathrm{d}}^2}{\epsilon_{\mathrm{m}} + \epsilon_{\mathrm{d}}} \right)^{1/2},$$

$$k_z^{\mathrm{m}} = \frac{\omega}{c} \left(\frac{-\epsilon_{\mathrm{m}}^2}{\epsilon_{\mathrm{m}} + \epsilon_{\mathrm{d}}} \right)^{1/2}. \tag{7.62}$$

Since $(\epsilon_{\mathrm{m}} + \epsilon_{\mathrm{d}})$ goes to zero at ω_{sp}, the decay constants increase with ω and diverge on approaching ω_{sp}. The field decay length l_z is equal to $1/|k_z|$, and this implies that the polaritons become more localized as ω approaches ω_{sp}. Note that it is usually the case that $|k_z^{\mathrm{m}}| > |k_z^{\mathrm{d}}|$, and hence that the plasmon extends further in the dielectric than in the metal, as illustrated in Fig. 7.15. At optical frequencies, l_z^{d} and l_z^{m} are typically a few hundred or few tens of nanometres, respectively. (See Exercise 7.20.)

Two other branches of plasmonics deal with the enhancement of radiative efficiencies and the increased transmission of light through sub-wavelength apertures. In the first case, the radiative efficiency of an atom in a dielectric can be enhanced by placing it close to a metal surface, and hence exploiting the large electric field amplitude that is present at the interface. In the second case, it has been demonstrated that plasmonic effects can enhance the transmission of light through periodic arrays of sub-wavelength-size holes in an optically thick metallic film. Details of these and other applications may be found in the Further Reading.

One of the aims of the research field of plasmonics is to propagate electromagnetic waves in a metal as surface plasmon polaritons. With values of l_z^{m} being in the sub-100 nm range, the waves can be confined to dimensions smaller than the wavelength of light, and we then enter the realm of **nanophotonics**, which is not accessible to conventional optics due to diffraction limits. The distance that the plasmon modes can propagate along the surface is determined by the imaginary part of ϵ_{m}. This can be several tens of micrometres (see Exercise 7.20), which is more than adequate for the applications that are being considered.

An issue that has to be addressed in plasmonics is the way to couple light to the polaritons. It is apparent from Fig. 7.16(b) that the polariton modes always lie below the light line. This means that polaritons can never transform directly into light: they are non-radiative modes. By the same token, it is not possible to couple light directly from the dielectric into the polariton modes, since it is never possible to match the wave vectors. Therefore, in order to couple external light into the surface plasmon polaritons, techniques must be used to change the wave vector of the light. One way to do this is to use a grating. In fact, it has been known for a long time that the reflectivity of metallic ruled gratings can drop significantly when one of the diffracted orders propagates parallel to the surface. This effect, which is called **Wood's anomaly**, is now known to be caused by the excitation of surface plasmon polaritons in the metal. Details of how the coupling to polariton modes is achieved in practice may be found in the references cited for Further Reading.

Colloids can be solids, liquids, or gases. The particle size in a colloid should be smaller than the wavelength of light. An interesting application of metal colloids is in the production of stained glass. The colouration of medieval stained glass is typically caused by gold, silver or copper nanoparticles incorporated into the glass during the melt process.

A striking example of surface plasmon effects is to be found in considering the optical properties of metal colloids. Metal colloids are made by dispersing a large number of very small metallic particles (i.e. 'nanoparticles') throughout a homogeneous medium such as glass or water. It has been known from antiquity that the colour of metal colloids is different from that of the bulk metal, and this phenomenon is now known to be caused by the resonant excitation of surface plasmons in the metal nanoparticles. However, in contrast to surface plasmon polaritons, the surface fields in the metallic nanoparticles do not propagate, since they

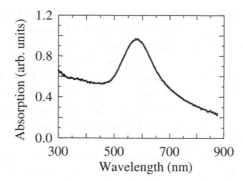

Fig. 7.17 Absorption spectrum for a thin film of gold nanoparticles embedded within an organic dielectric with $\epsilon_d \approx 2.5$. The film was grown on a glass substrate, and the spectrum was measured at room temperature. The particle size was 6–7 nm. Unpublished data from M.R. Sugden and T.R. Richardson.

are confined to the nanoparticles. They are therefore known as **localized surface plasmons**.

A simple explanation for the change of the optical properties of colloidal metals compared to the bulk can be given in terms of the polarizability. If the particle radius a is much smaller than the wavelength of the light, then it can shown that the polarizability is given by:

$$\alpha = 4\pi a^3 \frac{\epsilon_m - \epsilon_d}{\epsilon_m + 2\epsilon_d}, \qquad (7.63)$$

which has a resonance when $\epsilon_m = -2\epsilon_d$. In an undamped plasma, the resonance occurs at $\omega_p/\sqrt{3}$ when the dielectric is air (see Exercise 7.21), and is independent of the particle size. However, the resonance in real metals is shifted by interband absorption, and does depend somewhat on the size. Figure 7.17 shows the absorption spectrum of a thin film of gold nanoparticles embedded within an organic dielectric with $\epsilon_d \approx 2.5$. A strong plasmonic absorption peak centred at 580 nm (2.1 eV) is clearly resolved in the data. When the nanoparticles are suspended in water, the resonance occurs at higher frequencies owing to the lower permittivity of the dielectric. (See Exercise 7.22.) The absorption resonance is in fact in the green spectral region, and the colloid appears red, instead of the usual golden colour of the bulk metal. Similar effects can be observed for other metals.

The derivation of eqn 7.63, which has a similar form to the Clausius–Mossotti relationship given in eqn 2.35, may be found, for example, in Maier (2007). The general treatment of the interaction of light with conducting metal spheres is called **Mie theory**: see Born and Wolf (1999).

7.6   Negative refraction

Throughout this chapter we have been studying the optical properties of materials that have negative values of ϵ_r. We now wish to consider briefly the properties of materials that also have negative values of the relative magnetic permeability, μ_r. As we shall see, this possibility leads to the striking concept of a *negative* refractive index, which is a subject that has attracted much interest in recent years.

The general relationship between the refractive index of a medium and μ_r follows directly from Maxwell's equations. Equation A.29 shows that the speed of light in a medium is equal to $c/\sqrt{\epsilon_r \mu_r}$, and so we can write:

$$\tilde{n} = \sqrt{\epsilon_r \mu_r}. \qquad (7.64)$$

In everything that we have been considering so far in this book, we have been assuming that $\mu_r = 1$, so that eqn 7.64 reduces to $\tilde{n} = \sqrt{\epsilon_r}$. In fact, $(\mu_r - 1)$ characterizes the magnetic response of the medium, and magnetic dipoles respond to oscillating electromagnetic fields in much the same way that electric dipoles do. However, the natural resonant frequencies are low ($\sim$GHz at most), and so the magnetic response is usually negligible at optical frequencies. It is therefore important to clarify at the start that there are no known natural materials that have μ_r significantly different from unity at optical frequencies. The materials that we shall be considering here are therefore purely artificial.

The four possible general combinations of values of ϵ_r and μ_r are depicted schematically in Fig. 7.18. In quadrant I, both ϵ_r and μ_r are positive. This is the usual situation in a transparent optical medium, and the solutions to Maxwell's equations are travelling waves with a phase velocity determined by a refractive index n that is both real and positive. In quadrant II, μ_r is positive, but ϵ_r is negative. This is the scenario that we have considered at length in this chapter, since it applies to the case of a metal below its plasma frequency. In these conditions, $\tilde{n}$ is purely imaginary. This means that the waves decay exponentially in the medium: there are no propagating solutions, and incoming waves from the air are reflected. A similar situation would occur in quadrant IV, where ϵ_r is positive, but μ_r is negative. The final case to consider is that which corresponds to quadrant III in which *both* ϵ_r and μ_r are negative. The properties of materials that fall into quadrant III were first considered theoretically by Veselago in 1968.

Veselago's main conclusion was that a medium with both ϵ_r and μ_r negative would be transparent, and have a *negative* refractive index. Not surprisingly, this gives rise to many unusual properties. The $\boldsymbol{k}$ vector of the wave and the direction of energy flow are in opposite directions. Since the energy flow is determined by the Poynting vector $\boldsymbol{\mathcal{E}} \times \boldsymbol{H}$, the reversal of the direction of energy flow relative to $\boldsymbol{k}$ is consistent with reversing the direction of the magnetic field. This means that $\boldsymbol{\mathcal{E}}$, $\boldsymbol{H}$ and $\boldsymbol{k}$ now form a left-handed system instead of the usual right-handed arrangement, and so materials with $n < 0$ are sometimes called 'left handed'.

Even more striking effects occur when light is refracted on entering the negative index medium. If the incoming ray has an angle of incidence θ_i, then the angle of refraction θ_r is given by Snell's law:

$$\frac{\sin \theta_i}{\sin \theta_r} = n. \tag{7.65}$$

When n is negative, θ_i and θ_r have opposite signs, as shown in Fig. 7.19(a). This leads to the possibility that the medium can behave as a lens, as shown in Fig. 7.19(b). An object to the left of the medium is brought to a focus to the right, after passing through an intermediate focus. The trajectory of the rays only depends on the thickness of the medium, and there are therefore no aberrations. Hence the medium is said to behave as a *perfect* lens. Since the rays emerge exactly as they would from the

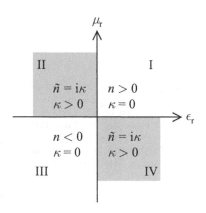

Fig. 7.18 Real and imaginary parts of the complex refractive index for four different combinations of values of ϵ_r and μ_r. The negative refractive index regime occurs in quadrant III, where both ϵ_r and μ_r are negative.

source, the medium is invisible.

Interesting though these properties might be, there would be no point considering them if it were not possible to obtain a medium that has $n < 0$. Metals such as silver or gold have negative values of ϵ_r at most frequencies, and so the issue becomes that of obtaining the negative value of μ_r. As mentioned above, there are no known natural materials that possess this property at useful frequencies. Hence the negative refractive index must be engineered by creating an artificial structure called a **metamaterial**. It was not until the mid-1990s that this subject became of widespread interest, following work by Pendry in which the first practical designs for metamaterials with a negative refractive index were proposed.

The principle behind a metamaterial is to create a metallic structure designed to behave like a magnetic dipole resonator. Figure 7.19(c) illustrates one of the standard designs considered in the literature, namely an array of split rings. In this case the magnetic response is determined by the design of the constituent units (i.e. the split rings), which must be smaller than the wavelength of the electromagnetic waves, but are still much larger than the underlying atoms. It is for this reason that the medium is called a *meta*material.

We have seen in Chapter 2 that the refractive response of an electric-dipole resonator is negative above the natural frequency ω_0. (See, for example, Fig. 2.6.) Magnetic resonators behave in a similar way, and so we can expect to obtain $\mu_r < 0$ in the frequency region above their resonant frequency. The difficulty is that the value of ω_0 depends on the size of the structure. A split-ring array with a period of $\sim 10\,\text{mm}$ has negative refraction at GHz frequencies, but much smaller structures must be used for optical frequency experiments. For this reason, the underlying principles of negative index materials are usually tested first at microwave frequencies. Further details of the design of metamaterials and progress in obtaining negative refraction at optical frequencies may be found in the works cited for Further Reading.

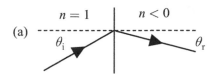

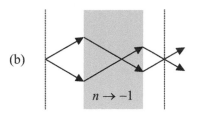

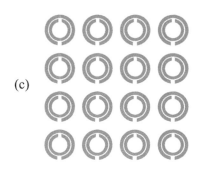

Fig. 7.19 (a) Negative refraction in a medium with $n < 0$. (b) Perfect lensing for $n < 0$. (c) Split ring design for producing a negative refractive index. In (a) and (b) the arrows give the direction of the Poynting vector.

Chapter summary

- Free electron effects are observed in metals and doped semiconductors. They can be modelled by the classical dipole oscillator model with no restoring force term. This approach is called the Drude–Lorentz model.

- The free electron plasma reflects strongly up to the plasma frequency, which depends on the electron density. The damping rate of the oscillations is determined by the momentum scattering time deduced from electrical conductivity measurements.

- Metals reflect strongly due to the plasma reflectivity effect. At frequencies above the plasma frequency, the metals become transparent. This effect is called the ultraviolet transparency of metals.

- Interband transitions are possible in metals from states below the Fermi energy to empty levels above it. The interband absorption can reduce the reflectivity from the value predicted by the Drude–Lorentz model, and must therefore be considered to obtain a good fit to experimental reflectivity data.

- Doped semiconductors reflect at frequencies in the infrared due to the free electrons and holes generated by the doping process. Free carrier absorption can be observed at frequencies above the plasma frequency but below the fundamental absorption edge at the band gap.

- P-type semiconductors show an additional absorption mechanism in the infrared due to intervalence band transitions.

- Doped semiconductors show sharp infrared absorption lines due to impurity transitions at low temperatures. At room temperature, the impurity states broaden the fundamental absorption edge.

- Plasma oscillations can occur at the plasma frequency. The quantized oscillations are called plasmons. These can be observed by electron energy-loss spectroscopy in metals, or by Raman scattering in doped semiconductors.

- Surface plasmons correspond to localized electromagnetic fields at the interface between a metal and a dielectric. Propagating surface plasmon polariton modes can be excited by coupling light to the metal with a grating or prism.

- Materials in which both the relative permittivity and magnetic permeability are negative are characterized by a negative refractive index. The artificial structures that show these effects are called metamaterials.

Further reading

The properties of electromagnetic waves in a conducting medium are covered in many electromagnetism and optics textbooks, for example Bleaney and Bleaney (1976), Born and Wolf (1999), or Hecht (2001).

The free carrier model of metals is covered in Singleton (2001). It is also covered in Ashcroft and Mermin (1976), Burns (1985), or Kittel (2005).

Free carrier reflectivity and absorption in semiconductors has been reviewed by Pidgeon (1980), and is also covered by Yu and Cardona (1996). Yu and Cardona give further details about intervalence band and impurity absorption.

The properties of plasmons are treated in depth in Maier (2007). Bulk plasmons are covered in Kittel (2005), while the classic text on surface plasmons is Raether (1988). Review articles on the research fields of plasmonics and nanophotonics may be found in Barnes et al. (2003), Lal et al. (2007), Maier and Atwater (2005), Murray and Barnes (2007), or Ebbeson et al. (2008).

The concept of negative refraction was proposed in Veselago (1968). Introductory reviews on the subject may be found in Pendry (2004) or Pendry & Smith (2004).

A more detailed review is given in Ramakrishna (2005). Progress on obtaining negative refraction at optical frequencies is reviewed in Shalaev (2007).

Exercises

(7.1) Derive a relationship between the Fermi energy E_F of a metal and its plasma frequency ω_p.

(7.2) The ionosphere reflects radio waves with frequencies up to about 3 MHz, but transmits waves with higher frequencies. Estimate the free electron density in the ionosphere.

(7.3) Estimate the skin depth of radio waves of frequency 200 kHz in sea water, which has an average electrical conductivity of about $4\,\Omega^{-1}\mathrm{m}^{-1}$. Hence discuss the difficulties you might encounter when attempting to communicate with a submerged submarine using radio waves.

(7.4) Cesium metal is found to be transparent to electromagnetic radiation of wavelengths below 440 nm. Calculate a value for the electron effective mass using the data given in Table 7.1.

(7.5) The momentum scattering time of silver is 4.0×10^{-14} s at room temperature. Calculate the dielectric constant at 500 nm, neglecting interband absorption effects. Hence estimate the reflectivity of a silver mirror at this wavelength. See Table 7.1 for the plasma frequency of silver.

(7.6) Estimate the fraction of light with wavelength 1 µm that is transmitted through a 20 nm thick gold film at 77 K, where the DC electrical conductivity is $2 \times 10^8\,\Omega^{-1}\mathrm{m}^{-1}$. The plasma frequency and electron density of gold are given in Table 7.1.

(7.7) Figure 7.20 shows the measured reflectivity of gold in the wavelength range 100–1000 nm. Account qualitatively for the shape of the spectrum, and deduce the energy gap between the d bands and the Fermi energy. Use the data to explain the characteristic colour of gold.

(7.8) What is the value of the dielectric constant of a medium that has zero reflectivity? Use eqn 7.22 to show that the reflectivity of a lightly damped doped semiconductor is zero at the angular frequency given in eqn 7.25.

(7.9) Use the data shown in Fig. 7.7 to deduce the value of the electron effective mass of InSb at each carrier density. Take $\epsilon_{\mathrm{opt}} = 15.6$.

(7.10) The absorption coefficient at room temperature of an n-type sample of InAs with a doping level of $1.4 \times 10^{23}\,\mathrm{m}^{-3}$ is found to be $500\,\mathrm{m}^{-1}$ at 10 µm. Estimate the momentum scattering time, given that the electron effective mass is $0.023\,m_0$ and the refractive index is 3.5.

(7.11)* A laser beam operating at 632.8 nm with an intensity of $10^6\,\mathrm{W\,m}^{-2}$ is incident on a sample of pure InP at room temperature. The absorption coefficient at this wavelength is $6 \times 10^6\,\mathrm{m}^{-1}$, and the carrier lifetime is 1 ns. Estimate the free carrier absorption coefficient at the wavelength of a CO_2 laser (10.6 µm), where the refractive index is 3.1. The effective mass and momentum scattering time for the electrons are $0.08\,m_0$ and 2×10^{-13} s, while the equivalent values for the holes are $0.6\,m_0$ and 5×10^{-14} s.

(7.12)* Consider the intervalence band processes illustrated in Fig. 7.9 for a heavily doped p-type sample of GaAs containing $1 \times 10^{25}\,\mathrm{m}^{-3}$ acceptors.

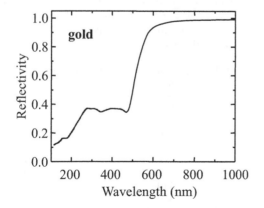

Fig. 7.20 Reflectivity of gold in the wavelength range 100–1000 nm. Data from Lide (1996).

*Exercises marked with an asterisk are more difficult.

The valence band parameters for GaAs are given in Table D.2.

(a) Work out the Fermi energy in the valence band on the assumption that the holes are degenerate. What are the wave vectors of the heavy and light holes at the Fermi energy?

(b) Calculate the upper and lower limits of the photon energies for the three absorption processes labelled (1), (2), and (3) in Fig. 7.9, namely the lh → hh, the SO → lh and the SO → hh transitions.

(7.13) Figure 7.11 shows the infrared absorption spectrum of n-type silicon, which has a dielectric constant of 16. Two series of lines labelled np_0 and $np_\pm$ are identified in the data.

(a) Show that the np_0 series is consistent with eqn 7.30, and deduce a value for the electron effective mass for these transitions.

(b) Show that the $np_\pm$ series follows the following formula:

$$h\nu = \frac{R_0^*}{1^2} - \frac{R_\pm^*}{n^2},$$

stating the values of R_0^* and $R_\pm^*$ deduced from the data.

(7.14) It is found that the infrared absorption spectrum of a lightly doped n-type semiconductor with a relative permittivity of 15.2 consists of a series of sharp lines at low temperatures. The energies of the lines are given by:

$$E(n) = R^*(1 - 1/n^2)$$

where R^* is 2.1 meV and n is an integer greater than 1. Explain why the energies of the absorption lines are almost independent of the type of impurity atoms used for the doping, and deduce a value for the electron effective mass.

(7.15) The fundamental absorption edge of a semiconductor shifts from 5.26 μm to 5.44 μm when doped with acceptors. Deduce a value for the ground state acceptor level energy relative to the valence band.

(7.16) The beam from an argon ion laser operating at 514.5 nm is incident on an n-type GaAs sample. A peak is observed in the intensity of the scattered light at 534.3 nm. Explain this observation, and estimate the electron density, given that $m_e^* = 0.067\, m_0$ and $n = 3.3$.

(7.17) Calculate the doping density at which the plasmons in n-type GaAs have the same wave number as the longitudinal optic phonon at 297 cm^{-1}. Take $m_e^* = 0.067\, m_0$ and $n = 3.3$.

(7.18) In a metal, the frequency at which $\epsilon_r = 0$ varies slightly with the wave vector $k \equiv |\mathbf{k}|$, and for small k we have:

$$\omega(k) \approx \omega_p \left(1 + \frac{3v_F^2 k^2}{10\omega_p^2}\right),$$

where v_F is the Fermi velocity of the electrons. Consider a metal with electron density $N = 10^{29}\,\mathrm{m}^{-3}$ and lattice constant $a = 4\,\text{Å}$. Calculate the relative size of the term in k^2 for $k = 0.1\,\pi/a$, i.e. 10% of the size of the Brillouin zone.

(7.19) In an electron energy-loss experiment on a metal, two series of peaks are observed that obey eqn 7.48 with $\hbar\omega_p = 10.3$ eV and 15.3 eV. Use the data in Table 7.1 to determine which metal is being investigated, and account for the two series of peaks.

(7.20) A surface plasmon mode with a frequency corresponding to a vacuum wavelength of 600 nm is excited at the interface between silver and air. The relative permittivity of silver at this frequency is given approximately by $\epsilon_m = -18 + i$.

(a) Calculate the decay constants k_z^d and k_z^m, and hence deduce the field decay lengths in the direction normal to the surface in both the air and in the metal.

(b) The propagation length in the direction parallel to the surface is defined as the distance over which the intensity drops by a factor of e^{-1}. Calculate the imaginary part of k_x, and hence deduce the propagation length for silver at 600 nm.

(7.21) Show that the resonance of the polarizability of colloidal metal nanoparticles that obey eqn 7.7 occurs at an angular frequency given by $\omega_p/\sqrt{1 + 2n^2}$ in a medium with a refractive index of n. Evaluate this frequency for the case where the dielectric is air.

Table 7.3 Complex relative permittivity of gold between 500 and 550 nm. Adapted from Raether (1988).

Wavelength (nm)	ϵ_1	ϵ_2
500	−2.3	3.6
510	−3.0	3.1
520	−3.7	2.7
530	−4.4	2.4
540	−5.2	2.2
550	−6.0	2.0

(7.22) (a) The complex relative permittivity of gold in the range 500–550 nm is given in Table 7.3. Use this information to estimate the resonance wavelength of colloidal gold nanoparticles in water, which has a refractive index of 1.33.

(b) Use the data for gold given in Table 7.1 to compare the result obtained in part (a) with the prediction of the previous exercise. Account for any difference.

8

Molecular materials

8.1 Introduction to organic materials	214
8.2 Optical spectra of molecules	216
8.3 Conjugated molecules	227
8.4 Organic opto-electronics	232
8.5 Carbon nanostructures	235
Chapter summary	243
Further reading	244
Exercises	245

In this chapter we consider the optical properties of electronic materials based on carbon–carbon bonds. In principle, this covers a huge range of compounds, and it is therefore necessary to restrict our attention to those that have the most interesting optical properties.

The bulk of the chapter is concerned with organic opto-electronic materials. These have grown considerably in importance since the 1990s following the development of organic light-emitting diodes and photovoltaic devices. The final part of the chapter gives a brief survey of the optical properties of carbon nanostructures. This is another subject that has developed rapidly in recent years, mainly as a result of the wider availability of high-quality carbon nanotube and single layer graphene samples.

8.1 Introduction to organic materials

The chemistry of organic molecules is based on the covalent bonds between the carbon atoms. Carbon has an electronic configuration of $1s^2\,2s^2\,2p^2$, with four valence electrons in the $n = 2$ atomic shell. In the diamond structure, each carbon atom forms four single covalent bonds with its nearest neighbour. In organic compounds, by contrast, there may be single, double, or triple bonds between adjacent carbon atoms. In molecules with double or triple bonds, the valence electrons are divided between σ and π bonds. It is easiest to see how this works by means of specific examples.

Consider the ethylene ($H_2C=CH_2$) molecule shown in Fig. 8.1(a). Each carbon atom is bonded to two hydrogen atoms, and has a double bond with the other carbon atom. The two 2s electrons hybridize with one of the 2p electrons to form three sp^2 bonds. These are the σ bonds and are arranged in a plane at an angle of about 120° to each other. The other 2p electron forms a π orbital derived from the $2p_z$ atomic orbital, with wave function lobes above and below the plane defined by the nuclei of the carbon and hydrogen atoms. The electrons in these π orbitals are called **π electrons**. The overlap of the π orbitals produces the second bond between the two carbon atoms.

Consider now the benzene molecule (C_6H_6) shown in Fig. 8.1(b). This is also a planar molecule, with the six carbon atoms arranged as a hexagon. Each carbon atom forms σ bonds with one hydrogen atom and its two adjacent carbon atoms. The π electrons now form a ring orbital above and below the plane of the hexagon. The chemical structure of

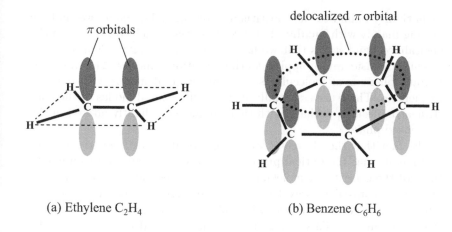

(a) Ethylene C$_2$H$_4$

(b) Benzene C$_6$H$_6$

Fig. 8.1 (a) The ethylene molecule (C$_2$H$_4$). The carbon and hydrogen atoms lie within a plane defined by the σ bonds shown by the thick black lines. The π orbitals lie above and below this plane. (b) The benzene molecule (C$_6$H$_6$). The π electrons form a delocalized ring orbital above and below the plane of the hexagon defined by the six carbon atoms.

benzene is traditionally drawn as a hexagon with alternating double and single bonds between the carbon atoms. (See for example Fig. 8.12(a).) In reality, however, the π electrons are shared equally between the two bonds on either side of each carbon atom. Organic molecules like benzene with alternating multiple and single bonds are said to be **conjugated**.

The π electrons in conjugated compounds are able to spread out in large delocalized orbitals. For example, the formation of the ring orbital in benzene allows the π electrons to spread out much more than the electrons in the σ bonds. On applying the concepts of quantum confinement developed in Section 6.1, we expect that the spreading of the π electron wave functions will lead to a reduction in the energy. This is indeed the case, as is demonstrated by the fact that the lowest electronic transition of ethylene occurs at around 6.9 eV, whereas the equivalent transition in benzene occurs at 4.6 eV. If we use larger molecules with more delocalized π electrons, we can reduce the confinement energy further, and push the transition energies down into the visible spectral region. This is why the transitions of the π electrons in conjugated molecules are the focus of interest in this chapter.

The benzene ring is an example of a **cyclic conjugated** molecule. The name follows from the fact that the electron wave functions must have cyclic periodicity around the closed ring. There are many other cyclic conjugated molecules that can be formed, and the optical properties of a few of these will be considered in Section 8.3.1. It is also possible to form **linear conjugated** molecules, in which the π electrons delocalize along a chain rather than into a ring. The conjugated polymers that are considered in Section 8.3.2 are good examples of this type of conjugation.

Solids based on conjugated molecules are formed by condensation of neutral organic compounds, and are held together by van der Waals interactions. These interactions are relatively weak compared to those within the molecule itself, which originate from the strong covalent bonds between the atoms. This is exemplified by the low melting point of organic solids and their generally soft structure. A consequence of the relatively weak intermolecule binding is that the electronic states re-

Single crystals of some conjugated organic materials have been prepared, but in most cases the samples are amorphous, having been prepared from the solution as thin films on glass substrates. It is not appropriate to apply traditional solid-state physics based on periodic crystalline structure to these amorphous materials.

main tightly bound to the constituent molecules. Therefore, we shall be dealing mainly with **localized** electronic states, which contrast with the delocalized band states that we have been considering in Chapters 3–7. A similar consideration applies to the vibrational modes. The most important phonons of molecular solids are localized modes with discrete frequencies. These are essentially just the vibrational modes of the constituent molecules, perhaps with their frequencies slightly altered in the condensed phase.

The fact that the electronic and vibrational states tend to be localized means that the optical properties of the solids are quite similar to those of the constituent molecules. In many cases the solid state merely provides a convenient way to incorporate large densities of molecules into an opto-electronic device, without necessarily introducing substantially new physics. This makes it clear that we need to understand the optical properties of isolated molecules first before we can properly understand the properties of molecular solids. The next section therefore gives a review of the electronic states and optical transitions of simple molecules.

An important aspect of molecular spectra is the strong coupling between the electronic and vibrational states, which means that the optical transitions are **vibronic** in character. This is equally true for isolated molecules and for the solids that we consider here. The basic principles of vibronic transitions in simple isolated molecules are discussed in Sections 8.2.2–8.2.4 below. This will provide us with a good basis for understanding the physics of molecular solids, and will also serve as a useful introduction for the other types of vibronic systems that are considered in Chapter 9.

8.2 Optical spectra of molecules

The optical properties of molecules are generally divided into three spectral regions:

- The far-infrared spectra: wavelength $\lambda > 100\,\mu m$.
- The infrared spectra: $\lambda \sim 1 - 100\,\mu m$.
- The visible and ultraviolet spectra: $\lambda < 1\,\mu m$.

These three spectral regions correspond respectively to transitions between the rotational, vibrational, and electronic states of the molecule. In this chapter we are concerned only with the visible/ultraviolet spectra, and so we restrict our attention to electronic transitions, with emphasis on conjugated molecules.

8.2.1 Electronic states and transitions

In order to understand the electronic spectra of molecules, we must first learn the terminology of the electronic states and transitions. Molecular electronic states can be arranged in order of increasing energy in much the same way as for atoms. Figure 8.2 gives a schematic diagram of how this looks for a typical conjugated molecule. The electrons fill up the molecular orbitals until they are all accounted for. The highest filled energy level is called the **HOMO** level (highest occupied molecular orbital). In the case of a conjugated molecule, this will be a π orbital, because the electrons in the σ bonds are very tightly bound. The first energy level above the HOMO state is called the **LUMO** level (lowest unoccupied molecular orbital). This will be an excited configuration of the π orbitals, and is labelled a π^* state. The lowest energy transition therefore involves the promotion of a π electron to a π^* state, and is thus labelled a $\pi \rightarrow \pi^*$ transition. Transitions involving the σ states occur at much higher energies, because it takes a large amount of energy to break a σ bond.

The electrons in the ground state of a molecule are all paired off in bonds with their spins anti-parallel. This means that the ground state HOMO level has a spin quantum number S equal to 0. The excited states, however, can either have $S = 0$ or $S = 1$. This is because the excitation process puts an unpaired electron in the excited state and leaves an unpaired electron in the HOMO state. According to the rules of addition of angular momenta, the two spin 1/2 electrons can combine to give a total spin of either 0 or 1. (See Appendix C.) This point is illustrated in Fig. 8.3. The **multiplicity** of the spin states is equal to $(2S+1)$, because there are $(2S+1)$ degenerate M_S levels. Hence the $S = 0$ states are known as **singlets**, while the $S = 1$ states are called **triplets**. Triplets tend to have lower energies than their singlet counterparts.

The separation of the electronic levels into singlet and triplet states has very important consequences for the optical spectra. Each molecule will have a series of singlet excited states labelled S$_1$, S$_2$, S$_3$, ..., in addition to its singlet ground state which is labelled S$_0$. There will be a similar series of triplet excited states labelled T$_1$, T$_2$, T$_3$, ... Since photons carry no spin, they can only excite transitions between electronic states of the same spin. Therefore, transitions from the S$_0$ ground state to the triplet excited state are not allowed. The main optical absorption edge therefore corresponds to the S$_0 \rightarrow$ S$_1$ singlet–singlet transition. The emission spectrum is likewise dominated by the S$_1 \rightarrow$ S$_0$ transition.

The singlet excited states have short lifetimes of order 1–10 ns due to the dipole-allowed transitions to the S$_0$ ground state. The lowest triplet state, on the other hand, has a long radiative lifetime because of the low probability for the T$_1 \rightarrow$ S$_0$ transition. The different time scales for the singlet–singlet and triplet–singlet transitions are conveniently distinguished by describing the emission processes as **fluorescence** and **phosphorescence**, respectively. As mentioned in Section B.3, this distinction is based on whether the emission is fast or slow, with the dividing line drawn at around 10^{-7} to 10^{-8} s in molecules. A schematic diagram of the two types of emission processes is given in Fig. 8.12(b).

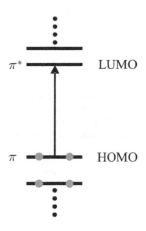

Fig. 8.2 Schematic electronic energy level diagram for a conjugated molecule. The lowest energy transition takes place between the HOMO (highest occupied molecular orbital) and the LUMO (lowest unoccupied molecular orbital) states. This is a $\pi \rightarrow \pi^*$ transition.

We might have expected the probability for triplet $\leftrightarrow$ singlet transitions to be exactly zero due to the spin selection rule. However, spin–orbit coupling can mix a small amount of singlet character into the triplet states and allows some probability for the transitions.

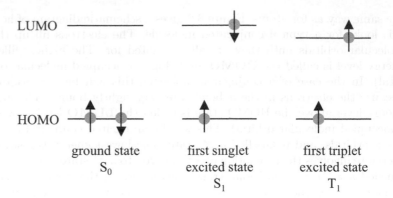

Fig. 8.3 The unpaired electrons of an excited molecule can either have their spins anti-parallel or parallel. The states with anti-parallel spins have spin quantum number $S = 0$ and are called singlets after their multiplicity. The states with parallel spins have $S = 1$ and are called triplets. The ground state is always a singlet.

8.2.2 Vibronic coupling

When a molecule makes an electronic transition, it is possible that it can also change its vibrational state. This leads to the concept of **vibronic** (i.e. *vibr*ational–*electronic*) transitions. The basic physics of the vibrational–electronic transitions can be understood with reference to Fig. 8.4. The diagram shows absorption and emission transitions between two electronic states of a molecule with energies of E_1 and E_2. For simplicity, we assume that the lower state is the S_0 ground state, and the upper state is a singlet exited state with allowed electric-dipole transitions from S_0.

Isolated molecules can, of course, also change their rotational state during an electronic transition, but since the rotational energies are so small, these considerations are not important here.

The atoms in a molecule can vibrate about their bonds, which gives the molecule vibrational energy in addition its electronic energy. Hence we must associate a series of vibrational levels with each electronic state, as shown in Fig. 8.4. Quantum mechanics tells us that the energy of a vibrational oscillation of angular frequency Ω is equal to $(n + 1/2)\hbar\Omega$, where n is the number of quanta excited. Thus the energy of the molecule in the ground-state level when n_1 quanta of frequency Ω_1 are excited is given by:

$$E = E_1 + (n_1 + 1/2)\hbar\Omega_1. \tag{8.1}$$

In the same way, the energy of the molecule in the excited electronic state with n_2 quanta of frequency Ω_2 excited is given by:

It will frequently be the case that the vibrational frequencies of the upper and lower states are very similar, and so it will be a reasonable assumption to set $\Omega_2 = \Omega_1$. In this case, we just denote the common vibrational frequency as Ω.

$$E = E_2 + (n_2 + 1/2)\hbar\Omega_2. \tag{8.2}$$

The subscripts on Ω allow for the possibility that the vibrational frequencies are different for the two electronic states.

We consider an optical transition in which an electron is promoted from the ground state to the excited state by absorbing a photon, as indicated by process (1) in Fig. 8.4. We assume that the molecule is initially in the lowest vibrational level of the ground state. This is reasonable because the energies of the vibrational quanta are typically of order $\sim 0.1\,\text{eV}$, and so there will be very few quanta excited at room temperature, where $k_B T \sim 0.025\,\text{eV}$. (See, for example, Exercise 8.3.) On applying conservation of energy to the transition with $n_1 = 0$, we find that:

$$\hbar\omega_a = \big(E_2 + (n_2 + 1/2)\hbar\Omega_2\big) - \big(E_1 + \hbar\Omega_1/2\big)$$
$$= \hbar\omega_0 + n_2\hbar\Omega_2 , \qquad (8.3)$$

where ω_a is the angular frequency of the absorbed photon and

$$\hbar\omega_0 = E_2 - E_1 + \tfrac{1}{2}\hbar(\Omega_2 - \Omega_1) . \qquad (8.4)$$

This vibrational–electronic process causes the electron to jump to the excited electronic state, and simultaneously creates vibrational quanta. Since n_2 can only take integer values, the absorption spectrum will in principle consist of a series of discrete lines with energies given by eqn 8.3. In practice, these discrete lines are often broadened into a continuum. (See the discussion of the experimental data in Section 8.2.5.)

The absorption transition leaves the molecule in the excited electronic state with a large amount of vibrational energy. This excess vibrational energy is rapidly lost in radiationless relaxation processes, as indicated by the wiggly arrow labelled (2) in Fig. 8.4. The relaxation process occurs by spreading the vibrational energy of the individual excited molecule throughout the rest of the system. (See the discussion in Section 8.2.4 below.) Ultimately, the excess vibrational energy ends up as heat.

Once the molecule has relaxed to the bottom of the excited state, it then returns to the ground state by emitting a photon at energy $\hbar\omega_e$, as shown by process (3) in Fig. 8.4. This leaves the molecule in an excited vibrational level of the ground state. The frequency of the photon is given by:

$$\hbar\omega_e = (E_2 + \hbar\Omega_2/2) - (E_1 + (n_1 + 1/2)\hbar\Omega_1)$$
$$= \hbar\omega_0 - n_1\hbar\Omega_1. \qquad (8.5)$$

Thus the emission spectrum consists of a series of vibrational–electronic lines with frequencies given by eqn 8.5. The molecule finally returns to the $n_1 = 0$ level of the ground state by losing the excess vibrational quanta in further radiationless relaxation processes, as shown by step (4) in Fig. 8.4.

On comparing eqns 8.3 and 8.5, we see that the absorption occurs at a higher energy than the emission, except for the cases when no vibrational quanta are excited during the electronic transitions. This is a very common phenomenon, and should be contrasted with atomic transitions, in which the absorption and emission frequencies coincide. The difference in energy between the absorption and emission spectra is called the **Stokes shift**.

8.2.3 Molecular configuration diagrams

The vibrational–electronic spectra of molecules can be understood in more detail with the aid of **configuration diagrams**. These are diagrams that show the electronic energy of a molecule as a function of the **configuration coordinates**. In order to understand how these diagrams work, we first consider the simplest type of molecule, namely a

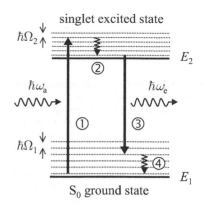

Fig. 8.4 Schematic diagram of the vibrational–electronic transitions in a molecule. The four processes indicated are respectively: (1) absorption; (2) non-radiative relaxation; (3) emission, and (4) non-radiative relaxation.

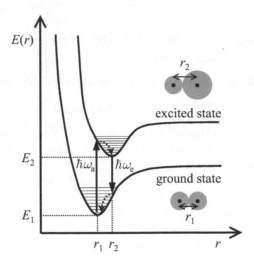

Fig. 8.5 Energy-level diagram for the ground state and an excited state of a simple diatomic molecule, as a function of the separation r between the two nuclei. Vibrational–electronic absorption and emission transitions at energies of $\hbar\omega_a$ and $\hbar\omega_e$ are indicated. The schematic 'dumb-bell' diagrams of the molecule, with the radius of one of the atoms increasing in the excited state, illustrate the point that the equilibrium separations of the nuclei in the two electronic states are different.

The validity of the Born–Oppenheimer approximation is demonstrated experimentally by the fact that the vibrational oscillations of the molecule (which are caused by displacements of the nuclei from their equilibrium positions) occur at frequencies in the infrared spectral range ($\sim 10^{13}$ Hz), whereas the electronic transitions occur in the visible and ultraviolet spectral regions (10^{14}–10^{15} Hz).

diatomic molecule, in which there is only one vibrational mode, namely the stretching of the bond between the two atoms. The vibrational configuration of the molecule can therefore be given a direct physical interpretation as the internuclear separation.

The electronic energy of a diatomic molecule is usually calculated by applying the **Born–Oppenheimer approximation**. This says that the electronic and nuclear motions are independent, and means that we can draw graphs of the electronic energy as a function of the internuclear separation. The approximation is valid because the nuclei are much heavier than the electrons, and therefore move on a far slower time scale.

Figure 8.5 shows a schematic configuration diagram of a typical diatomic molecule. The diagram shows the energy of the ground state and one of the excited states as a function of the separation r between the two nuclei. If the states are bound, there must be a minimum energy for some value of r. The position of the minimum in the ground state is labelled r_1 and corresponds to the equilibrium separation of the nuclei in the unexcited molecule. The minimum at r_2 is the mean separation of the nuclei when the molecule is in the excited electronic state. In general, r_1 and r_2 are not the same.

We can understand why the minima occur at different positions by discussing the behaviour of the simplest diatomic molecule, namely hydrogen. Consider the ground state of the H_2 molecule. When $r = \infty$, the atoms are independent of each other, and the ground-state energy is that of two separate hydrogen atoms each in the 1s level. As r decreases from ∞, the total energy of the system must decrease due to the cohesive energy of the H–H covalent bond. However, if r becomes too small, the energy will increase again due to the electron–electron and proton–proton repulsion. Therefore, the energy of the system must go through a minimum at some value of r, labelled r_1, and then increase strongly for smaller r. For the H_2 molecule, $r_1 = 0.074$ nm, and corresponds to the equilibrium separation of the nuclei in the ground state.

Now consider the energy of the first optically accessible excited state of the H_2 molecule. At $r = \infty$, this corresponds to one atom being in the 1s level and the other in the 2p level. The energy of the system will at first decrease with r due to the attractive forces between the atoms. We then go through another minimum labelled r_2 in Fig. 8.5 as the repulsive forces for small r become significant. In general, r_2 will not be equal to r_1 because the minimum energy is obtained when we maximize the cross-attractions between the electron of one atom and the proton of the other, while minimizing the sum of the proton–proton and electron–electron repulsions. This process obviously depends on the overlap of the electronic wave functions, which will be different for the orbitals of the $1s^2$ ground state and those of the 1s 2p excited state. In the case of H_2, the energy minimum occurs at $r_2 = 0.13$ nm for the 1s 2p state, which is substantially larger than r_1.

The difference between r_1 and r_2 can be given a very simple interpretation with the aid of the schematic 'dumb-bell' pictures of the molecule, which are included in Fig. 8.5. The equilibrium separation of the nuclei roughly corresponds to the point where the atomic orbitals begin to touch and bond together. In the ground state, both atoms have the same radius, but in the excited state one of them has a larger radius. It is thus obvious that r_2 will be larger than r_1.

The dependence of the electronic energy on the position coordinates shown in Fig. 8.5 is typical of other molecules, and can be used as a starting point for the discussion of the vibrational modes. The vibrational motion of the molecule will be determined by the shape of the $E(r)$ curve. Although the detailed functional form of $E(r)$ is complicated, it can be shown that for small displacements from the minimum position, the curve can always be approximated by a parabola. (See, for example, Exercise 8.5.) Therefore, near the minimum we can write:

$$E(x) = E_{\text{min}} + \tfrac{1}{2}\mu\Omega^2 x^2 \,, \tag{8.6}$$

where μ is the reduced mass of the molecule, and x is the displacement from the equilibrium position. In the case of the ground state, $x = r - r_1$, while for the excited state, $x = r - r_2$. Equation 8.6 describes a simple harmonic oscillator of frequency Ω. The quantized vibrations about the equilibrium position associate a series of uniformly spaced energy levels with each electronic state as shown in Fig. 8.5. The separation of the vibrational levels will be different for each electronic state because the value of Ω depends on the curvature of the $E(r)$ curve (see Exercise 8.5), which will differ from state to state.

8.2.4   The Franck–Condon principle

The optical transitions between the coupled vibrational–electronic levels of a molecule can be understood by invoking the **Franck–Condon principle**. The Franck–Condon principle, like the Born–Oppenheimer approximation, is a consequence of the fact that electrons are much lighter than nuclei.

Optical transitions from the $1s^2$ ground state to the 1s 2p excited state of H_2 occur at 11.3 eV, which is slightly larger than the Lyman α line in atomic hydrogen: see Exercise 8.4.

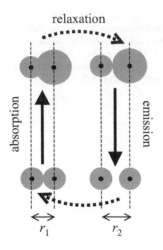

relaxation

absorption

emission

r_1 r_2

Fig. 8.6 Schematic representation of the processes that occur during the absorption and emission of photons by vibronic transitions in a molecule. r_1 and r_2 are the equilibrium separations of the nuclei in the ground and excited states respectively. One of the atoms has a larger radius in the excited molecule because the atom itself is in an excited state.

The steps that take place when photons are absorbed and re-emitted by a molecule according to the Franck–Condon principle are illustrated schematically in Fig. 8.6. The molecule starts in the ground state with a mean nuclear separation of r_1. The absorption of the photon promotes an electron to the excited state without altering r. The transition thus leaves the molecule in the excited state with a mean nuclear separation of r_1 instead of the equilibrium separation r_2. The separation of the nuclei rapidly relaxes to r_2 before re-emitting a photon. After the photon is emitted, the molecule is left in the ground state with a mean nuclear separation of r_2. Further rapid relaxation processes occur to complete the cycle and bring the molecule back to its equilibrium separation in the ground state. These four steps correspond to the four processes indicated in the simplified energy-level diagram shown in Fig. 8.4.

The 'rapid relaxation' processes that accompany the optical transitions need some clarification. If we think of the nuclear vibrations as analogous to those of a spring, we can see that the transition leaves the molecular spring in a compressed or extended state at time $t = 0$. We know that in this situation the spring will immediately begin to vibrate at its own natural frequency for $t > 0$. This is equivalent to instigating the oscillations of a specific vibrational mode in one particular molecule. However, the molecule may have other vibrational modes, and it can also interact with the other molecules that surround it. The relaxation of the vibrational energy created during the transition thus involves the spreading of the localized vibrational energy of a particular molecule to the other modes of the molecule and to the other molecules. This is a more technical way of saying that the excess energy ends up as heat. The vibrational relaxation typically occurs in less than 1 ps in a solid, which is much faster than the ~ 1 ns taken to re-emit a photon.

The Franck–Condon principle implies that we represent the optical transitions by vertical arrows in configuration diagrams, as shown in Fig. 8.5. The absorption of a photon puts the molecule in an excited vibrational state as well as an excited electronic state. The excess vibrational energy is lost very rapidly through non-radiative relaxation processes, as indicated by the dotted lines in Fig. 8.5. The frequencies of the photons absorbed and emitted are given by eqns 8.3 and 8.5. These describe a series of sharp lines with equal energy spacing.

In more complicated molecules with many degrees of freedom, the vibrational motion is described in terms of the normal modes of the coupled system. These vibrational modes are usually represented by a generalized coordinate Q, which has the dimensions of length. The Born–Oppenheimer approximation allows us to produce configuration diagrams in which we plot the electronic energy as a function of Q. Figure 8.7 is an example of such a configuration diagram. In general, the ground state and excited state have approximately parabolic minima at different values of the configuration coordinate. The optical transitions are indicated by vertical arrows, as prescribed by the Franck–Condon principle. The absorption and emission spectra consist of a series of lines with frequencies given by eqns 8.3 and 8.5, as shown in the right-hand

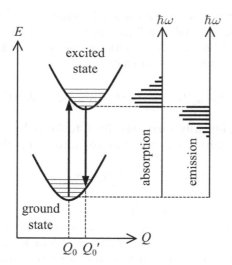

Fig. 8.7 Configuration diagram for two electronic states in a molecule. Vibrational–electronic transitions are indicated by the vertical arrows, together with a schematic representation of the absorption and emission spectra. The probability amplitudes for the relevant vibrational levels for the absorption transition are shown in Fig. 8.8.

side of the figure.

The relative intensities of the manifold of vibronic transitions can be calculated in the Franck–Condon approximation. The matrix element for an electric-dipole transition from an initial state Ψ_1 to a final state Ψ_2 is given by (cf. eqn B.27):

$$M_{12} = \langle 2| - e\boldsymbol{r} \cdot \boldsymbol{\mathcal{E}_0}|1\rangle \equiv -e \int d\xi_1 \cdots \int d\xi_N \ \Psi_2^* \, \boldsymbol{r} \cdot \boldsymbol{\mathcal{E}} \, \Psi_1 \,, \qquad (8.7)$$

where $\boldsymbol{r}$ is the position vector of the electron, $\boldsymbol{\mathcal{E}}$ is the electric field of the light wave, and $\xi_1, \cdots, \xi_N$ represent the coordinates for all the relevant internal degrees of freedom of the molecule. For the coupled vibrational–electronic states that we are considering here, the total wave function will be a product of an electronic wave function that depends only on the electron coordinate $\boldsymbol{r}$, and a vibrational wave function that depends only on the configuration coordinate Q. We thus write the vibronic wave function for an electronic state i and a vibrational level n as:

$$\Psi_{i,n}(\boldsymbol{r}, Q) = \psi_i(\boldsymbol{r}) \, \varphi_n(Q - Q_0) \,. \qquad (8.8)$$

The vibrational wave function $\varphi_n(Q - Q_0)$ is just the wave function of a simple harmonic oscillator centred at Q_0, the equilibrium configuration for the ith electronic state. In general, we have to assume that the equilibrium positions for different electronic states are not the same. We thus denote the equilibrium positions of the ground and excited states as Q_0 and Q_0' respectively, as indicated in Fig. 8.7. The difference between Q_0 and Q_0' is conveniently quantified by the dimensionless **Huang–Rhys parameter** S, defined by:

$$S = \frac{\frac{1}{2}\mu\Omega^2(Q_0' - Q_0)^2}{\hbar\Omega} = \frac{(Q_0' - Q_0)^2}{2(\hbar/\mu\Omega)} \,, \qquad (8.9)$$

where μ is the mass of the vibrational oscillator and Ω its angular frequency. Note that $\hbar/\mu\Omega$ is the mean squared amplitude for the zero-point motion of the oscillator.

In an electronic transition, the position vector that appears in M_{12} is the electron coordinate, because we are specifically considering the interaction with the dipole moment of the electron. The light will, of course, also interact with the dipole moment of the nucleus, but these transitions occur at much lower frequencies, and can be ignored in the Franck–Condon approximation.

The Hunag–Rhys parameter for a particular transition is usually determined by analysis of the shape of the vibronic spectra. See, for example, Exercise 8.6.

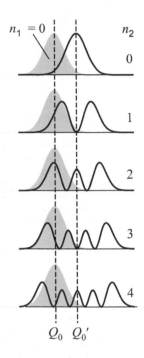

Let us consider a vibronic transition in which the vibrational quantum numbers of the initial and final states are n_1 and n_2 respectively. On inserting the vibronic wave functions from eqn 8.8 into the matrix element given in eqn 8.7, we find that:

$$M_{12} \propto \iint \psi_2^*(\boldsymbol{r}) \varphi_{n_2}^*(Q - Q_0') \, x \, \psi_1(\boldsymbol{r}) \varphi_{n_1}(Q - Q_0) \, \mathrm{d}^3 \boldsymbol{r} \, \mathrm{d}Q, \quad (8.10)$$

where we have arbitrarily taken the light to be polarized along the x axis. The matrix element can be separated into two parts:

$$M_{12} \propto \int \psi_2^*(\boldsymbol{r}) \, x \, \psi_1(\boldsymbol{r}) \, \mathrm{d}^3 \boldsymbol{r} \times \int \varphi_{n_2}^*(Q - Q_0') \varphi_{n_1}(Q - Q_0) \, \mathrm{d}Q. \quad (8.11)$$

The first factor is the electric-dipole moment for the electronic transition, which we are assuming to be non-zero. The second is the overlap of the initial and final vibrational wave functions. From Fermi's golden rule (eqn B.14), we know that the transition rate is proportional to the square of the matrix element. Hence the intensity I of the vibronic transition is given by:

$$I \propto |\langle n_2, Q_0' | n_1, Q_0 \rangle|^2 \equiv \left| \int_0^\infty \varphi_{n_2}^*(Q - Q_0') \, \varphi_{n_1}(Q - Q_0) \, \mathrm{d}Q \right|^2. \quad (8.12)$$

The square of the vibrational overlap integral that appears here is called the **Franck–Condon factor**.

To see how the Franck–Condon factor works in practice, we need to look at the probability densities (i.e. the square of the wave functions) for the vibrational levels involved in the transition. In the absorption transition shown in Fig. 8.7, the molecule starts in the $n_1 = 0$ level of the ground state, which has its equilibrium position at Q_0, and finishes in the n_2th level of the excited state, which is centred at Q_0'. The relevant wave functions are shown in Fig. 8.8. We see that we have a good overlap for several transitions, with the largest Franck–Condon factor for $n_2 \sim 2$. Hence we would expect the intensity to be largest for the $n_2 = 2$ level, as indicated by the schematic absorption spectrum shown in Fig. 8.7. Since harmonic oscillator wave functions with large n peak near the classical turning points at the edge of the potential well, we can give an approximate rule of thumb that the Franck–Condon factor will be largest for the levels with the edge of their classical potential well close to Q_0.

The reverse argument can be applied to the emission transitions. This leads to the schematic emission spectrum shown on the right of Fig. 8.7. In the simple model, we expect the emission spectrum to be the 'mirror' of the absorption spectrum when reflected about the centre frequency $\hbar \omega_0$. This is called the **mirror symmetry rule**.

Fig. 8.8 Probability densities for the vibrational levels involved in the absorption transition shown in Fig. 8.7. The initial wave function has been shaded. The molecule starts in the $n_1 = 0$ level of the ground state, and finishes in the n_2th level of the excited state. Q_0 and Q_0' are the equilibrium positions for the ground and excited states respectively.

8.2.5 Experimental spectra

The electronic transitions for very small molecules usually occur in the ultraviolet spectral region. Figure 8.9 shows the absorption spectrum of

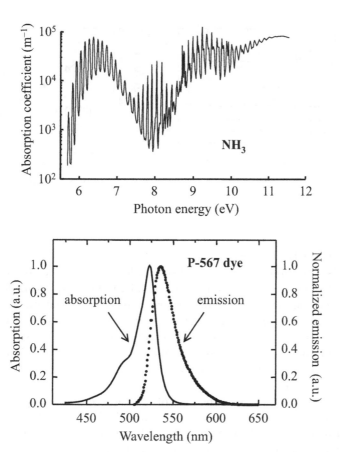

Fig. 8.9 Ultraviolet absorption of ammonia (NH₃) at standard temperature and pressure. After Watanabe (1954), © American Institute of Physics, reprinted with permission.

Fig. 8.10 Absorption and emission spectrum of pyrromethene 567 in benzene solution. After Gorman et al. (2000), © Excerpta Medica Inc., reprinted with permission.

ammonia (NH_3). Ammonia is a colourless gas at room temperature, with an absorption edge at 5.7 eV (217 nm). For photon energies above 5.7 eV a series of discrete lines are observed which exhibit the general behaviour shown schematically in Fig. 8.7. The spacing of the lines corresponds to the out-of-plane bending vibrational mode of the molecule with a quantized energy of approximately 0.114 eV. The progression peaks at $n_2 \sim 6$. Further progressions of vibrational–electronic transitions can be discerned in the data starting at 7.3 eV and 8.6 eV. These correspond to electronic transitions to higher singlet excited states of the molecule. It is evident that the data agree very well with the general predictions of the configuration diagram model, although the detailed interpretation can be quite complicated due to the overlapping of the different vibronic bands.

Figure 8.10 shows the absorption and emission spectra of the laser dye, pyrromethene 567, in benzene solution. Pyrromethene 567 is a large organic molecule with strong electronic transitions in the visible spectral region. The smaller transition energy compared to ammonia is, to a first approximation, just a consequence of the fact that the molecule is larger, and hence that the quantum confinement of the electrons is smaller. The Stokes shift of the emission is apparent in the data, as is approximate

We do not show spectra for the simplest molecules like H_2 here because the first electronic excited state is above the ionization limit of the molecule, and the vibronic lines are not well resolved. Ammonia shows discrete vibronic lines because it is a small, rigid molecule with very well-defined vibrational frequencies.

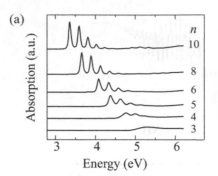 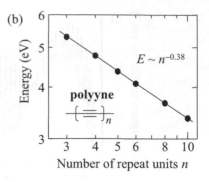

Fig. 8.11 Absorption data for polyyne molecules end-capped with trisopropy-lsilyl (TIPS) in hexane solution at room temperature. The size of the π orbital is determined by the number of carbon–carbon repeat units n. (a) Absorption spectra. (b) Log–log plot of the variation of the HOMO–LUMO energy gap E with n. After Slepkov et al. (2004), © American Institute of Physics, reprinted with permission.

Table 8.1 Hydrocarbon polyyne molecules.

	n	Bonding
C_2H_2	1	H–C≡C–H
C_4H_2	2	H–C≡C–C≡C–H
C_6H_2	3	H–C≡C–C≡C–C≡C–H
⋮	⋮	⋮
$C_{2n}H_2$	n	H–(C≡C–)$_n$H

mirror symmetry between the absorption and emission spectra.

No discrete vibronic lines are observed for the laser dye because the large molecule possesses many vibrational modes of differing frequencies, which generate overlapping progressions of lines that fill out into a continuum. Furthermore, the thermal motion of the molecules and also collisions with the benzene solvent broaden the transitions so that the individual lines cannot be resolved. We therefore obtain continuous absorption and emission bands that follow the envelope of the vibrational–electronic progressions.

The correlation between the size of a molecule and its optical transition energies is shown particularly clearly in Fig. 8.11. Here, absorption data is presented for a series of polyyne molecules. The polyynes are linear conjugated molecules based on alternating singlet and triplet carbon–carbon bonds. The simplest type of polyynes are hydrocarbons, where the series starts with acetylene (C_2H_2), and proceeds through diacetylene (C_4H_2), then triacetylene (C_6H_2), etc. (See Table 8.1.) In the particular case considered for Fig. 8.11, the hydrogen atoms at the end of the carbon chains have been replaced by 'TIPS' molecular units, i.e. triisopropylsilyl: $[(CH_3)_2CH]_3Si$. This change is made for purely practical reasons related to the molecular stability and solubility, and does not affect the basic argument.

The electrons in the π-bonds of the polyyne molecules can spread out along the carbon chain, and hence the size of the molecular orbitals increases with the number, n, of carbon–carbon repeat units. From the simple ideas of quantum confinement developed in Section 6.1, we expect the electronic energies to decrease as the molecular orbitals enlarge. We therefore expect the HOMO–LUMO transition energy to decrease with n. This prediction is indeed borne out by the experimental data. The absorption spectra in Fig. 8.11(a) show a clear red shift as n increases, and the straight line fit to the log–log plot in Fig. 8.11(b) indicates that the transition energies scale as $n^{-0.38}$. Another interesting feature of the spectra is the clarity of the vibronic progression for $n \geq 4$ compared to Fig. 8.10. This is a consequence of the rigid nature of the linear carbon molecule, which leads to very well defined vibrational frequencies.

8.3   Conjugated molecules

Having discussed the spectra of isolated molecules, we can now apply this knowledge to conjugated organic solids. Organic chemistry is capable of producing an enormous variety of conjugated molecules, and so we focus here only on those that are interesting for opto-electronic devices. There are two main classes of molecules that are usually employed in these applications, namely small molecules and conjugated polymers. The optical properties of both types are considered separately in the following subsections, and then the opto-electronic devices that are made from them are described in Section 8.4.

8.3.1   Small conjugated molecules

The definition of 'small' when applied to a conjugated molecule is somewhat imprecise. Therefore, in this text we adopt a practical working definition that it simply means that the conjugated molecule is not a polymer. Furthermore, we restrict our attention to the larger 'small molecules' that have electronic transition energies in and around the visible spectral region. We use anthracene as an example to illustrate the physics, and then consider the Alq_3 molecule that has considerable technological importance.

Anthracene ($C_{14}H_{10}$) is an example of an aromatic hydrocarbon. Aromatic hydrocarbons are carbon–hydrogen compounds containing benzene rings in their structure. The name derives from the strong aroma of the liquids and gases, but here we focus on crystalline solids. Anthracene crystals are held together by van der Waals interactions, which means that the covalent bonding within the molecule is much stronger than the interactions between the adjacent molecules in the crystal. We therefore expect the electronic states to be strongly localized, and the spectra of the crystals to be fairly similar to those of anthracene in solution.

The optical properties of molecules containing benzene rings are determined by their large delocalized π orbitals, as explained in Section 8.2.1. The absorption edge of benzene itself occurs well into the ultraviolet spectral region at 4.6 eV (267 nm), but larger aromatic hydrocarbons have transitions at lower energies. (See the discussion of the polyyne data in Section 8.2.5.) Anthracene has three benzene rings, as shown in Fig. 8.12(a), and $\pi \rightarrow \pi^*$ transitions in the violet spectral region are possible from the S_0 ground state to the first singlet excited state.

Figure 8.12(b) gives a simplified level diagram for the first three electronic states of the anthracene molecule and the transitions that can occur between them. Diagrams of this type are called **Jablonski diagrams**. As explained in Section 8.2.1, the states are classified by their spin quantum number S. The ground state is a singlet and is labelled S_0. The first singlet excited state (S_1) occurs at 3.3 eV, which is 1.5 eV above the first triplet excited state (T_1) at 1.8 eV. In Section 8.2.1 we explained how singlet–triplet transitions have a very low probability due to the spin selection rule. The absorption and emission spectra are therefore

The excitonic absorption spectrum of another aromatic hydrocarbon, namely pyrene ($C_{16}H_{10}$), is discussed in Section 4.5.3. Single crystals of simple aromatic hydrocarbons such as anthracene and pyrene can be produced, but this is not generally the case for other organic solids, which often tend to be prepared as amorphous thin films.

Fig. 8.12 (a) Chemical structure of the anthracene molecule ($C_{14}H_{10}$). (b) Jablonski diagram for anthracene. Absorption and fluorescence transitions can occur between the S_0 ground state and the S_1 excited state. Phosphorescent transitions from the T_1 state to the ground state are spin-forbidden and occur on a slow time scale. Electrons in the S_1 state have a small probability of transferring non-radiatively to the T_1 state by intersystem crossing.

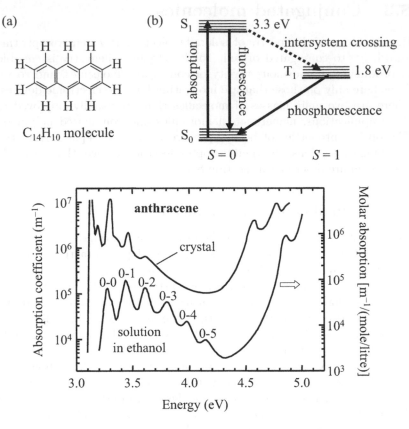

Fig. 8.13 Absorption spectrum of anthracene. The absorption spectrum of single crystals at 90 K is compared to that of a dilute molecular solution in ethanol. The vibronic transitions of the solution are labelled by the vibrational quantum numbers of the ground and excited states. After Wolf (1958), reprinted with permission.

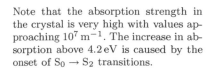

Note that the absorption strength in the crystal is very high with values approaching 10^7 m^{-1}. The increase in absorption above 4.2 eV is caused by the onset of $S_0 \rightarrow S_2$ transitions.

dominated by $S_0 \leftrightarrow S_1$ transitions. Phosphorescent $T_1 \rightarrow S_0$ transitions are only observed by using special techniques. (See below.)

Figure 8.13 compares the absorption spectrum of a dilute solution of molecular anthracene in ethanol with that of anthracene crystals at 90 K. At higher temperatures, the vibrational structure is less well resolved due to the thermal broadening of the vibronic lines. The absorption edge of the solution occurs at the energy for the $S_0 \rightarrow S_1$ transition at 3.3 eV. About six vibrational–electronic lines with approximately equal spacing are resolved, with the largest intensity for the transition with $n_2 = 1$. Similar vibrational structure is observed for the crystal. Three main absorption peaks with approximately the same spacing as in the solution are resolved. These correspond to transitions involving localized vibrations of the anthracene molecules. Additional lines are also observed due to coupling to new vibrational modes present in the crystal. Furthermore, the absorption edge occurs at a slightly lower energy than in the solution.

The S_1 excited state has a lifetime of 27 ns due to the dipole-allowed $S_1 \rightarrow S_0$ fluorescent transition. The fluorescence spectrum of the crystal consists of a broad vibronic band running from about 3.2 eV (390 nm) to 2.3 eV (530 nm). Prominent vibronic peaks occur at 3.05 eV, 2.93 eV 2.76 eV and 2.61 eV. These compare very favourably with the energies

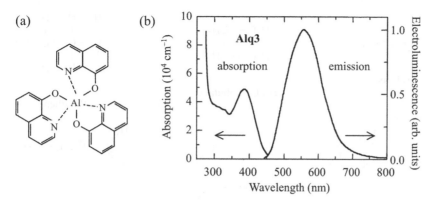

(a) (b)

Fig. 8.14 (a) Chemical structure of aluminium *tris*-quinolate (Alq3). (b) Absorption and emission spectra of Alq3 at room temperature. The emission spectrum was obtained from an Alq3 electroluminescent diode. Data from Garbuzov et al. (1996), © Elsevier, and Tang & VanSlyke (1987), © American Institute of Physics, reprinted with permission.

expected by applying the mirror symmetry rule to the absorption spectrum shown in Fig. 8.13. (See Exercise 8.12.)

Phosphorescent transitions from the T_1 state at 1.8 eV can be observed by techniques of **delayed fluorescence**. This involves exciting the crystal with a laser pulse, and then observing the emission at long times afterwards. Only the S_1 states are populated by the laser pulse due to the spin selection rule. Most of the molecules return directly to the ground state by radiative transitions, thereby generating prompt fluorescent emission within the first 27 ns after the pulse has arrived. However, there is a small probability for non-radiative $S_1 \rightarrow T_1$ **intersystem crossing**. The T_1 state has a long lifetime of 24 ms due to the low probability for radiative emission. The weak phosphorescent emission from the T_1 states thus persists as an 'afterglow' for 24 ms after the pulse has excited the sample.

The effects described for anthracene are typical of many molecular materials. The absorption and emission spectra are determined by transitions to singlet excited states. The triplet states are only observed in emission experiments as weak phosphorescence. Each electronic transition shows a manifold of vibronic peaks with energy separation determined by the vibrational frequencies of the system.

Figure 8.14 shows the absorption and emission spectra of aluminium *tris*-quinolate (or *tris* (8-hydroxyquinoline) aluminium), which is commonly known as Alq3. This is an important small molecule that is widely used in organic light emitting diodes. (See Section 8.4.) The $S_0 \rightarrow S_1$ absorption band peaks at 385 nm, and the emission is Stokes shifted into the green spectral region, peaking around 560 nm. The molecule is not as rigid as some others, and its vibrational lines are not resolved at room temperature.

Triplet–singlet transitions can occur by spin–orbit coupling, which causes a mixing between singlet and triplet states. Since spin–orbit interactions scale roughly as Z^2, where Z is the atomic number, the triplet–singlet transition rate is increased if heavy elements are present. In fact, it has been shown that the incorporation of heavy metals such as iridium, palladium, or platinum into the molecular structure leads to a strong enhancement of the phosphorescence.

8.3.2 Conjugated polymers

Polymers are long-chain molecules composed of repeated sequences of individual molecular units based on carbon–carbon bonds. The name **polymer** is the logical sequence that starts with **monomer** (single molecule), then progresses to **dimer** (double molecule), and finally ends

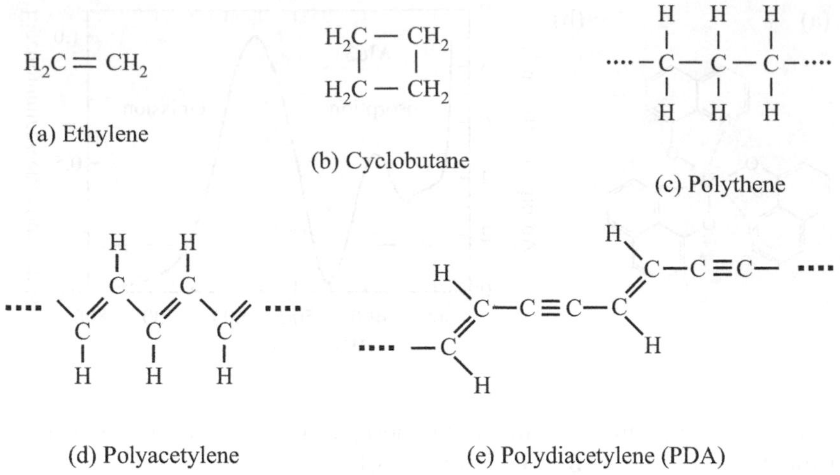

$$H_2C = CH_2$$

(a) Ethylene

$$\begin{array}{c} H_2C - CH_2 \\ |\qquad| \\ H_2C - CH_2 \end{array}$$

(b) Cyclobutane

$$\cdots - \underset{\underset{H}{|}}{\overset{\overset{H}{|}}{C}} - \underset{\underset{H}{|}}{\overset{\overset{H}{|}}{C}} - \underset{\underset{H}{|}}{\overset{\overset{H}{|}}{C}} - \cdots$$

(c) Polythene

(d) Polyacetylene

(e) Polydiacetylene (PDA)

Fig. 8.15 Chemical structures. (a) Ethylene (C_2H_4) monomer. (b) Ethylene dimer: $[C_2H_4]_2 \equiv C_4H_8$ (cyclobutane). (c) Polyethylene (polythene): $[CH_2]_n$. (d) Polyacetylene. (e) Polydiacetylene (PDA).

at polymer (many molecule). This progression is illustrated for ethylene in Figs 8.15(a–c). The dimer is $[C_2H_4]_2$, which is cyclobutane (C_4H_8). The polymer is polyethylene, or polythene for short, with a formula of $[CH_2]_n$, where n is a large number.

Polymers can be subdivided into two generic types, namely conjugated or saturated. As explained in Section 8.2.1, this division is based on whether there are alternating single–double bonds along the polymer backbone or not. The polythene structure shown in Fig. 8.15(c) is an example of a saturated polymer. In saturated polymers like polythene, all the electrons are incorporated into σ bonds, and are therefore very tightly bound. Their transitions are at high energies in the ultraviolet spectral region, and are not of particular interest here. This contrasts with many conjugated polymers, which have $\pi \to \pi^*$ transitions in the visible spectral region, and can therefore be described as **light-emitting polymers**.

Figure 8.15 also includes the chemical structures of two conjugated polymers. Figure 8.15(d) shows the simplest conjugated polymer, namely polyacetylene. This polymer is formed by combining many acetylene molecules (bonding $HC \equiv CH$) into a long chain with alternating single and double bonds between the carbon atoms:

$$\cdots = CH - CH = CH - \cdots .$$

The description of the molecule with alternating single and double bonds is only schematic. In reality, the spare electron of the double bond is shared equally between both bonds in a delocalized π orbital.

Figure 8.15(e) shows the chemical structure of a slightly more complicated conjugated polymer, namely polydiacetylene (PDA), which incorporates both double and triple bond π electrons. PDA is one of the most widely studied conjugated polymers because it is can form high-quality crystals at room temperature. This is not the case for most other conjugated polymers, for which only amorphous samples formed by spin-

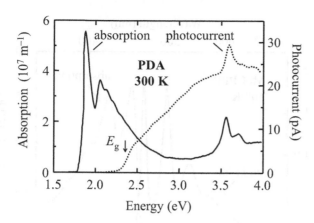

Fig. 8.16 Absorption spectrum of polydiacetylene (PDA) crystals at room temperature. The photocurrent spectrum of the same crystal is shown for comparison. After Möller & Weiser (1999), © Excerpta Medica Inc., reprinted with permission.

coating from solution are available. The absorption spectrum of a PDA single crystal is shown in Fig. 8.16. The spectrum has a broad $S_0 \to S_1$ absorption band starting at 1.8 eV. The band shows clear substructure, with two well-resolved vibronic peaks at 1.9 eV and 2.1 eV. The peaks around 3.6 eV are caused by $S_0 \to S_2$ transitions.

The optical spectra of conjugated polymers like PDA are strongly affected by excitonic effects. In Chapter 4 we discussed how the Coulomb interaction tends to bind electrons and holes together to form excitons, and how this has a strong effect on the optical spectra. Excitonic effects are present in molecular materials because the unpaired electron left in the ground state by an optical transition can be regarded as a hole, in the same way that unfilled states in the otherwise full valence band of a semiconductor are considered as holes. This hole acts like a positive charge because it represents the absence of a negative electron. Since the electronic states of molecular materials are strongly localized, the excitons that are formed are of the tightly-bound, or Frenkel, type. (See Section 4.5, especially Section 4.5.3.) These Frenkel excitons are observed as bound states below the band edge of the polymer.

The binding energy of the excitons in PDA can be determined by comparing the absorption and photocurrent spectra of the same crystal. The photocurrent spectrum of PDA included in Fig. 8.16 shows a threshold at 2.4 eV, which is about 0.5 eV above the absorption edge. This indicates that the absorption line at 1.9 eV is excitonic in character, because the neutral, tightly-bound excitons do not contribute to the conductivity, and the photon energy must exceed the HOMO–LUMO band gap at 2.4 eV before free electrons and holes are available to produce a photocurrent. The measurements therefore indicate that the binding energy of the exciton is 0.5 eV.

Figure 8.17 shows the absorption and emission spectrum of a thin film sample of the conjugated polymer called 'MeLPPP' at room temperature. The $S_0 \to S_1$ transition lies in the green-blue spectral region, and is thus important for opto-electronic applications, as will be described in the next section. The approximate mirror symmetry between the absorption and emission is evident in the data, with four vibrational lines

This is a different situation to the photocurrent spectra of free excitons shown in Figs 4.5 and 6.15. The weakly bound free excitons in GaAs can easily dissociate after formation to produce free electrons and holes. The Frenkel excitons in molecular materials, by contrast, do not easily dissociate into free electrons and holes due to their much larger binding energy.

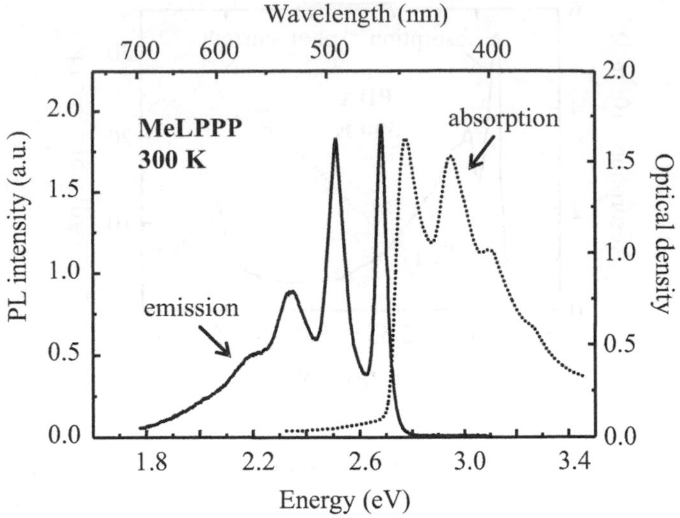

Fig. 8.17 Absorption and emission spectra of thin films of the ladder polymer MeLPPP at room temperature. MeLPPP is a methyl-substituted ladder-type polymer based on poly(paraphenylene). After Hertel et al. (1999), © American Institute of Physics, reprinted with permission.

clearly distinguishable.

The clarity of the vibrational structure in Fig. 8.17 compared, for example, to Fig. 8.10, is striking. This improved vibronic resolution is caused by the rigidity of the MeLPPP molecule, and also by the reduction of the thermal agitation in a thin film compared to a solution. Furthermore, it is noteworthy that the vibronic structure is more clearly resolved in the emission spectrum than in the absorption. This is a consequence of the disorder in the polymer. The vibronic lines are inhomogeneously broadened by random variations in the band gap energy along the polymer back bone. The absorption spectrum averages over all of these energy states, but the emission spectrum samples only a subset of them, because the excitons have time to migrate to the lowest energy states before they emit the photon.

8.4 Organic opto-electronics

It is clear from Figs 8.14 and 8.17 that the electronic transitions of conjugated molecules such as Alq3 and MeLPPP can lie in the visible spectral region ($\sim 1.7 - 3$ eV). These visible-emitting molecules can be used in a number of important opto-electronic applications. We discuss here briefly two of the most important of these, namely organic light-emitting diodes (O-LEDs) and organic photovoltaic devices.

As mentioned at the start of Section 8.3, there are two general approaches to O-LED technology, one based on small molecules and the other on light-emitting polymers. The study of electroluminescence in small molecules like anthracene goes back many years, but the efficiencies were generally so low and the operating voltages so high that practical devices could not be made. The first low-voltage small molecule O-LED was demonstrated in 1987, and the first polymer O-LED was reported in 1990, the latter building on previous work on the development of conducting polymers. The small molecule O-LED used alu-minium *tris*-quinolate (Alq3) in the active region and gave efficient green emission for voltages below 10 V, while the polymer device used poly-phenylenevinylene (PPV) as the active layer, and gave bright emission in the green-yellow spectral region when a voltage of about 15 V was applied. Since then, much research has been done to develop new small molecules and polymers that can emit over the whole visible spectral re-gion, and to reduce the operating voltage. At the present time, it is not clear whether small molecules or light-emitting polymers will ultimately give the best results.

The operating principle of an O-LED is much the same as that of its in-organic counterpart. Electrons and holes are injected from opposite sides of the diode, and they recombine by emitting light in the active region at the centre of the structure. Figure 8.18 gives a schematic diagram of the layer sequence used in an ideal organic LED (cf. Fig. 5.10(a)). The light-emitting layer is sandwiched between electron and hole trans-porting layers, and electrons and holes are injected from the cathode and anode respectively. The cathode is typically made of aluminium, magnesium, or calcium, while the anode is usually made from the alloy indium–tin oxide (ITO), which has the advantage of being both an effi-cient hole injector and also transparent at visible wavelengths. For this reason, ITO effectively behaves as a type of conducting glass. The whole structure is grown on a non-conducting glass substrate, and the light generated in the luminescent layer is emitted through the substrate.

The O-LED shown in Fig. 8.18 contains three different organic ma-terials separated by two heterojunctions. It is therefore an example of a double heterostructure device. This arrangement is the optimal one because it allows different materials to be used for each of the three functionalities that are required, namely transport of holes from the an-ode, emission in the active layer, and transport of electrons from the cathode. It is important to understand that the charge transport mech-anisms that occur in organic devices are different to those in inorganic electronic materials. Semiconductors like silicon and GaAs have delocal-ized band states which give rise to high electron and hole mobilities, but this is not generally the case for organic materials. Instead, the electrons and holes must move by hopping between localized states on particular molecules, and this means that the mobilities are generally much smaller. The efficient injection and transport of carriers into the active region is therefore an important issue in organic devices, and the best perfor-mance is usually achieved by using optimized materials for the electron and hole transport layers—hence double heterostructure devices.

In practice, many organic opto-electronic devices use a simpler layer sequence than the double-heterostructure one shown in Fig. 8.18. The electroluminescence spectrum for Alq3 shown in Fig. 8.14 is an example of the performance that can be achieved from a single heterostructure device. In this case the device had just two organic layers, namely Alq3 and diamine, with a single heterojunction between them. The Alq3 layer served as both the electron transporter and light emitter, while the di-

Alan Heegar, Alan MacDiarmid, and Hideki Shirakawa were awarded the No-bel prize for chemistry in 2000 for the discovery and development of conduc-tive polymers.

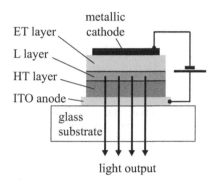

Fig. 8.18 Schematic layer sequence for a double heterostructure organic LED (O-LED). The hole-injecting anode is usually made from indium–tin oxide (ITO), while metals such as aluminium, magnesium or calcium are commonly used for the electron-injecting cath-ode. The holes and electrons drift through the hole-transporting (HT) and electron-transporting (ET) layers respectively, and recombine in the lu-minescent (L) layer.

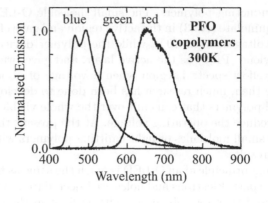

Fig. 8.19 Emission spectra of three polyfluorene (PFO) copolymers at room temperature. The shift in the emission wavelength is achieved by incorporating different substitutional units to the polymer backbone. Data from D.G. Lidzey on materials provided by Cambridge Display Technology Ltd.

The absorption spectrum of the polyfluorene-based polymer 'F8', which corresponds to the 'blue' polymer in Fig. 8.19, is shown in Fig. 1.6. The absorption peaks in the ultraviolet at 380 nm, and the emission is Stokes-shifted into the blue spectral region.

amine served as the hole transporter. The recombination occurred in the Alq3 layer within a distance of about 30 nm from the heterojunction. This short recombination distance is a consequence of the low diffusivity of holes in Alq3.

An example of the results that can be obtained from polymer devices is given in Fig. 8.19, which shows the emission spectra for three copolymers of polyfluorene. It can be seen that the whole visible spectrum can be covered by appropriate choice of the polymer structure, which opens the possibility for making full-colour organic displays. This technology is under active development by several opto-electronics companies.

The somewhat inferior electrical performance of organic devices compared to inorganic ones has to be offset against the low intrinsic cost of the active materials and the ease of fabrication of the devices. In contrast to inorganic opto-electronic devices, organic light-emitting diodes do not need to be crystalline. Instead, it is often sufficient to spin-coat thin film layers of amorphous materials onto the substrate. We have already noted that the optical properties of molecular materials are not markedly different from those of the individual molecules. Therefore, the absence of long range order has only a small effect on the main features of the optical spectra. This means that the preparation of the materials is much easier and cheaper, making them an attractive alternative to crystalline inorganic semiconductor LEDs for many applications. However, despite the rapid improvements in O-LED performance, all attempt to develop an electrically-injected organic laser have so far failed, although much progress has been made in optically-pumped lasers.

The other major opto-electronic application of organic materials is in photovoltaic devices, i.e. solar cells. It is apparent from the photocurrent spectrum of PDA shown in Fig. 8.16 that a current can be generated when light is incident on the device, and this provides a mechanism for converting solar light into electrical energy. The development of organic solar cells is a very active research field at the present time. Although the performance is still inferior to that of the best crystalline inorganic devices, efficiencies of several percent have been achieved, which compares favourably with the efficiency of typical amorphous silicon devices ($\sim 10\%$). Given the low cost and ease of fabrication of organic materials

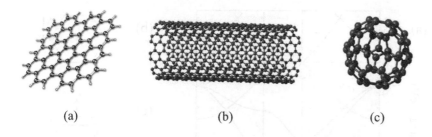

(a)   (b)   (c)

Fig. 8.20 Carbon nanostructures. (a) Graphene. (b) Carbon nanotube. (c) Buckminsterfullerene (C_{60}). The images in (b) and (c) are from http://homepage.mac.com/jhgowen/ research/nanotube_page/nanotube. html.

compared to inorganic ones, the use of organic materials in photovoltaic applications appears to be very promising, although there are still long term issues about the chemical stability of the molecules that have to be resolved.

8.5   Carbon nanostructures

8.5.1   Introduction

Carbon underpins the whole of organic chemistry. It is not surprising, then, that elemental carbon is itself highly interesting. Carbon comes in many different forms, with two of the most widely known allotropes being diamond and graphite. Diamond is a three-dimensional insulator crystal with an indirect band gap in the ultraviolet spectral region at 5.5 eV. Graphite, by contrast, is a quasi two-dimensional semimetal. Its structure consists of atomic layers of carbon atoms held together by π bonds. The individual layers of carbon atoms are called **graphene** layers, and graphite's electrical conductivity arises from the motion of the electrons in the delocalized π orbitals in the direction parallel to the layers.

The carbon atoms in a graphene layer are arranged in a hexagonal pattern, as illustrated in Fig. 8.20(a). With a thickness of just one atomic layer, graphene is the ultimate two-dimensional material. Individual graphene layers were first isolated in 2004, and since then there has been a great deal of interest in investigating their physical properties. A brief summary of the optical properties of graphene is given in Section 8.5.2 below.

A graphene layer can be rolled up to form a **nanotube**, as shown in Fig. 8.20(b). The π electrons can easily flow up and down the tube, but their motion in the direction perpendicular to the tube axis is constrained by the diameter of the tube. Carbon nanotubes are therefore one-dimensional materials. Their optical properties are considered in Section 8.5.3 below.

The final carbon nanostructure that we consider here is the **C_{60} molecule** shown in Fig. 8.20(c). The truncated icosahedron cage is called the **buckminsterfullerene** or **bucky ball** structure. Note that the structure contains both hexagons and pentagons, with 20 of the former and 12 of the latter. The optical properties of bucky balls are

The optical properties of a specific type of defect in diamond called the NV centre have attracted much interest in recent years. See Section 9.2.2.

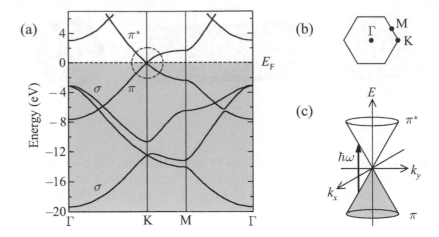

Fig. 8.21 Band structure of graphene. (a) Full band structure. (b) High symmetry points of the hexagonal Brillouin zone. (The lattice structure is shown in Fig. 8.23.) (c) An optical transition near the K point, which corresponds to the circled region of (a). Shading indicates that the states are occupied. The data in (a) are from Machón et al. (2002), © American Physical Society, reprinted with permission.

considered in Section 8.5.4 below.

8.5.2   Graphene

Graphene is a two-dimensional material with many interesting physical properties. Its band structure is shown in Fig. 8.21(a), with the notation for the high symmetry points of the Brillouin zone given in Fig. 8.21(b). The top of the valence band occurs at the K point, where there is no energy gap to the π^* states in the conduction band. Figure 8.21(c) shows an enlargement of the band structure near the K point. Note that the bands are linear at this point.

The linear band dispersion of graphene at the K point gives rise to many striking properties. The most obvious one is that all the conduction electrons have the same velocity of $\sim c/300$, irrespective of their energy. (See eqn 2.25.) This contrasts with the usual behaviour in which $E \propto k^2$, and the velocity increases as E increases. The exception to this is when the particle has a negligibly small rest mass. This can be seen from the Einstein energy:

$$E^2 = m^2 c^4 + c^2 p^2 \,, \tag{8.13}$$

where m is the rest mass and p is the linear momentum. The second term dominates when m is negligible, and the energy is linear in p, implying $E \propto k$. The linear dispersion therefore implies that the conduction electrons behave like relativistic particles with negligible mass, and must therefore be treated by the Dirac equation of relativistic quantum mechanics. For this reason, the K point of the Brillouin zone of graphene is known as the Dirac point.

The relativistic properties of the electrons in graphene have many fascinating implications. Here, we concentrate just on the optical properties. These are governed by optical transitions between the valence and conduction bands at the Dirac point, as shown in Fig. 8.21(c). Since the energy gap is zero, transitions are possible for all photon frequencies. In a conventional two-dimensional material, the transition rate is independent of frequency on account of the constant density of states. (See Section 6.4.2.) This argument, which is based on a parabolic E–k dispersion, clearly does not apply to graphene. Nevertheless, graphene does show similar behaviour, with the absorption being independent of the energy at optical frequencies. The interesting aspect is that the absorption rate is governed only by the fine structure constant $\alpha = e^2/\hbar c$, with the absorbance of a single layer being equal to $\pi\alpha = 2.3\%$. We thus have a simple solid state material that clearly illustrates quantum electrodynamical effects.

These predictions for graphene have been confirmed by experiment. Figure 8.22(a) shows the transmission spectrum of a single layer of graphene in the visible spectral region. The data show that the absorbance is indeed independent of the frequency, and takes a constant value of $\pi\alpha = 2.3\%$ per graphene layer. This implies that the transmission of a multilayer sample will be equal to $1 - \pi\alpha N$, where N is the number of graphene layers, which is clearly demonstrated by the data for multiple layers shown in Fig. 8.22(b). The absorbance of 2.3% per layer might seem small at first thinking, but is in fact very strong, given that the graphene layer is only one atom thick.

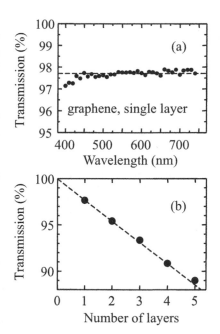

Fig. 8.22 (a) Transmission of a single layer of graphene in the visible spectral region. The dashed line is the transmission expected for a constant absorbance of $\pi\alpha$. (b) Variation of the transmission with the number of graphene layers. After Nair et al. (2008), © AAAS, reprinted with permission. The slight drop in the transmission at short wavelengths in (a) is related to the low-frequency tail of a broad absorption peak at 4.6 eV. See Kravets *et al.* (2010) and Mak *et al.* (2011).

8.5.3 Carbon nanotubes

A carbon nanotube can be considered as a rolled-up sheet of graphene. There are many different ways to do this, and there are therefore a great variety of nanotube structures. Consider the graphene honeycomb lattice shown in Fig. 8.23. The fundamental lattice vectors $\boldsymbol{a_1}$ and $\boldsymbol{a_2}$ of the structure are shown. In a nanotube, the graphene sheet is rolled up so that one of the translation vectors of the lattice becomes the circumference. We can thus define the circumference vector of the nanotube as:

$$\boldsymbol{c} = n_1 \boldsymbol{a_1} + n_2 \boldsymbol{a_2}, \tag{8.14}$$

where n_1 and n_2 are integers, and the tube axis is perpendicular to $\boldsymbol{c}$. This circumference vector is usually called the **chiral vector** and is denoted (n_1, n_2). The diameter of a nanotube is given by (see Exercise 8.18):

$$d = \frac{|\boldsymbol{c}|}{\pi} = \frac{a_0}{\pi}\sqrt{n_1^2 + n_1 n_2 + n_2^2}, \tag{8.15}$$

where $a_0 = |\boldsymbol{a_1}| = |\boldsymbol{a_2}| = 0.2461\,\text{nm}$ is the length of the basis vectors.

Three different types of circumference vectors are indicated in Fig. 8.23. Those with $n_2 = 0$ and $n_2 = n_1$ are called 'zigzag' and 'armchair' nanotubes respectively, and all the remainder are simply called 'chiral'. The chiral angle θ is defined as the angle between the chiral vector and the zigzag direction, and is given by (see Exercise 8.18):

$$\theta = \tan^{-1}\left(\frac{\sqrt{3}n_2}{2n_1 + n_2}\right). \tag{8.16}$$

Armchair nanotubes thus have chiral angles of $30°$.

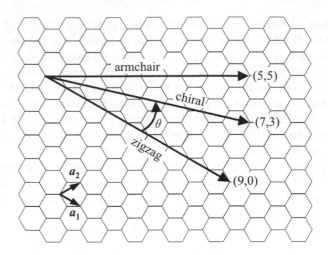

Fig. 8.23 Definition of the lattice vectors a_1 and a_2 for the graphene lattice, and the chiral vectors for a nanotube. The chiral angle θ is the angle between the chiral vector and the 'zigzag' direction.

In discussing the properties of carbon nanotubes, it is important to distinguish between **single-walled nanotube** (SWNT) and **multi-wall nanotube** (MWNT) structures. As the names suggest, these correspond respectively to nanotubes composed of a single cylinder with a unique chiral vector, and those composed of several concentric cylinders with differing chiral vectors. Much progress has been made in recent years in techniques to isolate individual SWNTs, making the study of nanotubes with well-defined chiral vectors possible.

The electronic properties of nanotubes follow from their chiral structure. We have seen that graphene is a semimetal on account of its zero energy gap at the K point of the Brillouin zone. (See Fig. 8.21.) Nanotubes, by contrast, can be either metallic or semiconducting. The nanotube is metallic if (see Exercise 8.19):

$$n_1 - n_2 = 3m, \qquad (8.17)$$

where m is an integer (positive, negative, or zero). In all other cases the nanotube is a semiconductor with a finite energy gap between the conduction and valence bands. It is therefore apparent that one-third of nanotubes are metallic, and two thirds semiconducting.

Multi-wall nanotubes will typically contain some metallic and some semiconducting tubes, and will therefore usually be highly conducting.

The electrons in a nanotube are free to move along the axis (usually defined as the z direction), but experience two-dimensional cylindrical confinement in the perpendicular directions. We thus have an almost ideal one-dimensional system, which can be treated as a **quantum wire**. (See Section 6.1.) The wave functions of the electrons are of the form:

$$\Psi(x, y, z) = \frac{1}{\sqrt{L}} \, \psi_{ij}(x, y) \, e^{ik_z z}, \qquad (8.18)$$

where k_z is the wave vector along the tube axis, L is the normalization length, and (i, j) are indices that identify the quantum-confined circumferential states of the tube. The energy of the electrons is therefore given by:

$$E(k_z, n) = E_n + \frac{\hbar^2 k_z^2}{2m^*}, \qquad (8.19)$$

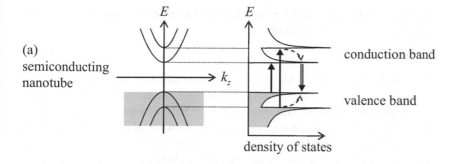

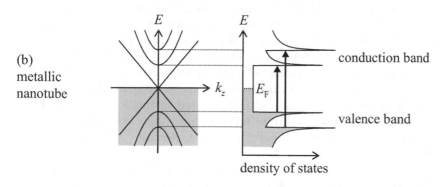

Fig. 8.24 Energy bands and corresponding density of states for (a) semiconducting nanotubes, and (b) metallic nanotubes. The shading indicates the states that are occupied. E_F is the Fermi energy. The first two absorption transitions at the energies of the van Hove singularities are indicated by the upward arrows in both cases. The downward arrow in (a) indicates an emission transition, with the dashed lines representing relaxation processes.

where n is an integer that specifies the quantum-confined state. The density of states per unit length for each band is given by (see Exercise 8.20):

$$g_n(E) = \frac{\sqrt{2m^*}}{\pi \hbar}(E - E_n)^{-1/2}, \qquad (8.20)$$

as appropriate for a 1-D material. We thus expect **van Hove singularities** in the density of states at the energies of each quantized level.

See Section 3.5 for an explanation of van Hove singularities.

   Figure 8.24 illustrates the band structure for semiconducting and metallic nanotubes together with their density of states. For each confined state we have a parabolic band, with a van Hove singularity at the energy threshold. In semiconducting nanotubes, there is an energy gap between the highest filled state in the valence band and the lowest empty state in the conduction band, as shown in part (a). The magnitude of this gap varies with the tube diameter and lies at about $0.8\,\text{eV}$ (1500 nm) for a tube with a diameter of 1 nm. (See Fig. 8.25.) Metallic nanotubes have the additional linear band derived from the K point of the Brillouin zone of graphene. (See Fig. 8.21 and its discussion.) Since the band passes through the origin, there is no gap between the top of the valence band and the bottom of the conduction band. There is therefore a continuum of states between the quantum-confined levels, as shown in part (b) of Fig. 8.24.

   Optical transitions can occur between states in the valence band and the conduction band. The selection rules dictate that the quantum number n of the electron and hole states must be identical, and conservation of momentum requires that k_z is unchanged. Owing to the van Hove sin-

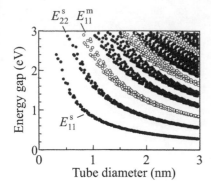

Fig. 8.25 Kataura plot of the calculated energies of the confined states versus tube diameter. The solid and open circles correspond to semiconducting and metallic nanotubes respectively. The first three energy states are labelled according to the notation of eqn 8.21, with additional superscripts to identify semiconducting (s) and metallic (m) nanotubes. See Kataura et al. (1999). Data from Dr S. Maruyama, www.photon.t.u-tokyo.ac.jp/~maruyama/nanotube.html.

gularities at the threshold for each band, the transition rate is enhanced at photon energies that satisfy

$$\hbar\omega = E_n^c - E_n^v \equiv E_{nn}, \qquad (8.21)$$

where the superscripts identify the conduction and valence bands respectively. The E_{11} and E_{22} transitions are illustrated for both semiconducting and metallic nanotubes in Figs 8.24(a) and (b) respectively. Optical transitions are, of course, possible at other photon energies, but the transitions at the frequencies that satisfy eqn 8.21 are expected to stand out from the continuum on account of their higher transition rate.

Fluorescence can be observed in semiconducting nanotubes when electrons excited in the conduction band recombine with holes in the valence band. This is typically done by exciting electrons and holes into a higher band by photo-excitation. The electrons and holes then relax by phonon emission to the lowest bands, and emit photons with energies given by $\hbar\omega = E_{11}$. This process is illustrated in Fig. 8.24(a) for the case where the electrons and holes are initially excited in the $n = 2$ bands. Fluorescence is not observed from metallic nanotubes because the hole in the valence band is very rapidly refilled by electrons from the occupied states above it.

Figure 8.25 shows a plot of the energy gap defined by eqn 8.21 as a function of the tube diameter. Such a diagram is called a 'Kataura' plot. The solid and open circles correspond to semiconducting and metallic nanotubes respectively. As we would expect for a quantum confinement effect, the magnitude of the energy gaps decrease as the tube diameter increases, varying roughly as $1/d$. For any particular tube, there is a series of energy gaps that correspond to increasing values of n. The fundamental band gap (E_{11}^s) of the semiconducting tubes moves into the visible spectral region for tube diameters smaller than about 5 nm. Note that the zero gap states of the metallic tubes are not shown in Fig. 8.25.

The calculated energy gaps shown in Fig. 8.25 can be compared to experimental data obtained by a variety of techniques, including absorption, fluorescence, and photoluminescence excitation spectroscopy, or resonant Raman scattering. The experimental data generally agree well with the theoretical calculations. For example, Fig. 8.26(a) shows the emission spectra of an ensemble of semiconducting SWNTs suspended in D_2O. Several strong peaks corresponding to the E_{11} transition of nanotubes with different diameters are clearly observed. Three specific peaks are identified, namely:

- 915 nm (1.355 eV): $(n_1, n_2) = (9, 1)$, $d = 0.75$ nm;
- 955 nm (1.298 eV): $(n_1, n_2) = (8, 3)$, $d = 0.77$ nm;
- 1023 nm (1.212 eV): $(n_1, n_2) = (7, 5)$, $d = 0.82$ nm.

As expected, the emission energy decreases with increasing tube diameter. Furthermore, other experiments demonstrate that the E_{11} absorption and emission lines occur at very similar energies, which confirms the physical picture shown in Fig. 8.24.

When individual SWNTs are isolated, the spectrum consists of just a single strong line, as shown in Fig. 8.26(b). The emission lines have a Lorentzian shape with a width of 23 meV, which is very close to $k_B T$ at room temperature. Note that there is a slight shift of the (7,5) peak in Fig. 8.26(b) compared to Fig. 8.26(a). This is a manifestation of the fact that the ensemble spectrum represents the weighted average of an inhomogeneous distribution of many different nanotubes, each of which has a Lorentzian spectrum as in part (b), but with slightly differing transition energies. The peak energy of a single SWNT can therefore lie anywhere within the inhomogeneous linewidth of the ensemble. A similar phenomenon was discussed for quantum dots in connection with Fig. 6.22(a).

The discussion above makes no consideration of excitonic effects. The exciton binding energy E_X in a carbon nanotube is much larger than in a typical bulk III–V semiconductor due to its reduced dimensionality. The binding energy varies inversely with the diameter, and is given roughly by $E_X \sim 0.3/d$ when E_X is measured in eV and d in nm. This implies that $E_X \approx 0.4$ eV for $d = 0.8$ nm. As explained in Chapter 4, the dominant excitonic transition occurs at $(E_G - E_X)$, where E_G is the band gap and E_X is the exciton binding energy. This implies that the actual band gaps are somewhat larger than the energies measured in the absorption and emission experiments, since the observed transitions are excitonic in nature.

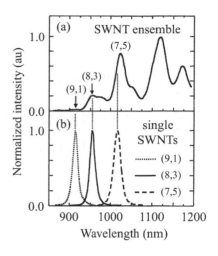

Fig. 8.26 Emission spectra from semiconducting single-wall carbon nanotubes (SWNTs) at room temperature. (a) SWNT suspension in D_2O. (b) Individual SWNTs on a glass substrate. The labels indicate the chiral vectors (n_1, n_2) for the nanotubes. After Hartschuh et al. (2003), © AAAS, reprinted with permission.

8.5.4 Carbon bucky balls

The discovery of the C_{60} molecule in 1985 by Curl, Kroto, and Smalley was recognized by the Nobel prize for chemistry in 1996. The molecule has the 'bucky ball' or 'fullerene' structure shown in Fig. 8.20(c). This structure is similar to a soccer ball, and is named after the architect R. Buckminster Fuller, who is noted for the design of geodesic domes.

The most important features of the optical spectra of C_{60} are summarized in the Jablonski diagram shown in Figure 8.27(a). The states are labelled by their parity, with 'g' (German *gerade*) and 'u' (German *ungerade*) indicating 'even' and 'odd' respectively. The HOMO ground state (S_0) has even parity, as does the LUMO first singlet excited state (S_1) at 1.85 eV, making electric-dipole transitions from S_0 to S_1 forbidden by the parity selection rule. The first allowed transition is therefore to the S_2 state at 2.7 eV.

Once the molecule has been excited to a higher-lying singlet state, it relaxes very rapidly to the S_1 state by non-radiative relaxation processes. There are then two possible decay routes back to the ground state, namely radiative transitions or intersystem crossing to the T_1 triplet level, followed by non-radiative or phosphorescent decay. Electric dipole transitions from S_1 to S_0 should again be forbidden, but vibronic coupling in the excited state and crystal disorder can create a small admixture between even and odd states, thereby partially lifting the parity selection rule. The resulting radiative lifetime is typically around 1.8 μs.

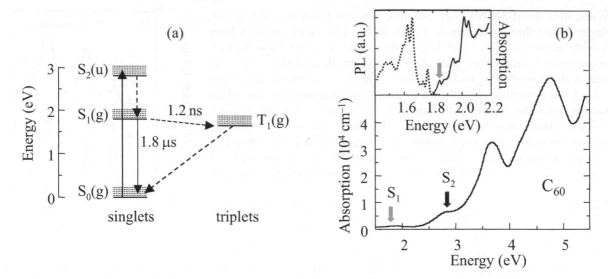

Fig. 8.27 (a) Jablonski diagram for C_{60}. The labels 'g' and 'u' denote the parity of the states. The solid arrows indicate optical transitions, while the dashed arrows indicate non-radiative relaxation processes. The time constants indicated are typical values and can vary somewhat from sample to sample. (b) Absorption spectrum of a C_{60} thin film at room temperature. The inset shows the absorption and normalized photoluminescence (PL) (dotted line) spectra for crystalline C_{60} at 10 K. The grey and black arrows indicate the onsets of the $S_0 \to S_1$ and $S_0 \to S_2$ transitions, respectively. Data taken from Ren et al. (1991), © American Institute of Physics, and Schlaich et al. (1995), © Elsevier, reprinted with permission.

Note that the time constants quoted here are only typical values, and can vary significantly between samples. The intersystem crossing rate in C_{60} is relatively fast because the S_1 and T_1 levels are nearly degenerate.

This long radiative lifetime has to be compared to the shorter inter-system crossing time of 1.2 ns. The $T_1 \to S_0$ transitions are strongly forbidden and hence have a very low probability, effectively making the decay route via T_1 a non-radiative channel. It is therefore apparent from eqn 5.5 that the radiative efficiency is very low, with typical values being around 10^{-3} for crystalline C_{60} at low temperatures.

Figure 8.27(b) shows the absorption spectrum of a solid C_{60} thin film at room temperature. The absorption at the HOMO → LUMO gap of 1.85 eV is weak, as expected for dipole-forbidden transitions, and the first strong feature is observed at 2.7 eV, which corresponds to the $S_0 \to S_2$ transition. The strong line at 3.6 eV corresponds to a transition from an odd parity lower level in the valence band to the even parity S_1 level, while the line at 4.7 eV arises from $S_0 \to S_n$ transitions, where S_n is the next odd parity singlet excited state above S_2.

The inset in Fig. 8.27 shows the photoluminescence (PL) spectrum of crystalline C_{60} at 10 K and the detailed absorption spectrum at the HOMO–LUMO band gap. As mentioned above, electric-dipole transitions between the HOMO and LUMO states are forbidden by the parity selection rule. This means that the transitions must occur either by higher-order processes (e.g. electric quadrupole) or by a mechanism that destroys the parity of the states. Examples of the second type of mechanism include crystal disorder and vibronic coupling. These both lead to distortions of the icosahedral symmetry of the C_{60} molecule—static in the former case and dynamic in the latter—and hence to a loss of the inversion symmetry that defines the parity.

Both the absorption and PL spectra in the inset of Fig. 8.27 exhibit strong vibronic sub-structure, together with additional peaks caused by the fine structure of the S_1 level. The relative intensity of the vibronic peaks in the PL spectrum is found to vary significantly from sample to sample on account of the strong sensitivity to defects and crystalline disorder. The purely electronic 0–0 vibronic line occurs at 1.846 eV and is identified by the grey arrow. The fact that this line is absent from the PL spectrum clearly demonstrates that vibronic coupling (i.e. coupling to phonons) is an important factor in allowing the radiative emission to occur.

Carbon bucky balls have potential applications as optical limiting devices in the spectral region 500–800 nm. The absorption rate is small at low powers, since the photon energy lies below the $S_0 \rightarrow S_2$ threshold at 2.7 eV. As the power is increased, the photo-excited electrons transfer rapidly to the lowest triplet state by intersystem crossing and accumulate there on account of the long lifetime of the T_1 level. These electrons can then absorb light in the same spectral region by making transitions to odd parity triplet excited states. A new absorption channel thus opens up at high intensities, which thereby limits the transmission of high power pulses. Such an optical limiter is potentially useful for making safety goggles to protect the eye from intense laser pulses.

The lifetime of the T_1 level is very long ($\gtrsim 50$ ms) because the transitions to the S_0 level violate both the spin and parity selection rules.

Chapter summary

- The optical spectra of molecular solids are determined by localized electronic and vibrational states that are closely related to the states of the isolated molecules.

- Transitions between electronic states are vibronic in character. Vibrational quanta are simultaneously excited during the transition.

- The vibronic spectra can be understood by using configuration diagrams. The Franck–Condon principle says that the configuration coordinates do not change during a vibronic transition, and so the transitions are represented by vertical arrows on configuration diagrams.

- Radiationless relaxation occurs within the vibronic bands after the transition has occurred. The emission spectra are red-shifted with respect to the absorption spectra.

- The excited states of molecules can be divided into singlets and triplets. The ground state is always a singlet. Transitions to singlet excited states are spin-allowed, and dominate the absorption and emission spectra. Singlet–triplet transitions are spin-forbidden and have very low probabilities.

- Emission from singlet states is called fluorescence, while triplet–singlet emission is called phosphorescence.

- Conjugated organic molecules have delocalized π orbitals. Larger molecules have smaller transition energies due to the reduced confinement of the π electrons.

- Conjugated molecules are generally categorized as being either small molecules or polymers. Many of these have transitions in the visible spectral region and can be used to make light-emitting diodes. Their absorption properties can also be exploited in photovoltaic devices.

- Graphene is a two-dimensional layer of carbon atoms, and has a frequency-independent absorption coefficient determined by the fine structure constant.

- Carbon nanotubes can be treated as one-dimensional systems (i.e. quantum wires) and can be either metallic or semiconducting depending on their chirality. Van Hove peaks are observed in their optical spectra at the energy thresholds for new confined states. Only semiconducting nanotubes fluoresce.

- C_{60} molecules are commonly called bucky balls or fullerenes. Their radiative efficiency is low because electric-dipole transitions are forbidden from the lowest singlet excited state.

Further reading

The electronic states of simple molecules are discussed in many introductory quantum mechanics texts, for example, Gasiorowicz (1996). More detailed discussions of molecular spectra can be found in Banwell and McCash (1994) or Haken and Wolf (1995). Klessinger and Michl (1995) give an advanced treatment of the photophysics of organic molecules.

An introduction to molecular crystals may be found in Wright (1995), while Pope and Swenberg (1999) give an authoritative treatment of the optical properties of organic crystals, polymers, and fullerenes. An overview of the properties of excitons in nanoscale systems may be found in Scholes and Rumbles (2006).

A useful collection of articles on organic electronic materials may be found in Farchioni and Grosso (2001). Reviews on organic electroluminescent devices are given in Friend et al. (1999) and Mueller (2000). An interesting collection of articles on organic electronic devices, including lasers and solar cells, may be found in Forrest and Thompson (2007). Organic photovoltaic devices are also described in depth in Brabec et al. (2008).

Reviews of the properties of graphene are given in Geim and Novoselov (2007), Geim and MacDonald (2007), and Castro Neto et al. (2009). An overview of the properties of fullerenes and nanotubes may be found in Dresselhaus et al. (1996). More recent accounts on nanotubes and graphene may be found in Reich et al. (2004), Dresselhaus et al. (2007), or Saito and Zettl (2008). A discussion of the possible applications of nanotubes in electronics and photonics may be found in Avouris (2009).

Exercises

(8.1) The Schrödinger equation for a one-dimensional harmonic oscillator is given by:

$$-\frac{\hbar^2}{2m}\frac{\mathrm{d}^2\Psi}{\mathrm{d}x^2} + \tfrac{1}{2}m\Omega^2 x^2\Psi = E\Psi.$$

Show that the following wave functions are solutions, stating the value of a and the energies of the three states.

$$\Psi_1 = C_1\,e^{-x^2/2a^2},$$

$$\Psi_2 = C_2\,x\,e^{-x^2/2a^2},$$

$$\Psi_3 = C_3\left(1 - \frac{2x^2}{a^2}\right)e^{-x^2/2a^2}.$$

(8.2) Consider the π electrons along a conjugated polymer as a one-dimensional system of length d defined by the total length of the molecule. This allows us to use the infinite potential well model described in Section 6.3.2 to estimate the electron energy. Use this approximation to find the value of d required to give the lowest energy transition at 500 nm. Hence estimate the number of repeat units within a polymer that emits at this wavelength, given that the carbon–carbon bond length is about 0.1 nm.

(8.3) The three vibrational modes of the carbon dioxide molecule have frequencies of 2×10^{13} Hz, 4×10^{13} Hz and 7×10^{13} Hz. For each mode, calculate the ratio of the number of molecules with one vibrational quantum excited to those with none when the temperature is 300 K.

(8.4) Calculate the energy difference between two pairs of isolated hydrogen atoms, one of which has both atoms in the 1s state, and the other has one atom in the 1s state and the other in the 2p state. Account qualitatively for the difference between this value and the measured transition energy of 11.3 eV between the ground state and the first electronic excited state of the H_2 molecule.

(8.5) The potential energy $U(r)$ of two neutral molecules separated by a distance r is sometimes described by the Lennard–Jones potential:

$$U(r) = \frac{A}{r^{12}} - \frac{B}{r^6}, \tag{8.22}$$

where A and B are positive fitting constants.
(a) Justify the r^{-6} dependence of the attractive part of the potential.
(b) Sketch the form of $U(r)$ and show that the energy has a minimum at $r = r_0 = (2A/B)^{1/6}$.
(c) Write down the Taylor expansion of $U(r)$ for small displacements about r_0 and hence show that the form of the potential is parabolic near the minimum. Calculate the angular frequency for harmonic oscillations about this minimum in terms of A, B and the reduced mass μ of the molecule.

(8.6) Consider a vibronic transition from the ground state to an excited state at $T = 0$.
(a) Explain why the absorption spectrum would be expected to be of the form:

$$I(\hbar\omega) \propto \sum_{n=0}^{\infty} |\langle n, Q_0'|0, Q_0\rangle|^2 \delta(\hbar\omega - \hbar\omega_0 - n\hbar\Omega),$$

where Q_0 and Q_0' are the equilibrium co-ordinates of the ground and excited states respectively, $|\langle n, Q_0'|0, Q_0\rangle|^2$ is a Franck–Condon factor, $\hbar\omega_0$ is the energy of the zero-phonon line, Ω is the vibrational frequency of the excited state, and $\delta(x)$ is the Dirac delta function.[1]
(b) The overlap factor that appears in the expression for $I(\hbar\omega)$ can be evaluated exactly and is given by:

$$|\langle n, Q_0'|0, Q_0\rangle|^2 = \frac{\exp(-S)S^n}{n!},$$

where S is the Huang–Rhys parameter for the transition. Sketch the spectra that would be expected for the cases: (i) $S = 0$, (ii) $S = 1$, and (iii) $S = 5$.

(8.7) The absorption spectrum of benzene is found to consist of a series of lines with wavelengths given by 267 nm, 261 nm, 254 nm, 248 nm, 243 nm, 238 nm, and 233 nm. Estimate the energy of the S_1 excited state, and the dominant vibrational frequency of the molecule.

(8.8) Use the data shown in Fig. 8.9 to draw a schematic configuration diagram for the ground state and the first two singlet excited states of the ammonia molecule.

(8.9) Explain why spin–orbit coupling can allow radiative transitions between singlet and triplet states.

---

[1]The Dirac delta function is defined with $\delta(x) = 0$ for $x \neq 0$, and $\int_{-\infty}^{+\infty} \delta(x)\,\mathrm{d}x = 1$. It can be considered as the $\xi \to 0$ limit of a function that is zero everywhere except in the range $-\xi/2 \to +\xi/2$, where it takes the value of $1/\xi$.

(8.10) Under certain circumstances it is possible to observe weak emission at 760 nm from the pyrromethene dye studied in Fig. 8.10. This emission is found to have a lifetime of 0.3 ms. Suggest a possible explanation for this result.

(8.11) Use the data shown in Fig. 8.13 to estimate the energy of the dominant vibrational modes of anthracene crystals and molecules in solution.

(8.12) Apply the mirror symmetry rule to the absorption spectrum of anthracene crystals shown in Fig. 8.13 to deduce the shape of the emission spectrum.

(8.13) Repeat Exercise 8.12 for the PDA absorption spectrum shown in Fig. 8.16.

(8.14) The absorption edge of crystalline anthracene ($C_{14}H_{10}$) occurs at 400 nm, but photoconductivity experiments show a different threshold at 295 nm. Deduce the value of the binding energy of the ground-state Frenkel exciton of anthracene.

(8.15) In Raman scattering, photons are shifted to lower energy by emitting vibrational quanta as they pass through the sample. The angular frequency of the down-shifted photon is equal to $(\omega - \Omega)$, where ω and Ω are the frequencies of the incoming photon and the vibrational mode. By referring to the data given in Fig. 8.17, estimate the wavelength of the Raman-shifted photons generated when a helium neon laser operating at 633 nm is incident on a sample of MeLPPP.

(8.16) Equal numbers of electrons and holes are injected into two identical samples of a molecular material. In one case, the electrons and holes are injected optically; in the other, they are injected electrically. Explain why the luminescence from the electrically excited sample is expected to be four times weaker than that from the optically excited one. (Hint: this is related to the formation of triplet excitons.)

(8.17) A polymer light-emitting diode emits at 550 nm at an operating current of 10 mA.
(a) Explain why the maximum quantum efficiency we might expect from the device is only 25%.
(b) Calculate the total optical power emitted on the assumption that the internal quantum efficiency is 25%.
(c) If the operating voltage is 5 V, what is the power conversion efficiency? Would you expect to obtain this efficiency in a practical device ?

(8.18) (a) Consider the lattice translation vectors of graphene shown in Fig. 8.23. Show that $a_0 = \sqrt{3}a$, where $a_0 = |\boldsymbol{a_1}| = |\boldsymbol{a_2}|$, and a is the carbon–carbon distance (0.142 nm).
(b) Prove equation 8.15.
(c) Prove eqn 8.16.

(8.19) (a) Explain why the $\boldsymbol{k}$ states that are available to electrons in a carbon nanotube with chiral vector $\boldsymbol{c}$ must satisfy the condition:

$$\boldsymbol{k} \cdot \boldsymbol{c} = 2\pi m \,,$$

where m is an integer.
(b) The metallic nature of graphene arises from the zero energy gap at $\boldsymbol{k} = (\boldsymbol{k_1} - \boldsymbol{k_2})/3$, i.e. at the K point of the Brillouin zone. ($\boldsymbol{k_1}$ and $\boldsymbol{k_2}$ are the reciprocal lattice vector of graphene.) A nanotube will therefore be metallic if the K point is one of its allowed $\boldsymbol{k}$ vectors. Show that this condition is only fulfilled when the chiral vector (n_1, n_2) satisfies:

$$n_1 - n_2 = 3m \,,$$

where m is an integer.

(8.20) By considering a nanotube as a 1-D quantum wire with electron energies given by eqn 8.19, show that the density of states is given by eqn 8.20.

(8.21) Calculate the radiative efficiency of a C_{60} molecule with the time constants given in Fig. 8.27(a).

Luminescence centres

9

In this chapter we consider the physics of optically active defects and impurities in crystalline host materials. The electronic states of the defects are strongly coupled to the phonons of the crystal by the electron–phonon interaction. We are thus dealing with a vibronic system, with optical properties analogous to the vibrational–electronic spectra of the molecular materials studied in Chapter 8. We therefore begin by reviewing the physics of vibronic transitions, and then focus on two general categories of luminescence centres, namely colour centres and paramagnetic ion impurities. These materials are widely used in solid-state lasers and phosphors.

9.1	**Vibronic absorption and emission**	**247**
9.2	**Colour centres**	**250**
9.3	**Paramagnetic impurities in ionic crystals**	**255**
9.4	**Solid-state lasers and optical amplifiers**	**261**
9.5	**Phosphors**	**264**
	Chapter summary	**266**
	Further reading	**267**
	Exercises	**268**

9.1 Vibronic absorption and emission

The electronic states of the impurity atoms doped into a crystal couple strongly to the vibrational modes of the host material through the electron–phonon interaction. This gives rise to continuous **vibronic** bands that are conceptually different from the electronic bands studied in the band theory of solids. The electronic states are localized near specific lattice sites in the crystal, and the continuous spectral bands arise from coupling the discrete electronic states to a continuous spectrum of vibrational (phonon) modes. This contrasts strongly with interband transitions which involve continuous bands of delocalized electronic states.

The basic processes involved in the vibrational–electronic transitions in molecular materials were described in Sections 8.2.2–8.2.4. The principles developed there form a good starting point for the more general vibronic systems that we consider here. There are, however, two additional aspects of the physics that need to be discussed.

(1) We shall be dealing with the optical transitions from a low density of luminescent dopant ions or defects within an optically inert crystal. The interaction with the crystal host therefore has a strong effect on the spectra.

(2) We must consider the coupling of the electronic states to a continuous spectrum of vibrational modes, rather than the discrete modes of a molecule. The density of states for the vibrational modes is determined by the phonon dispersion curves.

The formation of vibronic bands is depicted schematically in Fig. 9.1. Figure 9.1(a) shows the optical transitions between the ground state of an isolated atom (e.g. a dopant ion) at energy E_1 and one of its electronic

In principle, molecular crystals also have continuous phonon bands. In practice, however, the vibronic spectra are usually dominated by the localized vibrational modes associated with the internal vibrations of the molecule itself rather than delocalized phonon modes of the whole crystal.

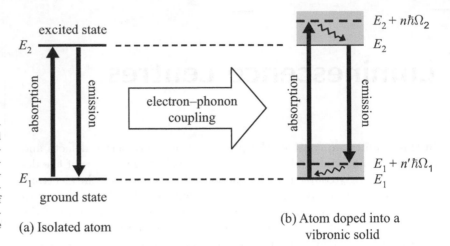

Fig. 9.1 (a) Optical transitions between the ground state and an excited state of an isolated atom. (b) Absorption and emission transitions in a vibronic solid, in which the electron–phonon interaction couples each electronic state to a continuous band of phonons. Note that energies of the electronic states might also be shifted in the solid: see Section 9.3.1.

excited states at energy E_2. If this atom is inserted into a crystalline host material, the electronic levels can couple to the vibrations of the lattice through the electron–phonon interaction. At this stage, we do not wish to enter into the microscopic details of how such an interaction might occur, but merely consider the possibility that the coupling might be present. The presence of the coupling associates a continuous band of phonon modes with each electronic state, as shown in Fig. 9.1(b).

Optical transitions can occur between the vibronic bands if the selection rules permit them. We first consider an absorption transition. Before the photon is incident, the electron will be at the bottom of the ground-state band. The absorption of a photon simultaneously puts the electron in an excited electronic state and creates one or more phonons, as shown in Fig. 9.1(b). Conservation of energy requires that:

$$\hbar\omega_a = (E_2 + n\hbar\Omega_2) - E_1 = (E_2 - E_1) + n\hbar\Omega_2\,, \tag{9.1}$$

where $\hbar\omega_a$ is the energy of the photon, n is an integer, and Ω_2 is the phonon angular frequency. Equation 9.1 shows that absorption is possible for a band of energies from $(E_2 - E_1)$ up to the maximum energy allowed by the electron–phonon coupling.

After the photon has been absorbed, the electron relaxes to the bottom of the upper band by non-radiative processes. The system then returns to the ground state band by a vibronic transition of energy:

$$\hbar\omega_e = E_2 - (E_1 + n'\hbar\Omega_1) = (E_2 - E_1) - n'\hbar\Omega_1\,, \tag{9.2}$$

where Ω_1 is the phonon angular frequency, and n' is an integer. Once the electron is in the ground-state band, it relaxes to the bottom of the band by non-radiative transitions, dissipating the excess vibrational energy as heat in the lattice.

In the relaxation process, the vibrational energy of the phonons excited during the optical transition rapidly spreads throughout the other phonon modes in the crystal and ultimately becomes heat.

On comparing eqns 9.1 and 9.2, we see that in a vibronic system the emission generally occurs at a lower energy than the absorption. This red shift is called the **Stokes shift**. It is apparent from Fig. 9.1(b) that

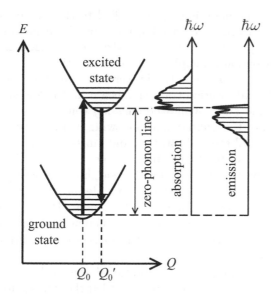

Fig. 9.2 Configuration diagram for the ground state and one of the excited electronic states of a vibronic solid. The optical transitions are indicated by the vertical arrows. The right-hand side of the figure shows the general shape of the absorption and emission spectra that would be expected.

the Stokes shift arises from the vibrational relaxation that takes place within the vibronic bands. This contrasts with isolated atoms in which the absorption and emission lines occur at the same frequency.

The Stokes shift between absorption and emission can be understood in more detail by using configuration diagrams. The concept of configuration diagrams was introduced in Section 8.2.3 in the context of the vibrational–electronic spectra of molecules. This model carries over directly to the discussion of the optical transitions in a vibronic solid. The electronic energy of the optically active species is a function of the vibrational configuration of the system as shown schematically in Fig. 9.2. This diagram shows the energy of two electronic states of a vibronic system as a function of Q, the configuration coordinate. We have assumed that the electronic states are bound, and they therefore have a minimum energy for some value of Q. In general, the equilibrium positions for the two states will occur at different values of the configuration coordinate. Therefore we label the positions of the minima for the ground state and excited states as Q_0 and Q_0' respectively. The difference between Q_0 and Q_0' is quantified by the Huang–Rhys parameter for the transition. (See eqn 8.9.)

The basic physical processes involved in the optical transitions of a vibronic solid are similar to those in a molecule, and we only give a brief summary here. More details can be found in Section 8.2.3. The energy of the electronic ground state can be expanded as a Taylor series about the minimum at Q_0 as follows:

$$E(Q) = E(Q_0) + \frac{\mathrm{d}E}{\mathrm{d}Q}(Q - Q_0) + \frac{1}{2}\frac{\mathrm{d}^2 E}{\mathrm{d}Q^2}(Q - Q_0)^2 + \cdots . \quad (9.3)$$

Since we are at a minimum, we know that $\mathrm{d}E/\mathrm{d}Q$ must be zero. Hence the $E(Q)$ curve will be approximately parabolic for small displacements

On first encountering configuration diagrams, it is quite confusing to understand exactly what the configuration coordinate represents physically. In the case of molecules discussed in Section 8.2.3, it is easy to see that Q corresponds to the amplitude of one of the normal modes of the vibrating molecule. In a vibronic solid, Q might, for example, represent the average separation of the dopant ion from the cage of neighbouring ions in the host lattice. In this case, the vibrations would correspond to a breathing mode in which the environment pulsates radially about the optically active ion. This is equivalent to a localized phonon mode of the whole crystal. In general there will a large number of vibrational modes in a solid, and the configuration coordinate can represent the amplitude of any one of these modes or perhaps a linear combination of several of them.

from Q_0. The same analysis can be applied to the excited state. This means that to first order we have harmonic oscillator potentials with a series of equally spaced energy levels as sketched in Fig. 9.2.

The Franck–Condon principle discussed in Section 8.2.4 tells us that optical transitions are represented by vertical arrows on the configuration diagram. The absorption transition begins in the lowest vibrational level of the ground state, while the emission commences at the lowest vibrational level of the excited state following non-radiative relaxation. This gives rise to vibronic absorption and emission bands as shown in the right-hand side of the figure. In principle, the absorption and emission bands for a particular vibrational mode should consist of a series of discrete lines similar to those observed in molecules, each corresponding to the creation of a specific number of phonons. However, in practice the electronic states can couple to many different phonon modes with a whole range of frequencies, and thus the spectra usually fill out to form continuous bands.

The transitions from the lowest vibrational level of the ground state to the lowest level of the excited state are called the **zero-phonon lines**. Since there are no phonons involved, we have $n = n' = 0$ in eqns 9.1 and 9.2, so that the absorption and emission lines occur at the same frequency. In the absorption spectrum there will be a band of vibronic transitions with energies larger than that of the zero-phonon line, while in the emission spectra there will be a corresponding band with lower energy. The shapes of the absorption and emission bands depend on the overlap of the vibrational wave functions as determined by the Franck–Condon factor given in eqn 8.12. In general, the peak occurs away from the zero-phonon line on account of the difference between Q_0 and Q_0'. As with molecules, we would expect mirror symmetry between the emission and absorption about the zero-phonon line. In the sections that follow, we apply these general principles to the optical spectra of colour centres and luminescent impurities.

9.2 Colour centres

Insulator crystals like diamond have large band gaps and should therefore be colourless. However, it is not uncommon to come across imperfect crystals that contain vacancies in the lattice and are coloured (e.g. pink diamonds). The defects that cause the colouration are called **colour centres** or **F-centres**, where the F stands for *Farbe*, the German word for colour. In this section we consider two important examples of colour centres, namely F centres in alkali halides and nitrogen vacancy defects in diamond.

9.2.1 F-centres in alkali halides

The alkali halides are colourless insulators with band gaps in the ultraviolet spectral region. (See Table 4.3.) Figure 9.3(a) gives a schematic representation of an F-centre in an alkali halide crystal. The F-centre

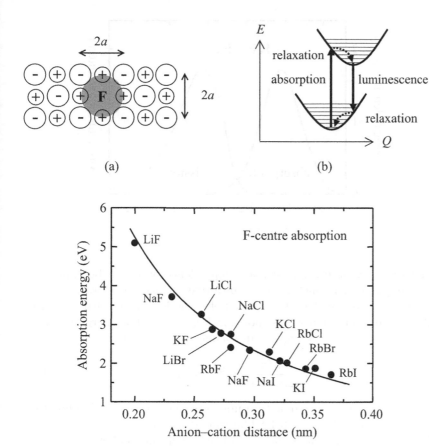

Fig. 9.3 (a) An F-centre in an alkali halide crystal. The centre consists of an electron trapped at an anion vacancy. The shaded region represents the wave function of the electron. (b) Configuration diagram corresponding to the vibronic transitions of the trapped electron in an F-centre.

Fig. 9.4 Energy (E) of the peak absorption in the F-band for several face-centred cubic alkali halide crystals. The energies are plotted against the anion–cation distance a. The solid line is a fit with $E \propto 1/a^2$. After Baldacchini (1992), reprinted with permission from Plenum Publishers.

consists of an electron trapped at an anion (i.e. negative ion) vacancy. The anion vacancies are typically created by introducing an excess of the metal ion. This might be done, for example, by heating the crystal in alkali vapour and then cooling it quickly. Alternatively, the vacancies can be produced by irradiation with X-rays or by electrolysis. The absence of the negative ion acts like a positive hole that can attract an electron. The trapped electron is in a bound state with characteristic energy levels.

Optical transitions between the bound states of the trapped electron cause the colouration of the crystals. The trapped electrons couple to the vibrations of the host crystal and this gives rise to vibronic absorption and emission. The processes that take place are illustrated in the generic configuration diagram shown in Fig. 9.3(b). These transitions are known as F-bands.

Experimental measurements on the F-centres in alkali halides indicate that the frequency of the F-band absorption is approximately proportional to a^{-2} where a is the cation–anion distance in the host crystal. This is clearly evident in the data shown in Fig. 9.4, which plots the energy of the peak in the F-band absorption as a function of a. The solid line is a fit to the data with the energy proportional to $1/a^2$.

Most of the alkali halide crystals have the face-centred cubic sodium-chloride structure, in which the cube edge dimension is equal to $2a$. The exceptions are CsCl, CsBr, and CsI, which have simple cubic structures.

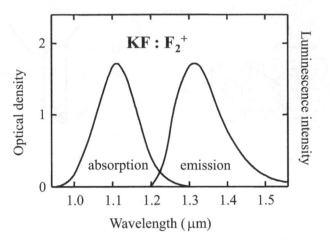

Fig. 9.5 Absorption and emission bands of the F_2^+ centre in KF. After Mollenauer (1985), © Excerpta Medica Inc., reprinted with permission.

This approximate inverse-square dependence on a can be explained by a simple model which gives an intuitive understanding of the basic physics. We assume that the trapped electron is confined inside a rigid cubic box of dimension $2a$, as shown in Fig. 9.3(a). The energy levels of an electron of mass m_0 trapped in such a box are given by:

$$E = \frac{\hbar^2 \pi^2}{2m_0(2a)^2}(n_x^2 + n_y^2 + n_z^2),\qquad (9.4)$$

where n_x, n_y and n_z are quantum numbers that specify the bound electronic states. (See Exercise 9.1.) The ground state has $n_x = n_y = n_z = 1$, while the first excited state has one of the quantum numbers equal to 2. The lowest energy transition thus occurs at a photon energy given by:

$$h\nu = \frac{3h^2}{8m_0}\frac{1}{(2a)^2}.\qquad (9.5)$$

Detailed analysis reveals that the energy of the F-band absorption actually scales as $a^{-1.84}$ rather than a^{-2}. This empirical formula is called the Mollwo–Ivey relationship.

The model therefore predicts the approximate a^{-2} dependence of the F-band absorption energy, but overestimates the transitions energies somewhat. (See Exercise 9.2.) A more realistic approach would calculate how the electron wave function tries to maximize its overlap with the positive ions while minimizing the overlap with the negative ones.

The simple electron in a box model can also explain the microscopic origin of the coupling between the trapped electrons and the vibrations of the host crystal. A displacement of the neighbouring ions from their equilibrium positions would alter the size of the box in which the electron is trapped. This in turn would alter the electronic energy through eqn 9.4. Such a displacement of the ions could be caused by a vibration of the crystal. Hence the vibrations are coupled to the electronic energy levels, and we have a vibronic system.

The transition energies in the F_2^+ centres are lower than those of F-centres because the electron can move over two lattice sites, and hence the box in which the electron is confined is larger. (See Exercise 9.4.)

Figure 9.5 shows the absorption and emission bands of a slightly more complicated type of colour centre, namely the F_2^+ centre in KF. The emission bands of this F-centre are in the near-infrared spectral region, and the crystals can be used to make tunable lasers, as will be discussed further in Section 9.4. The F_2^+ centre consists of a single electron trapped at two adjacent anion vacancies. Since the centre consists of one electron and two holes, it has a net positive charge of one unit. The Stokes shift and the mirror symmetry between the absorption and emission are clearly evident in the data.

9.2.2   NV centres in diamond

The nitrogen vacancy (NV) centre in diamond has been widely studied in recent years on account of its possible application in quantum information processing (QIP) and magnetometry. Figure 9.6 shows a schematic diagram of the defect. It consists of a substitutional nitrogen atom in the diamond crystal with a vacancy at an adjacent lattice site. These defects occur in natural diamonds, but are normally studied in a more controlled way in synthetic crystals. The substitutional nitrogen impurities may be naturally present in the crystal, and the vacancies are then introduced, for example, by irradiation with electrons, protons, or neutrons. Alternatively, both nitrogen impurities and vacancies can be introduced in high-purity crystals by nitrogen implantation techniques. In both cases, the NV centres are formed when the vacancies migrate to the nitrogen sites during annealing.

NV centres are normally labelled with a superscript that identifies their charge state. The neutral centre is labelled NV^0, while NV^- designates a centre that contains a single trapped electron. The NV^- centre is the most common charge state, and is our focus of attention here. The trapped electron in an NV^- centre interacts with the unpaired electron from each of the three adjacent carbon atoms and also with the two unpaired electrons from the nitrogen atom, forming a six-electron system. These six electrons can be considered as two holes in a filled $n = 2$ atomic shell, giving rise to resultant spin states with $S = 0$ (singlets) or $S = 1$ (triplets).

The level scheme for the NV^- centre is shown in Fig. 9.7(a). The ground state is a triplet labelled ^3A. This label refers to the symmetry, with the superscript indicating the spin multiplicity, i.e. $(2S+1)$. Electric dipole allowed optical transitions are possible to the first triplet excited state labelled ^3E, which lies 1.945 eV above the ground state. Dipole allowed transitions back to the ground state are then possible, with a lifetime of 13 ns. Both the ground and excited states are broadened by vibronic coupling, which leads to broad absorption and emission bands.

Figure 9.7(b) shows the absorption spectrum of NV^- centres at 80 K, together with the emission spectrum at 2 K. The luminescence spectrum is measured by exciting the crystal within the vibronic absorption band, for example with an argon ion laser operating at 514 nm (2.41 eV). The zero-phonon line at 1.945 eV (637.2 nm) is well resolved in both spectra, with the vibronic absorption and emission showing approximate mirror symmetry about that line. The quantum efficiency at low temperatures is less than 100% on account of a singlet excited state labelled ^1A between the ^3E and ^3A levels. Electrons in the ^3E band can transfer non-

There are several other defect centres in diamond that are interesting for applications in QIP. For example, the NE8 centre, which consists of a nickel atom surrounded by four nitrogen atoms within a diamond crystal, is a very good single-photon source at 802 nm. See Gaebel et al. (2004) for further details.

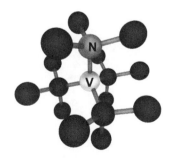

Fig. 9.6 Schematic representation of an NV centre in diamond, which consists of a substitutional nitrogen atom with an adjacent vacancy. Image courtesy of P. Neumann and F. Reinhard.

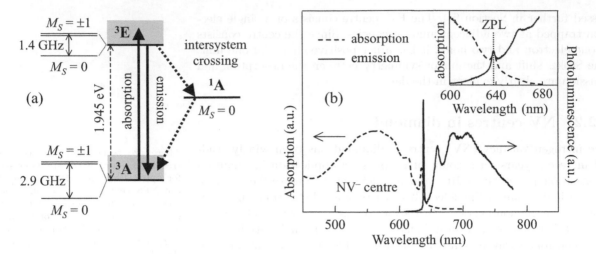

Fig. 9.7 (a) Level diagram for the NV^- centre in diamond. Both the ground state and the triplet excited state are split into a singlet and doublet by the crystal field, as indicated on the expanded scale to the left. (b) Absorption spectrum at 80 K, and emission spectrum at 2 K. The inset shows the corresponding spectra at room temperature near the zero-phonon line (ZPL) in more detail. Data from Mita (1996), © American Physical Society, Jelezko et al. (2002), © American Institute of Physics, and Acosta et al. (2009), © American Physical Society, reprinted with permission.

radiatively to the ^1A level by intersystem crossing. The time constant for this transfer process is around 30 ns for the $M_S = \pm 1$ sub-levels of the ^3E state. The ^1A level is metastable on account of its long (~ 300 ns) lifetime, which reflects the spin-forbidden nature of the transitions back to the ground state. Hence the intersystem decay route via the ^1A level is effectively non-radiative.

The appeal of NV^- centres in QIP is based on a number of factors.

(1) The defect density can be controlled to a level where only one is present within a focused laser spot, making it possible to study the properties of *single* NV^- centres.

(2) The ^3A ground state is paramagnetic (i.e. $S \neq 0$). The $M_S = 0$ and $M_S \pm 1$ sub-levels are split by the crystal field as shown in Fig. 9.7(a), and form a two-level quantum system that can serve as a quantum bit (or 'qubit').

(3) The state of the qubit can be controlled coherently by electron spin resonance (ESR) at the frequency which corresponds to the energy splitting of the sub-levels, namely 2.9 GHz. The coherence time of the spin can be extremely long (up to ~ 2 ms at room temperature), and this permits many operations to be performed before dephasing occurs.

Optical pumping is a technique of atomic physics by which a spin polarization in the ground state is created. The technique involves repeated excitation to an excited state followed by relaxation back to the ground state, and relies on selection rules that favour relaxation to some magnetic sub-levels in preference to others.

(4) The initial state of the system can be defined by optical pumping, and the final state read-out by optical spectroscopy.

The full details of how all of this is done may be found in the literature cited for further reading. We restrict ourselves here to a brief discussion of the fourth point, which concerns the optical properties of the defect.

We adopt a notation in which we use a subscript to indicate the value of M_S, so that, for example, $^3A_0$ refers to the $M_S = 0$ sub-level of the ^3A state.

The ability to produce a well-defined initial spin state is an important requirement for QIP with the NV^- centre. This is not easily achieved on account of the small energy splitting of the $^3A_0$ and $^3A_{\pm 1}$ sub-levels, and would normally require extremely low temperatures. (See Exercise 9.5.) However, the selection rules for optical transitions make it possible to use optical pumping methods to populate the $^3A_0$ sub-level selectively. The selection rules dictate that M_S is conserved during an optical transition. Thus if the system starts in the $^3A_0$ sub-level, it will return to the same sub-level after excitation to the ^3E excited state. However, if the system starts in the $^3A_{\pm 1}$ sub-levels and is excited to the $^3E_{\pm 1}$ states, there is a finite probability that it can relax to the $^3A_0$ sub-level if intersystem crossing via the $^1A_0$ state occurs. Hence a non-thermal population of the $^3A_0$ sub-level can be created after a few cycles of the optical excitation–relaxation process.

The other aspect of the optical properties of the NV^- centre that makes it useful for QIP is that it is possible to read out the M_S value of the ground state by optical spectroscopy. This method relies on the M_S-conserving nature of the optical excitation process, and the fact that the probability for $^3E \rightarrow {}^1A$ intersystem crossing is much larger for the $^3E_{\pm 1}$ sub-levels than for the $^3E_0$ sub-level. The $^3E_{\pm 1}$ sub-levels therefore have a smaller radiative efficiency, which means that the M_S value of the ground state can be determined by exciting the system to the ^3E band with a laser and measuring the emission intensity. If the system is initialized in the $^3A_0$ sub-level by optical pumping, a change in the M_S value induced by ESR techniques can then be detected by measuring a decrease in the fluorescence intensity.

Intersystem crossing from the $^3E_0$ excited state to the $^1A_0$ intermediate state has a very low probability. The relative probability of intersystem crossing from $^3E_{\pm 1} \rightarrow {}^1A_0$ compared to radiative relaxation from $^3E_{\pm 1} \rightarrow {}^3A_{\pm 1}$ is about 40%. See Manson et al. (2006) for a discussion of the relative decay rates.

9.3 Paramagnetic impurities in ionic crystals

In this section we discuss the optical transitions of paramagnetic metal ions doped into ionic crystals. We focus on ions from the transition metal and rare earth series of the periodic table. These have optically active unfilled 3d or 4f shells respectively, as listed in Table 9.1. They are naturally present in certain minerals, but are deliberately doped into synthetic crystals for technological applications. The optical transitions of these doped crystals are the basis for many solid-state lasers, and are also widely used in phosphors for fluorescent lighting and cathode ray tubes.

Table 9.1 Atomic number Z and electronic configuration of the atoms from the transition metal and rare-earth series of the periodic table.

Series	Z	Configuration
Transition metal	21–30	[Ar] $3d^n\,4s^2$
Rare earth	58–70	[Xe] $4f^n\,6s^2$

9.3.1 The crystal-field effect and vibronic coupling

Metal ions doped as impurities in an ionic crystal substitute at the cation (i.e. positive ion) lattice sites. For example, when Cr_2O_3 is doped into an Al_2O_3 crystal to form ruby, the Cr^{3+} ions directly substitute for the Al^{3+} ions. The impurities will normally be present at a low density, so that the interactions between neighbouring dopants are negligible due to their large separation. Hence, the main effect that we need to consider is the perturbation of the electronic levels of the dopant ions due to the crystalline environment in which they are placed.

The optical properties of free ions in the gas phase are characterized by sharp emission and absorption lines with wavelengths determined by their energy levels. When the same ions are doped into a crystalline host, the optical properties are modified by the interactions with the crystal. If the interaction is weak, the emission and absorption spectra will remain as discrete lines, but perhaps with their frequency slightly shifted and certain degeneracies lifted. On the other hand, if the interaction is strong, the frequencies of the transitions will be quite different from those of the isolated ions, and the spectra may be broadened into a continuum. We shall see below that the 4f series dopants are generally weakly coupled to the crystal, while the 3d series tend to be strongly coupled.

A positive ion doped into a crystal finds itself surrounded by a regular matrix of anions (i.e. negative ions). For example, the Cr^{3+} ions in ruby are surrounded by six O^{2-} ions arranged in an octahedral arrangement, as depicted in Fig. 9.8. These negative ions produce an electric field at the cation site, which perturbs the atomic levels of the ion. This interaction is known as the **crystal-field effect**.

The shift of the energy levels of the dopant ion caused by the crystal field can be calculated by perturbation theory. The calculation starts with the gross structure of the free ion with the electrons arranged in the principal atomic shells. It then proceeds by adding on perturbations in order of diminishing size. The details of these perturbation calculations are beyond the scope of this text, and at this level we are just able to make four qualitative remarks.

(1) The crystal-field coupling can be considered to consist of two different contributions. The first arises from the **static** crystal field. This is the perturbation to the energy levels caused by the electric field of the crystal when all of the ions are at their time-averaged equilibrium positions. The second is the **dynamic** effect. This refers to the additional perturbation caused by displacing the neighbouring anions from their equilibrium positions, which alters the electric field experienced by the dopant ion and hence alters the perturbation to its energy levels.

(2) The lifting of the degeneracies of the atomic levels of the free ion due to the *static* field is determined by the symmetry of the crystalline environment. (See, for example, Exercises 9.8 and 9.9.) A useful analogy can be made here with the case of a free atom in a magnetic field. The free atom is spherically symmetric, which implies that the magnetic levels are degenerate. The application of an external field defines a preferred axis, and the levels split by the Zeeman effect. The same is true for the ions doped in the crystal. The magnetic levels of the free ion are degenerate, but are split in the crystal because the crystal defines axes so that not all directions are equivalent. This point is illustrated in Fig. 1.8.

Fig. 9.8 The octahedral crystal environment. The cation dopant is surrounded by six equidistant anions located at the corners of an octahedron.

Group theory provides an extremely powerful tool for working out the lifting of degeneracies by the static crystal field. For example, it tells us that the five-fold degenerate 3d orbitals of a free transition metal ion are split into a doublet and triplet by an octahedral crystal field. However, it cannot tell us which of the levels is at the higher energy, or the size of the splitting. This requires detailed numerical modelling of the interaction between the cation and the electric field of the neighbouring anions: see Exercise 9.9.

(3) The *dynamic* crystal-field effect is the origin of the vibronic coupling in these systems. Vibrations of the crystal cause the ions to be displaced from their equilibrium positions and therefore alter the electric field experienced by the dopant ion. This in turn alters the perturbation of the electronic levels, and thus couples the vibrations to the electronic levels of the system. This is equivalent to an electron–phonon interaction. One way to look at this is to consider the phonon as acting like an amplitude modulation on the crystal field. This induces side bands at the phonon frequency on the electronic levels through the crystal-field effect. In some cases it is possible to resolve distinct side bands in the optical spectra that correspond to specific phonon frequencies, but more often than not, the side bands form a continuum due to the continuous distribution of frequencies of the phonon modes.

(4) The magnitude of the crystal-field effects for the transition metal and rare-earth ions are very different. This is a consequence of the electronic configurations of the optically-active electrons. (See Table 9.1.) Transition-metal ions are formed when the outermost 4s electrons of the neutral atoms are removed. The 3d orbitals therefore lie on the outside of the ion and have a large radius. Rare-earth ions, by contrast, are formed when the outermost 6s electrons are removed. They have a relatively small radius (see Exercise 9.7), and are partly shielded from external fields by the electrons in the filled 5s and 5p shells. This means that the transition-metal ions are much more sensitive to the crystal field than the rare earths.

These four points apply to a wide range of paramagnetic ions in crystalline hosts. In the following subsections, we discuss the properties of the 3d and 4f series ions separately, starting with the rare earths.

9.3.2 Rare-earth ions

The rare-earth ions occur in the periodic table after lanthanum (element 57) and are therefore alternatively known as **lanthanides**. The ions commonly occur in divalent or trivalent forms. For example, the neutral europium atom has an electronic configuration of [Xe] $4f^7 6s^2$, and ions are formed by losing the 6s electrons first and then one of the 4f electrons, giving configurations of [Xe] $4f^7$ (Eu^{2+}) and [Xe] $4f^6$ (Eu^{3+}). We restrict our attention here mainly to the trivalent ions, which have the electronic configurations listed in Table 9.2.

The magnitude of the crystal-field effect in rare-earth ions is relatively small, due to the screening of the optically-active levels (see point 4 above). Furthermore, the spin–orbit coupling is quite large because it roughly varies as Z^2, and Z is in the range 58–70. This means that the crystal-field effects are smaller than the spin–orbit coupling. Therefore, in treating the crystal-field effects by perturbation theory, we must apply the spin–orbit interaction first.

The spin–orbit interaction splits the gross structure of the free ions into fine structure terms defined by the quantum numbers $|L, S, J\rangle$, denoted in spectroscopic notation as $^{2S+1}L_J$. (See Appendix C.) The crys-

In one electron atoms the spin–orbit interaction scales as Z^4, but this reduces to an approximate Z^2 dependence when there is screening by other electrons.

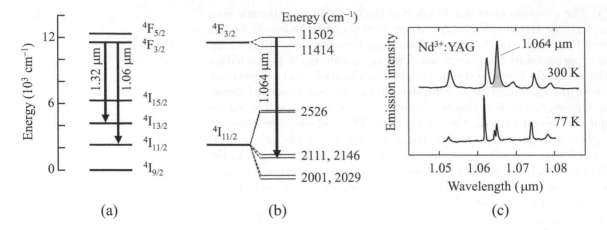

Fig. 9.9 (a) Energy-level diagram for Nd^{3+} ions in a YAG crystal. The energies are given in wave number units. ($1\,cm^{-1} \equiv$ 1.240×10^{-4} eV.) (b) Fine structure of the $^4F_{3/2} \rightarrow {}^4I_{11/2}$ transition. (c) Emission spectrum for the $^4F_{3/2} \rightarrow {}^4I_{11/2}$ transition at 77 K and 300 K. The laser transition at $1.064\,\mu m$ has been highlighted by shading. After Koningstein and Geusic (1964), © American Physical Society, reprinted with permission.

Table 9.2 Electronic configurations of trivalent rare-earth ions. Z is the atomic number.

Ion	Z	Configuration
Ce^{3+}	58	[Xe] $4f^1$
Pr^{3+}	59	[Xe] $4f^2$
Nd^{3+}	60	[Xe] $4f^3$
Pm^{3+}	61	[Xe] $4f^4$
Sm^{3+}	62	[Xe] $4f^5$
Eu^{3+}	63	[Xe] $4f^6$
Gd^{3+}	64	[Xe] $4f^7$
Tb^{3+}	65	[Xe] $4f^8$
Dy^{3+}	66	[Xe] $4f^9$
Ho^{3+}	67	[Xe] $4f^{10}$
Er^{3+}	68	[Xe] $4f^{11}$
Tm^{3+}	69	[Xe] $4f^{12}$
Yb^{3+}	70	[Xe] $4f^{13}$

tal field then perturbs these states, shifting their energies slightly and causing new splittings. However, the size of these shifts is much smaller than the spin–orbit splittings, and so the optical spectra of the dopant ions are generally fairly similar to those of the free ions.

As an example of these effects, we can consider the optical spectra of Nd^{3+} ions doped into an yttrium aluminium garnet ($Y_3Al_5O_{12}$ or 'YAG') crystal. We choose this example because Nd:YAG crystals form the gain medium in one of the most important solid-state lasers. The electronic configuration of Nd^{3+} is [Xe] $4f^3$. Hund's rules tell us that the ground state has $S = 3/2$, $L = 6$, and $J = 9/2$, which is a $^4I_{9/2}$ level. Above this ground state there is a progression of excited states. Figure 9.9(a) shows the first five excited states without the crystal-field fine structure. Two important transitions are identified, namely the $^4F_{3/2} \rightarrow {}^4I_{13/2}$ line at $1.32\,\mu m$ and the $^4F_{3/2} \rightarrow {}^4I_{11/2}$ line at $1.06\,\mu m$. Lasing has been demonstrated for both transitions, although the $1.06\,\mu m$ line is the more important.

The $^4F_{3/2} \rightarrow {}^4I_{13/2}$ and $^4F_{3/2} \rightarrow {}^4I_{11/2}$ laser transitions in Nd^{3+} have $\Delta J = 5$ and 4 respectively, and are therefore forbidden for the free ion. In the crystal, the selection rules are not so strict, and this provides a small probability for the transitions to occur. The mechanism that relaxes the selection rules is the crystal-field interaction, which can mix states of different J. Since the states arise from the same electronic configuration of $4f^3$, they have the same parity, and the transitions must proceed by magnetic-dipole processes. However, the absence of local inversion symmetry at the Nd^{3+} sites means that the parity is no longer well-defined in the crystal, and hence that there is also some probability for electric-dipole processes. The upper state has a long lifetime of $230\,\mu s$ at 300 K on account of the selection rules. This long lifetime is beneficial for energy storage, and explains why Nd^{3+} lasers are capable of giving such high pulse energies.

Figure 9.9(b) shows the crystal-field fine structure for the $^4F_{3/2} \rightarrow$ $^4I_{11/2}$ transition at 1.06 μm. The octahedral symmetry of the YAG crystal field lifts the degeneracy of the M_J states of the free ion, with states of the same $|M_J|$ having the same energy. Thus the upper $^4F_{3/2}$ level, which has four degenerate M_J states in the free ion corresponding to $M_J = -3/2, -1/2, +1/2$ and $+3/2$, is split by the crystal field into two sub-levels identified by $M_J = \pm 3/2$ and $M_J = \pm 1/2$. Similarly, the lower $^4I_{11/2}$ level splits into six sub-levels. The size of the crystal-field splittings is of order 100 cm^{-1}, which is approximately an order of magnitude smaller than the spin–orbit splitting.

In atomic physics, the splitting of levels by an electric field is called the Stark effect. The energy shift normally depends only on $|M_J|$ rather than the absolute value of M_J. Therefore, since the crystal-field splitting is caused by electric fields, it is not surprising that the shift is insensitive to the sign of M_J.

Figure 9.9(c) shows experimental data for the emission spectrum of the $^4F_{3/2} \rightarrow {}^4I_{11/2}$ transition at 77 K and 300 K. The spectrum consists of sharp lines rather than a continuum, which demonstrates the weak nature of the vibronic coupling. Transitions involving most of the sub-levels of the upper and lower levels are clearly identifiable in the spectra. The laser transition at 1.064 μm is identified in the 300 K spectrum, and the states involved are indicated in Fig. 9.9(b).

The emission lines in Fig. 9.9(c) are broader at 300 K than at 77 K. This is a consequence of the stronger electron–phonon coupling at the higher temperature. The linewidth of the 1.064 μm emission line is 120 GHz at 300 K. As we shall see in Section 9.4, this broadening is very beneficial for making short pulse lasers.

9.3.3 Transition-metal ions

The transition metals are found in the fourth row of the periodic table and have atomic numbers from 21–30, with electronic configurations of [Ar] $3d^n 4s^2$. The divalent ions are formed by losing the outermost 4s electrons, and higher valencies are possible if one or more of the 3d electrons are also lost. Table 9.3 lists some of the more common ions that are found in solid-state crystals.

A characteristic aspect of the physics of transition-metal ions is the strong interaction with the crystal field, and hence the strong vibronic coupling. As mentioned in point 4 of Section 9.3.1, this is caused by the relatively large radius of the 3d orbitals and the fact that they are unshielded by outer filled shells. This makes their electronic states very sensitive to the crystalline environment. A striking example of this is that Cr^{3+} ions are responsible for both the red colour of ruby and the green colour of emerald. The change in colour arises from the shift in the energy levels in changing the host crystal from sapphire (Al_2O_3) in ruby to beryl ($Be_3Al_2Si_6O_{18}$) in emerald. This can be contrasted with the behaviour of rare earth dopants when the crystal host is changed. For example, the 1.064 μm transition of Nd:YAG only shifts to 1.053 μm when the host crystal is changed to YLF (YLiF$_4$).

On comparing the transition metals with the rare earths, the crystal-field effect is stronger and the spin–orbit interaction is smaller. The latter

Table 9.3 Electronic configurations of common transition-metal ions.

Ion	Configuration
Ti^{3+}, V^{4+}	[Ar] $3d^1$
V^{3+}, Cr^{4+}	[Ar] $3d^2$
V^{2+}, Cr^{3+}	[Ar] $3d^3$
Cr^{2+}, Mn^{3+}	[Ar] $3d^4$
Mn^{2+}, Fe^{3+}	[Ar] $3d^5$
Fe^{2+}	[Ar] $3d^6$
Co^{2+}	[Ar] $3d^7$
Ni^{2+}	[Ar] $3d^8$
Cu^{2+}	[Ar] $3d^9$

The transmission spectrum of ruby is given in Fig. 1.7. The red colour is caused by the two strong absorption bands in the green/yellow and blue spectral regions.

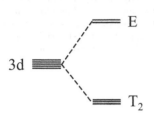

Fig. 9.10 Splitting of the degenerate 3d levels of a Ti^{3+} ion in an octahedral crystal environment.

The details of the level schemes become more complicated if there is more than one 3d electron present and/or the crystal environment has lower than octahedral symmetry.

point is a consequence of the Z^2 dependence of the spin–orbit interaction. This means that in treating the crystal-field effect by perturbation theory, we should consider the crystal-field interactions first, and then apply the spin–orbit coupling afterwards. This means that the character of the states is very different to those of the free ion.

As an example we can consider a transition-metal ion dopant in the octahedral crystalline environment shown in Fig. 9.8. We take the simplest case in which the metal only has a single 3d electron, as in Ti^{3+}, which has an electronic configuration of $3d^1$. The octahedral crystal field interacts with the degenerate 3d levels of the free ion and splits them into a doublet and a triplet, as shown in Fig. 9.10. This splitting can be deduced by group theory, and can also be worked out explicitly by calculating the perturbation due to the crystal field. (See Exercise 9.9.)

The nomenclature used for the crystal-field-split levels in Fig. 9.10 is taken from group theory, as was the case with the NV^- centre. The doublet is labelled as an E state, and the triplet as a T_2 state. These states are sometimes further specified by their spin multiplicity and their parity. Thus the doublet is a ^2E_g state, with $^2T_{2g}$ for the triplet. The superscript prefix refers to the spin multiplicity (i.e. that the single electron has two spin states), while the subscript refers to the parity, with 'g' being short for *gerade*, the German word for 'even'. (Odd parity states are labelled 'u', which is short for *ungerade*.)

Figure 9.11 shows the absorption and emission spectra of Ti^{3+} ions doped into the octahedral sapphire (Al_2O_3) host at 300 K. The spectra correspond to transitions between the T_{2g} ground-state level and the E_g excited state. Since the upper and lower levels both have even parity, electric-dipole transitions should be forbidden. However, the introduction of the Ti^{3+} impurities slightly distorts the octahedral environment of the host, and mixes in states of odd parity, giving some transition probability. This gives an upper state radiative lifetime of 3.9 µs, which is shorter than the non-radiative lifetime at 300 K. The luminescent efficiency at 300 K is therefore high (see eqn 5.5 and Exercise 5.4), which explains why Ti:sapphire crystals make good lasers.

The experimental data show clearly that the absorption and emission spectra consist of continuous bands rather than sharp lines. This is a consequence of the strong vibronic broadening of the ground and the excited states. The Stokes shift of the emission is also apparent in the data, together with the approximate mirror symmetry of the emission and absorption about the zero phonon wavelength of ~ 630 nm.

The general shape of the spectra shown in Fig. 9.11 is typical of many other transition-metal-ion-doped crystals. The crystal field splits the atomic levels derived from the 3d states, and then the strong coupling to phonons broadens these states into continuous vibronic bands. This gives rise to continuous vibronic absorption and emission bands, which are particularly useful for making tunable lasers, as discussed in the next section.

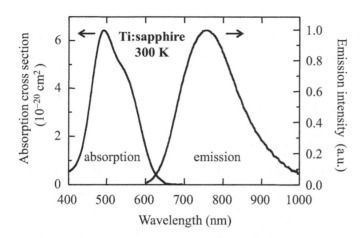

Fig. 9.11 Absorption and emission spectra for Ti^{3+} ions doped into sapphire (Al_2O_3) at 300 K. After Moulton (1986), reprinted with permission.

9.4   Solid-state lasers and optical amplifiers

Many important solid-state lasers use transition-metal ions or rare-earth ions as the gain medium. For example, the first laser ever demonstrated used ruby (Cr^{3+} doped into Al_2O_3) as the active material. The lasers can be generally classified as having either a fixed or tunable wavelength. The emission spectra of rare-earth ions usually consist of very specific wavelengths, and generally fall into the first category. Transition-metal ions, on the other hand, often show broad emission bands, which give rise to the possibility for tunable laser operation over a very wide range of wavelengths.

A key requirement for laser operation is that there should be **population inversion** between the upper and lower laser levels. (See eqn B.13 in Appendix B.) By this we mean that the population of the upper level exceeds that of the lower level. This ensures that the rate of stimulated emission exceeds the rate of absorption, and hence that there is optical amplification (i.e. gain) in laser crystal.

Population inversion is achieved by 'pumping' atoms into the upper laser level by a variety of mechanisms. Figure 9.12 indicates how this is done for the 1.064 μm line of the Nd:YAG laser. The upper laser level is the $^4F_{3/2}$ state. This level is populated by first pumping electrons from the ground state to excited states such as the $^4F_{5/2}$ level identified in Fig. 9.9(a). Alternatively, the upper laser level can be populated by pumping to other excited states not shown in Fig. 9.9(a). Some of these are broadened into bands by vibronic coupling, and can thus absorb a wide range of frequencies, which makes them easier to pump. The electrons in the higher excited states rapidly relax to the upper laser level by non-radiative decay. This gives rise to population inversion with respect to the $^4I_{11/2}$ state, and if a suitable cavity is provided, lasing can occur. Rapid non-radiative decay to the $^4I_{9/2}$ state ensures that the electrons do not accumulate in the lower laser level and reduce the population inversion.

Nd:YAG lasers have traditionally been pumped by bright flash lamps.

The population inversion scheme in ruby is considered in Exercise 9.12. The publication date of this book (2010) coincides with the 50th anniversary of the ruby laser. Ruby contrasts with most other lasers in that it is a **three-level** rather than a **four-level** system. The difference depends on whether the lower laser level is the ground state or an excited state. In the former case, the lower laser level has a high initial occupancy, and more than 50% of the atoms must be pumped to the upper level to obtain population inversion. This is not the case for four-level lasers.

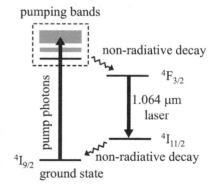

Fig. 9.12 Population inversion scheme for the 1.064 μm transition in a Nd:YAG laser. The pump photons are typically obtained from a flash lamp or diode laser.

Table 9.4 Common solid-state lasers based on rare-earth or transition-metal ions. The lasers operate at room temperature unless stated otherwise.

Laser	Active ion	Configuration	Host	Wavelength (nm)
Ti:sapphire	Ti^{3+}	$3d^1$	sapphire (Al_2O_3)	700–1100
Ruby	Cr^{3+}	$3d^3$	sapphire (Al_2O_3)	694
Alexandrite	Cr^{3+}	$3d^3$	beryl ($BeAl_2O_4$)	700–820
Cr:LiSAF	Cr^{3+}	$3d^3$	LiSAF ($LiSrAlF_6$)	780–1010
Cr:LiCAF	Cr^{3+}	$3d^3$	LiCAF ($LiCaAlF_6$)	720–840
Cr:forsterite	Cr^{4+}	$3d^2$	forsterite (Mg_2SiO_4)	1150–1350
Co:MgF$_2$	Co^{2+}	$3d^7$	magnesium fluoride (MgF_2)	1500–2500 at 77 K
Nd:YAG	Nd^{3+}	$4f^3$	yttrium aluminium garnet (YAG: $Y_3Al_5O_{12}$)	1064
Nd:glass	Nd^{3+}	$4f^3$	phosphate glass	1054
Nd:YLF	Nd^{3+}	$4f^3$	yttrium lithium fluoride (YLF: $LiYF_4$)	1047 and 1053
Nd:vanadate	Nd^{3+}	$4f^3$	yttrium vanadate (YVO_4)	1064
Yb:YAG	Yb^{3+}	$4f^{13}$	yttrium aluminium garnet (YAG: $Y_3Al_5O_{12}$)	1030 at 100 K
Erbium fibre	Er^{3+}	$4f^{11}$	optical fibre	1530–1560

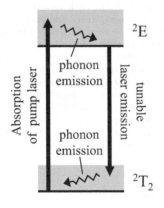

Fig. 9.13 Level diagram for the tunable vibronic emission in a Ti:sapphire laser.

However, the transition from the ground state to the $^4F_{5/2}$ state conveniently matches the optimum emission wavelength of GaAs quantum well diode lasers around 800 nm. (See Section 6.6.) This has given rise to a new generation of Nd:YAG lasers pumped by high power semiconductor lasers, which are much more efficient and stable than their counterparts that use flash lamps.

The population inversion mechanism in Ti:sapphire lasers follows the general procedure shown in Fig. 9.13. Electrons are pumped from the ground state of the $^2T_2$ band to an excited level within the ^2E band. These electrons relax to the bottom of the ^2E band by phonon emission, and this creates population inversion with respect to the vibronic levels of the $^2T_2$ band, which are rapidly depleted by further phonon emission. Lasing can then occur over a broad range of wavelengths within the emission band shown in Fig. 9.11.

Ti:sapphire lasers can be pumped by argon ion lasers, whose emission lines at 488 nm and 514 nm match well to the absorption bands of the Ti:sapphire crystal shown in Fig. 9.11. Alternatively, frequency-doubled Nd lasers (e.g. Nd:YAG, Nd:YLF, or Nd:YVO$_4$) operating around 532 nm can be used. In the latter case, the 532 nm radiation is obtained by doubling the frequency of the 1064 nm laser line using the techniques of nonlinear optics discussed in Chapter 11. It might at first seem counterintuitive to use one laser to pump another, but it actually makes sense because it is an efficient way to convert the discrete frequencies of a fixed-wavelength, high-power laser to continuously tunable radiation.

Table 9.4 lists a number of important solid-state lasers based on transition-metal or rare-earth ions. As is apparent from the table, it is possible to cover a wide range of frequencies in the visible and near-infrared spectral regions by using these sources. Of the lasers listed, the Nd^{3+} lasers have found the most widespread applications in industrial and medical environments, due to their high power output and rugged

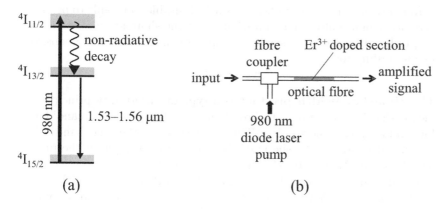

(a)   (b)

Fig. 9.14 The erbium-doped fibre amplifier. (a) Level scheme. The $^4I_{11/2}$ band is 1.27 eV above the ground state, and is suitable for pumping with 980 nm diode lasers. Rapid non-radiative relaxation occurs to the bottom of the $^4I_{13/2}$ band. This creates population inversion for the $^4I_{13/2} \to {}^4I_{15/2}$ vibronic transition, and hence gain between 1.53 μm and 1.56 μm. (b) Schematic diagram of the fibre amplifier. The 980 nm pump laser is coupled into the erbium-doped section by means of a fibre coupler.

structure.

The last gain medium listed in Table 9.4, namely the erbium-doped optical fibre, has become increasingly important for use in telecommunications systems. The level scheme for the Er^{3+} ion is shown in Fig. 9.14(a). The $^4I_{11/2}$ band of the Er^{3+} ions is 1.27 eV above the ground state, which makes it suitable for pumping with 980 nm diode lasers. Rapid non-radiative relaxation occurs to the bottom of the $^4I_{13/2}$ band, where the electrons accumulate due to the long lifetime of the state (11 ms). This creates population inversion with respect to the ground state band, generating optical gain for the $^4I_{13/2} \to {}^4I_{15/2}$ vibronic band between 1.53 μm and 1.56 μm.

The erbium ions are doped into a section of optical fibre, which is pumped by a 980 nm semiconductor diode laser introduced by a fibre coupler, as shown schematically in Fig. 9.14(b). Lasing can occur if mirrors are placed around the gain medium, but usually there is no cavity and the gain of the erbium ions is used to amplify signals. The gain peaks around 1.55 μm, which is one of the preferred wavelengths for silica fibre systems. Amplification factors of about 10^3 can be obtained with a few metres of erbium fibre.

We mentioned in Section 9.2 that colour centres can also be used as laser crystals. The population inversion mechanism follows the same general scheme as for the Ti:sapphire laser shown in Fig. 9.13. The electrons are first pumped to an intermediate state within the upper band, from where they relax by phonon emission. They then return to the lower band by emitting the laser photon, and finally relax to the ground state by further phonon emission. Colour centre lasers are mainly used for spectroscopic studies in the infrared spectral region. For example, laser operation has been demonstrated for the F_2^+ centre in KF between 1.22 μm and 1.50 μm, which covers most of the emission bands of this crystal. (See Fig. 9.5.) The laser is conveniently pumped by the 1.064 μm line of a Nd:YAG laser, which matches well to the absorption band between 1.0 μm and 1.2 μm. With other combinations of host crystals and colour centres, it is possible to cover a wide range of wavelengths in the infrared spectral region between 1 μm and 4 μm.

The fibre losses are very small at 1.55 μm. Nevertheless, in long distance systems (e.g. transatlantic) the signals in the fibre must still be amplified at regular intervals to compensate for the small (but non-zero) losses.

In many modern laser applications it is desirable to be able to produce very short light pulses. The duration Δt of the shortest pulses that can be produced by a laser is set by the spectral width $\Delta \nu$ of the emission line according to:

$$\Delta \nu \Delta t \sim 1. \tag{9.6}$$

This **time–bandwidth product** is a type of uncertainty principle. It means that laser crystals with broad emission lines are good candidates for producing very short pulses. The precise value of the time–bandwidth product depends on the shape of the pulse. For example, if the pulses are Gaussian, $\Delta \nu \Delta t = 0.441$. (See Exercise 9.13.)

We mentioned in connection with Fig. 9.9(c) that the linewidth of the 1064 nm line in Nd:YAG is about 120 GHz at 300 K, which leads to pulses as short as a few picoseconds. However, the shortest pulses at present are produced by Ti:sapphire lasers. The extremely broad spectral width of the emission band ($\Delta \nu \sim 10^{14}$ Hz) makes it possible to generate pulses shorter than 10 fs, which permits many interesting studies of dynamical effects in physics, chemistry, and biology. These ultrafast lasers are typically pumped by diode-pumped Nd:YAG lasers, and, as such, represents a *tour de force* of present-day solid-state optical technology, combining semiconductor quantum wells (Chapter 6) with nonlinear optics (Chapter 11) and solid-state laser science (the present chapter).

9.5   Phosphors

The term **phosphor** covers a wide range of solids that emit visible light when excited either by a beam of electrons or by short wavelength photons. In this section we briefly discuss the applications of phosphors in lighting applications.

The traditional lighting application for phosphors is in fluorescent tubes. This technology was developed around the time of the Second World War and quickly established itself for general lighting applications due to its greater efficiency compared to incandescent lamps. The tubes contain mercury vapour at a low pressure, and the inside of the glass is coated with the phosphor. An electrical discharge excites the mercury atoms, which emit ultraviolet radiation at 254 nm and 185 nm. This ultraviolet light is then absorbed by the phosphor, and re-emitted in the visible spectral region.

For many years fluorescent lighting was dominated by halophosphate phosphors incorporating Sb^{3+} and Mn^{2+} dopants. However, the advent of rare-earth phosphors in 1975 revolutionized the technology. By using a blend of three rare-earth dopants, one emitting in the blue, one in the green, and one in the red, it is possible to make highly efficient tubes with a very good white-light colour balance.

Figure 9.15 shows the emission spectrum of a tricolour lamp blended to give a colour balance equivalent to a black-body source at 4000 K. The lamp incorporates a carefully selected mixture of $BaMgAl_{10}O_{17}:Eu^{2+}$, $CeMgAl_{11}O_{19}:Tb^{3+}$, and $Y_2O_3:Eu^{3+}$. The Eu^{2+} ($4f^7$) ions emit in the

Phosphors also find widespread application in the cathode ray tubes used in oscilloscopes and older computer monitors. In these devices an electron beam strikes a screen coated with a suitable phosphor which then emits light by the cathodoluminescence process. (See Section 5.4.4.)

Similar phosphors are used for the red, green, and blue pixels in older colour televisions and computer monitors.

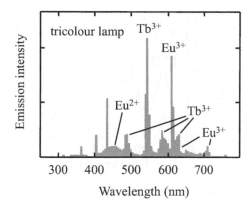

Fig. 9.15 Emission spectrum of a tricolour fluorescent lamp with a colour balance equivalent to a black-body source at 4000 K. The main emission lines from the blend of Eu^{2+}, Eu^{3+} and Tb^{3+} phosphors in the lamp are identified. The sharp lines at 405 nm and 436 nm originate from the mercury discharge. There is also a mercury line at 545 nm which is very close to the main Tb^{3+} emission line. After Smets (1992), reprinted with permission from Plenum Publishers.

blue at 450 nm, the Tb^{3+} $(4f^8)$ ions in the green at 550 nm, and the Eu^{3+} $(4f^6)$ ions in the red at 610 nm. These emission lines are clearly visible in the spectrum of the tricolour lamp, together with other weaker emission lines from the phosphors and also the mercury lines at 405 nm, 436 nm, and 545 nm. These tricolour lamps are much more efficient than the older halophosphates, and also offer a much better colour balance.

In recent years phosphors have found an important new application in phosphor-converted LEDs (light-emitting diodes). In these devices a phosphor (or blend of phosphors) is combined with a short wavelength semiconductor LED to produce white light. These white-light LEDs lie at the basis of the rapidly-developing solid-state lighting industry.

Figure 9.16(a) shows a schematic diagram of a white-light LED. The device consists of a short wavelength semiconductor LED based on nitride materials (see Section 5.4) combined with appropriate phosphors. There are several common strategies available for making white-light LEDs in this way:

The purpose of the reflector cup in the LED is to increase the output in the forward direction by reflecting the light that is emitted downwards.

- use a GaN device emitting in the ultraviolet (UV) to excite a blend of three phosphors (i.e. red, green, and blue) as in a tricolour lamp;

- use a blue-emitting GaInN alloy together with a yellow phosphor (e.g. Ce^{3+}:YAG);

- use a blue-emitting GaInN alloy together with a blend of green and red phosphors.

The first method offers the best colour control, but is less efficient, since the UV LED technology is less well developed, and more energy is lost in the colour conversion process. (See Exercise 9.18.) In the second and third approaches, the nitride LED serves as both the blue emitter and the excitation source for the phosphors. The devices based on just a single phosphor are simpler to make, but offer less good colour rendition compared to those incorporating two phosphors.

Figure 9.16(b) shows the emission spectrum of a white-light LED based on a blue-emitting GaInN quantum well chip. The device incorporated a blend of two phosphors labelled I and II. Both phosphors are based on the Eu^{2+} ion. The phosphors have absorption bands in the

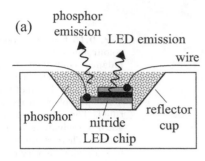
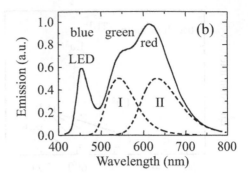

Fig. 9.16 (a) Schematic diagram of a phosphor-converted LED. (b) Emission spectrum of a white-light LED. The device incorporated a blue-emitting GaInN quantum well LED together with a green phosphor (I, $SrSi_2O_2N_2:Eu^{2+}$) and red phosphor (II, $Sr_2Si_5N_8:Eu^{2+}$). After Mueller-Mach et al. (2005), reprinted with permission.

The Eu^{2+} phosphors emit light by dipole-allowed $5d \rightarrow 4f$ transitions. The excited-state 5d shells have a relatively large radius and are therefore highly susceptible to crystal-field effects and vibronic coupling. They are therefore broadened into bands, with their energy depending strongly on the properties of the host crystal.

blue spectral region that overlap the LED output, and their emission wavelength is controlled by the choice of host material. In the case of phosphor I, the host was $SrSi_2O_2N_2$, and the emission was in the green spectral region, while for phosphor II, the host was $Sr_2Si_5N_8$ and the emission was in the red. The combined emission spectrum of the LED shows a peak in the blue spectral region around 450 nm from the nitride quantum well, together with a broad emission band in the green–red spectral range from the phosphors. The equivalent colour temperature of the device was 3200 K.

White-light LEDs offer several advantages over fluorescent tubes in lighting applications. Firstly, they operate at low voltages, which makes them compatible with small battery-operated devices (e.g. mobile phones, flashlights.) Secondly, they contain no environmentally hazardous elements such as mercury. Finally, the overall energy conversion efficiency is already comparable to that achieved in fluorescent tubes and is predicted to increase significantly in the future. As a result, these phosphor-converted LEDs are already finding very widespread application and are expected to form the basis for the next generation of general lighting sources.

Chapter summary

- Luminescence centres are optically active defects and impurities within crystalline hosts. The electronic states of the centres are localized at the defect or impurity from which they arise.

- The electronic states couple to the lattice vibrations of the host crystal through the electron–phonon interaction. Optical transitions between the states are vibronic and involve the simultaneous excitation of phonons.

- The vibronic coupling leads to broad absorption and emission bands in many materials. The emission occurs at a lower energy than the absorption. This red-shift of the emission bands is called the Stokes shift.

- Colour centres (F-centres) consist of an electron trapped at a vacancy in an insulator crystal. Vibronic transitions between the bound states of the electron give rise to broad absorption and emission bands. Nitrogen vacancy (NV) centres offer very promising prospects for applications in quantum information processing.

- The energy levels of paramagnetic ions doped into ionic crystals are perturbed by the crystal field of their local environment. In the rare-earth ions, the crystal-field effects are quite small, but in the transition-metal ions, the crystal-field effects are very large.

- The optical spectra of rare-earth ions tend to consist of discrete lines. The crystal-field effect causes small splittings of transitions that are degenerate in the free ions.

- The optical spectra of transition-metal ions consist of broad vibronic bands. The emission is Stokes-shifted with respect to the absorption.

- Paramagnetic ions and colour centres can be used as the gain medium in solid-state lasers. Rare-earth ion lasers tend to operate on discrete wavelengths, while the transition-metal ions and colour centres give rise to tunable laser wavelengths. Erbium ions doped into optical fibres can be used as optical amplifiers at $1.55\,\mu\mathrm{m}$.

- Rare-earth ion phosphors are frequently used as the light-emitting material in fluorescent lighting and cathode ray tubes. They can also be combined with short wavelength semiconductor LEDs to produce white-light LEDs for applications in solid-state lighting.

Further reading

The basic physics of colour centres is covered in Ashcroft & Mermin (1976), Burns (1985), or Kittel (2005), and the crystal-field effect is explained in more detail in Blundell (2001). A good introduction to luminescent impurities is given by Elliott & Gibson (1974), while authoritative treatments of vibronic systems may be found in Henderson & Imbusch (1989) or Hayes & Stoneham (1985).

Reviews on colour centres in diamond and the read-out of spins by optical spectroscopy may be found respectively in Jelezko & Wrachtrup (2006) and Jelezko & Wrachtrup (2004). A detailed account of the decay dynamics of NV^- centres may be found in Manson et al. (2006). An account of the present state of the art in coherent manipulations of single spins in NV centres in diamond may be found in Balasubramanian et al. (2009).

Detailed information about solid-state lasers may be found in Henderson & Bartram (2000), Silfvast (2004), or Svelto (1998). A collection of review papers on colour centres, transition-metal ions, and phosphors may be found in Di Bartolo (1992). Reviews on white-light LEDs and solid-state lighting may be found in Narukawa (2004), Shur & Žukauskas (2005), or Schubert et al. (2006).

Exercises

(9.1) A colour centre may be modelled as an electron of mass m_0 confined to move in a cubic box with a cube edge length of $2a$. On the assumption that the potential barriers at the edge of the box are infinite, solve the Schrödinger equation for the electron and hence derive eqn 9.4.

(9.2) The solid line in Fig. 9.4 is a fit to the data with $E = 0.21/a^2$, where E is measured in eV and a in nm. How does this fit compare to the prediction of eqn 9.5 ?

(9.3) The anion–cation distance in KBr is 0.33 nm. Estimate the energy of the F-band absorption peak in this crystal.

(9.4) An electron is trapped in a hard rectangular box with square ends orientated along the z axis. Calculate the energy of the electron if the length of the box is $2b$ and its cross-sectional area is b^2. Hence explain why we might expect the transitions of an F_2^+ centre to be at about half the energy of the equivalent F centre. Does this model fit the experimental data for KF given in Figs 9.4 and 9.5?

(9.5) (a) Calculate the relative occupation of the $M_S = 0$ and $M_S = \pm 1$ sub-levels of the ^3A ground state of an NV$^-$ centre in thermal equilibrium at 2 K.
(b) Calculate the temperature that would need to be reached to initialize 80% of the electrons into the $M_S = 0$ level.

(9.6) Use the data in Fig. 9.7(b) to determine the energy of the dominant phonon mode that interacts with the NV$^-$ centre.

(9.7) The expectation value of the radius of an electron in a hydrogenic atom is given by

$$\langle r \rangle = \frac{n^2 a_H}{Z} \left(\frac{3}{2} - \frac{l(l+1)}{2n^2} \right),$$

where Z is the atomic number, a_H is the Bohr radius of hydrogen, n is the principal quantum number and l is the orbital quantum number. Use this result to argue that:
(a) The radius of the 3d orbitals in a transition-metal ion is larger than that of the 4f orbitals in a rare-earth ion.
(b) The 3d orbitals of a transition-metal ion are the outermost orbitals of the atom, whereas the 4f orbitals of a rare-earth ion are not.

(9.8)* Consider the interaction between an electron in an outer p orbital with the electric field of a crystalline host environment.
(a) Explain why the p_x, p_y, and p_z orbitals are degenerate if the ion is placed in an octahedral crystal, as sketched in Fig. 9.8.
(b) Explain why the p states split into a singlet and a doublet if the crystal has uniaxial symmetry, that is, if the ions of the crystal host are closer along the z axis than in the x and y directions.
(c) State whether the energy of the singlet is higher or lower than that of the doublet if the nearest neighbour ions are negative.

(9.9)* In this exercise we consider the splitting of the 3d levels of a transition-metal ion in an octahedral crystalline environment. We assume that the cation is located at the origin and is surrounded by six anions of charge q located at $(\pm a, 0, 0)$, $(0, \pm a, 0)$, and $(0, 0, \pm a)$, as shown in Fig. 9.8. In this case, the potential near the origin is of the form:

$$V(\mathbf{r}) = \frac{6\alpha}{a} + \frac{35\alpha}{4a^5}\left(x^4 + y^4 + z^4 - \frac{3}{5}r^4 \right),$$

where $\alpha = q/4\pi\epsilon_0$ and $\mathbf{r} \equiv (x, y, z)$ is the position vector relative to the origin. This can be rewritten in spherical polar co-ordinates as:

$$V(\mathbf{r}) = \frac{6\alpha}{a} + \frac{7\alpha r^4}{2a^5}\left(C_{4,0} + \sqrt{5/14}(C_{4,4} + C_{4,-4}) \right),$$

where $C_{l,m}(\theta, \phi)$ is a spherical harmonic function.
(a) The five m states of the 3d orbitals can be written as:

$$\psi_m(\mathbf{r}) \equiv |m\rangle = f(r)\, C_{2,m},$$

where

$$
\begin{aligned}
C_{2,0} &= \left(3\cos^2\theta - 1 \right)/2, \\
C_{2,\pm 1} &= \mp (3/2)^{1/2} \cos\theta \sin\theta\, e^{\pm i\phi}, \\
C_{2,\pm 2} &= (3/8)^{1/2} \sin^2\theta\, e^{\pm i2\phi}.
\end{aligned}
$$

Use the fact that $|m\rangle \propto e^{im\phi}$ to show that the matrix elements:

$$\langle m|V|m'\rangle \equiv \iiint \psi_m^* \, V \, \psi_{m'} \, d^3\mathbf{r}$$

are zero unless $m' = m$ or $m' = m \pm 4$.

---

*Exercises marked with an asterisk are more challenging.

(b) By writing:

$$\langle \pm 2 | V | \pm 2 \rangle = A$$
$$\langle \pm 1 | V | \pm 1 \rangle = B$$
$$\langle 0 | V | 0 \rangle = C$$
$$\langle m | V | m \pm 4 \rangle = D,$$

show that the eigenstates of the system are as follows:

Energy	Wave function
$A + D$	$(|2\rangle + |-2\rangle)/\sqrt{2}$
$A - D$	$(|2\rangle - |-2\rangle)/\sqrt{2}$
B	$|\pm 1\rangle$
C	$|0\rangle$

(c) Symmetry (or explicit calculation) demands that $A + D = C$ and $A - D = B$. This implies that the system splits into a doublet and triplet, the states of which are labelled dγ and dϵ respectively. Show that these states take the following form in Cartesian co-ordinates:

$$\psi_{\mathrm{d}\gamma} \propto (2z^2 - x^2 - y^2) \text{ and } (x^2 - y^2),$$
$$\psi_{\mathrm{d}\epsilon} \propto xy, yz, \text{ and } zx.$$

(d) By considering the shapes of the dγ and dϵ wave functions, explain why the doublet has the higher energy in a d^1 electronic configuration, but the lower energy in a d^9 configuration. *Hint*: a d^9 configuration can be considered as a single hole in a filled d shell.

(9.10) Explain why the intensity of the 1.064 μm line of the Nd:YAG crystal is greater at 300 K than at 77 K. (See Fig. 9.9(c).)

(9.11) Explain why population inversion between two levels gives rise to optical gain at the energy difference between the two levels.

(9.12) The level scheme for the 694.3 nm line of a ruby laser is shown in Fig. 9.17. The lower laser level (level 0) is the ground state, and the upper level is an excited state (level 2). Ruby has strong absorption bands in the green/blue spectral regions (see Fig. 1.7), and these are used as intermediate pumping bands (level 1) to produce population inversion between levels 2 and 0.
(a) Explain why lasing is not possible unless more than 50% of the atoms in the ground state have been promoted to the upper laser level.
(b) In a particular laser, a flash lamp pumps 60% of the atoms from the ground state to the upper

laser level, which then emits a short laser pulse. Calculate the maximum energy of this pulse if the laser rod has a volume of 10^{-6} m^3 and the doping density of the Cr^{3+} ions in the crystal is 1×10^{25} m^{-3}.

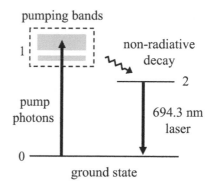

Fig. 9.17 Level diagram for a ruby laser. Lasing occurs between levels 2 and 0, after atoms have been pumped to the upper level via level 1.

(9.13) A laser emits pulses with a Gaussian time dependence of the form $I(t) = I_0 \exp(-t^2/\tau^2)$. The centre frequency of the laser light is ω_0.
(a) By considering the Fourier transform of the electric field, show that the pulses have a spectrum of the form $I(\omega) = I(\omega_0) \exp[-\tau^2(\omega - \omega_0)^2]$.
(b) Hence show that the time–bandwidth product of the pulses, namely $\Delta\nu\Delta t$, where $\Delta\nu$ and Δt are the full width at half maximum of the pulse in the frequency and time domains respectively, is equal to $2\ln 2/\pi$.

(9.14) The linewidth of the 1.054 μm transition of Nd^{3+} in a phosphate glass host is 7.5×10^{12} Hz. Suggest a possible explanation for why this is about 60 times larger than that of the 1.064 μm line in Nd:YAG crystal. Estimate the duration of the shortest pulses that can be obtained from a Nd:glass laser.

(9.15) Explain why the radiative lifetime for the E$_{\mathrm{g}} \rightarrow$ T$_{2\mathrm{g}}$ transition in titanium-doped sapphire is in the microsecond range. Would you classify this emission as fluorescence or phosphorescence?

(9.16) The radiative lifetime of the upper laser level of Co:MgF$_2$ is 1.8 ms. The measured excited state lifetime decreases from 1.4 ms at 77 K to 0.06 ms at 300 K. Account for the temperature dependence of the excited state lifetime, and explain why the operating temperature for the Co:MgF$_2$ laser is 77 K and not 300 K.

(9.17) A titanium-doped sapphire laser operating at 800 nm is pumped by an argon ion laser at 514 nm. Calculate the maximum possible power output if the pump power is 5 W, stating the assumptions you make. What happens to the energy that is not emitted as laser light?

(9.18) A white-light LED contains a phosphor emitting at 650 nm. Calculate the energy conversion efficiency for the phosphor when it is excited (a) by an ultraviolet LED operating at 350 nm, and (b) by a blue LED operating at 450 nm.

Phonons

10

In this chapter we turn our attention to the interaction between light and the phonons in a solid. Phonons are vibrations of the atoms in a crystal lattice, and have resonant frequencies in the infrared spectral region. This contrasts with the optical properties of bound electrons, which occur at visible and ultraviolet frequencies.

The main optical properties of phonons can be explained to a large extent by classical models. We therefore make extensive use of the classical dipole oscillator model developed in Chapter 2. This will allow us to understand why polar solids reflect and absorb light strongly within a band of infrared frequencies. We then introduce the concepts of polaritons and polarons, before moving on to discuss the physics of inelastic light scattering. We shall see how Raman and Brillouin scattering techniques give us complementary information to infrared reflectivity data, which is why they are so extensively used in phonon physics. Finally we briefly discuss why phonons have a finite lifetime, and how this affects the reflectivity and inelastic scattering spectra.

We assume that the reader is familiar with the basic physics of phonons, which is covered in all introductory solid-state physics texts. A partial list of suitable preparatory reading is given under Further Reading at the end of the chapter.

10.1	Infrared active phonons	271
10.2	Infrared reflectivity and absorption in polar solids	273
10.3	Polaritons	281
10.4	Polarons	282
10.5	Inelastic light scattering	285
10.6	Phonon lifetimes	290
Chapter summary		292
Further reading		292
Exercises		293

10.1 Infrared active phonons

The atoms in a solid are bound to their equilibrium positions by the forces that hold the crystal together. When the atoms are displaced from their equilibrium positions, they experience restoring forces, and vibrate at characteristic frequencies. These vibrational frequencies are determined by the phonon modes of the crystal.

The resonant frequencies of the phonons occur in the infrared spectral region, and the modes that interact directly with light are called **infrared active** (**IR active**). Detailed selection rules for deciding which phonon modes are IR active can be derived by using group theory. At this level we just discuss the general rules based on the dispersion of the modes, their polarization, and the nature of the bonding in the crystal.

The phonon modes of a crystal are sub-divided into two general categories:

- acoustic or optical;
- transverse or longitudinal.

The group-theory approach is beyond the scope of this book, although we shall give some simple arguments based on symmetry when we consider inelastic light scattering in Section 10.5.

The phonon dispersion curves for real crystals are more complicated than those shown in Fig. 10.2 because the longitudinal and transverse polarizations tend to have different frequencies.

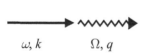

$$\omega, k \qquad \Omega, q$$

Fig. 10.1 Lattice absorption process by an infrared active phonon. The straight arrow represents the photon that is absorbed, while the wiggly arrow represent the phonon that is created.

It will come as no surprise to realize that it is the 'optical' rather than the acoustic modes that are directly IR active. These optically active phonons are able to absorb light at their resonant frequency. The basic process by which a photon is absorbed by the lattice and a phonon is created is represented in Fig. 10.1. Conservation laws require that the photon and the phonon must have the same energy and momentum. We shall see below that this condition can only be satisfied for the optical modes.

Figure 10.2 shows the generic dispersion curves for the acoustic and optical phonons in a simple crystal. The angular frequency Ω of the acoustic and optical phonons is plotted against the wave vector q in the positive half of the first Brillouin zone. At small wave vectors the slope of the acoustic branch is equal to v_s, the velocity of sound in the medium, while the optical modes are essentially dispersionless near $q = 0$.

The figure also shows the dispersion of the light waves in the crystal, which have a constant slope of $v = c/n$, where n is the refractive index. The refractive index has been highly exaggerated here in order to make the dispersion of the photon noticeable on the same scale as the phonon dispersion. The requirement that the photon and phonon should have the same frequency and wave vector is satisfied when the dispersion curves intersect. Since $c/n \gg v_s$, the only intersection point for the acoustic branch occurs at the origin, which corresponds to the response of the crystal to a static electric field. The situation is different for the optical branch: there is an intersection at finite ω, which is identified with the circle in Fig. 10.2. Since the optical branch is essentially flat for small q, the frequency of this resonance is equal to the frequency of the optical mode at $q = 0$.

Photons couple to phonons through the driving force exerted on the atoms by the AC electric field of the light. Since electromagnetic waves are transverse, they only apply driving forces to the transverse vibrations of the crystal, which means that they only couple to the transverse optic (TO) phonon modes. Furthermore, there will only be an interaction if the atoms are charged. This means that the crystal must have some ionic character in order for its TO phonons to be optically active.

The ionicity of a solid arises from the way the crystal binding occurs. An ionic crystal consists of an alternating sequence of positive and negative ions held together by their mutual Coulomb attraction. Covalent crystals, by contrast, consist of neutral atoms with the electrons shared equally between the neighbouring nuclei. This means that none of the optical phonons of purely covalent solids like silicon are IR active. Most other materials fall somewhere between these two limits. For example, the bond in a III–V semiconductor is only partly covalent, and the shared electrons lie slightly closer to the group V atoms than to the group III atoms, which gives the bond a partly ionic character. The bonds with an ionic character are called **polar** bonds to stress the point that the asymmetric electron cloud between the atoms creates a dipole that can interact with electric fields. Provided the bond has some polar character, its phonons can be IR active.

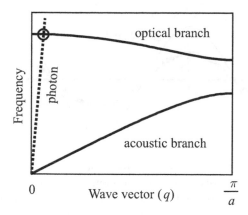

Fig. 10.2 Dispersion curves for the acoustic and optical phonon branches in a typical crystal with a lattice constant of a. The dispersion of the photon modes in the crystal is shown by the dotted line.

The conclusions of this section are summarized in Table 10.1.

10.2 Infrared reflectivity and absorption in polar solids

Table 10.1 Infrared activity of the phonon modes in polar and non-polar crystals. LA: longitudinal acoustic, TA: transverse acoustic, LO: longitudinal optic, TO: transverse optic.

Mode	Polar crystal	Non-polar crystal
LA	no	no
TA	no	no
LO	no	no
TO	yes	no

Experimental data show that polar solids absorb and reflect light very strongly in the infrared spectral region when the frequency is close to resonance with the TO phonon modes. We have come across several examples of this already. For example, the transmission spectra of sapphire and CdSe given in Fig. 1.4 show that there are spectral regions in the infrared where no light is transmitted. This is a consequence of lattice absorption.

The aim of this section is to account for this result by modelling the interaction of photons with TO phonons. To do this we shall make extensive use of the classical oscillator model developed in Chapter 2, especially Section 2.2. This will allow us to calculate the frequency dependence of the complex dielectric constant $\tilde{\epsilon}_r(\omega)$, from which we shall be able to determine the important optical properties such as the reflectivity and absorption.

10.2.1 The classical oscillator model

The interaction between electromagnetic waves and a TO phonon in an ionic crystal is most easily treated by considering a linear chain, as illustrated in Fig. 10.3. The chain consists of a series of unit cells, each containing a positive ion (black circle) and a negative ion (grey circle). The waves are taken to be propagating along the chain in the z direction. We are dealing with a transverse mode, and so the displacement of the atoms is in the x or y directions. Furthermore, in an optic mode the different atoms within each unit cell move in opposite directions, with a fixed ratio between their displacements which is not necessarily equal to unity.

We are interested in the interaction between a TO phonon mode with

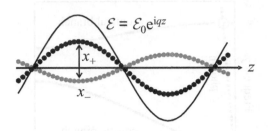

Fig. 10.3 Interaction of a TO phonon mode propagating in the z direction with an electromagnetic wave of the same wave vector. The black circles represent positive ions, while the grey circles represent the negative ions. The solid line represents the spatial dependence of the electric field of the electromagnetic wave.

The data for SiO_2 glass shown in Fig. 2.7 illustrates the connection between the infrared absorption in solids and that of the constituent molecules quite well. The glass is amorphous, and therefore does not have long range order with delocalized phonon modes. The absorption in the range 10^{13} to 10^{14} Hz is basically caused by the vibrational absorption of the SiO_2 molecules themselves, although the frequencies are not necessarily exactly the same in the solid as in the free molecule.

$q \approx 0$ and an infrared light wave of the same frequency and wave vector. This means that we are considering phonons with a very long wavelength of $\sim 10\,\mu m$ matched to that of an infrared photon. This phonon wavelength is huge compared to the size of a unit cell in a crystal, which is usually less than 10^{-9} m. The size of the atoms has been highly exaggerated in Fig. 10.3 to make the physics of the interaction clearer. In fact, the real size of the atoms is tiny compared to the wavelength, and there will be thousands of unit cells within one period of the wave.

The solid line in the figure represents the spatial dependence of the AC electric field of the infrared light wave. At resonance, the wave vector of the photon and the phonon are the same. This means that the driving force exerted by the light on the positive and negative ions is in phase with the lattice vibration. At the same time, the anti-parallel displacements of the oppositely charged atoms generate an AC electric field in phase with the external light. This implies that there is a strong interaction between the TO phonon mode and the light wave when the wave vectors and frequencies match.

For long wavelength TO modes with $q \approx 0$, the motion of the atoms in different unit cells is almost identical, and we therefore need to concentrate on what is happening within the unit cell itself. This enables us to see that there is a close connection between the TO phonons at $q = 0$ and the vibrational modes of the molecules from which the crystal is formed. We can therefore make use of some of the principles developed in molecular physics, for example: the selection rules for deciding whether a particular phonon mode is IR or Raman active. (See Section 10.5.2.)

The interaction between the TO phonon and the light wave can be modelled by writing down the equations of motion for the displaced ions. The displacements of the positive and negative ions in a TO mode are in opposite directions and are given the symbols x_+ and x_- respectively, as indicated in Fig. 10.3. The appropriate equations of motion are:

$$m_+ \frac{d^2 x_+}{dt^2} = -K(x_+ - x_-) + q\mathcal{E}(t) \qquad (10.1)$$

$$m_- \frac{d^2 x_-}{dt^2} = -K(x_- - x_+) - q\mathcal{E}(t) \qquad (10.2)$$

where m_+ and m_- are the masses of the two ions, K is the restoring constant of the medium, and $\mathcal{E}(t)$ is the external electric field due to the light wave. The effective charge per ion is taken to be $\pm q$.

By dividing eqn 10.1 by m_+ and eqn 10.2 by m_-, and then subtracting, we obtain:

$$\frac{d^2}{dt^2}(x_+ - x_-) = -\frac{K}{\mu}(x_+ - x_-) + \frac{q}{\mu}\mathcal{E}(t),\qquad(10.3)$$

where μ is the reduced mass given by:

$$\frac{1}{\mu} = \frac{1}{m_+} + \frac{1}{m_-}.\qquad(10.4)$$

By putting $x = x_+ - x_-$ for the relative displacement of the positive and negative ions within their unit cell, we can recast eqn 10.3 in the simpler form:

$$\frac{d^2 x}{dt^2} + \Omega_{TO}^2 x = \frac{q}{\mu}\mathcal{E}(t),\qquad(10.5)$$

where we have written Ω_{TO}^2 for K/μ. Ω_{TO} represents the natural vibrational frequency of the TO mode at $q=0$ in the absence of the external light field. Note that the displacement of the charged atoms creates an electric dipole of magnitude qx. We thus have a vibrational dipole, as discussed in Section 2.1.2.

Equation 10.5 represents the equation of motion for undamped oscillations of the lattice driven by the forces exerted by the AC electric field of the light wave. In reality, we should have incorporated a damping term to account for the finite lifetime of the phonon modes. The physical significance of the phonon lifetime will be discussed further in Section 10.6. At this stage, we simply introduce a phenomenological damping rate γ, and rewrite eqn 10.5 as

$$\frac{d^2 x}{dt^2} + \gamma\frac{dx}{dt} + \Omega_{TO}^2 x = \frac{q}{\mu}\mathcal{E}(t).\qquad(10.6)$$

This now represents the response of a damped TO phonon oscillator to a resonant light wave.

Equation 10.6 is identical in form to eqn 2.5 in Chapter 2, with m_0 replaced by μ, ω_0 by Ω_{TO} and $-e$ by q. Therefore, we can use all the results derived in Section 2.2 to model the response of the medium to a light field of angular frequency ω with $\mathcal{E}(t) = \mathcal{E}_0 e^{-i\omega t}$. In particular, we can go directly to the formula for the frequency dependence of the dielectric constant without repeating all the steps in the derivation. By adapting the symbols appropriately in eqn 2.14, we immediately write down:

$$\epsilon_r(\omega) = 1 + \chi + \frac{Nq^2}{\epsilon_0\mu}\frac{1}{(\Omega_{TO}^2 - \omega^2 - i\gamma\omega)},\qquad(10.7)$$

where $\epsilon_r(\omega)$ is the complex dielectric constant at angular frequency ω. χ represents the non-resonant susceptibility of the medium, and N is the number of unit cells per unit volume.

Equation 10.7 can be tidied up by introducing the static and high-frequency dielectric constants ϵ_{st} and ϵ_∞ respectively. In the limits of low and high frequency, we obtain from eqn 10.7:

$$\epsilon_{st} \equiv \epsilon_r(0) = 1 + \chi + \frac{Nq^2}{\epsilon_0\mu\Omega_{TO}^2},\qquad(10.8)$$

Care should be taken here not to confuse q for charge and q for phonon wave vector. It is usually obvious from the context which meaning is intended. For a strongly ionic crystal such as NaCl, q would just be equal to $\pm e$. However, for crystals with polar covalent bonds such as the III–V compounds, q will represent an effective charge which is determined by the asymmetry of the electron cloud within the bond.

and

$$\epsilon_\infty \equiv \epsilon_{\mathrm{r}}(\infty) = 1 + \chi. \tag{10.9}$$

In principle, we should consider the local field corrections discussed in Section 2.2.4 here. This is an unnecessary complication at this level which does not add much to the main conclusions. We therefore neglect local field effects, and base our discussion on eqn 10.10.

Thus we can write:

$$\epsilon_{\mathrm{r}}(\omega) = \epsilon_\infty + (\epsilon_{\mathrm{st}} - \epsilon_\infty)\frac{\Omega_{\mathrm{TO}}^2}{(\Omega_{\mathrm{TO}}^2 - \omega^2 - \mathrm{i}\gamma\omega)}. \tag{10.10}$$

This is our main result, which will be used in the next subsections to derive the infrared optical coefficients. As discussed in Section 2.2.2, and in particular in connection with Fig. 2.6, we should understand '$\omega = \infty$' in a relative sense here. ϵ_∞ represents the dielectric constant at frequencies well above the phonon resonance, but below the next natural frequency of the crystal due, for example, to the bound electronic transitions in the visible/ultraviolet spectral region.

10.2.2   The Lyddane–Sachs–Teller relationship

Before working out the frequency dependence of the infrared reflectivity, it is useful to investigate one rather striking implication of eqn 10.10. Suppose we have a lightly damped system so that we can set $\gamma = 0$. Then at a certain frequency which we label ω', eqn 10.10 tells us that the dielectric constant can fall to zero. The condition for this to happen is:

$$\epsilon_{\mathrm{r}}(\omega') = 0 = \epsilon_\infty + (\epsilon_{\mathrm{st}} - \epsilon_\infty)\frac{\Omega_{\mathrm{TO}}^2}{(\Omega_{\mathrm{TO}}^2 - \omega'^2)}. \tag{10.11}$$

This can be solved to obtain:

$$\omega' = \left(\frac{\epsilon_{\mathrm{st}}}{\epsilon_\infty}\right)^{1/2}\Omega_{\mathrm{TO}}. \tag{10.12}$$

What does $\epsilon_{\mathrm{r}} = 0$ mean physically? We encountered another system for which $\epsilon_{\mathrm{r}} = 0$ when we discussed plasma oscillations in Section 7.5.1. We saw there that a dielectric medium can support longitudinal waves at frequencies that satisfy $\epsilon_{\mathrm{r}}(\omega) = 0$. We can understand this by considering Gauss's law. If the medium has no free charges, the total charge density is zero, and we have (see eqn A.10):

$$\nabla \cdot \boldsymbol{D} = \nabla\cdot(\epsilon_{\mathrm{r}}\epsilon_0\boldsymbol{\mathcal{E}}) = 0, \tag{10.13}$$

Transverse and longitudinal electromagnetic waves must satisfy $\nabla \cdot \boldsymbol{\mathcal{E}} = 0$ and $\nabla \times \boldsymbol{\mathcal{E}} = 0$ respectively. As discussed in Section 7.5.1, separate solutions are possible for the two different types of wave. The longitudinal modes can only exist at frequencies for which $\epsilon_{\mathrm{r}} = 0$.

where we have made use of eqn A.3 to relate the electric displacement $\boldsymbol{D}$ to the electric field $\boldsymbol{\mathcal{E}}$ in a dielectric medium. If $\epsilon_{\mathrm{r}} \neq 0$, it follows that $\nabla \cdot \boldsymbol{\mathcal{E}} = 0$, which implies that the waves are transverse. This is the usual case that applies when considering the propagation of electromagnetic waves in a dielectric medium. However, if $\epsilon_{\mathrm{r}} = 0$, we can satisfy eqn 10.13 with waves for which $\nabla \cdot \boldsymbol{\mathcal{E}} \neq 0$, i.e. longitudinal waves.

The longitudinal electromagnetic waves that we are considering here are generated by longitudinal optical (LO) phonons. The LO phonon modes generate longitudinal electromagnetic waves just as TO phonons generate transverse electromagnetic waves. The waves at $\omega = \omega'$ therefore correspond to LO phonons, and we identify ω' with the frequency of

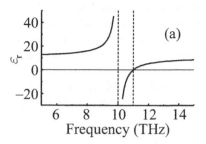
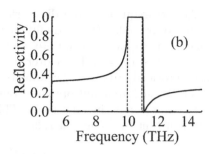

Fig. 10.4 Frequency dependence of the dielectric constant and reflectivity for a crystal with $\nu_{TO} = 10\,\text{THz}$, $\nu_{LO} = 11\,\text{THz}$, $\epsilon_{st} = 12.1$, and $\epsilon_\infty = 10$. The curves have been calculated from eqns 10.15 and 10.16. Phonon damping is ignored in this calculation.

the LO mode at $q = 0$, namely Ω_{LO}. This allows us to rewrite eqn 10.12 in the following form:

$$\frac{\Omega_{LO}^2}{\Omega_{TO}^2} = \frac{\epsilon_{st}}{\epsilon_\infty}. \qquad (10.14)$$

This result is known as the **Lyddane–Sachs–Teller (LST) relationship**. The validity of the relationship can be checked by comparing the values of Ω_{LO}/Ω_{TO} deduced from neutron or Raman scattering experiments with those calculated from eqn 10.14 using known values of the dielectric constants. Some results are given in Table 10.2. It is apparent that the agreement is generally very good.

An interesting corollary of the LST relationship is that it implies that the LO phonon and TO phonon modes of non-polar crystals are degenerate. This follows because there is no infrared resonance, and therefore $\epsilon_{st} = \epsilon_\infty$. This is indeed the case for the purely covalent crystals of the group IV elements, namely diamond (C), silicon, and germanium.

Table 10.2 Comparison of the measured ratio Ω_{LO}/Ω_{TO} for several materials to the value predicted by the Lyddane–Sachs–Teller relationship. Data from Madelung (1996).

Crystal	Ω_{LO}/Ω_{TO}	$(\epsilon_{st}/\epsilon_\infty)^{1/2}$
Si	1	1
GaAs	1.07	1.08
AlAs	1.12	1.11
BN	1.24	1.26
ZnSe	1.19	1.19
MgO	1.81	1.83
AgF	1.88	1.88

10.2.3 Reststrahlen

Having discussed the properties of the system at the special frequency of $\omega = \Omega_{LO}$, we can now calculate the infrared optical constants. It is easier to understand the general behaviour if we assume that the damping term is small. We thus set $\gamma = 0$ in eqn 10.10, and discuss the properties of a material with a dielectric constant that has the following frequency dependence:

$$\epsilon_r(\nu) = \epsilon_\infty + (\epsilon_{st} - \epsilon_\infty)\frac{\nu_{TO}^2}{(\nu_{TO}^2 - \nu^2)}. \qquad (10.15)$$

We have divided all the angular frequencies by 2π here, so that we can compare the predictions to experimental data, which are usually presented against frequency (ν) rather than angular frequency (ω). We shall discuss the effect of including the damping term when we compare our model to the experimental data in connection with Fig. 10.5.

Figure 10.4(a) plots the frequency dependence of the dielectric constant $\epsilon_r(\nu)$ calculated from eqn 10.15 for a polar crystal with the following parameters: $\nu_{TO} = 10\,\text{THz}$, $\nu_{LO} = 11\,\text{THz}$, $\epsilon_{st} = 12.1$, and $\epsilon_\infty = 10$. ($1\,\text{THz} = 10^{12}\,\text{Hz}$.) These figures are quite close to those that would be found in a typical III–V semiconductor. Note that the phonon frequencies have been chosen to satisfy the LST relationship given in eqn 10.14.

At low frequencies the dielectric constant is just equal to ϵ_{st}. As ν increases from 0, $\epsilon_r(\nu)$ gradually increases until it diverges when the resonance at ν_{TO} is reached. Between ν_{TO} and ν_{LO}, ϵ_r is negative. Precisely at $\nu = \nu_{LO}$, $\epsilon_r = 0$. Thereafter, ϵ_r is positive, and gradually increases asymptotically towards the value of ϵ_∞.

The most important optical property of a polar solid in the infrared spectral region is the reflectivity. This can be calculated from the dielectric constant by using eqn 1.29:

We can see from eqn 1.20 that $\sqrt{\epsilon_\infty}$ corresponds to the refractive index of the medium at frequencies well above the optical phonon resonances. This will be the refractive index measured at near-infrared and visible frequencies below the band gap of the material.

$$R = \left| \frac{\tilde{n} - 1}{\tilde{n} + 1} \right|^2 = \left| \frac{\sqrt{\epsilon_r} - 1}{\sqrt{\epsilon_r} + 1} \right|^2. \tag{10.16}$$

Figure 10.4(b) plots the reflectivity calculated from eqn 10.16 for the dielectric constant shown in Fig. 10.4(a). At low frequencies the reflectivity is $(\sqrt{\epsilon_{st}} - 1)^2/(\sqrt{\epsilon_{st}} + 1)^2$. As ν approaches ν_{TO}, R increases towards unity. In the frequency region between ν_{TO} and ν_{LO}, $\sqrt{\epsilon_r}$ is imaginary, so that R remains equal to unity. R drops rapidly to zero as ν increases above ν_{LO} (see Exercise 10.2), and then increases gradually towards the high-frequency asymptote of $(\sqrt{\epsilon_\infty} - 1)^2/(\sqrt{\epsilon_\infty} + 1)^2$.

We see from this analysis that the reflectivity is equal to 100% in the frequency region between ν_{TO} and ν_{LO}. This frequency region is called the **Reststrahl band**. 'Reststrahl' is the German word for 'residual ray', with 'Reststrahlen' being its plural, i.e. 'residual rays'. Light cannot propagate into the medium in the Reststrahl band.

Figure 10.5 shows experimental data for the reflectivity of InAs and GaAs in the infrared spectral region. InAs has TO and LO phonon frequencies at $218.9\,\mathrm{cm}^{-1}$ and $243.3\,\mathrm{cm}^{-1}$ respectively, while for GaAs we have $\nu_{TO} = 273.3\,\mathrm{cm}^{-1}$ and $\nu_{LO} = 297.3\,\mathrm{cm}^{-1}$. We see that the reflectivity is very high for frequencies between the TO and LO phonon frequencies in both materials, and there is a sharp dip in the reflectivity just above the LO phonon resonance.

Experimental infrared spectra are frequently plotted against the wave number $\bar{\nu} \equiv 1/\lambda$. The wave number is effectively a frequency unit, with $1\,\mathrm{cm}^{-1}$ equivalent to $2.998 \times 10^{10}\,\mathrm{Hz}$.

On comparing these results with the prediction shown in Fig. 10.4(b), we see that the general agreement between the model and the experimental data is very good. The main difference is that in both materials the maximum reflectivity in the Reststrahl band is less than 100%. This reduction in the reflectivity is caused by ignoring the damping term. (See Example 10.1 and Exercise 10.4.) The damping also broadens the edge so that there is only a minimum in R just above ν_{LO} rather than a zero.

The magnitude of γ can be found by fitting the experimental data to the full dependence given in eqn 10.10. The values of γ obtained in this way are around 10^{11} to $10^{12}\,\mathrm{s}^{-1}$, which implies that the optical phonons have a lifetime of about 1–10 ps. ($1\,\mathrm{ps} = 10^{-12}\,\mathrm{s}$.) The physical significance of this short lifetime will be discussed in Section 10.6.

10.2.4   Lattice absorption

When we introduced the classical oscillator model in Section 2.2 of Chapter 2, we made the point that we expect high absorption coefficients

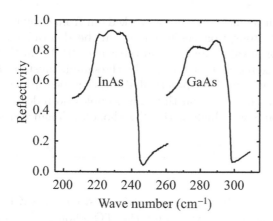

Fig. 10.5 Infrared reflectivity of InAs and GaAs at 4.2 K. A wave number of $1\,\mathrm{cm}^{-1}$ is equivalent to a frequency of 2.998×10^{10} Hz. After Hass (1967), © Academic Press, reprinted with permission.

whenever the frequency matches the natural resonances of the medium. The reader might therefore be wondering why we have been concentrating on calculating the reflectivity rather than the absorption due to the TO phonon resonances.

This question is further prompted by recalling the analogy between the infrared absorption of polar solids and that of isolated molecules. In both cases we are basically treating the interaction of photons with quantized vibrational modes. In molecular physics we usually discuss this in terms of the infrared absorption spectrum. The absorption spectra show strong peaks whenever the frequency coincides with the infrared active vibrational modes and the molecule can absorb a photon by creating one vibrational quantum. This is directly analogous to the process for solids shown in Fig. 10.1 in which a photon is absorbed and a phonon is created.

The answer to these questions is that the lattice does indeed absorb very strongly whenever the photon is close to resonance with the TO phonon. As stressed in Chapter 2, the fundamental optical properties of a dielectric—the absorption, refraction, and reflectivity—are all related to each other because they are all determined by the complex dielectric constant. The distinction between absorption and reflection is merely a practical one. Polar solids have such high absorption coefficients in the infrared that unless the crystal is less than $\sim 1\,\mu$m thick, no light at all will be transmitted. This is clearly seen in the transmission spectra of Al_2O_3 and CdSe shown in Fig. 1.4. For this reason, it is only sensible to consider lattice absorption in thin film samples. In thick crystals, we must use reflectivity measurements to determine the vibrational frequencies. This contrasts with molecular physics, where we are usually dealing with low density gases, which give rise to much smaller absorption coefficients.

The absorption coefficients expected at the resonance with the TO phonon can be calculated from the imaginary part of the dielectric constant. At $\omega = \Omega_{\mathrm{TO}}$ we have from eqn 10.10:

$$\epsilon_{\mathrm{r}}(\Omega_{\mathrm{TO}}) = \epsilon_\infty + \mathrm{i}(\epsilon_{\mathrm{st}} - \epsilon_\infty)\frac{\Omega_{\mathrm{TO}}}{\gamma} . \qquad (10.17)$$

The extinction coefficient κ can be worked out from ϵ_r by using eqn 1.26, and then the absorption coefficient α can be determined from κ via eqn 1.19. Typical values for α are in the range 10^6–10^7 m^{-1}. (See Example 10.1 and Exercise 10.6.) This is why the sample must be thinner than $\sim 1\,\mu$m in order to perform practical absorption measurements. Infrared absorption measurements on thin film samples do indeed confirm that the absorption is very high at the TO phonon resonance frequency.

Example 10.1

The static and high-frequency dielectric constants of NaCl are $\epsilon_{st} = 5.9$ and $\epsilon_\infty = 2.25$ respectively, and the TO phonon frequency ν_{TO} is 4.9 THz.

(a) Calculate the upper and lower wavelengths of the Reststrahl band.

(b) Estimate the reflectivity at 50 μm, if the damping constant γ of the phonons is $10^{12}\,$s^{-1}.

(c) Calculate the absorption coefficient at 50 μm.

Solution

(a) The Reststrahl band runs from ν_{TO} to ν_{LO}. We are given ν_{TO}, and we can calculate ν_{LO} from the LST relationship (eqn 10.14). This gives

$$\nu_{LO} = \left(\frac{\epsilon_{st}}{\epsilon_\infty}\right)^{1/2} \times \nu_{TO} = \left(\frac{5.9}{2.25}\right)^{1/2} \times 4.9\,\text{THz} = 7.9\,\text{THz}\,.$$

Therefore the Reststrahl band runs from 4.9 THz to 7.9 THz, or 38 μm to 61 μm.

(b) At 50 μm we are in middle of the Reststrahl band. We therefore expect the reflectivity to be high. We insert the values for ϵ_{st}, ϵ_∞, γ and $\Omega_{TO} = 2\pi\nu_{TO}$ into eqn 10.10 with $\omega = 2\pi\nu$ ($\nu = 6$ THz) to find:

$$\epsilon_r = 2.25 + 3.65\frac{(4.9)^2}{(4.9)^2 - 6^2 - i(1)(6)/2\pi} = -5.0 + 0.57i\,.$$

We then obtain the real and imaginary parts of the refractive index from eqns 1.25 and 1.26:

$$n = \frac{1}{\sqrt{2}}\left(-5.0 + [(-5.0)^2 + (0.57)^2]^{1/2}\right)^{1/2} = 0.13\,,$$

and

$$\kappa = \frac{1}{\sqrt{2}}\left(+5.0 + [(-5.0)^2 + (0.57)^2]^{1/2}\right)^{1/2} = 2.2.$$

We finally substitute these values of n and κ into eqn 1.29 to find the reflectivity:

$$R = \frac{(n-1)^2 + \kappa^2}{(n+1)^2 + \kappa^2} = \frac{(-0.87)^2 + (2.2)^2}{(1.13)^2 + (2.2)^2} = 0.91.$$

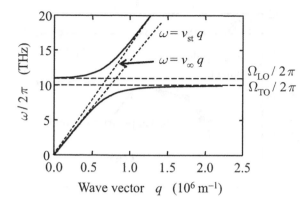

Fig. 10.6 Polariton dispersion predicted from eqn 10.18 with ϵ_r given by eqn 10.15. The curves are calculated for a crystal with $\nu_{TO} = 10$ THz, $\epsilon_{st} = 12.1$, and $\epsilon_\infty = 10$. The asymptotic velocities v_{st} and v_∞ are equal to $c/\sqrt{\epsilon_{st}}$ and $c/\sqrt{\epsilon_\infty}$ respectively.

This value is close to the measured reflectivity of NaCl in the Reststrahl band at room temperature.

(c) We can calculate the absorption coefficient α from the extinction coefficient by using eqn 1.19. We have already worked out that $\kappa = 2.2$ in part (b). Hence we find:

$$\alpha = \frac{4\pi\kappa}{\lambda} = \frac{4\pi \times 2.2}{50 \times 10^{-6}} = 5.5 \times 10^5 \, \text{m}^{-1}.$$

This shows that the light would be absorbed in a thickness of about $2 \, \mu\text{m}$.

10.3 Polaritons

The dispersion curves of the photons and TO phonons were discussed in broad terms in connection with Fig. 10.2. We now wish to consider the circled intersection point in Fig. 10.2 in more detail. As we shall see, the two dispersion curves do not actually cross each other. This is a consequence of the strong coupling between the TO phonons and the photons when their frequencies and wave vectors match. This leads to the characteristic anticrossing behaviour that is observed in many coupled systems.

The coupled phonon–photon waves are called **phonon polaritons**. As the name suggests, these classical waves are mixed modes that have characteristics of both polarization waves (the TO phonons) and the photons. The dispersion of the polaritons can be deduced from the relationship:

A different type of polariton—namely a surface plasmon polariton—was considered in Section 7.5.2.

$$\omega = vq = \frac{c}{\sqrt{\epsilon_r}} q, \tag{10.18}$$

where the second part of the equation comes from eqn A.29, with $\mu_r = 1$. The resonant response of the polar solid is contained implicity in the frequency dependence of ϵ_r.

Figure 10.6 shows the polariton dispersion calculated for a lightly damped medium. The dielectric constant is given by eqn 10.15, and is

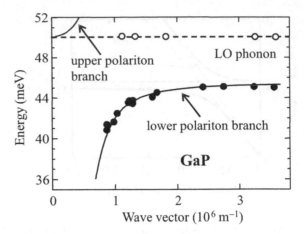

Fig. 10.7 Dispersion of the TO and LO phonons in GaP measured by Raman scattering. The solid lines are the predictions of the polariton model with $h\nu_{TO} = 45.5\,\text{meV}$, $\epsilon_\infty = 9.1$ and $\epsilon_{st} = 11.0$. After Henry and Hopfield (1965), © American Physical Society, reprinted with permission.

plotted for the same parameters as in Fig. 10.4(a). At low frequencies the dielectric constant is equal to ϵ_{st}, and the dispersion of the modes is given by $\omega = cq/\sqrt{\epsilon_{st}}$. As ω approaches Ω_{TO}, the dielectric constant increases, and the velocity of the waves decreases, approaching zero at Ω_{TO} itself. For frequencies in the Reststrahl band between Ω_{TO} and Ω_{LO}, the dielectric constant is negative. No modes can propagate, and all the photons that are incident on the medium are reflected. For frequencies above Ω_{LO}, ϵ_r is positive again and propagating modes are possible once more. The velocity of the waves gradually increases with increasing frequency, approaching a value of $c/\sqrt{\epsilon_\infty}$ at high frequencies.

The dispersion of the polariton modes has been measured for a number of materials. Figure 10.7 shows the measured dispersion of the TO and LO phonons in GaP at small wave vectors. The results were obtained by Raman scattering techniques. (See Section 10.5.2.) The experimental data reproduce very well the polariton dispersion model indicated in Fig. 10.6. The solid line is the calculated polariton dispersion, which gives a very accurate fit to the experimental points for the transverse modes. Note that the LO phonons do not show any dispersion here because they do not couple to the light waves.

10.4   Polarons

So far in this chapter we have been considering the direct interaction between a light wave and the phonons in a crystal. As we have seen, this gives rise to strong absorption and reflection in the infrared spectral region. The optical phonons can, however, contribute indirectly to a whole host of other optical properties that depend primarily on the electrons through the **electron–phonon coupling**. In this section we will consider the **polaron** effect, which is one of the most important examples of this.

Consider the motion of a free electron through a polar solid, as shown in Fig. 10.8. The electron will attract the positive ions that are close

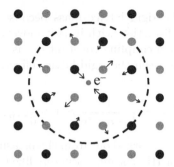

Fig. 10.8 Schematic representation of a polaron. A free electron moving through an ionic lattice attracts the positive (black) ions, and repels the negative (grey) ones. This produces a local distortion of the lattice within the polaron radius shown by the dashed circle.

to it, and repel the negative ones. This produces a local displacement of the lattice in the immediate vicinity of the electron. The lattice distortion accompanies the electron as it moves through the crystal. The electron with its local lattice distortion is equivalent to a new elementary excitation of the crystal, and is called a polaron.

The polaron effect can be conceived in terms of an electron surrounded by a cloud of virtual phonons. We think of the electron absorbing and emitting phonons as it moves through the crystal. These phonons produce the local lattice distortion. The displacement of the ions is in the same direction as the electric field of the electron, and we are therefore dealing with longitudinal optic phonons.

The strength of the electron–phonon interaction in a polar solid can be quantified by the dimensionless coupling constant $\alpha_{\rm ep}$, which is given by:

$$\alpha_{\rm ep} = \frac{1}{137} \left(\frac{m^* c^2}{2h\nu_{\rm LO}} \right)^{1/2} \left[\frac{1}{\epsilon_\infty} - \frac{1}{\epsilon_{\rm st}} \right], \qquad (10.19)$$

where $1/137$ is the fine structure constant from atomic physics. The mass m^* that appears here is the usual effective mass deduced from the curvature of the band structure (cf. eqn D.6):

$$m^* = \hbar^2 \left(\frac{{\rm d}^2 E}{{\rm d}k^2} \right)^{-1}. \qquad (10.20)$$

The polaron theory can be applied equally to electrons or holes by taking the appropriate effective masses in the formulæ.

Values for $\alpha_{\rm ep}$ for three binary compound semiconductors, namely GaAs, ZnSe, and AgCl, are given in Table 10.3. We see that the coupling constant increases from GaAs (0.06) through ZnSe (0.40) to AgCl (2.2). This is because the ionicity increases as we go from the III–V semiconductor, in which the bonding is partly covalent, to the I–VII compound, which is highly ionic. In a non-polar crystal such as silicon, $\epsilon_\infty = \epsilon_{\rm st}$, and $\alpha_{\rm ep} = 0$. There is therefore no polaron effect.

The effective mass given by eqn 10.20 is calculated by assuming that the lattice is rigid. However, the concept of a rigid lattice is only a theoretical one, and any experiment we perform to measure m^* will actually measure the polaron mass m^{**} instead. This is because it is not possible to hold the lattice rigid as the electron moves. The polaron

It can be shown that the average number of virtual LO phonons that move with the electron is equal to $\alpha_{\rm ep}/2$, where $\alpha_{\rm ep}$ is defined in eqn 10.19. We do not consider the longitudinal acoustic modes here because they do not produce a polarization in the medium: the positive and negative ions move in the same direction, and this produces no electric-dipole moment.

Table 10.3 Electron–phonon coupling constant $\alpha_{\rm ep}$ calculated from eqn 10.19 for GaAs, ZnSe, and AgCl. The figures for ZnSe are for the cubic crystal structure. Data from Madelung (1996).

	GaAs	ZnSe	AgCl
$m_{\rm e}^*/m_0$	0.067	0.13	0.30
ϵ_∞	10.9	5.4	3.9
$\epsilon_{\rm st}$	12.4	7.6	11.1
$\nu_{\rm LO}$ (THz)	8.5	7.6	5.9
$\alpha_{\rm ep}$	0.06	0.40	2.2

mass is larger than the rigid lattice mass because the electron has to drag the local lattice distortion with it as it moves.

If the electron–phonon coupling constant α_{ep} is small, we can give an explicit relationship between the rigid lattice effective mass m^* and the polaron mass m^{**}:

$$\frac{m^{**}}{m^*} = \frac{1}{1 - \alpha_{ep}/6} \approx 1 + \frac{1}{6}\alpha_{ep}. \tag{10.21}$$

Values of m^* are actually worked out from the measured values of m^{**} by applying eqn 10.21. For III–V semiconductors like GaAs with $\alpha_{ep} < 0.1$, m^{**} only differs from m^* by about 1%. The polaron effect is thus only a small correction. This correction becomes more significant for II–VI compounds (e.g. $\sim 7\%$ for ZnSe). With highly ionic crystals like AgCl, the small α_{ep} approximation is not valid. The actual polaron mass of AgCl is $0.43m_0$, which is about 50% larger than the rigid lattice value.

An example of an experiment to measure the effective mass is **cyclotron resonance**. In this technique, we measure the infrared absorption in the presence of a magnetic field B. As discussed in Section 3.3.6, the electron energy is quantized in terms of the cyclotron energy:

$$E_n = (n + 1/2)\hbar\omega_c, \tag{10.22}$$

where n is an integer, and

$$\omega_c = \frac{eB}{m^*}. \tag{10.23}$$

Optical transitions with $\Delta n = \pm 1$ can take place between the ladder of levels defined by eqn 10.22. We therefore observe absorption at a wavelength λ given by:

$$\frac{hc}{\lambda} = \frac{e\hbar B}{m^*}. \tag{10.24}$$

This absorption usually occurs in the far-infrared spectral region, and the effective mass can be deduced from the values of λ and B at resonance. In a typical experiment, we use a fixed wavelength source from an infrared laser and find the value of B which gives the maximum absorption. For example, the cyclotron resonance occurs at about 6.1 T in GaAs $(m^* = 0.067m_0)$ for the 118 μm line from a methanol laser. The effective mass we find this way is the polaron mass m^{**}, not the value determined by the curvature of the bands given by eqn 10.20.

It can be shown that, in addition to the change of the mass, the polaron effect causes a reduction in the band gap by an amount:

$$\Delta E_g = -\alpha_{ep}\,\hbar\Omega_{LO}. \tag{10.25}$$

With a III–V material like GaAs, this again produces only a relatively small effect: $\Delta E_g \sim -0.1\%$. In practice, when we measure E_g by optical spectroscopy, we always measure the polaron value.

Another important parameter of the polaron is its radius, r_p, which specifies how far the lattice distortion extends. This is depicted schematically in Fig. 10.8 by the dashed circle drawn around the electron that

A particularly clear manifestation of the electron–phonon coupling can be observed in cyclotron resonance experiments when $B = m^*\Omega_{LO}/e$, so that $\omega_c = \Omega_{LO}$. The degenerate electron and phonon modes anticross with each other as the field is swept through this condition, and the cyclotron resonance line splits into a doublet. The magnitude of the splitting is directly proportional to the electron–phonon coupling constant α_{ep}. This effect was first observed in n-type InSb.

causes the lattice distortion. If α_{ep} is small, we can give an explicit formula for r_p:

$$r_p = \left(\frac{\hbar}{2m^*\Omega_{LO}} \right)^{1/2}. \tag{10.26}$$

This gives $r_p = 4.0\,\text{nm}$ for GaAs and $3.1\,\text{nm}$ for ZnSe. Both values are significantly larger than the unit cell size ($\sim 0.5\,\text{nm}$), which is important because the theory used to derive eqns 10.21–10.26 assumes that we can treat the medium as a polarizable continuum. This approximation is only valid if the radius of the polaron is very much greater than the unit cell size. A polaron that satisfies this criterion is called a **large polaron**. In highly ionic solids such as AgCl and the alkali halides, α_{ep} is not small and the polaron radius is comparable to the unit cell size. In this case we have a **small polaron**. The mass and radius have to be calculated from first principles.

The small polaron effect in highly ionic crystals leads to **self-trapping** of the charge carriers. The local lattice distortion is very strong, and the charge carrier can get completely trapped in its own lattice distortion. The carrier effectively digs itself into a pit and cannot get out of it. This is particularly the case for the holes in alkali halide crystals. The only way they can move is by **hopping** to a new site. The electrical conductivity of most alkali halide crystals is limited by this thermally activated hopping process at room temperature.

Self-trapping effects are also important when considering Frenkel excitons. As discussed in Section 4.5, these are bound electron-hole pairs localized at individual atom or molecule sites within the lattice. The self-trapping of either the electron or hole can exacerbate the tendency for the exciton to localize, thereby instigating the transition from Wannier (free) to Frenkel exciton behaviour. The ground-state excitons observed in many alkali halide, rare gas, and organic crystals are of the self-trapped Frenkel type.

Polaronic hopping effects are also important in the conduction processes in organic semiconductors like polydiacetylene.

10.5 Inelastic light scattering

Inelastic light scattering describes the phenomenon by which a light beam is scattered by an optical medium and changes its frequency in the process. It contrasts with elastic light scattering, in which the frequency of the light is unchanged. The interaction process is illustrated in Fig. 10.9. Light incident with angular frequency ω_1 and wave vector k_1 is scattered by an excitation of the medium of frequency Ω and wave vector q. The scattered photon has frequency ω_2 and wave vector k_2. Inelastic light scattering can be mediated by many different types of elementary excitations in a crystal, such as phonons, magnons, or plasmons. In this chapter we are concerned exclusively with phonon processes.

Inelastic light scattering from phonons is generally classified by the type of phonons that are involved:

We presented a Raman scattering spectrum from plasmons in n-type GaAs in Fig. 7.13.

- **Raman scattering**. This is inelastic light scattering from optical phonons.

- **Brillouin scattering**. This is inelastic light scattering from acoustic phonons.

The physics of the two processes is essentially the same, but the experimental techniques differ. We therefore consider the general principles first, and then consider the details of each technique separately.

10.5.1   General principles of inelastic light scattering

Inelastic light scattering can be subdivided into two generic types:

- **Stokes scattering**;
- **Anti-Stokes scattering**.

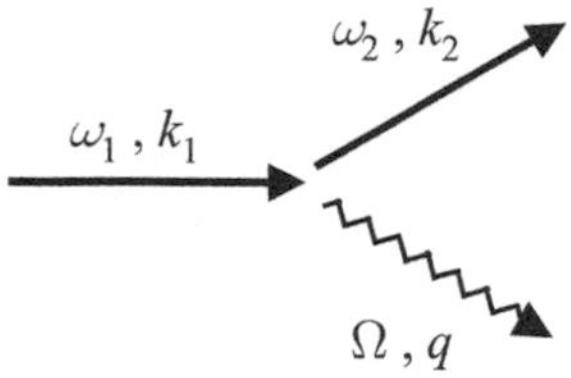

Fig. 10.9 An inelastic light scattering process. The straight arrows represent photons, while the wiggly arrow represents the phonon. The process shown corresponds to Stokes scattering in which the photon is shifted to lower frequency.

Stokes scattering corresponds to the emission of a phonon (or some other type of material excitation), while anti-Stokes scattering corresponds to phonon absorption. The interaction shown in Fig. 10.9 is thus a Stokes process. Conservation of energy during the interaction requires that:

$$\omega_1 = \omega_2 \pm \Omega, \tag{10.27}$$

while conservation of momentum gives:

$$\boldsymbol{k}_1 = \boldsymbol{k}_2 \pm \boldsymbol{q}. \tag{10.28}$$

The $+$ signs in eqns 10.27 and 10.28 correspond to phonon emission (Stokes scattering), while the $-$ signs correspond to phonon absorption (anti-Stokes scattering). Thus the light is shifted down in frequency during a Stokes process, and up in frequency in an anti-Stokes event.

Anti-Stokes scattering will only be possible if there are phonons present in the material before the light is incident. The probability for anti-Stokes scattering therefore decreases on lowering the temperature as the phonon populations decrease. This means that the probability for anti-Stokes scattering from optical phonons is very low at cryogenic temperatures. On the other hand, Stokes scattering does not require a phonon to be present and can therefore occur at any temperature. The full quantum mechanical treatment shows that the ratio of anti-Stokes to Stokes scattering events is given by:

$$\frac{I_{\text{anti-Stokes}}}{I_{\text{Stokes}}} = \exp\left(-\hbar\Omega/k_{\text{B}}T\right). \tag{10.29}$$

This will be the ratio of the intensities of the anti-Stokes and Stokes lines observed in the Raman or Brillouin spectra.

The frequencies of the phonons involved can be deduced from the frequency shift of the scattered light by using eqn 10.27. Thus the main use of inelastic light scattering is to measure phonon frequencies. This means that inelastic light scattering can give complementary information to that obtained from the infrared spectra. For example, infrared reflectivity measurements tell us nothing about the acoustic phonons, but we can measure the frequencies of some of the acoustic modes by Brillouin scattering experiments. We shall consider this complementarity in more detail when we discuss the selection rules for Raman scattering in subsection 10.5.2 below.

The maximum phonon frequency in a typical crystal is about 10^{12} to 10^{13} Hz. This is almost two orders of magnitude smaller than the frequency of a photon in the visible spectral region. Equation 10.27 therefore tells us that the maximum frequency shift for the photon will be around 1%. The magnitude of the wave vector of the photon is directly proportional to its frequency, and we can therefore make the approximation:

$$|\boldsymbol{k}_2| \approx |\boldsymbol{k}_1| = \frac{n\omega}{c}, \tag{10.30}$$

where n is the refractive index of the crystal and ω is the angular frequency of the incoming light.

We know from eqn 10.28 that $|\boldsymbol{q}| = |\boldsymbol{k}_1 - \boldsymbol{k}_2|$. The maximum possible value of $|\boldsymbol{q}|$ thus occurs for the **back-scattering geometry** in which the outgoing photon is emitted in the direction back towards the source. In this case, we have:

$$q \approx |\boldsymbol{k} - (-\boldsymbol{k})| \approx 2\,\frac{n\omega}{c}. \tag{10.31}$$

By inserting typical values into eqn 10.31, we conclude that the maximum value of q that can be accessed in an inelastic light scattering experiment is of order $10^7\,\mathrm{m}^{-1}$. This is very small compared to the size of the Brillouin zone in a typical crystal ($\sim 10^{10}\,\mathrm{m}^{-1}$). Inelastic light scattering is thus only able to probe small wave vector phonons.

Raman and Brillouin scattering are generally weak processes, and we therefore expect that the scattering rate will be small. This is because we are dealing with a higher-order interaction than for linear interactions such as absorption. Figure 10.9 shows us that three particles are present in the Feynman diagram for inelastic light scattering rather than the two for absorption. (See Fig. 10.1.) Therefore, a higher-order perturbation term must be involved. This means that we usually have to employ very sensitive detectors to observe the signals even when using a powerful laser beam as the excitation source.

The inelastic light scattering efficiency can be enhanced by many orders of magnitude by exploiting plasmonic effects. As discussed in Section 7.5.2, the electric field amplitude is enhanced substantially at the surface of metallic structures by surface plasmons, and this greatly increases the scattering probability. This effects is known as **surface-enhanced Raman scattering**. See Maier (2007) for further details.

10.5.2 Raman scattering

C.V. Raman was awarded the Nobel prize in 1930 for his discovery of inelastic light scattering from molecules. The process which now carries his name refers to scattering from high frequency excitations such as the vibrational modes of molecules. In the present context of phonon physics, it refers specifically to inelastic light scattering from optical phonons.

Optical phonons are essentially dispersionless near $q = 0$. We argued above that inelastic light scattering can only probe the phonon modes with $q \approx 0$. Therefore, Raman scattering gives little information about the dispersion of optical phonons, and its main use is to determine the frequencies of the LO and TO modes near the Brillouin zone centre.

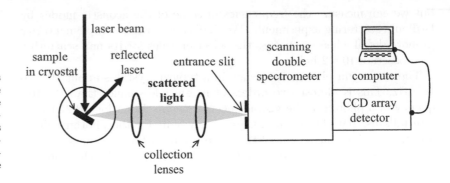

Fig. 10.10 Experimental apparatus used to record Raman spectra. The sample is excited with a laser, and the scattered photons are collected and focused into a spectrometer. The signals are recorded using a sensitive photon-counting detector such as a photomultiplier tube or a charge coupled device (CCD).

For example, when Raman techniques are used to measure polariton dispersion curves (see Section 10.3, and especially Fig. 10.7), we are only probing a very small portion of the Brillouin zone near $q = 0$.

The complementarity of infrared reflectivity and inelastic light scattering measurements become more apparent when we consider the selection rules for deciding whether a particular optical phonon is Raman active or not. These rules are not the same as those for determining whether the mode is IR active. The full treatment requires the use of group theory. However, a simple rule can be given for crystals that possess inversion symmetry. In these centrosymmetric crystals, the vibrational modes must either have even or odd parity under inversion. The odd parity modes are IR active, while the even parity modes are Raman active. Thus the Raman active modes are not IR active, and vice versa. This is called the **rule of mutual exclusion**, and is a well-known result in molecular physics. In non-centrosymmetric crystals, some modes may be simultaneously IR and Raman active.

As an example of these rules, we can compare silicon and GaAs. Silicon has the diamond structure with inversion symmetry, while GaAs has the non-centrosymmetric zinc-blende structure. The TO modes of silicon are not IR active, but they are Raman active, while the TO modes of GaAs are both Raman and IR active.

The observation of a Raman spectrum requires specialized apparatus to overcome the difficulties that are inherent to the technique. We pointed out above that the signal is relatively weak, which means that we have to use an intense source such as a laser to produce a sizeable scattering rate. However, the frequency shift of the scattered photons is quite small. We thus need to resolve a weak Raman signal which is very close in wavelength to the elastically scattered light from the laser.

Figure 10.10 shows a basic experimental arrangement that can be used to measure Raman spectra. The sample is excited with a suitable laser, and the scattered light is collected and focused onto the entrance slit of a scanning spectrometer. The number of photons emitted at a particular wavelength is registered on a photon-counting detector and then the results are stored on a computer for analysis. Photomultiplier tubes have traditionally been employed as the detector in this application,

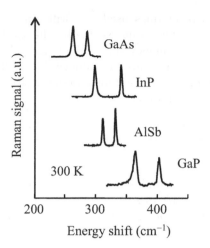

Fig. 10.11 Raman spectra for the TO and LO phonons of GaAs, InP, AlSb, and GaP at 300 K using a Nd:YAG laser at 1.06 μm. The spectra are plotted against the wave number shift: $1\,\mathrm{cm}^{-1}$ is equivalent to an energy shift of 0.124 meV. The LO mode is the one at higher frequency. After Mooradian (1972), © Excerpta Medica Inc., reprinted with permission.

but modern arrangements now tend to use array detectors made with charge coupled devices (CCD arrays). By orientating the sample appropriately, the reflected laser light can be arranged to miss the collection optics. However, this still does not prevent a large number of elastically scattered laser photons entering the spectrometer, and this could potentially saturate the detector. To get around this problem, a high-resolution spectrometer with good stray light rejection characteristics is used.

One way to achieve good stray light rejection is to use a double spectrometer, which is essentially two spectrometers in tandem.

Figure 10.11 shows the Raman spectrum obtained from four III–V semiconductor crystals at 300 K. The laser source was a Nd:YAG laser operating at 1.06 μm, and a double monochromator with a photomultiplier tube were used to detect the signal. Two strong lines are observed for each crystal. These correspond to the Stokes-shifted signals from the TO and LO phonons, with the LO phonons at the higher frequency. The values obtained from this data agree very well with those deduced from infrared reflectivity measurements. (See Exercise 10.13.)

10.5.3 Brillouin scattering

L. Brillouin gave a theoretical discussion of the scattering of light by acoustic waves in 1922. The technique named after him now refers to inelastic light scattering from acoustic phonons. Its main purpose is to determine the dispersion of these acoustic modes.

The frequency shift of the photons in a Brillouin scattering experiment is given by (see Exercise 10.14):

$$\delta\omega = v_{\mathrm{s}} \frac{2n\omega}{c} \sin\frac{\theta}{2}, \tag{10.32}$$

where ω is the angular frequency of the incident light, n is the refractive index of the crystal, v_{s} is the velocity of the acoustic waves, and θ is the angle through which the light is scattered. Measurements of $\delta\omega$ therefore allow the velocity of the sound waves to be determined if the refractive index is known.

The experimental techniques used for Brillouin scattering are more sophisticated than those for Raman scattering due to the need to be able to detect much smaller frequency shifts. Single-mode lasers must be used to ensure that the laser linewidth is sufficiently small, and a scanning Fabry–Perot interferometer is used instead of a grating spectrometer to increase the spectral resolution.

Example 10.2

When light from an argon ion laser operating at 514.5 nm is scattered by optical phonons in a sample of AlAs, two peaks are observed at 524.2 nm and 525.4 nm. What are the values of the TO phonon and LO phonon energies?

Solution

We can work out the energies of the phonons by using eqn 10.27. The photons have been red-shifted, and thus we are dealing with a Stokes process. For the 524.2 nm line we therefore have:

$$\Omega = \omega_1 - \omega_2 = 2\pi c(1/\lambda_1 - 1/\lambda_2) = 6.8 \times 10^{13} \, \text{rad/s} \, .$$

For the 525.4 nm line we find $\Omega = 7.6 \times 10^{13} \, \text{rad/s}$. The higher frequency phonon is the LO mode. Hence we find $\hbar\Omega_{\text{TO}} = 45 \, \text{meV}$ and $\hbar\Omega_{\text{LO}} = 50 \, \text{meV}$.

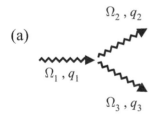

(a)

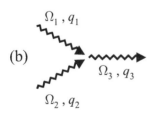

(b)

Fig. 10.12 Three-phonon interaction processes. Each wiggly arrow represents a phonon. These processes are caused by anharmonicity in the crystal.

10.6   Phonon lifetimes

The discussion of the phonon modes as classical oscillators in Section 10.2 led us to introduce a phenomenological damping constant γ. This damping term is needed to explain why the reflectivity in the Reststrahl band is less than unity. Analysis of the experimental data led us to conclude that γ is typically in the range 10^{11} to $10^{12} \, \text{s}^{-1}$. This very rapid damping is a consequence of the finite lifetime τ of the optical phonons. Since γ is equal to τ^{-1}, the data implies that τ is in the range 1–10 ps.

The very short lifetime of the optical phonons is caused by **anharmonicity** in the crystal. Phonon modes are solutions of the equations of motion with the assumption that the vibrating atoms experience harmonic restoring forces (i.e. forces that are linear in the displacement). In reality, this is only an approximation that is valid for small displacements. In general, the atoms sit in a potential well of the form:

$$U(x) = C_2 x^2 + C_3 x^3 + C_4 x^4 + \cdots \, . \tag{10.33}$$

An example of how interatomic interactions lead to a potential of this form is considered in Exercise 10.16.

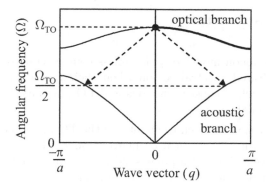

Fig. 10.13 Decay of an optical phonon into two acoustic phonons by a three-phonon interaction of the type shown in Fig. 10.12(a).

The term in x^2 in eqn 10.33 is the harmonic term. This leads to simple harmonic oscillator equations of motion with a restoring force $-\mathrm{d}U/\mathrm{d}x$ proportional to $-x$. The terms in x^3 and higher are the **anharmonic** terms, which are derived from restoring forces that vary non-linearly with x (e.g. $F \propto -x^2$). These anharmonic terms allow phonon–phonon scattering processes. For example, the term in x^3 allows interactions involving three phonons. Figure 10.12 illustrates two possible permutations for a three-phonon process.

Figure 10.12(a) shows a three-phonon interaction in which one phonon is annihilated and two new phonons are created. This type of anharmonic interaction is responsible for the fast decay of the optical phonons. We can see why this is so by referring to the generic phonon dispersion curve for the first Brillouin zone shown in Fig. 10.13. Lattice absorption or Raman scattering creates optical phonons with $q \approx 0$. Three-phonon processes allow these phonons to decay into two acoustic phonons as indicated in Fig. 10.13. Momentum and energy can be conserved if the two acoustic phonons have opposite wave vectors, and their frequency is half that of the optical phonon. With more complex dispersion relationships, and also the possibility for higher-order processes, many other types of decay can contribute to the short lifetime of the optical phonons.

The lifetime of the optical phonons can be deduced from Raman data in two different ways. Firstly, the spectral width of the Raman line is affected by lifetime broadening. Provided that other sources of broadening are smaller, the linewidth in frequency units is expected to be $(2\pi\tau)^{-1}$. Thus measurements of the linewidth give a value for τ independently of the reflectivity data. Secondly, τ can be measured directly by time-resolved Raman spectroscopy. The lifetime of the LO phonons in GaAs has been determined in this way to be 7 ps at 77 K. This value agrees with the linewidth measured in the conventional Raman spectrum. It is also similar to the lifetime of the TO phonons deduced from reflectivity measurements.

Chapter summary

- The TO phonon modes of polar solids couple strongly to photons in the infrared spectral region when their frequencies and wave vectors match. Acoustic phonons and LO phonons do not couple directly to light waves.

- The interaction between the light and the TO phonon can be modelled by using the classical oscillator model. This model explains why the reflectivity of a polar solid is very high for frequencies in the Reststrahl band between ν_{TO} and ν_{LO}.

- The reflectivity in the Reststrahl band is 100% for an undamped system, but damping due to the finite phonon lifetime reduces the reflectivity in real crystals.

- The frequencies of the TO and LO phonon modes are related to each other by the Lyddane–Sachs–Teller relationship given in eqn 10.14.

- The lattice absorbs strongly at the TO phonon frequency. The absorption can be measured directly in thin film samples.

- The strongly coupled phonon–photon waves at frequencies near the Reststrahl band are described as phonon polariton modes.

- The electron–phonon coupling in polar crystals leads to polaron effects. Polarons are charge carriers surrounded by a local lattice distortion. The phonon cloud around the electron or hole increases its mass. Polaronic effects are strong in ionic crystals like the alkali halides.

- Raman and Brillouin scattering are inelastic light scattering processes from optical and acoustic phonons respectively. Energy and momentum must be conserved in the scattering process.

- Stokes and anti-Stokes inelastic light scattering processes correspond to phonon emission and absorption respectively. Anti-Stokes scattering from optical phonons is very improbable at low temperatures.

- Optical phonons have short lifetimes due to the possibility of decay into two acoustic phonons by anharmonic interactions.

Further reading

Introductory reading on phonons may be found in practically any solid-state physics text, for example: Ashcroft and Mermin (1976), Burns (1985), Ibach and Luth (2003), or Kittel (2005).

The theory of polaritons and polarons is described in more detail in Madelung (1978). Pidgeon (1980) and Seeger (1997) discuss cyclotron resonance experiments in detail. The properties of self-trapped excitons are covered by Song and Williams (1993), while Pope and Swenberg (1999) discuss polaronic hopping transport, especially in organic semiconductors.

A classic text on the infrared physics of molecules and

solids is Houghton and Smith (1966). The techniques of inelastic light scattering are described in detail by Mooradian (1972) or Yu and Cardona (1996). The study of phonon dynamics by ultra-fast laser techniques is described by Shah (1999).

Exercises

(10.1) State, with reasons, which of the following solids would be expected to show strong infrared absorption: (a) ice, (b) germanium, (c) solid argon at 4 K, (d) ZnSe, (e) SiC.

(10.2) Show that the reflectivity of an undamped polar solid falls to zero at a frequency given by

$$\nu = \left(\frac{\epsilon_{st} - 1}{\epsilon_\infty - 1} \right)^{1/2} \nu_{TO},$$

where ϵ_{st} and ϵ_∞ are the low- and high-frequency dielectric constants, and ν_{TO} is the frequency of the TO phonon mode at the Brillouin zone centre.

(10.3) The static and high-frequency dielectric constants of LiF are $\epsilon_{st} = 8.9$ and $\epsilon_\infty = 1.9$ respectively, and the TO phonon frequency ν_{TO} is 9.2 THz. Calculate the upper and lower wavelengths of the Reststrahl band.

(10.4) Estimate the reflectivity in the middle of the Reststrahl band for a crystal with $\nu_{TO} = 10$ THz, $\epsilon_{st} = 12.1$, and $\epsilon_\infty = 10$, when the damping constant γ is (a) 10^{11} s^{-1} and (b) 10^{12} s^{-1}.

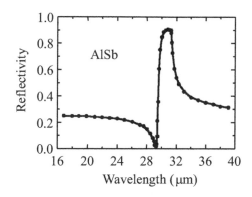

Fig. 10.14 Infrared reflectivity of AlSb. After Turner and Reese (1962), © American Physical Society, reprinted with permission.

(10.5) Figure 10.14 shows the measured infrared reflectivity of AlSb. Use this data to estimate:

(a) the frequencies of the TO and LO phonons of AlSb near the Brillouin zone centre;
(b) the static and high-frequency dielectric constants, ϵ_{st} and ϵ_∞;
(c) the lifetime of the TO phonons.
Are the experimental values found in parts (a) and (b) consistent with the Lyddane–Sachs–Teller relationship?

(10.6) Estimate the absorption coefficient at the TO phonon frequency in a typical polar solid with a damping constant γ of (a) 10^{11} s^{-1} and (b) 10^{12} s^{-1}. Take $\nu_{TO} = 10$ THz, $\epsilon_{st} = 12.1$, and $\epsilon_\infty = 10$.

(10.7) Explain qualitatively why the reflectivity of NaCl in the middle of the Reststrahl band is observed to decrease from 98% at 100 K to 90% at 300 K.

(10.8) The static and high-frequency dielectric constants of InP are $\epsilon_{st} = 12.5$ and $\epsilon_\infty = 9.6$ respectively, and the TO phonon frequency ν_{TO} is 9.2 THz. Calculate the wave vector of a polariton mode with a frequency of 8 THz. (Ignore phonon damping.)

(10.9) In an infrared absorption experiment on n-type CdTe, the cyclotron resonance condition is satisfied at 3.4 T for the 306 μm line from a deuterated methanol laser. Calculate (a) the polaron mass, and (b) the rigid lattice electron effective mass, given that $\epsilon_\infty = 7.1$, $\epsilon_{st} = 10.2$, and $\nu_{LO} = 5.1$ THz.

(10.10) Discuss the qualitative differences you would expect between the Raman spectrum observed from diamond to that shown for the III–V crystals in Fig. 10.11.

(10.11) In an inelastic light scattering experiment on silicon using an argon ion laser at 514.5 nm, Raman peaks are observed at 501.2 nm and 528.6 nm. Account for the origin of the two peaks, and estimate their intensity ratios if the sample temperature is 300 K.

(10.12) NaCl is a centrosymmetric crystal. Would you expect the TO phonon modes to be IR active, or

Raman active, or both?

(10.13) Use the data in Fig. 10.11 to deduce the energies in meV of the TO and LO phonons of GaAs, InP, AlSb, and GaP at 300 K. How do the values for GaAs obtained from this data relate to the infrared reflectivity data given in Fig. 10.5?

(10.14) A photon of angular frequency ω is scattered inelastically through an angle θ by an acoustic phonon of angular frequency Ω. By considering the conservation of momentum in the process, show that Ω is given by:

$$\Omega = v_s \frac{2n\omega}{c} \sin \frac{\theta}{2},$$

where v_s and n are the velocity of sound and the refractive index in the medium respectively. (You may assume that $\omega \gg \Omega$.) Hence justify eqn 10.32.

(10.15) A Brillouin scattering experiment is carried out on a crystal with a refractive index of 3 using light from a laser with a wavelength of 488 nm. The scattered photons are found to be down-shifted in frequency by 10 GHz when observed in the back-scattering geometry with $\theta = 180°$. Calculate the speed of sound in the crystal.

(10.16)* The potential energy per molecule of an ionic crystal with a nearest neighbour separation of r may be approximated by the following form:

$$U(r) = \frac{\beta}{r^{12}} - \frac{\alpha e^2}{4\pi\epsilon_0 r},$$

where α is the Madelung constant of the crystal, and β is a fitting parameter.
(a) Account for the functional form of $U(r)$.
(b) Show that $U(r)$ has a minimum value when $r = r_0$, where $r_0^{11} = 48\beta\pi\epsilon_0/\alpha e^2$.
(c) Expand U(r) as a Taylor series about r_0, and hence show that the potential takes the form given by eqn 10.33 for small displacements about r_0, stating the value of the constant C_3 in terms of α and r_0.

(10.17) High-resolution Raman experiments on a GaAs crystal indicate that the LO phonon line has a spectral width of $0.85\,\text{cm}^{-1}$. Use this value to estimate the lifetime of the LO phonons, on the assumption that the spectrum is lifetime-broadened.

*Exercises marked with an asterisk are more challenging.

Nonlinear optics

11

Practically everything we have been describing so far in this book falls into the realm of linear optics, where it is assumed that properties such as the refractive index, absorption coefficient, and reflectivity are independent of the optical power. This approximation is only valid at low power levels. With a high-power laser, it is possible to enter a different realm of behaviour called **nonlinear optics**. In this subject we consider the consequences of allowing the electric susceptibility, and all the properties that follow from it, to vary with the strength of the electric field of the light beam.

Nonlinear optics is a subject in its own right that has grown in importance as the applications of lasers have become more common. A text such as this would be incomplete without some mention of the types of phenomena that can occur in solid-state materials. The objective here is to give a brief introduction to the subject, based predominantly on the classical dipole oscillator model developed in Chapter 2. It is hoped that this may form a basis for further reading in more comprehensive treatments. A partial list of introductory nonlinear optics texts may be found in the Bibliography.

11.1	The nonlinear susceptibility tensor	295
11.2	The physical origin of optical nonlinearities	298
11.3	Second-order nonlinearities	305
11.4	Third-order nonlinear effects	317
	Chapter summary	326
	Further reading	327
	Exercises	328

11.1 The nonlinear susceptibility tensor

The optical properties of materials are described through the real and imaginary parts of the dielectric constant ϵ_r. The dielectric constant is derived from the polarization $\boldsymbol{P}$ of the medium according to:

$$\begin{aligned} \boldsymbol{D} &= \epsilon_0 \boldsymbol{\mathcal{E}} + \boldsymbol{P}\,, \\ &= \epsilon_0 \epsilon_r \boldsymbol{\mathcal{E}}\,. \end{aligned} \tag{11.1}$$

In linear optics, we assume that $\boldsymbol{P}$ depends linearly on the electric field $\boldsymbol{\mathcal{E}}$ of the light wave, so that we can write:

$$\boldsymbol{P} = \epsilon_0 \chi \boldsymbol{\mathcal{E}}\,, \tag{11.2}$$

where χ is the electric susceptibility. By combining eqns 11.1 and 11.2 we derive the usual relationship between ϵ_r and χ, namely:

$$\epsilon_r = 1 + \chi\,. \tag{11.3}$$

In nonlinear optics we consider the possibility that the relationship between $\boldsymbol{P}$ and $\boldsymbol{\mathcal{E}}$ is more general than that given in eqn 11.2. We start by considering a nonlinear medium in which the polarization is parallel to the electric field, so that we do not need to consider the vector nature of $\boldsymbol{P}$ and $\boldsymbol{\mathcal{E}}$ at this stage. We split the polarization P into the first-order linear response $P^{(1)}$, plus a whole series of nonlinear terms of increasing order according to:

$$P^{\text{nonlinear}} = P^{(1)} + P^{(2)} + P^{(3)} + \cdots , \qquad (11.4)$$

where $P^{(n)}$ is the nth-order polarization.

In analogy with eqn 11.2, we now introduce the **nonlinear susceptibility** $\chi^{\text{nonlinear}}$, and the nth-order nonlinear susceptibility, $\chi^{(n)}$. These are defined by the following equations:

$$
\begin{aligned}
P^{\text{nonlinear}} &= \epsilon_0 \chi^{\text{nonlinear}} \mathcal{E} \\
&= \epsilon_0 \chi^{(1)} \mathcal{E} + \epsilon_0 \chi^{(2)} \mathcal{E}^2 + \epsilon_0 \chi^{(3)} \mathcal{E}^3 + \cdots ,
\end{aligned}
\qquad (11.5)
$$

where $\mathcal{E}$ is the magnitude of the applied field.

The various terms in eqns 11.4 and 11.5 correspond directly with each other so that

$$
\begin{aligned}
P^{(1)} &= \epsilon_0 \chi^{(1)} \mathcal{E} , & (11.6) \\
P^{(2)} &= \epsilon_0 \chi^{(2)} \mathcal{E}^2 , & (11.7) \\
P^{(3)} &= \epsilon_0 \chi^{(3)} \mathcal{E}^3 . & (11.8)
\end{aligned}
$$

$$\vdots$$

By comparing eqns 11.3 and 11.5, we see that:

$$
\begin{aligned}
\epsilon_r^{\text{nonlinear}} &= 1 + \chi^{\text{nonlinear}} , \\
&= 1 + \chi^{(1)} + \chi^{(2)} \mathcal{E} + \chi^{(3)} \mathcal{E}^2 + \cdots ,
\end{aligned}
\qquad (11.9)
$$

where $\chi^{(1)}$ is just the normal linear susceptibility. Equation 11.9 implies that the dielectric constant depends on the electric field through the nonlinear susceptibilities. Since the optical power is proportional to $\mathcal{E}^2$, this means that ϵ_r also depends on the optical power. Hence properties like the refractive index and absorption coefficient become power-dependent in nonlinear materials.

The different-order nonlinear susceptibilities give rise to a whole host of nonlinear effects. The majority of these phenomena can be attributed to either the $\chi^{(2)}$ or $\chi^{(3)}$ terms in the polarization. These are either called **second-order** or **third-order nonlinear** effects as appropriate. Some of these will be discussed in Sections 11.3 and 11.4.

The well-defined axes of crystalline materials make it necessary to consider that the nonlinear response of the medium may depend on the directions in which the fields are applied. For example, we could apply two optical fields in different directions and then generate a nonlinear polarization along a third direction. This type of behaviour can be described by generalizing eqns 11.7 and 11.8 to allow for the anisotropic response of the medium. For example, the components of the second-order nonlinear polarization $\boldsymbol{P}^{(2)}$ can be written in the following form:

$$P_i^{(2)} = \epsilon_0 \sum_{j,k} \chi_{ijk}^{(2)} \mathcal{E}_j \mathcal{E}_k \,. \tag{11.10}$$

The quantity $\chi_{ijk}^{(2)}$ that appears here is the **second-order nonlinear susceptibility tensor**, and the subscripts i, j, and k correspond to the Cartesian coordinate axes x, y, and z. It will usually be convenient to define these axes so that they coincide with the principal axes of the crystal whenever this is possible.

Equation 11.10 shows that there are nine different contributions for each component of $\boldsymbol{P}^{(2)}$. For example, the term with $\chi_{xyz}^{(2)}$ gives the nonlinear polarization generated along the x axis when one optical field is applied along the y axis and another along the z axis. We can also generate a nonlinear polarization along the x axis by applying two fields along x via the $\chi_{xxx}^{(2)}$ term, and so on for all nine possible permutations of j and k. At first sight it might therefore appear that we have to measure 27 different quantities in order to quantify the second-order nonlinear response of an anisotropic medium completely. Fortunately, this is not usually the case, because the high degree of symmetry found in crystals requires that many of the terms are zero, and many of the others are the same. This point is developed further in Section 11.3.2.

The third-order nonlinear response of an anisotropic medium is also described by a tensor relationship. We can generalize eqn 11.8 by writing the components of the third-order nonlinear polarization as follows:

$$P_i^{(3)} = \epsilon_0 \sum_{j,k,l} \chi_{ijkl}^{(3)} \mathcal{E}_j \mathcal{E}_k \mathcal{E}_l \,, \tag{11.11}$$

where $\{i,j,k,l\} \in \{x,y,z\}$, and $\chi_{ijkl}^{(3)}$ is the third-order nonlinear susceptibility tensor. $\chi_{ijkl}^{(3)}$ is a fourth-rank tensor with 81 components. As with the second-order nonlinear susceptibility, symmetry may require that many of these terms are the same or zero.

It is not necessary that $\mathcal{E}_k$ and $\mathcal{E}_l$ in eqn 11.10 should be derived from different light beams. In many cases there will only be a single beam incident on the crystal, and $\mathcal{E}_k$ and $\mathcal{E}_l$ will just be the components of the electric field resolved along the appropriate axes.

Example 11.1

Potassium dihydrogen phosphate (KDP) is a uniaxial crystal with a four-fold axis of rotation about the z axis. The tetragonal $\overline{4}2m$ symmetry class of the crystal demands that the only non-zero components of the second-order nonlinear susceptibility tensor are the ones with i, j, and k all different, namely $\chi_{xyz}^{(2)}$, $\chi_{yxz}^{(2)}$, $\chi_{xzy}^{(2)}$ $\chi_{zxy}^{(2)}$, $\chi_{yzx}^{(2)}$, and $\chi_{zyx}^{(2)}$. Furthermore, symmetry also requires that

$$\chi_{xyz}^{(2)} = \chi_{yxz}^{(2)} = \chi_{xzy}^{(2)} = \chi_{yzx}^{(2)} \,, \tag{11.12}$$

and

$$\chi_{zxy}^{(2)} = \chi_{zyx}^{(2)} \,. \tag{11.13}$$

Determine the direction of the nonlinear polarization when a powerful laser beam is propagating along the optic axis.

Solution

If the laser is propagating in the z direction, then the electric field of the light will be polarized along the x or y directions. The nonlinear polarization is therefore given by eqn 11.10 with $\mathcal{E}_z = 0$. This gives:

$$P_i^{(2)} = \epsilon_0 \left(\chi_{ixx}^{(2)} \mathcal{E}_x \mathcal{E}_x + \chi_{ixy}^{(2)} \mathcal{E}_x \mathcal{E}_y + \chi_{iyx}^{(2)} \mathcal{E}_y \mathcal{E}_x + \chi_{iyy}^{(2)} \mathcal{E}_y \mathcal{E}_y \right).$$

If $i = x$ or $i = y$, then all the terms on the right-hand side are zero because $\chi_{ijk}^{(2)}$ is zero unless i, j and k are all different. This means that the nonlinear polarization vector is given by:

$$
\begin{aligned}
P_x^{(2)} &= 0 \\
P_y^{(2)} &= 0 \\
P_z^{(2)} &= \epsilon_0 \left(\chi_{zxy}^{(2)} \mathcal{E}_x \mathcal{E}_y + \chi_{zyx}^{(2)} \mathcal{E}_y \mathcal{E}_x \right).
\end{aligned}
$$

We therefore conclude that the nonlinear polarization is pointing along the optic axis, irrespective of the direction of the polarization of the input laser beam.

11.2   The physical origin of optical nonlinearities

The discussion in the previous section gave no indication as to why a particular material should be nonlinear or not. The magnitude of the electric field that binds an electron to an atom is typically around 10^{10}–$10^{11}\,\mathrm{V\,m^{-1}}$. (See Exercise 11.1.) It might therefore be expected that nonlinear effects would become important when the electric field of the light is comparable to this value. From the relationship between the intensity of a light beam and its electric field given by eqn A.44 in Appendix A, namely:

$$I = \frac{1}{2} c \epsilon_0 n \mathcal{E}^2 , \tag{11.14}$$

we see that we need optical intensities around $10^{19}\,\mathrm{W\,m^{-2}}$ to produce fields of this magnitude. Intensities as high as this can just about be achieved with very powerful lasers, but in fact the nonlinear effects set in at much lower intensity levels. This is because we can produce a sizeable macroscopic result by adding together the very small nonlinear effects in a very large number of atoms. This only works if the nonlinear phenomena in all the individual atoms are in phase with each other. This effect is called 'phase matching', and is discussed in Section 11.3.3.

The approach taken to explaining the microscopic origin of optical nonlinearities depends on whether the frequency is close to one of the natural transition frequencies of the atoms or not. If it is, then we are dealing with a **resonant** nonlinear effect, while if it is not, we are considering a **non-resonant** nonlinearity. These two situations are discussed separately below, starting with the non-resonant nonlinearities. It turns out that the non-resonant effects can be explained in terms of the classical oscillator model by introducing anharmonic terms. On the other hand, we need to use a quantum model to account properly for resonant effects.

11.2.1 Non-resonant nonlinearities

In Chapter 2 we explained how we can calculate the response of a medium to electromagnetic waves by assuming that it consists of a series of oscillators with characteristic resonant frequencies. In the near-infrared, visible, or ultraviolet spectral ranges we are normally considering the response due to the electrons. We have been assuming that these are bound to the atoms by harmonic restoring forces such that the displacement induced by the driving field of the light wave is linear. As with most oscillatory systems, this will only be true for small displacements. If the system is driven hard by the strong field of an intense laser beam, the displacements will be large, and it may no longer be valid to assume that the displacement varies linearly with the driving field.

We can account for the non-resonant nonlinear effects by assuming that the electron is bound in an **anharmonic** potential well of the form:

$$U(x) = \frac{1}{2}m_0\omega_0^2 x^2 + \frac{1}{3}m_0 C_3 x^3 + \frac{1}{4}m_0 C_4 x^4 + \cdots, \qquad (11.15)$$

where ω_0 is the natural resonant frequency and $x = 0$ corresponds to the equilibrium position of the electron. It is assumed that $\omega_0^2 \gg C_3 x \gg C_4 x^2 \ldots$, so that it makes sense to carry out the power series expansion, and that the harmonic term dominates for small displacements. The power series expansion is a simplification of the more complicated functional forms that would appear in a real atom. (See Exercise 10.16 for a worked example for the vibrational potential energy.)

We concentrate here on the second-order effects, and consider only the x^3 term in eqn 11.15. The restoring force for displacements from the equilibrium position is given by:

$$F(x) = -\frac{dU}{dx} = -\left(m_0\omega_0^2 x + m_0 C_3 x^2\right). \qquad (11.16)$$

This shows that the strength of the restoring force now depends on the direction of the displacement: the electron experiences a stronger force for positive displacements than for negative displacements. If we drive the electron with the AC electric field of a light wave, the displacements will be smaller during the positive part of the cycle than for the negative part. Since the dipole moment per unit volume of the medium is equal to $-Nex$, the polarization will likewise be asymmetric in the field direction. Hence the relationship between P and $\mathcal{E}$ will not be linear and will involve powers of $\mathcal{E}$ greater than one.

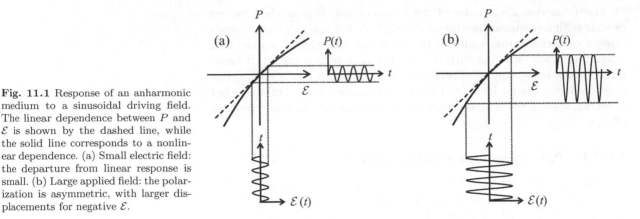

Fig. 11.1 Response of an anharmonic medium to a sinusoidal driving field. The linear dependence between P and $\mathcal{E}$ is shown by the dashed line, while the solid line corresponds to a nonlinear dependence. (a) Small electric field: the departure from linear response is small. (b) Large applied field: the polarization is asymmetric, with larger displacements for negative $\mathcal{E}$.

The nonlinear relationship between P and $\mathcal{E}$ is sketched in Fig. 11.1. For small fields the departure from the linear response is negligible. Hence the polarization closely follows the applied field, as shown in Fig. 11.1(a). However, if the magnitude of the applied field is increased, the response becomes asymmetric, with larger displacements for negative fields. This point is illustrated in Fig. 11.1(b), which shows how the application of a sinusoidal electric field gives a distorted output if the material is nonlinear. It is well known from electrical circuit theory that the distorted output can be described by including higher harmonics. In the case shown in Fig. 11.1(b), the output contains a second harmonic wave with 20% of the amplitude of the fundamental.

This simple discussion shows that the inclusion of the anharmonic term generates a signal at twice the frequency of the applied wave. We can see from eqn 11.7 that this is equivalent to a second-order nonlinearity, because if $\mathcal{E}(t) = \mathcal{E}_0 \sin \omega t$, then

$$
\begin{aligned}
P^{(2)}(t) &= \epsilon_0 \chi^{(2)} \mathcal{E}_0^2 \sin^2 \omega t \\
&= \frac{1}{2} \epsilon_0 \chi^{(2)} \mathcal{E}_0^2 (1 - \cos 2\omega t) \, .
\end{aligned} \tag{11.17}
$$

Thus if $\chi^{(2)}$ is non-zero, the medium generates a wave at 2ω when driven at frequency ω. This is the same conclusion as that derived from the consideration of the anharmonic term in the potential, and shows that the two treatments are equivalent.

The relationship between C_3 and $\chi^{(2)}$ can be made more precise by finding an approximate solution to the equation of motion of the electron when driven by an AC electric field at frequency ω. To do this, we proceed as in Section 2.2, but now include the anharmonic term in the restoring force. The equation of motion is thus:

$$
m_0 \frac{\mathrm{d}^2 x}{\mathrm{d}t^2} + m_0 \gamma \frac{\mathrm{d}x}{\mathrm{d}t} + m_0 \omega_0^2 x + m_0 C_3 x^2 = -e\mathcal{E} \, , \tag{11.18}
$$

where γ accounts for damping. $\mathcal{E}$ is the driving field of the electromagnetic wave, which is assumed to have the following time dependence:

$$\mathcal{E}(t) = \mathcal{E}_0 \cos \omega t = \frac{1}{2}\mathcal{E}_0 \left(e^{i\omega t} + e^{-i\omega t}\right) . \qquad (11.19)$$

We have seen above that the inclusion of the C_3 term leads to a response at frequency 2ω in addition to the one at ω. Therefore we write the time dependence of the electron displacement as:

$$x(t) = \frac{1}{2} \left(X_1 e^{-i\omega t} + X_2 e^{-2i\omega t} + \text{c.c.}\right) , \qquad (11.20)$$

where 'c.c.' stands for complex conjugate. We assume that the nonlinear term is small, so that $X_1 \gg X_2$.

On substituting eqn 11.20 into eqn 11.18, we obtain:

$$
\begin{aligned}
(-\omega^2 - i\omega\gamma + \omega_0^2)&(X_1 e^{-i\omega t} + \text{c.c.})/2 \\
+ (-4\omega^2 - 2i\omega\gamma + \omega_0^2)&(X_2 e^{-2i\omega t} + \text{c.c.})/2 \\
+ C_3(X_1^2 e^{-2i\omega t} + 2X_1^* X_2 e^{-i\omega t} &+ \cdots + \text{c.c.})/4 \qquad (11.21) \\
&= \frac{-e\mathcal{E}_0}{2m_0} \left(e^{-i\omega t} + \text{c.c.}\right) ,
\end{aligned}
$$

where it is assumed that the anharmonic term is small and the ellipsis represents the higher-order cross terms at frequencies other than ω and 2ω. For eqn 11.21 to hold at all times, the coefficients of $e^{\pm i\omega t}$ and $e^{\pm 2i\omega t}$ must be the same on both sides of the equation. We are assuming that the nonlinear response is small, and so we can neglect the term at frequency ω generated from the anharmonic part of the potential. Hence we obtain:

$$X_1 = \frac{-e\mathcal{E}_0}{m_0} \frac{1}{(\omega_0^2 - \omega^2) - i\gamma\omega} . \qquad (11.22)$$

This is exactly the same result as eqn 2.9 in Section 2.2, which is hardly surprising, since it represents the linear response of the system. The polarization at frequency ω follows directly:

$$
\begin{aligned}
P(\omega, t) &= -Nex(\omega, t) \\
&= -Ne\left(X_1 e^{-i\omega t} + \text{c.c.}\right)/2 \qquad (11.23) \\
&= \epsilon_0 \chi(\omega)\mathcal{E}(t) ,
\end{aligned}
$$

where the third line is just the standard definition of the linear susceptibility as in eqn 11.2. By combining eqns 11.19, 11.22, and 11.23, we obtain the usual result for the linear susceptibility:

$$\chi(\omega) = \frac{Ne^2}{m_0\epsilon_0[(\omega_0^2 - \omega^2) - i\gamma\omega]} . \qquad (11.24)$$

We now solve for X_2 to find the nonlinear response by equating the coefficients of $e^{-2i\omega t}$ in eqn 11.21. This gives:

$$(-4\omega^2 - 2i\omega\gamma + \omega_0^2)\frac{X_2}{2} + \frac{C_3}{4}X_1^2 = 0 , \qquad (11.25)$$

It is important to keep track of all the conjugate terms in this model, for otherwise we can lose some of the important cross terms. We focus on the negative frequency terms in order to keep consistency with Section 2.2.

from which we obtain, using eqns 11.22 and 11.24:

$$X_2 = -\frac{C_3 X_1^2}{2(\omega_0^2 - 4\omega^2 - 2i\omega\gamma)}$$

$$= -\frac{C_3 e^2 \mathcal{E}_0^2}{2m_0^2(\omega_0^2 - \omega^2 - i\gamma\omega)^2(\omega_0^2 - 4\omega^2 - 2i\omega\gamma)} \quad (11.26)$$

$$= -\frac{m_0 C_3 \epsilon_0^3 \chi(\omega)^2 \chi(2\omega)}{2N^3 e^4} \mathcal{E}_0^2.$$

The polarization at frequency 2ω is given by

$$P(2\omega, t) = -Nex(2\omega, t) = -Ne(X_2 e^{-2i\omega t} + \text{c.c.})/2. \quad (11.27)$$

Now in the case we are considering where the polarization at frequency 2ω is generated by nonlinear conversion of the driving field at frequency ω, $P(2\omega)$ will also be given by eqns 11.7 and 11.19 as:

$$P(2\omega, t) = \epsilon_0 \chi^{(2)} \mathcal{E}(t)^2 = \epsilon_0 \chi^{(2)} \left(\frac{\mathcal{E}_0}{2}\right)^2 (e^{-2i\omega t} + \text{c.c.}). \quad (11.28)$$

Thus by combining eqns 11.26–11.28 we obtain the final result (Miller's rule):

We have restricted our attention here to the second-order nonlinearity, but it is obvious that the derivation can be generalized by including higher-order anharmonic terms, which would then explain the origin of higher-order nonlinearities.

$$\chi^{(2)} = \frac{m_0 C_3 \chi(\omega)^2 \chi(2\omega)\epsilon_0^2}{N^2 e^3}. \quad (11.29)$$

This shows that the second-order nonlinear susceptibility is directly proportional to C_3, the anharmonic term in the equation of motion. Equation 11.29 is reasonably successful in predicting the dispersion of $\chi^{(2)}$ in a large number of crystals. This is because it is found empirically that the anharmonic constant C_3 does not vary very much from material to material.

Equation 11.29 tells us that $\chi^{(2)}$ increases as ω approaches ω_0, through the frequency dependence of $\chi(\omega)$ given by eqn 11.24. This effect is known as resonance enhancement. In a lightly damped system, the classical treatment breaks down as we get closer to the resonant frequency due to the divergence in $\chi(\omega)$. Hence we have to adopt a different approach if we are close to resonance. This is discussed in the next subsection.

11.2.2 Resonant nonlinearities

The off-resonant nonlinear effects discussed in the previous section are all 'virtual' processes. This means that no real transitions take place because the photon energy does not coincide with any of the transition frequencies of the atoms. The situation is obviously completely different if the laser frequency is in resonance with an atomic transition. In this case, the atoms can absorb photons and make transitions to excited states as the beam propagates through the medium.

The absorption rate is normally determined by the matrix element for the transition and the density of states according to Fermi's golden rule (eqn B.14). This allows us to determine the absorption coefficient for a

particular material at a particular frequency. All this presupposes that
the intensity of the light beam on the sample is small. If the intensity is
high, the absorption coefficient becomes intensity dependent. Since the
absorption coefficient is related to the dielectric constant, this means
that the dielectric constant is intensity dependent. In other words, we
are dealing with an optical nonlinearity.

The intensity dependence of the absorption rate can be understood
through the Einstein B coefficients discussed in Section B.1 of Appendix B. Consider the propagation of an intense laser beam of frequency
ν through an absorbing medium. Figure 11.2 illustrates the simplest case
to discuss, namely a medium containing atoms with just two levels: level
1 at energy E_1 and level 2 at energy E_2, where $E_2 > E_1$. We assume
that there are N_1 atoms per unit volume in the lower level, and N_2 per
unit volume in level 2. The total number of atoms per unit volume is
N_0, where $N_0 = N_1 + N_2$.

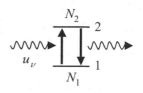

Fig. 11.2 Transitions induced by a resonant laser beam of energy density u_ν.
Photons are removed from the beam by
absorption transitions from level $1 \rightarrow 2$
and are added by stimulated emission
transitions from level $2 \rightarrow 1$.

We consider the case in which the laser is resonant with the atomic
transition frequency of the atoms, such that $h\nu = E_2 - E_1$. The laser
beam will be absorbed as it propagates through the medium by transitions upwards from level 1 to level 2. At the same time, photons will be
added to the beam by stimulated emission. The stimulated emission rate
is not normally considered when discussing the propagation through an
absorbing medium because it is assumed that N_2 is negligible. However,
if the laser intensity is large, we can no longer make this assumption,
because the absorption transitions will create a substantial population
in the upper level. This gives rise to a significant stimulated emission
rate, and effectively reduces the absorption coefficient.

The reduction of the absorption coefficient due to stimulated emission
can be modelled by considering an incremental beam slice of thickness
dz as illustrated in Fig. 11.3. The number of photons absorbed per unit
time in the incremental slice is given by:

$$\delta N_{\text{absorbed}} = B_{12} N_1 u_\nu g(\nu) \times A dz \,, \tag{11.30}$$

where u_ν is the energy density of the beam at position z, $g(\nu)$ is the
spectral lineshape function of the transition, and A is the beam area.
$A dz$ is thus the volume of the slice. The extra factor of $g(\nu)$ in eqn 11.30
compared to eqn B.5 arises from the difference between the beam en-
ergy density u_ν (units: J m^{-3}) and spectral energy density $u(\nu)$ (units:
$\text{J m}^{-3}\,\text{Hz}^{-1}$). The Einstein coefficients are traditionally defined when
considering the interaction of a broadband light source (e.g. black-body
radiation) with a narrow atomic absorption line. Here, by contrast, we
are considering a laser beam with a spectral linewidth that is smaller
than that of the transition. The inclusion of the spectral lineshape function is required for dimensional consistency and ensures that the transition rate is proportional to the shape of the absorption line.

The stimulated emission rate can be treated in the same way. We see
from eqn B.6 that the number of photons added per unit time to the
beam is given by:

$$\delta N_{\text{stimulated}} = B_{21} N_2 u_\nu g(\nu) \times A dz \,. \tag{11.31}$$

For a Lorentzian line, $g(\nu)$ takes the
form:

$$g(\nu) = \frac{\Delta\nu/2\pi}{(\nu - \nu_0)^2 + (\Delta\nu/2)^2} \,,$$

where ν_0 is the transition frequency and
$\Delta\nu$ its half width. Note that this is normalized such that:

$$\int_0^\infty g(\nu)\,\mathrm{d}\nu = 1 \,.$$

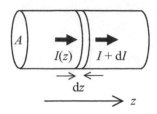

Fig. 11.3 Propagation of a laser beam of area A through an absorbing medium.

Hence the total reduction in the photon number per unit time is given by:

$$\delta N_{\text{total}} = \delta N_{\text{absorbed}} - \delta N_{\text{stimulated}}$$
$$= (B_{12}N_1 - B_{21}N_2)u_\nu g(\nu)A dz\,. \tag{11.32}$$

This is effectively the net absorption rate.

The intensity of the beam decreases as it propagates through the medium due to absorption. If we take the intensity at position z to be $I(z)$, and the change in the intensity in the incremental slice to be dI, we can write

$$A dI = -\delta N_{\text{total}} \times h\nu\,,$$
$$= -(B_{12}N_1 - B_{21}N_2)u_\nu g(\nu)h\nu A dz\,. \tag{11.33}$$

The left-hand side is the beam energy change per unit time in the incremental slice at z. The right-hand side is the change in the photon number per unit time given by eqn 11.32 multiplied by the energy of each photon. Conservation of energy requires that these two quantities must be the same.

Equation 11.33 can be simplified by noting from eqn A.43 in Appendix A that $I = cu_\nu/n$, where n is the refractive index of the medium. Hence we obtain:

$$\frac{dI}{dz} = -\frac{B_{12}(N_1 - N_2)g(\nu)h\nu n}{c}I\,, \tag{11.34}$$

The argument can easily be generalized to the case where the degeneracies are different.

where we have assumed that the degeneracies of the two levels are the same so that $B_{12} = B_{21}$ (cf. eqn B.10). Equation 11.34 can be compared to the standard definition of the absorption coefficient α given in eqn 1.3, which implies that

$$\frac{dI}{dz} = -\alpha I\,. \tag{11.35}$$

Hence by comparing eqns 11.34 and 11.35 we see that

$$\alpha = B_{12}(N_1 - N_2)g(\nu)h\nu n/c\,. \tag{11.36}$$

This shows that the absorption coefficient is proportional to the population difference between the lower and upper levels.

At low intensities we can assume that $N_1 \approx N_0$ and $N_2 \approx 0$. Equation 11.36 then just reduces to the usual result where the absorption coefficient is proportional to the number of atoms in the system. However, at high intensities, the laser pumps a large number of atoms into the upper level so that N_2 increases and N_1 decreases. Hence the absorption coefficient begins to decrease as the population difference between the upper and lower levels decreases.

Note that eqn 11.37 only applies to homogeneously broadened systems, and that the saturation intensity measured for a particular absorption line will depend on the detailed rate constants for that specific transition.

The decrease of the absorption with the laser power can be characterized by introducing the **saturation intensity** I_s. The absorption coefficient is found experimentally to depend on the intensity I according to the following relationship:

$$\alpha(I) = \frac{\alpha_0}{1 + I/I_s}\,, \tag{11.37}$$

where α_0 is the absorption measured in the linear regime when $I \ll I_\mathrm{s}$. A medium which shows the behaviour indicated by eqn 11.37 is called a **saturable absorber**.

At low intensity levels, eqn 11.37 can be expanded to obtain:

$$\alpha(I) = \alpha_0 - (\alpha_0/I_\mathrm{s})I\,. \tag{11.38}$$

This shows that the absorption decreases linearly with I. Now α is proportional to the imaginary part of ϵ_r (cf. eqns 1.19 and 1.24), and I is proportional to $\mathcal{E}^2$. Hence ϵ_r varies in proportion to $\mathcal{E}^2$, and from eqn 11.9 we see that this is equivalent to a $\chi^{(3)}$ process. In other words, the resonant nonlinearities due to saturable absorption are third-order nonlinear effects. (See Exercise 11.11.)

The analysis of the saturable absorber above applies primarily to the discrete absorption lines found in atomic systems. However, in solid-state materials we are often more interested in the saturation of an absorption band rather than a discrete line. For example, in Section 11.4.7 we present data for the saturable absorption of interband transitions and also of excitons.

In treating the nonlinear saturation of interband absorption, it is useful to adopt a slightly different approach which is based on the Pauli exclusion principle. The dependence of α on $(N_1 - N_2)$ in eqn 11.36 can be considered as a consequence of the Fermi–Dirac statistics of the electrons. For absorption to be possible, the lower level must contain an electron, while the upper level must be empty. Hence the absorption coefficient will obey

$$\alpha = \alpha_0(f_1 - f_2)\,, \tag{11.39}$$

where f_1 and f_2 are the Fermi occupancies of the lower and upper levels respectively. α_0 is the low power absorption when the lower level is full and the upper level empty: that is, when $f_1 = 1$ and $f_2 = 0$. The absorption at high powers is calculated by working out the filling of the levels after a large number of electrons and holes have been excited by the absorption of a laser pulse.

11.3   Second-order nonlinearities

In this section we discuss a few of the more important effects that are associated with the second-order nonlinear susceptibility $\chi^{(2)}$. We begin by considering the general principles of nonlinear frequency mixing, and the effects of the crystal symmetry on the nonlinear coefficients. We then introduce the concept of phase matching, which is crucially important for obtaining large nonlinear signals. Finally, we give a brief discussion of the linear electro-optic effect.

11.3.1   Nonlinear frequency mixing

The second-order nonlinear polarization is given by eqn 11.7. If the medium is excited by sinusoidal waves at frequencies ω_1 and ω_2 with amplitudes $\mathcal{E}_1$ and $\mathcal{E}_2$ respectively, then the nonlinear polarization will be equal to:

$$P^{(2)}(t) = \epsilon_0 \chi^{(2)} \times \mathcal{E}_1 \cos\omega_1 t \times \mathcal{E}_2 \cos\omega_2 t$$
$$= \epsilon_0 \chi^{(2)} \mathcal{E}_1 \mathcal{E}_2 \left[\cos\left(\omega_1 + \omega_2\right)t + \cos\left(\omega_1 - \omega_2\right)t\right]/2\,. \quad (11.40)$$

This shows that the second-order nonlinear response generates polarization waves at the sum and difference frequencies of the input fields according to:

$$\omega_{\text{sum}} = \omega_1 + \omega_2, \quad (11.41)$$
$$\omega_{\text{diff}} = \omega_1 - \omega_2. \quad (11.42)$$

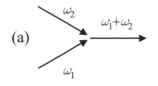

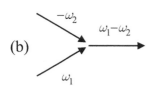

Fig. 11.4 Feynman diagrams for second-order nonlinear frequency mixing processes.

The medium then re-radiates at ω_{sum} and ω_{diff}, thereby emitting light at frequencies $(\omega_1 + \omega_2)$ and $(\omega_1 - \omega_2)$. This effect is called **nonlinear frequency mixing**. If the frequencies are the same, the sum frequency is twice the input frequency. This effect is called **frequency doubling** or **second-harmonic generation**, and has already been introduced in the discussion of eqn 11.17.

Nonlinear frequency mixing processes can be represented by Feynman diagrams as indicated in Fig. 11.4. Figure 11.4(a) shows the process for sum frequency mixing, while Fig. 11.4(b) represents difference frequency mixing. Conservation of energy applies at each vertex. The negative input frequency at ω_2 for the difference frequency mixing process in Fig. 11.4(b) reflects the fact that $\cos\omega t = (e^{+i\omega t} + e^{-i\omega t})/2$, so that we can represent real waves either with positive or negative frequencies on a Feynman diagram. In quantum-mechanical terms, we would say that the sum frequency mixing process annihilates two input photons at frequencies ω_1 and ω_2, with the creation of a new photon at frequency ω_{sum}, while difference frequency mixing annihilates one photon at frequency ω_1 and creates two photons, one at frequency ω_2 and the other at ω_{diff}. The creation of the photon at frequency ω_2 in the latter case is stimulated by the presence of a large number of existing photons at frequency ω_2 from the input field.

One of the most important uses for nonlinear optical processes is to generate new frequencies from fixed-wavelength lasers. The most common technique is frequency doubling. In this case, we have a single input beam, and the sum frequency mixing works by taking two photons from the input beam and generating a new photon at the doubled frequency.

Figure 11.5 shows a schematic experimental arrangement for generating the second, third, and fourth harmonics of a Nd:YAG laser operating at 1064 nm. The second harmonic at $1064/2 = 532$ nm is generated by frequency doubling of the fundamental. The second harmonic beam can then be doubled again with an additional nonlinear crystal to generate the fourth harmonic at 266 nm. The third harmonic can be generated by carrying out sum frequency mixing of the fundamental and the second harmonic beams in another nonlinear crystal. These techniques are standard procedures in modern laser physics.

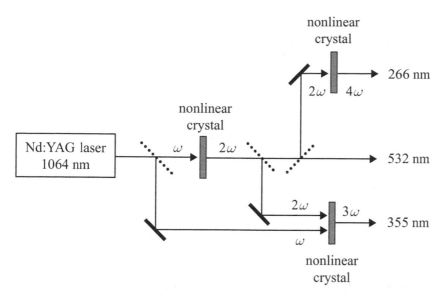

Fig. 11.5 Nonlinear frequency conversion of a Nd:YAG laser operating at 1064 nm. The beam is first doubled to 532 nm. This beam can then either be used as the output, or doubled again to 266 nm. Alternatively, the output at 532 nm can be mixed with the fundamental at 1064 nm to generate the third harmonic at 355 nm. The residual pump beams transmitted through the nonlinear crystals are separated from the harmonics by filters that are not shown in the diagram.

When the two input fields are at the same frequency, we see from eqn 11.42 that the difference frequency is zero. This effect is called **optical rectification**, and refers to the phenomenon by which a static electric field is produced from fields of optical frequencies. The **Pockels effect**, which is also known as the **linear electro-optic effect**, is the reverse of this process, and is discussed further in Section 11.3.4.

The sum frequency mixing process shown in Fig. 11.4(a) can work the other way round as well. In this case, we bring in a single input field at frequency ω and create two new photons at frequencies of ω_1 and ω_2, where $\omega_1 + \omega_2 = \omega$. This process is called **down conversion**. It is apparent that the output frequencies generated by down-conversion are not uniquely defined. Any combination of frequencies that satisfies the conservation of energy requirements can in principle be generated. However, the number of photons emitted at any particular frequency will only be large if the phase matching conditions discussed in Section 11.3.3 are satisfied.

Down conversion can be used to amplify a weak beam by a process called **parametric amplification**. If we introduce a weak 'signal' field at frequency ω_s in the presence of a strong pump field at frequency ω, it can generate an 'idler' field at $\omega_i = \omega - \omega_s$ by difference frequency mixing with the pump field. These new idler photons then generate more signal photons by further mixing with the pump field. The process then repeats itself. If the phase matching conditions are satisfied, it is possible to transfer power from the pump beam to the signal and idler beams. Furthermore, if the crystal is inside an optical cavity which is resonant with either ω_s or ω_i, then oscillation can occur. This process is called **parametric oscillation**, and can lead to the generation of intense beams at tunable frequencies even though we started from a fixed-wavelength laser.

A very interesting aspect of down-conversion processes is that the photons are always created in pairs. This means that the photon statistics at frequency ω_1 are directly correlated with those at ω_2. This gives rise to a whole host of quantum optical effects. For example, the correlated photon pair can be used as the basis of a 'heralded' single-photon source: the detection of a photon in one beam 'heralds' the presence a single photon in the other.

Table 11.2 Nonlinear coefficients of a number of important nonlinear crystals. The nonlinear coefficients are all measured at 1064 nm. Data from Tang (1995) and Klein et al. (2003).

Crystal	Symmetry	Transmission range (nm)	Nonlinear coefficient (pm/V)
KDP (KH$_2$PO$_4$)	$\overline{4}$2m	200–1500	$d_{36} = 0.39$ $d_{14} = 0.4$
KTP (KTiOPO$_4$)	mm2	350–4400	$d_{31} = 6.5$ $d_{32} = 5.0$ $d_{33} = 14$ $d_{24} = 7.6$ $d_{15} = 6.1$
BBO (β-BaB$_2$O$_4$)	3m	190–2500	$d_{22} = 2.1$ $d_{31} = 0.26$
LBO (LiB$_3$O$_5$)	mm2	160–2600	$d_{32} = 1.2$
LiNbO$_3$	3m	400–5000	$d_{31} = -4.8$ $d_{33} = -30$ $d_{22} = 2.3$

11.3.3   Phase matching

Nonlinear effects are generally small, and we therefore need a long length of the nonlinear medium to obtain a useful nonlinear conversion efficiency. For this to work, we need that the phases of the nonlinear waves generated throughout the whole crystal are all the same so that the fields add together coherently. When this is achieved, we are in a regime called **phase matching**. As we shall see below, phase matching does not normally occur, and can only be achieved if the nonlinear crystal is orientated in a very precise direction.

We can see why phase matching is an important issue by considering a simple example. Suppose we wish to use a nonlinear crystal to double the frequency of a Nd:YAG laser from 1064 nm to 532 nm, as shown schematically in Fig. 11.5. All materials are dispersive to some extent, and this means that the refractive index at 532 nm will be different to that at 1064 nm. Therefore, the second harmonic waves at 532 nm will propagate with a different phase velocity to the fundamental at 1064 nm. This means that the second harmonic waves generated at the front will arrive at the back of the crystal at a different time to the fundamental, and so the 532 nm waves generated at the back of the crystal will be out of phase with those from the front.

The phase mismatch introduced by the frequency doubling process can be calculated from the wave vectors of the two waves. If the beams are travelling in the z direction, then the nonlinear waves will propagate as $\exp(\mathrm{i}k^{(2\omega)}z)$, where $k^{(2\omega)}$ is the wave vector at frequency 2ω. On the other hand, the fundamental beam propagates as $\exp(\mathrm{i}k^{(\omega)}z)$, where $k^{(\omega)}$ is the wave vector at frequency ω. Now since $P^{(2)} \propto \mathcal{E}^2$, the nonlinear polarization at a given point in the medium will be created with a phase of $[\exp(\mathrm{i}k^{(\omega)}z)]^2 = \exp(\mathrm{i}2k^{(\omega)}z)$. Hence the phase difference $\Delta\Phi$ between the nonlinear waves created at a distance z into the crystal and those created at the front of the crystal is given by

$$\Delta\Phi = (k^{(2\omega)} - 2k^{(\omega)})z\,. \tag{11.46}$$

We introduce the coherence length l_c for the nonlinear process as the distance over which the phase mismatch becomes equal to 2π:

$$(k^{(2\omega)} - 2k^{(\omega)}) \times l_\mathrm{c} = 2\pi\,. \tag{11.47}$$

This can be rewritten in terms of the refractive indices $n^{2\omega}$ and n^ω at the two frequencies as:

$$\frac{2\omega}{c}\left[n^{2\omega} - n^\omega\right]l_\mathrm{c} = 2\pi\,. \tag{11.48}$$

Hence

$$l_\mathrm{c} = \frac{\pi c}{\omega[n^{2\omega} - n^\omega]} = \frac{\lambda}{2[n^{2\omega} - n^\omega]}\,, \tag{11.49}$$

where λ is the vacuum wavelength of the fundamental beam. Taking a typical example with $\lambda = 1\,\mu\mathrm{m}$ and $n^{2\omega} - n^\omega \sim 10^{-2}$ we find that $l_\mathrm{c} \sim 50\,\mu\mathrm{m}$.

Equation 11.49 shows us that only the waves emitted within a very short distance of the surface will add together coherently. This clearly greatly restricts the efficiency of the nonlinear conversion process, because only a very short length of the nonlinear crystal is actually useful. The situation would be completely different if we could somehow arrange that $n^{2\omega} = n^\omega$. In this case the nonlinear waves generated throughout the whole crystal would all have the same phase and would thus add together coherently. This is the phase-matching condition.

At first sight it might seem that there is no way to satisfy the condition $n^{2\omega} = n^\omega$ in any material with normal dispersion properties. However, this neglects the fact that the anisotropic crystals that are used for nonlinear mixing are birefringent. This opens new possibilities to balance the dispersion against the birefringence. For example, consider a uniaxial crystal with normal dispersion and negative birefringence, such that $n^{2\omega} > n^\omega$ and $n_\mathrm{e} < n_\mathrm{o}$. (See Sections 2.4 and 2.5.1.) In such a crystal it is possible to obtain phase matching by propagating the beam at frequency 2ω as an extraordinary ray, and the beam at ω as an ordinary ray. It is shown in Example 11.2 below that phase matching can then be achieved for a very specific orientation of the crystal.

The phase-matching condition can be given an intuitive physical interpretation if we notice that if $n^{2\omega} = n^\omega$, then $k^{(2\omega)} = 2k^{(\omega)}$. This corresponds to momentum conservation in the nonlinear process. In the more general case when a photon of wave vector $\boldsymbol{k}$ is generated by mixing two photons with wave vectors $\boldsymbol{k}_1$ and $\boldsymbol{k}_2$, the phase-matching condition can be written as

$$\boldsymbol{k} = \boldsymbol{k}_1 + \boldsymbol{k}_2 . \tag{11.50}$$

In down conversion where one photon is split into two output photons, the phase-matching condition of eqn 11.50 applies to each pair of photons created in the process.

Example 11.2

The ordinary and extraordinary refractive indices of a uniaxial crystal are n_o and n_e respectively. A laser beam is propagating at an angle θ to the optic (z) axis as shown in Fig. 2.13. The laser is linearly polarized along the x direction.

(a) Show that there is an angle θ at which the phase-matching condition can be met for second-harmonic waves polarized as extraordinary rays.
(b) Evaluate the phase-matching angle for potassium dihydrogen phosphate (KDP) at the wavelength of a Nd:YAG laser (1064 nm). The relevant refractive indices for KDP are: $n_o(1064\,\text{nm}) = 1.494$, $n_o(532\,\text{nm}) = 1.512$, and $n_e(532\,\text{nm}) = 1.471$.

The arrangement with the pump beam as an ordinary ray (o-ray) and the second harmonic as an extraordinary ray (e-ray) is called **type I phase matching**. In **type II phase matching**, one of the pump photons is an o-ray and the other is an e-ray, while the second harmonic is again an e-ray.

Solution
(a) The general condition for phase matching is that

$$n^{2\omega} = n^{\omega} . \tag{11.51}$$

The fundamental wave is polarized along the x axis, and so its refractive index is n_o^{ω}, irrespective of θ. The refractive index for the second harmonic waves polarized as extraordinary rays is given by the result of Exercise 2.16, namely:

$$\frac{1}{n(\theta)^2} = \frac{\sin^2 \theta}{n_e^2} + \frac{\cos^2 \theta}{n_o^2} , \tag{11.52}$$

where n_o and n_e are evaluated at 2ω. Hence the phase-matching condition given in eqn 11.51 is met when

$$\frac{1}{(n_o^{\omega})^2} = \frac{\sin^2 \theta}{(n_e^{2\omega})^2} + \frac{\cos^2 \theta}{(n_o^{2\omega})^2} . \tag{11.53}$$

(b) On inserting the appropriate refractive indices into eqn 11.53, we have:

$$\frac{1}{1.494^2} = \frac{\sin^2 \theta}{1.471^2} + \frac{\cos^2 \theta}{1.512^2} .$$

This is satisfied with $\theta = 41°$. To achieve phase matching we therefore hold the crystal in a gimbal mount and carefully rotate its orientation until the optic axis is at 41° to the direction of the laser beam.

11.3.4 Electro-optics

In electro-optics, a DC electric field is used to change the refractive index of an optical material. As we saw in Section 2.5.2, the refractive index change can be either linear in the field (the Pockels effect) or quadratic (the Kerr effect). In this subsection we focus on the linear electro-optic effect, leaving relevant discussion of the quadratic effect to Section 11.4.3.

It was pointed out in Section 11.3.1 that the linear electro-optic effect can be considered as a type of second-order nonlinearity in which the frequency of the driving field is equal to zero. This has an immediate consequence. Since the second-order susceptibility is zero if the crystal has inversion symmetry, then the linear electro-optic effect can only be observed in crystals that lack inversion symmetry. In fact, the same materials (e.g. KDP, LiNbO$_3$) are frequently used in both nonlinear optics and electro-optics.

The general treatment of the linear electro-optic effect starts from the index ellipsoid of the anisotropic medium. The index ellipsoid describes the variation of the refractive index with the direction of the electric field of the light, and takes the form:

$$\frac{x^2}{n_x^2} + \frac{y^2}{n_y^2} + \frac{z^2}{n_z^2} = 1\,, \tag{11.54}$$

where n_x, n_y, and n_z are the refractive indices measured for light polarized along the principal axes, namely $\hat{x}$, $\hat{y}$, and $\hat{z}$ respectively. The refractive index measured for light polarized along the (x, y, z) direction is then given by:

$$n = \sqrt{x^2 + y^2 + z^2}\,. \tag{11.55}$$

As mentioned in Section 2.5.2, the basic effect of the electric field is to change the anisotropy of the crystal. This effect can be quantified by re-writing the index ellipsoid in the following form:

$$\left(\frac{1}{n^2}\right)_1 x^2 + \left(\frac{1}{n^2}\right)_2 y^2 + \left(\frac{1}{n^2}\right)_3 z^2$$
$$+ 2\left(\frac{1}{n^2}\right)_4 yz + 2\left(\frac{1}{n^2}\right)_5 xz + 2\left(\frac{1}{n^2}\right)_6 xy = 1\,. \tag{11.56}$$

On comparing eqns 11.54 and 11.56, it is immediately apparent that at zero applied field we must have that:

$$\left(\frac{1}{n^2}\right)_1 = \frac{1}{n_x^2}\,, \quad \left(\frac{1}{n^2}\right)_2 = \frac{1}{n_y^2}\,, \quad \left(\frac{1}{n^2}\right)_3 = \frac{1}{n_z^2}\,, \tag{11.57}$$

and

$$\left(\frac{1}{n^2}\right)_4 = \left(\frac{1}{n^2}\right)_5 = \left(\frac{1}{n^2}\right)_6 = 0\,. \tag{11.58}$$

The changes in the index ellipsoid are determined by the components of the **electro-optic coefficient tensor** r_{ij}, defined by:

$$\Delta\left(\frac{1}{n^2}\right)_i = \sum_{j=1}^{3} r_{ij}\mathcal{E}_j\,. \tag{11.59}$$

See Born & Wolf (1999), Nye (1957) or Yariv (1997) for a discussion of the index ellipsoid (also called the 'indicatrix' in some older texts) and its use in linear optics. In uniaxial crystals with their optic axis along z, we set $n_x = n_y = n_o$ and $n_z = n_e$ in eqn 11.54, but in biaxial crystals, all three indices are different.

Written out explicitly, this becomes:

$$
\begin{pmatrix}
\Delta(1/n^2)_1 \\
\Delta(1/n^2)_2 \\
\Delta(1/n^2)_3 \\
\Delta(1/n^2)_4 \\
\Delta(1/n^2)_5 \\
\Delta(1/n^2)_6
\end{pmatrix}
=
\begin{pmatrix}
r_{11} & r_{12} & r_{13} \\
r_{21} & r_{22} & r_{23} \\
r_{31} & r_{32} & r_{33} \\
r_{41} & r_{42} & r_{43} \\
r_{51} & r_{52} & r_{53} \\
r_{61} & r_{62} & r_{63}
\end{pmatrix}
\begin{pmatrix}
\mathcal{E}_x \\
\mathcal{E}_y \\
\mathcal{E}_z
\end{pmatrix}.
\tag{11.60}
$$

As with the nonlinear optical coefficient tensor, symmetry requires that many of the tensor components are zero, and that some (or all) of the non-zero components have the same magnitude.

Let us consider the case of a KDP (KH_2PO_4) crystal, which has tetragonal $\bar{4}2m$ symmetry. With no field applied, KDP is a uniaxial birefringent crystal, with ordinary and extraordinary refractive indices of n_o and n_e respectively. (See Section 2.5.1.) It is shown in Example 11.3 below that the application of a DC field along the optic (i.e. z) axis rotates the principal axes of the crystal by 45°, as illustrated in Fig. 11.6(a). The modified refractive indices with respect to the new principal axes $(\hat{x}', \hat{y}', \hat{z}')$ are given by:

$$
\begin{aligned}
n_1(\mathcal{E}) &= n_o + n_o^3 r_{63}\mathcal{E}/2\,, \\
n_2(\mathcal{E}) &= n_o - n_0^3 r_{63}\mathcal{E}/2\,, \\
n_3(\mathcal{E}) &= n_e\,.
\end{aligned}
\tag{11.61}
$$

This shows that the field induces anisotropy in the x–y plane, where there was none before. The effect of the field is therefore to induce birefringence, as mentioned previously in Section 2.5.2.

Let us consider a ray propagating in the z direction incident on a KDP crystal. With no field applied, this is an ordinary ray: it experiences no birefringence, and emerges with its polarization unchanged. However, when the field is applied, the ray now experiences birefringence on account of the induced anisotropy in the x–y plane. This means that the output polarization will be different to the input polarization.

Let us assume that the input beam is linearly polarized along the y axis. The $\mathcal{E}$-vector can be resolved into equal components along the x' and y' axes, which experience refractive indices of $n_1(\mathcal{E})$ and $n_2(\mathcal{E})$ respectively. The phase difference between them is given by:

$$
\Delta\Phi(\mathcal{E}) = \frac{2\pi}{\lambda}|n_1 - n_2|L = \frac{2\pi}{\lambda}n_o^3 r_{63}\mathcal{E}L\,,
\tag{11.62}
$$

where L is the length of the crystal, and λ is the vacuum wavelength. In general, the output beam will be elliptically polarized. However, if $\Delta\Phi = \pi/2$, then the electro-optic crystal acts as a quarter wave plate and turns the beam into circularly polarized light. Alternative, if $\Delta\Phi = \pi$, then the electro-optic crystal acts as a half wave plate, and the beam emerges linearly polarized with its polarization rotated by 90°.

The modification of the polarization induced by the field can be used to make a light amplitude modulator, as shown in Fig. 11.6(b). The

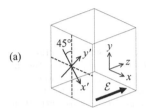

(a)

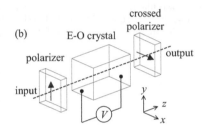

(b)

Fig. 11.6 (a) Rotated principal axes x' and y' for a crystal with tetragonal $\bar{4}2m$ symmetry when a field is applied along the z axis. (b) An electro-optic (E-O) modulator based on the crystal shown in part (a). This arrangement is often called a Pockels cell.

crystal is mounted between crossed polarizers, so that there is no output without a field applied. When a beam is incident with its polarization parallel to the axis of the input polarizer and a voltage is applied, the beam emerges from the electro-optic crystal with a different polarization. This means that some of the light now passes through the second polarizer. The output of the device is thus controlled by the voltage applied to the crystal, with the maximum output occurring when the voltage applied induces a π phase shift.

The principles that we have illustrated for the KDP crystal with tetragonal $\overline{4}2m$ symmetry can be applied to other crystals with different symmetry classes. Exercise 11.8 considers the case of trigonal crystals with 3m symmetry (e.g. LiNbO$_3$) and explains how they can be used as electro-optic phase modulators, while Exercise 11.9 considers crystals with cubic symmetry. Table 11.3 lists the values of the electro-optic coefficients for some of the crystals commonly used in modulators.

Table 11.3 Electro-optic coefficients r_{ij} of common electro-optic crystals at 633 nm. KDP and ADP are abbreviations for KH$_2$PO$_4$ and NH$_4$H$_2$PO$_4$ respectively. After Yariv (1997).

Crystal	Symmetry	r_{ij} (pm/V)
KDP	$\overline{4}2m$	$r_{41} = 8$
		$r_{63} = 11$
ADP	$\overline{4}2m$	$r_{41} = 23$
		$r_{63} = 7.8$
LiNbO$_3$	3m	$r_{13} = 9.6$
		$r_{22} = 6.8$
		$r_{33} = 31$
		$r_{51} = 33$
KNbO$_3$	2mm	$r_{13} = 28$
		$r_{42} = 380$
		$r_{51} = 105$
GaP	$\overline{4}3m$	$r_{41} = -0.97$

Example 11.3

Consider a uniaxial crystal of the tetragonal $\overline{4}2m$ class (e.g. KDP) with ordinary and extraordinary refractive indices equal to n_o and n_e respectively. When the optic axis is defined to be along the z direction, Neumann's principle requires that the electro-optic coefficient tensor of crystals with this symmetry class takes the form:

$$
r = \begin{pmatrix} 0 & 0 & 0 \\ 0 & 0 & 0 \\ 0 & 0 & 0 \\ r_{41} & 0 & 0 \\ 0 & r_{41} & 0 \\ 0 & 0 & r_{36} \end{pmatrix}. \tag{11.63}
$$

(a) Write down the modified index ellipsoid when an electric field of magnitude $\mathcal{E}$ is applied along the z axis.

(b) Show that the principal axes of the crystal with the field applied are rotated by $-45°$ about the z axis compared to the case with no field applied.

(c) Hence find the modified refractive indices of the crystal measured relative to the rotated principal axes.

Solution

(a) We have a uniaxial crystal with a refractive index of n_o for light polarized along the x or y directions, and n_e for light polarized along z. With no field applied, the index ellipsoid is therefore of the form:

$$
\frac{x^2 + y^2}{n_o^2} + \frac{z^2}{n_e^2} = 1.
$$

The modified ellipsoid in the presence of the field is found by substituting into eqn 11.60 with $\mathcal{E}_x = \mathcal{E}_y = 0$ and $\mathcal{E}_z = \mathcal{E}$. There is only one non-zero term, namely:

$$\Delta\left(\frac{1}{n^2}\right)_6 = r_{63}\mathcal{E}.$$

The modified index ellipsoid is therefore given by eqn 11.56 as:

$$\frac{x^2}{n_{\rm o}^2} + \frac{y^2}{n_{\rm o}^2} + \frac{z^2}{n_{\rm e}^2} + 2r_{63}\mathcal{E}xy = 1.$$

(b) We define new axes rotated by $-45°$ about the z axis such that:

$$
\begin{aligned}
x' &= (x-y)/\sqrt{2} \\
y' &= (x+y)/\sqrt{2} \\
z' &= z,
\end{aligned}
$$

which implies that $x = (x'+y')/\sqrt{2}$, $y = (-x'+y')/\sqrt{2}$ and $z = z'$. On substituting into the modified index ellipsoid we obtain:

$$\frac{x'^2 + y'^2}{n_{\rm o}^2} + \frac{z'^2}{n_{\rm e}^2} + r_{63}\mathcal{E}(-x'^2 + y'^2) = 1.$$

This can be written in the form:

$$\frac{x'^2}{n_1^2} + \frac{y'^2}{n_2^2} + \frac{z'^2}{n_3^2} = 1,$$

if we set:

$$
\begin{aligned}
1/n_1^2 &= 1/n_{\rm o}^2 - r_{63}\mathcal{E}, \\
1/n_2^2 &= 1/n_{\rm o}^2 + r_{63}\mathcal{E}, \\
1/n_3^2 &= 1/n_{\rm e}^2.
\end{aligned}
$$

This shows that x' and y' are the modified principal axes of the crystal with the field applied.

(c) It is immediately apparent from part (b) that $n_3(\mathcal{E}) = n_{\rm e}$. We can find $n_1(\mathcal{E})$ and $n_1(\mathcal{E})$ by writing:

$$
\begin{aligned}
n_1(\mathcal{E}) &= n_{\rm o}(1 - n_{\rm o}^2 r_{63}\mathcal{E})^{-1/2}, \\
n_2(\mathcal{E}) &= n_{\rm o}(1 + n_{\rm o}^2 r_{63}\mathcal{E})^{-1/2},
\end{aligned}
$$

and assuming that the field-induced changes are small compared to $n_{\rm o}$. The approximation $(1+x)^{-1/2} = (1 - x/2)$ is then valid, and so we conclude:

$$
\begin{aligned}
n_1(\mathcal{E}) &= n_{\rm o} + n_{\rm o}^3 r_{63}\mathcal{E}/2, \\
n_2(\mathcal{E}) &= n_{\rm o} - n_{\rm o}^3 r_{63}\mathcal{E}/2, \\
n_2(\mathcal{E}) &= n_{\rm e}.
\end{aligned}
$$

11.4   Third-order nonlinear effects

Third-order nonlinear effects are particularly important in isotropic media, such as gases, liquids, and glasses. This is because an isotropic medium possesses inversion symmetry, and all the components of $\chi^{(2)}_{ijk}$ must therefore be zero. (See Section 11.3.2.) Hence the lowest-order nonlinear susceptibility with non-vanishing components is $\chi^{(3)}$. In this section we start by giving an overview of third-order nonlinear phenomena, and then focus on isotropic media in more detail, mentioning optical fibres as a specific example. Finally we discuss resonant third-order nonlinear effects in semiconductors.

11.4.1   Overview of third-order phenomena

A third-order nonlinear polarization is generated when three input fields are applied to the nonlinear medium. If the input fields are at frequencies ω_1, ω_2, and ω_3, then the nonlinear polarization is given by eqn 11.8 or more generally by eqn 11.11. In the simplest case without considering the tensor aspect of the susceptibility, this gives:

$$P^{(3)}(t) = \epsilon_0 \chi^{(3)} \times \mathcal{E}_1 \cos \omega_1 t \times \mathcal{E}_2 \cos \omega_2 t \times \mathcal{E}_3 \cos \omega_3 t , \qquad (11.64)$$

where $\mathcal{E}_1$, $\mathcal{E}_2$, and $\mathcal{E}_3$ are the amplitudes of the three waves. Hence the frequency ω_4 of the nonlinear polarization must satisfy:

$$\omega_4 = \omega_1 + \omega_2 + \omega_3 , \qquad (11.65)$$

where the frequencies on the right-hand side can be either positive or negative. This reflects the fact that $\cos \omega t = (\mathrm{e}^{+\mathrm{i}\omega t} + \mathrm{e}^{-\mathrm{i}\omega t})/2$ and therefore contains both positive and negative frequency terms.

Figure 11.7 shows the Feynman diagrams for several third-order nonlinear processes. Figure 11.7(a) gives the diagram for the general process, with three input photons corresponding to the driving fields and one output photon corresponding to the nonlinear polarization. The output frequency is the sum of the input frequencies, as required by conservation of photon energy. Since four photons are involved, the general phenomenon is often called **four-wave mixing**. In the discussion below,

As with the second-order processes, the negative frequencies represent the creation of photons in the nonlinear process. This is a legitimate outcome because photons are bosons and hence the nonlinear interaction can stimulate the creation of photons at the input frequency as well as causing annihilation of input photons.

Table 11.4 Third-order nonlinear effects. The third column lists the frequencies of the light beams incident on the nonlinear medium, while the fourth gives the frequency of the output beam or the nonlinear polarization. A frequency of zero indicates a DC electric field. Note that there are several other third-order nonlinear effects such as two-photon absorption which have not been included in this table.

Effect	Alternative names	Input frequencies	Output frequencies
Generic four-wave mixing		$\omega_1, \omega_2, \omega_3$	$\lvert \pm\omega_1 \pm \omega_2 \pm \omega_3 \rvert$
Frequency tripling	Third-harmonic generation	ω	3ω
Optical Kerr effect	Degenerate four-wave mixing	ω	ω
	Nonlinear refraction		
	Self-phase modulation		
DC Kerr effect	Quadratic electro-optic effect	$\omega, 0$	ω
Stimulated four-wave mixing	Stimulated Raman scattering	ω, ω_s	ω_s
	Stimulated Brillouin scattering		

we concentrate on the three specific examples of four-wave mixing illustrated in Figs 11.7(b)–(d), namely: frequency tripling, the optical Kerr effect, and stimulated Raman scattering. There are, of course, many other important third-order nonlinear phenomena, but we do not have space to discuss them all here. The main effects that we consider are summarized in Table 11.4.

11.4.2　Frequency tripling

In practice it is usually easier to generate the third harmonic of a laser beam by two second-order processes according to the scheme shown in Fig. 11.5, rather than by using a single third-order conversion using the $\chi^{(3)}$ nonlinearity.

Figure 11.7(b) shows the Feynman diagram for frequency tripling. This is the equivalent of frequency doubling for a $\chi^{(2)}$ process. Three collinear fields at the same frequency are incident on the medium from a single input laser beam. With $\omega_1 = \omega_2 = \omega_3 = +\omega$, we find from eqn 11.65 that the output frequency is equal to 3ω. In other words, the nonlinear process directly generates the third harmonic of the fundamental. As with frequency doubling, the conversion efficiency will only be large if the phase-matching condition determined by momentum conservation is satisfied. (See Section 11.3.3.) Frequency tripling experiments are useful for the spectroscopic information they give about the magnitude of $\chi^{(3)}$ and its relation to the atomic transitions of the medium.

11.4.3　The optical Kerr effect and the nonlinear refractive index

Figure 11.7(c) shows the Feynman diagram for the optical Kerr effect. In this process we have just a single beam at frequency ω incident on the nonlinear medium, and the nonlinear interaction produces a third-order polarization at the same frequency as the input laser beam. This works by using $\omega_1 = \omega_2 = +\omega$ and $\omega_3 = -\omega$ in eqn 11.65. No phase-matching problems occur in this case because the nonlinear polarization is at the same frequency as the driving fields and thus the fields are in phase throughout the whole medium. Since the frequencies of all four photons are the same, the optical Kerr effect is sometimes called degenerate four-wave mixing.

One of the main consequences of the optical Kerr effect is that the refractive index begins to depend on the intensity of the beam. This can be seen by calculating the change of the dielectric constant produced by the light. From eqn 11.9 we see that the dielectric constant in a nonlinear medium with $\chi^{(2)} = 0$ is given by:

$$\epsilon_{\rm r}^{\rm nonlinear} = 1 + \chi^{(1)} + \chi^{(3)}\mathcal{E}^2 \,, \tag{11.66}$$

where $\chi^{(1)}$ is the linear susceptibility and $\mathcal{E}$ is the optical electric field amplitude. We split this into its linear and nonlinear parts by writing:

$$\epsilon_{\rm r}^{\rm nonlinear} = \epsilon_{\rm r} + \Delta\epsilon \,, \tag{11.67}$$

where

$$\epsilon_{\rm r} = 1 + \chi^{(1)} \tag{11.68}$$

and

$$\Delta\epsilon = \chi^{(3)}\mathcal{E}^2 \,. \tag{11.69}$$

$\epsilon_{\rm r}$ is the usual relative permittivity for the linear regime and $\Delta\epsilon$ is the change caused by the nonlinear process. In a non-absorbing medium, the refractive index n is equal to the square root of the relative permittivity (cf. eqn A.31). Hence we may write:

$$n = (\epsilon_{\rm r} + \Delta\epsilon)^{1/2} = \sqrt{\epsilon_{\rm r}} + \frac{\Delta\epsilon}{2\sqrt{\epsilon_{\rm r}}} \equiv n_0 + \Delta n \,, \tag{11.70}$$

where we have assumed that $\Delta\epsilon \ll \epsilon_{\rm r}$ in the second equality, and in the third we have split the refractive index into its linear part $n_0 = \sqrt{\epsilon_{\rm r}}$ and its nonlinear part Δn. On comparing eqns 11.69 and 11.70 we find that

$$n = n_0 + \frac{\chi^{(3)}\mathcal{E}^2}{2n_0} = n_0 + \frac{\chi^{(3)}}{n_0^2 c\epsilon_0}I \,, \tag{11.71}$$

where we have used the proportionality between I and $\mathcal{E}^2$ given by eqn 11.14 in the second identity.

We now introduce the **nonlinear refractive index** n_2 according to:

$$n(I) = n_0 + n_2 I \,. \tag{11.72}$$

By comparing eqns 11.71 and 11.72 we find that:

$$n_2 = \frac{1}{n_0^2 c\epsilon_0} \chi^{(3)} \,. \tag{11.73}$$

This shows that n_2 is directly proportional to $\chi^{(3)}$, and hence that third-order nonlinearities cause the refractive index to vary with the intensity.

Table 11.5 lists the measured values of the nonlinear refractive index for a number of materials. It is apparent that the nonlinear refractive index can be either positive or negative. For a given material, n_2 is

Table 11.5 Nonlinear refractive index n_2 of selected materials. E_g: band gap, λ: wavelength of measurement, $\hbar\omega$: photon energy, n_0: linear refractive index. Data from Sheik–Bahae et al. (1991) and DeSalvo et al. (1996).

Material	E_g (eV)	λ (nm)	$\hbar\omega/E_g$	n_0	n_2 (m^2 W^{-1})
GaAs	1.42	1064	0.87	3.47	-3.3×10^{-17}
CdTe	1.44	1064	0.81	2.84	-2.9×10^{-17}
AlGaAs	1.57	850	0.93	3.30	-2.5×10^{-17}
AlGaAs	1.57	830	0.95	3.30	-8.9×10^{-17}
AlGaAs	1.57	810	0.98	3.30	-3.3×10^{-16}
ZnTe	2.26	1064	0.52	2.79	1.3×10^{-17}
CdS	2.42	1064	0.48	2.34	5.0×10^{-18}
CdS	2.42	532	0.96	2.34	-6.1×10^{-17}
ZnSe	2.58	1064	0.45	2.48	2.9×10^{-18}
ZnSe	2.58	532	0.90	2.70	-6.2×10^{-18}
Fused silica (SiO$_2$)	7.8	1064	0.15	1.48	2.1×10^{-20}
Fused silica (SiO$_2$)	7.8	532	0.30	1.48	2.2×10^{-20}
Fused silica (SiO$_2$)	7.8	355	0.45	1.48	2.4×10^{-20}
Fused silica (SiO$_2$)	7.8	266	0.60	1.50	7.8×10^{-20}

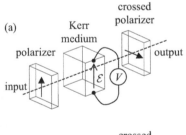

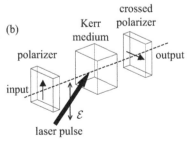

Fig. 11.8 (a) A Kerr cell based on the DC Kerr effect. (b) A Kerr gate based on the optical Kerr effect.

positive at low photon energies, and turns negative for $\hbar\omega \gtrsim 0.7E_g$, where E_g is the band gap. Note that $|n_2|$ is resonantly enhanced as the photon energy approaches E_g, as demonstrated by the data for AlGaAs.

It is apparent from eqn 11.71 that Δn is proportional to $\mathcal{E}^2$ in the optical Kerr effect. This is reminiscent of the DC Kerr effect considered in Section 2.5.2, and explains the name 'optical Kerr effect'. Both types of Kerr effect are examples of quadratic electro-optic effects. In the DC Kerr effect, $\mathcal{E}$ is an applied DC field, whereas in the optical Kerr effect, $\mathcal{E}$ is the electric field of the light. The DC Kerr effect can in fact be considered as a $\chi^{(3)}$ process in which $\omega_1 = \omega_2 = 0$ and $\omega_3 = \omega$. This contrasts with the *linear* electro-optic effect considered in Section 11.3.4, which is a $\chi^{(2)}$ process.

The analogy between the DC and optical Kerr effects is illustrated in Fig. 11.8. Fig. 11.8(a) shows a conventional Kerr cell based on the DC Kerr effect, which operates in a similar way to the Pockels cell shown in Fig. 11.6(b). The cell comprises a Kerr medium inserted between crossed polarizers. With no external field applied, the Kerr medium is isotropic, and no light is transmitted through the crossed polarizers. When the voltage is applied, birefringence is induced in proportion to $\mathcal{E}^2$ (see eqn 2.51), and the light that emerges from the Kerr medium is no longer vertically polarized. This means that some light is now transmitted through the second polarizer, allowing the cell to be used as an intensity modulator.

Fig. 11.8(b) illustrates a Kerr gate based on the optical Kerr effect. The principle of operation is the same as the Kerr cell shown in Fig. 11.8(a), except that now there is no applied voltage, and the birefringence is induced by an intense laser pulse via the optical Kerr effect. The gate therefore only transmits light while the laser pulse is impinging on the crystal. By using an ultrafast pulse from a mode-locked laser, it is possible to use this arrangement to produce a very fast (e.g. 1 ps) optical shutter.

11.4.4   Stimulated Raman scattering

Figure 11.7(d) shows the Feynman diagram for the **stimulated Raman effect**. A weak beam at frequency ω_s is incident on the medium together with a powerful pump beam at frequency ω. With $\omega_1 = +\omega$, $\omega_2 = -\omega$, and $\omega_3 = \omega_s$, we see from eqn 11.65 that the nonlinear wave is at frequency ω_s. Hence the presence of a field at ω_s generates more photons at the same frequency through nonlinear mixing with the pump field. The beam at frequency ω_s can then experience gain by the same sort of parametric mixing that was discussed in Section 11.3.1. The process becomes a Raman effect if we tune the frequency of ω_s such that $\omega - \omega_s = \Omega$, where Ω is the frequency of one of the vibrational modes of the medium. The nonlinear susceptibility is resonantly enhanced in these conditions because the two frequencies are strongly coupled together through the natural vibrations of the medium.

The spontaneous Raman scattering processes discussed in Section 10.5 can be related to the stimulated effects we are considering here by making an analogy with spontaneous and stimulated radiative emission. Spontaneous radiative emission can be considered to be a stimulated process that is triggered by a vacuum photon from the zero-point fluctuations of the quantized electromagnetic field. In the same way we can regard spontaneous Raman scattering as a stimulated nonlinear process instigated by vacuum photons. In fact, when we generate stimulated Raman beams by passing a laser through a suitable medium, there is usually no initial field at frequency ω_s to start the process. This field has to come from spontaneous Raman scattering, which itself is instigated by vacuum fluctuations. Hence the generation of a stimulated Raman beam is considered to start from the zero-point fluctuations of the field.

Stimulated Raman scattering was discovered very soon after the invention of the laser. In 1962 E.J. Woodbury and W.K. Ng observed that an intense beam was generated at 766 nm when passing a strong beam from a ruby laser at 694.3 nm through a nitrobenzene cell. Analysis of these results showed that the difference in the frequency of the two photons corresponded exactly with one of the vibrational modes of the molecule at 4.0×10^{13} Hz. The same phenomenon has subsequently been observed in many different liquids and gases, and also in solids.

In solids the scattering can be mediated either by the optical or the acoustic vibrational modes. In the former case it is the Raman-active LO and TO phonons at $q = 0$ that are involved, where q is the wave vector of the phonon, as discussed in Section 10.5.2. This gives rise to discrete frequency shifts analogous to those observed in molecules. In the latter case, it is the acoustic phonons that are involved, and the process is usually called **stimulated Brillouin scattering**. The frequency shift caused by stimulated Brillouin scattering depends on the angle through which the light is scattered, as determined by the energy and wave vector conservation laws. (See eqn 10.32.)

11.4.5   Isotropic third-order nonlinear media

The third-order nonlinear response is described by the nonlinear susceptibility tensor $\chi^{(3)}_{ijkl}$ defined in eqn 11.11. This has 81 elements, many of which are the same or zero in materials with a high degree of symmetry. In a completely isotropic medium such as a gas, there are 21 non-zero

Table 11.6 Non-zero components of the third-order nonlinear susceptibility in an isotropic medium.

$$\chi^{(3)}_{xxxx} = \chi^{(3)}_{yyyy} = \chi^{(3)}_{zzzz}$$

$$\chi^{(3)}_{xxyy} = \chi^{(3)}_{yyxx} = \chi^{(3)}_{xxzz} = \chi^{(3)}_{zzxx} = \chi^{(3)}_{yyzz} = \chi^{(3)}_{zzyy}$$

$$\chi^{(3)}_{xyxy} = \chi^{(3)}_{yxyx} = \chi^{(3)}_{xzxz} = \chi^{(3)}_{zxzx} = \chi^{(3)}_{yzyz} = \chi^{(3)}_{zyzy}$$

$$\chi^{(3)}_{xyyx} = \chi^{(3)}_{yxxy} = \chi^{(3)}_{xzzx} = \chi^{(3)}_{zxxz} = \chi^{(3)}_{yzzy} = \chi^{(3)}_{zyyz}$$

elements that are listed in Table 11.6. The interrelationships listed in the table suggest that there are four independent values, but this is not the case because it can be shown that the susceptibilities must also satisfy:

$$\chi^{(3)}_{xxxx} = \chi^{(3)}_{xxyy} + \chi^{(3)}_{xyxy} + \chi^{(3)}_{xyyx} . \tag{11.74}$$

This means that there are in fact only three independent elements. Furthermore, if we are well below the natural resonance frequencies of the atoms, it will also be true that:

$$\chi^{(3)}_{xxyy} = \chi^{(3)}_{xyxy} = \chi^{(3)}_{xyyx} = \frac{1}{3}\chi^{(3)}_{xxxx} . \tag{11.75}$$

This result is only valid at low frequencies, and is called **Kleinman symmetry**. In this limit, there are thus only two physically different third-order nonlinear susceptibilities.

Glasses are isotropic because they do not possess a crystal structure and therefore have no preferred axes. Doped glasses have optical transitions in the visible spectral region (see Section 1.4.5) and the Kleinman symmetry condition does not hold at frequencies close to the band gap of the dopant.

Glasses are perhaps the most important examples of isotropic optical materials. We usually use glasses at wavelengths where they are transparent. We are therefore well below the band gap in the ultraviolet spectral region, and the Kleinman symmetry condition given in eqn 11.75 will normally hold.

When an intense laser beam propagates through a glass, it can alter the refractive index by the optical Kerr effect in accordance with eqn 11.72. This produces a nonlinear phase shift given by:

$$\Delta\Phi^{\text{nonlinear}} = \frac{2\pi}{\lambda}\Delta n\, L = \frac{2\pi}{\lambda}\, n_2 I\, L , \tag{11.76}$$

where λ is the vacuum wavelength, L is the length of the medium and I is the intensity. The laser beam therefore alters its own phase, an effect called **self-phase modulation**.

11.4.6   Nonlinear propagation in optical fibres and solitons

Self-phase modulation effects can be observed very clearly in optical fibres, even though the nonlinear refractive index is very small. This is

Fig. 11.9 A group of runners on a mattress is an analogy of a soliton pulse in an optical fibre. The indentation of the group slows the faster runners and speeds the slower ones, thereby compensating for the tendency of the group to break up. In the same way, the nonlinear phase shift of an intense short pulse can compensate for the pulse broadening effect due to the dispersion of the optical fibre. After Mollenauer and Gordon (1994), reprinted with permission from Plenum Publishers.

because the beam is focused to a very small area in the fibre, so that the intensity is very high even at moderate power levels. Large phase shifts can then be achieved by using long lengths of fibre.

The subject of **solitons** is a very interesting aspect of the propagation of light pulses down an optical fibre in the nonlinear regime. A short laser pulse must necessarily contain a spread of frequencies in order to satisfy the Fourier transform limit given approximately by eqn 9.6. Since the glass that makes up the fibre is dispersive (see Fig. 2.10), the different frequency components of the pulse will experience slightly different refractive indices. This means that their velocities will be different, and the pulse will gradually broaden in time as it propagates down the fibre. (See Exercise 2.15.) This becomes a serious problem if we are trying to transmit a sequence of closely-spaced data pulses down the fibre. The soliton effect discovered by John Scott Russell in 1834 can eliminate this problem. Russell noticed that the bow wave of a Scottish canal barge did not disperse when the amplitude of the wave was large enough. He successfully explained his observation by realizing that the dispersion of the water wave was being balanced by the nonlinear effects due to the large amplitude. The same phenomenon can occur in optical fibres.

Solitons can be observed in optical fibres at frequencies where the dispersion is negative. This is because the n_2 of the fibre is positive in the near infrared spectral region, and so the dispersion must be negative if the two effects are to cancel. Most glasses have positive dispersion at optical frequencies due to the electronic absorption in the ultraviolet. However, as noted in Section 2.4, the dispersion of the SiO_2 used to make optical fibres is zero at $1.3\,\mu$m and becomes negative for longer wavelengths. This can be seen in the data shown in Fig. 2.7. This means that we are in the right regime to observe solitons at $1.55\,\mu$m, which is the preferred wavelength for telecommunication systems because the losses of the fibre are smallest.

The basic physics of soliton propagation can be understood in a simple way without recourse to heavy mathematics by reference to the cartoon given in Fig. 11.9. This shows a group of athletes crossing a mattress, with the faster runners at the front. The group creates a valley in the mattress that retards the faster runners but helps the slower ones. On a hard surface the group would break up with the faster runners leaving the slower ones behind, but the valley in the mattress opposes this effect and keeps the group together. In the right conditions the nonlinear phase shift of an intense pulse in an optical fibre can have a similar effect, creating a trapped light pulse similar to the trapped group of runners in the cartoon. The resulting pulse can travel indefinitely long distances along the fibre without significant broadening. This is a very useful property for long-distance optical fibre telecommunication systems.

Example 11.4

A laser pulse at 1.55 μm of intensity $10^{12}\,\mathrm{W\,m^{-2}}$ is propagating down an optical fibre of length 100 m. Calculate the nonlinear phase shift if the nonlinear refractive index is $2 \times 10^{-20}\,\mathrm{m^2\,W^{-1}}$.

Solution
We calculate the nonlinear phase shift from eqn 11.76 with $\lambda = 1.55 \times 10^{-6}\,\mathrm{m}$, $n_2 = 2 \times 10^{-20}\,\mathrm{m^2\,W^{-1}}$, and $I = 10^{12}\,\mathrm{W\,m^{-2}}$. This gives:

$$\Delta\Phi^{\mathrm{nonlinear}} = \frac{2\pi}{1.55 \times 10^{-6}} \times (2 \times 10^{-20}) \times 10^{12} \times 100 = 8.1 \,.$$

The nonlinear phase shift is thus 2.6π. This example shows that the nonlinear phase shift can be very large, even though the nonlinearity is small. This occurs because of the very large values of L that can be used in optical fibres.

11.4.7 Resonant nonlinearities in semiconductors

In Section 11.2.2 we explained how the absorption coefficient of an absorbing medium is expected to depend on the intensity at high power levels. This phenomenon has been studied extensively in semiconductors, with a particular emphasis on developing nonlinear switching devices for applications in optical information processing.

The simplest mechanism that can cause saturable absorption in a semiconductor is illustrated in Fig. 11.10. This shows the band diagram of a direct gap semiconductor when a large number of electrons have been excited from the valence band to the conduction band. The electrons are excited by absorption of an intense laser pulse with photon energy greater than the band gap. The electrons fill up the states at the bottom of the conduction band, leaving empty states at the top of the valence band. This blocks further interband absorption transitions at photon energies close to the band gap such as the one shown in the figure, because there are no electrons in the valence band to be excited, and furthermore, the destination states in the conduction band are full. Therefore, as we turn up the intensity of the exciting laser, we find that

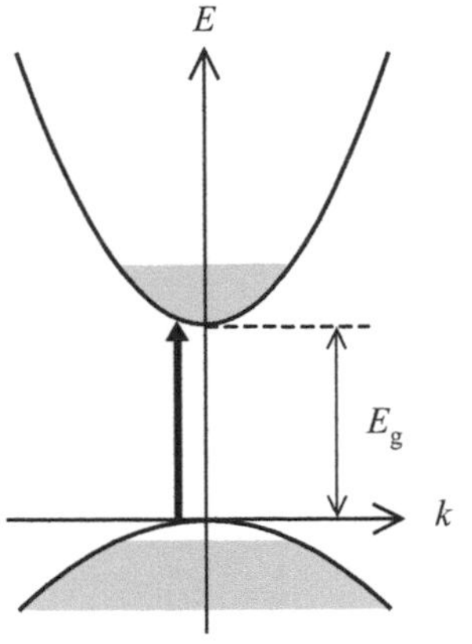

Fig. 11.10 Band-filling nonlinearity in an excited semiconductor. The interband absorption transition shown by the arrow in the diagram is blocked because there are no electrons in the valence band to absorb, and the destination states in the conduction band are filled.

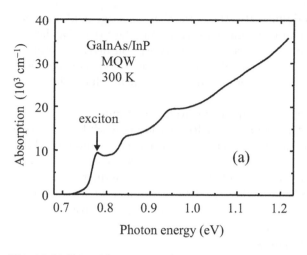
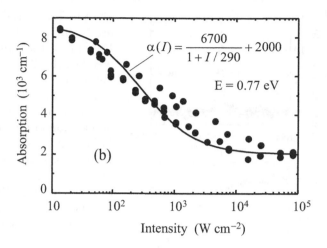

Fig. 11.11 Saturable excitonic absorption in a semiconductor multiple quantum well. The sample contained 30 Ga$_{0.53}$In$_{0.47}$As quantum wells of width 15.4 nm with InP barriers. The temperature was 300 K. Part (a) shows the linear absorption of the sample, while (b) shows the intensity dependence of the absorption at 0.77 eV. This photon energy corresponds to the first excitonic peak in the linear absorption spectrum, and is identified by the arrow in (a). The solid line in (b) is a fit to the data using eqn 11.77. After Westland et al. (1987) and Fox et al. (1987), © American Institute of Physics, reprinted with permission.

the absorption gradually saturates as the bands fill up. This effect is therefore called a **band-filling** nonlinearity.

Another mechanism that can cause the absorption to depend on the intensity is the saturation of excitons. This effect was described in Section 4.4. At high carrier densities the Coulomb force holding the excitons together is screened, and the electron and hole states near $k = 0$ required for the formation of excitons are filled. Both of these effects cause bleaching of the exciton absorption. If the carriers are excited by the absorption of a laser beam, the excitonic absorption will decrease with increasing laser intensity.

Figure 11.11 shows the intensity dependence of the exciton absorption in a GaInAs multiple quantum well sample at room temperature. The linear absorption spectrum is given in Fig. 11.11(a). The step-like spectrum expected for the 2-D quantum wells is clearly apparent in the data, together with the broadened peaks due to the excitons. Figure 11.11(b) shows the saturation of the absorption coefficient α at the first exciton peak as a function of intensity I. The intensity dependence of α is given approximately by:

$$\alpha(I) = \alpha_b + \frac{\alpha_0}{1 + I/I_s}, \tag{11.77}$$

with $\alpha_b = 2000\,\text{cm}^{-1}$, $\alpha_0 = 6700\,\text{cm}^{-1}$, and $I_s = 290\,\text{W cm}^{-2}$. The α_b term in eqn 11.77 represents an unsaturable background absorption, while the second term represents a saturable absorption of the form given in eqn 11.37.

The intensity dependence of α shown in Fig. 11.11(b) gives rise to a very large change in the refractive index. (See Exercise 11.15.) The nonlinear refractive index n_2 associated with the exciton saturation is $\sim 10^{-8}\,\text{m}^2\,\text{W}^{-1}$. This is many orders of magnitude larger than the values

Quantum wells are good materials to demonstrate excitonic nonlinearities because they show strong excitonic absorption even at room temperature. See Section 6.4.4 for further details.

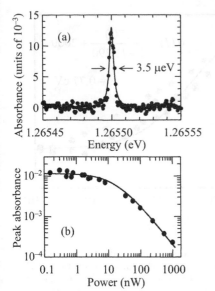

Fig. 11.12 (a) Excitonic absorption line of a single self-assembled InGaAs quantum dot at liquid helium temperature. (b) Saturation of the absorption as a function of resonant laser power. The fit is a saturation curve of the form given in eqn 11.78 with a saturation power of 18 nW. After Kroner et al. (2008), © Elsevier, reprinted with permission.

given in Table 11.5, and illustrates the point that resonant nonlinearities are much larger than non-resonant ones. However, the price that has to be paid for the large $|n_2|$ value is the high absorption at the operating wavelength. This restricts the length of the devices that can be used, and therefore puts a limit on the maximum nonlinear phase shift that can be obtained.

The saturable absorption of quantum wells has found an important application in ultrafast laser physics. Ultrafast lasers work by the mode locking principle, and one of the methods used to achieve mode locking is to incorporate a saturable absorber into the laser cavity. By using molecular beam epitaxy, it is possible to make a semiconductor saturable absorber mirror (SESAM), which consists of semiconductor quantum well layers grown on top of a dielectric mirror. Such mirrors are now routinely used as the saturable absorber in a number of commercial mode-locked solid-state lasers.

In recent years there has been a growing interest in investigating the nonlinear properties of semiconductor quantum dots. The optical properties of quantum dots were discussed in Section 6.8, with Section 6.8.3 focusing on self-assembled III–V dots. Figure 11.12(a) shows the excitonic absorption line of a single self-assembled InGaAs quantum dot at liquid helium temperatures. A very sharp line with a linewidth of only 3.5 μeV is observed at low excitation powers. Figure 11.12(b) shows the absorption strength (i.e. absorbance) measured when the dot is irradiated with a laser tuned to resonance with the exciton. The absorbance is observed to saturate strongly as a function of increasing laser power P. The fit is a saturation curve of the form (cf. eqn 11.37):

$$\alpha(P) = \frac{\alpha_0}{1 + P/P_\mathrm{s}},\tag{11.78}$$

with a saturation power P_s of 18 nW. The good fit to the data shows that the quantum dot approximates well to a saturable two-level atom system, which is the starting point for the analysis of nonlinear optics in atomic physics. This is one of the reasons why quantum dots are often described as 'solid-state atoms', and have interesting possibilities for applications in quantum optics.

Chapter summary

- Nonlinear optical effects become important at the high power levels available from lasers. Nonlinear effects cause the optical susceptibility, and all the properties that follow from it, to depend on the magnitude of the electric field of the light.

- Nonlinear effects are characterized by the nonlinear susceptibility tensor. All the components of the second-order nonlinear susceptibility tensor are zero if the medium possesses inversion symmetry. Some crystals lack inversion symmetry and have non-zero second-order nonlinear susceptibilities. The form of the tensor is determined by the symmetry class of the crystal.

- Non-resonant nonlinearities are caused by the anharmonicity of the restoring force on the bound electrons.

- Resonant nonlinearities are caused by stimulated emission from the upper level when its population becomes significant. The saturation of interband absorption can be understood in terms of the blocking of the transitions due to the Pauli exclusion principle.

- The nonlinear polarization in a second-order effect is proportional to the product of two electric fields. This gives rise to a series of second-order effects such as frequency doubling, sum frequency generation, and difference frequency generation.

- Strong second-order nonlinear signals are only achieved when the phase-matching condition is satisfied.

- The anisotropy induced by a DC electric field in an electro-optic crystal is called the Pockels effect, and can be used to make amplitude or phase modulators.

- The nonlinear polarization in a third-order effect is proportional to the product of three fields. Third-order effects are generally known as four-wave mixing.

- The third-order susceptibility is the lowest-order non-zero term in the nonlinear response of isotropic media such as gases, liquids, and glasses.

- Third-order effects give rise to nonlinear refraction, frequency tripling, and stimulated Raman scattering. Self-phase modulation describes the nonlinear phase shift induced by a beam through the nonlinear refractive index.

- The saturation of interband transitions and excitonic absorption in semiconductors gives rise to large third-order nonlinearities.

Further reading

More extensive introductory reading on nonlinear optics may be found in Yariv (1997) or Tang (1995). Butcher and Cotter (1990) give a comprehensive introduction to the theory of nonlinear optics.

A classic discussion of the effects of the crystal symmetry on the physical properties of materials is given by Nye (1957).

The optical nonlinearities due to excitons are reviewed in Chemla (1985). A more detailed review on excitonic nonlinearities in quantum wells is given by Schmitt-Rink et al. (1989).

Exercises

(11.1) Use the Bohr model of the atom to show that the electric field experienced by an electron bound in the nth shell of a hydrogenic atom of atomic number Z is equal to $(Z^3/n^4)e/4\pi\epsilon_0 a_H^2$, where a_H is the Bohr radius of hydrogen. Make a rough estimate of this value for the valence electrons of silicon.

(11.2) Estimate the magnitude of the electric field of the light wave for: (a) a 10 ns long pulse of energy 1 J from a Nd:glass laser with a beam diameter 5 mm; (b) a 1 mW continuous wave semiconductor laser focused into an optical fibre with a core area of $20\,\mu\text{m}^2$ and a refractive index of 1.45.

(11.3) A gas of atoms is subjected to a strong DC electric field. State with reasons whether you would expect to be able to observe any frequency doubling or not from the gas.

(11.4) Which of the following materials would you expect to have non-zero components in the second-order nonlinear susceptibility: (a) NaCl, (b) GaAs, (c) water, (d) glass, (e) crystalline quartz, (f) ZnS (wurtzite).

(11.5)* An intense laser is propagating through an absorbing medium as described in Section 11.2.2. The medium contains atoms with two non-degenerate levels 1 and 2, and the separation of the levels is resonant with the laser frequency. The atoms are initially in a state with $N_1 = N_0$ and $N_2 = 0$, and the laser beam is turned on at time $t = 0$.
(a) Explain why it is not possible to get population inversion between the two levels, no matter how intense the laser is.
(b) Show that if spontaneous emission is neglected, the time dependence of the population difference $\Delta N = N_1 - N_2$ between the two levels is given by:

$$\Delta N(t) = N_0 \exp\left(-2B_{12}u_\nu g(\nu)t\right),$$

where u_ν is the energy density of the beam, $g(\nu)$ is the spectral lineshape function, and B_{12} is the Einstein B coefficient for absorption. What does this relationship imply about the absorption coefficient?

(11.6) The nonlinear optical coefficient tensor of a crys-

tal with orthorhombic symmetry is given by

$$d = \begin{pmatrix} 0 & 0 & 0 & d_{14} & 0 & 0 \\ 0 & 0 & 0 & 0 & d_{25} & 0 \\ 0 & 0 & 0 & 0 & 0 & d_{36} \end{pmatrix}.$$

A laser beam is incident on the crystal. The beam travels in the x–y plane, and is polarized with its electric field in the x–y plane. Show that a second harmonic beam is produced which is polarized along the z axis. Show also that the magnitude of the second harmonic beam is a maximum if the incoming beam travels at 45° to the x axis.

(11.7) Calculate the phase-matching angle for KDP at the wavelength of the ruby laser (694 nm). The relevant refractive indices are: $n_o(694\,\text{nm}) = 1.506$, $n_o(347\,\text{nm}) = 1.534$, $n_e(347\,\text{nm}) = 1.490$.

(11.8) Trigonal crystals with 3m symmetry (e.g. LiNbO$_3$) are uniaxial, and have electro-optic coefficient tensors of the form:

$$r = \begin{pmatrix} 0 & -r_{22} & r_{13} \\ 0 & r_{22} & r_{13} \\ 0 & 0 & r_{13} \\ 0 & r_{51} & 0 \\ r_{51} & 0 & 0 \\ -r_{22} & 0 & 0 \end{pmatrix},$$

when the optic axis is defined to coincide with the z axis.
(a) Deduce the changes in the ordinary and extraordinary refractive indices (n_o and n_e respectively) when a DC electric field of magnitude $\mathcal{E}$ is applied along the optic axis.
(b) Find the phase change induced by the electric field for a crystal of length L when light of vacuum wavelength λ is propagating along the y axis with linear polarization in the z direction.
(c) Explain how such a device can be used as a phase modulator.

(11.9) The anisotropy induced by an electric field applied along the z axis for crystals with the cubic zinc-blende structure ($\bar{4}$3m symmetry) can be written in the following form:

$$
\begin{aligned}
n_{x'} &= n_0 + \tfrac{1}{2}n_0^3 r_{41}\mathcal{E}_z, \\
n_{y'} &= n_0 - \tfrac{1}{2}n_0^3 r_{41}\mathcal{E}_z, \\
n_z &= n_0,
\end{aligned}
$$

*Exercises marked with an asterisk are more challenging.

where, as with the case of the tetragonal crystal considered in Fig. 11.6(a), x' and y' are axes at $45°$ to the crystallographic axes x and y. n_0 is the refractive index at $\mathcal{E}_z = 0$, and r_{41} is the electro-optic coefficient.

(a) Show that when light of vacuum wavelength λ propagates through the crystal in the z direction, the phase difference introduced between the x' and y' polarizations is equal to

$$\Delta\Phi(\mathcal{E}) = \frac{2\pi}{\lambda} r_{41} n_0^3 V \,,$$

where V is the voltage applied across the crystal to produce the field along the z direction.

(b) Evaluate the voltage at which the phase change is equal to π for CdTe at $10.6\,\mu$m, where $r_{41} = 6.8\,$pm/V and $n_0 = 2.6$.

(11.10) The electro-optic coefficient r_{63} of KDP at $633\,$nm is equal to $11\,$pm/V, where $n_o = 1.5074$. Calculate the voltage that must be applied to change the transmission of the modulator shown in Fig. 11.6(b) to 50% at this wavelength.

(11.11) Explain why the imaginary part of the third-order nonlinear susceptibility must be non-zero in a saturable absorber medium.

(11.12) Show that the third-order nonlinear polarization produced by a linearly polarized laser beam in an isotropic medium is always parallel to the $\mathcal{E}$ vector of the laser.

(11.13) A laser beam of wavelength $1.55\,\mu$m propagates down an optical fibre of length $10\,$m with a core diameter of $5\,\mu$m. Calculate the power that must be launched into the fibre to produce a nonlinear phase shift of π, if the fibre has a nonlinear refractive index of $2 \times 10^{-20}\,$m$^2\,$W^{-1}.

(11.14)* A short laser pulse excites $10^{24}\,$m^{-3} electron-hole pairs into a sample of GaAs at $4\,$K. Calculate the shift in the absorption edge at the band gap due to the presence of the photoexcited carriers. The band parameters for GaAs are given in Table D.2. Ignore excitonic effects.

(11.15)* The $n = 1$ heavy-hole exciton of a GaAs multiple quantum well sample has a peak absorption coefficient of $8 \times 10^5\,$m^{-1} at $847\,$nm. By assuming that the exciton behaves like a classical dipole oscillator, estimate the change of the refractive index that can be obtained from this sample by completely saturating the absorption. The nonresonant refractive index is 3.5.

(11.16) (a) Use eqn 4.8 to calculate the saturation density for excitons in InP. (The relevant band structure parameters for InP are given in Table D.2, and the dielectric constant is 12.5.)

(b) If the carrier recombination time is $1\,$ns, and the absorption coefficient is $10^6\,$m^{-1}, estimate the saturation intensity when a laser is tuned to the exciton wavelength.

Appendix A

Electromagnetism in dielectrics

A.1 Electromagnetic fields and Maxwell's equations   330

A.2 Electromagnetic waves   333

Further reading   339

This appendix summarizes the principal results of electromagnetism that are used throughout the book. It is hoped that the reader will be familiar with this material. The main purpose of the appendix is to collect together the equations in a concise form for quick reference, and also to define the notation. SI units are used throughout. A short bibliography of suitable supplementary texts is given under Further Reading.

A.1 Electromagnetic fields and Maxwell's equations

The response of a dielectric material to an external electric field is characterized by three macroscopic vectors:

- the **electric field strength** $\mathcal{E}$;
- the **polarization** P;
- the **electric displacement** D.

The microscopic response of the material is determined primarily by the polarization. For this reason, the first task in all the examples treated by electromagnetism in this book is to calculate P. The dielectric constant ϵ_r is then determined from P, and the optical properties are deduced from ϵ_r.

The polarization is defined as the net dipole moment per unit volume. The application of a field produces a polarization by the forces exerted on the positive and negative charges of the atoms that are contained within the medium. If the molecules have permanent dipole moments, the field will apply a torque to these randomly orientated dipoles and tend to align them along the field direction. If there are no permanent dipoles, the field will push the positive and negative charges in opposite directions and induce a dipole parallel to the field. In either case, the end result is the same: the application of the field tends to produce many microscopic dipoles aligned parallel to the direction of the external field. This generates a net dipole moment within the dielectric, and hence a polarization.

The microscopic dipoles will all tend to align along the field direction, and so the polarization vector will be parallel to $\mathcal{E}$. This allows us to write:

$$P = \epsilon_0 \chi \mathcal{E}, \tag{A.1}$$

where ϵ_0 is the **electric permittivity** of free space and χ is the **electric susceptibility** of the medium. The value of ϵ_0 is $8.854 \times 10^{-12}\,\mathrm{F\,m^{-1}}$ in SI units.

Equation A.1 makes two assumptions that need a brief word of explanation.

(1) We have assumed that the medium is isotropic, even though we know that some materials are anisotropic. In particular, anisotropic crystals have preferred non-equivalent axes, and P will not necessarily be parallel to $\mathcal{E}$.

(2) We have assumed that P varies linearly with $\mathcal{E}$. This will not always be the case. In particular, if the optical intensity is very large, we can enter the realm of nonlinear optics, in which eqn A.1 is not valid.

A discussion of how to treat non-isotropic materials may be found in Section 2.5, while nonlinear optics is the subject of Chapter 11.

Both of these qualifications introduce unnecessary complications at this stage, and are not considered further in this appendix.

The electric displacement D of the medium is related to the electric field $\mathcal{E}$ and polarization P through:

$$D = \epsilon_0 \mathcal{E} + P. \tag{A.2}$$

This may be considered to be the definition of D. By combining eqns A.1 and A.2, we can write:

$$D = \epsilon_0 \epsilon_r \mathcal{E}, \tag{A.3}$$

where

$$\epsilon_r = 1 + \chi. \tag{A.4}$$

ϵ_r is the **relative dielectric constant** of the medium, and is an extremely important parameter in the understanding of the propagation of light through dielectrics.

In electrostatic problems we are frequently interested in calculating the spatial dependence of the electric field, and hence the electric potential V, from the free charge density ϱ. This calculation can be performed by using the Poisson equation:

$$\nabla^2 V = -\frac{\varrho}{\epsilon_r \epsilon_0}. \tag{A.5}$$

Poisson's equation is derived from Gauss's law of electrostatics:

$$\nabla \cdot \mathcal{E} = \frac{\varrho}{\epsilon_r \epsilon_0}. \tag{A.6}$$

We recall that the electric field strength is the gradient of the potential:

$$\mathcal{E} = -\nabla V. \tag{A.7}$$

Equation A.5 follows directly by substituting for $\mathcal{E}$ in eqn A.6 using eqn A.7. Once we know V, we can then calculate $\mathcal{E}$ from eqn A.7. This approach is useful for treating devices that have a fixed potential across them determined by an external voltage source.

The response of a material to external magnetic fields is treated in a similar way to the response of dielectrics to electric fields. The **magnetization** M of the medium is proportional to the **magnetic field strength** H through the **magnetic susceptibility** χ_M:

$$M = \chi_M H. \tag{A.8}$$

The **magnetic flux density** B is related to H and M through:

$$\begin{aligned} B &= \mu_0(H + M) \\ &= \mu_0(1 + \chi_M)H \\ &= \mu_0\mu_r H, \end{aligned} \tag{A.9}$$

where μ_0 is the magnetic permeability of the vacuum and $\mu_r = 1 + \chi_M$ is the **relative magnetic permeability**, of the medium. The value of μ_0 is $4\pi \times 10^{-7}\,\mathrm{H\,m^{-1}}$ in SI units.

The laws that describe the combined electric and magnetic response of a medium are summarized in **Maxwell's equations** of electromagnetism:

$$\nabla \cdot D = \varrho \tag{A.10}$$

$$\nabla \cdot B = 0 \tag{A.11}$$

$$\nabla \times \mathcal{E} = -\frac{\partial B}{\partial t} \tag{A.12}$$

$$\nabla \times H = j + \frac{\partial D}{\partial t}, \tag{A.13}$$

where ϱ is the free charge density, and j is the current density. The first of these four equations is Gauss's law of electrostatics (eqn A.6) written in terms of D rather than $\mathcal{E}$. The second is the equivalent of Gauss's law for magnetostatics with the assumption that free magnetic monopoles do not exist. The third equation combines the Faraday and Lenz laws of electromagnetic induction. The fourth is a statement of Ampere's law, with the second term on the right-hand side to account for the displacement current.

The second Maxwell equation naturally leads to the concept of the **vector potential**. This is defined through the equation

$$B = \nabla \times A. \tag{A.14}$$

It is apparent that the vector potential A automatically satisfies eqn A.11, because $\nabla \cdot (\nabla \times A) = 0$ for all A. By substituting for B in the third Maxwell equation (eqn A.12) using eqn A.14, we see that:

$$\nabla \times \mathcal{E} = -\frac{\partial}{\partial t}(\nabla \times A) = \nabla \times \left(-\frac{\partial A}{\partial t}\right). \tag{A.15}$$

The solution is

$$\mathcal{E} = -\frac{\partial A}{\partial t} + \mathrm{constant}, \tag{A.16}$$

where the constant is any vector whose curl is zero. If the scalar potential is V, then we can combine eqn A.16 with eqn A.7 by writing:

$$\mathcal{E} = -\frac{\partial \boldsymbol{A}}{\partial t} - \boldsymbol{\nabla} V \,. \qquad (A.17)$$

This works because $\boldsymbol{\nabla} \times \boldsymbol{\nabla} V = 0$. The more general definition of $\mathcal{E}$ given in eqn A.17 reduces to eqn A.7 when the magnetic field does not vary with time, and to

$$\mathcal{E} = -\frac{\partial \boldsymbol{A}}{\partial t} \qquad (A.18)$$

when the scalar potential is constant throughout space.

It is important to note that the definition of $\boldsymbol{A}$ in eqn A.14 does not define the vector potential uniquely. We can add any vector of the form $\boldsymbol{\nabla} \varphi$ to $\boldsymbol{A}$ without changing $\boldsymbol{B}$ because

$$\boldsymbol{\nabla} \times (\boldsymbol{A} + \boldsymbol{\nabla} \varphi) = \boldsymbol{\nabla} \times \boldsymbol{A} + \boldsymbol{\nabla} \times (\boldsymbol{\nabla} \varphi) = \boldsymbol{\nabla} \times \boldsymbol{A} \,. \qquad (A.19)$$

$\varphi(\boldsymbol{r})$ can be any scalar function of $\boldsymbol{r}$. For this reason, we have to give an additional definition, which specifies the **gauge** in which we are working. The **Coulomb gauge** is defined by

$$\boldsymbol{\nabla} \cdot \boldsymbol{A} = 0 \,. \qquad (A.20)$$

This gauge is convenient because it allows us to recover Poisson's equation (A.5) by taking the divergence of eqn A.17. The vector potential in the Coulomb gauge is used in the semi-classical treatment of the interaction of light with atoms discussed in Section B.2 of Appendix B.

A.2   Electromagnetic waves

Maxwell was able to show that eqns A.10–A.13 were consistent with wave-like solutions in a medium with no free charges or currents. To see this we first simplify eqns A.12 and A.13 by setting $\boldsymbol{j} = 0$ and eliminating $\boldsymbol{B}$ and $\boldsymbol{D}$ using eqns A.3 and A.9. This gives:

$$\boldsymbol{\nabla} \times \mathcal{E} = -\mu_0 \mu_{\mathrm{r}} \frac{\partial \boldsymbol{H}}{\partial t} \,, \qquad (A.21)$$

and

$$\boldsymbol{\nabla} \times \boldsymbol{H} = \epsilon_0 \epsilon_{\mathrm{r}} \frac{\partial \mathcal{E}}{\partial t} \,. \qquad (A.22)$$

We then take the curl of eqn A.21 and eliminate $\boldsymbol{\nabla} \times \boldsymbol{H}$ using eqn A.22. This gives:

$$\boldsymbol{\nabla} \times (\boldsymbol{\nabla} \times \mathcal{E}) = -\mu_0 \mu_{\mathrm{r}} \epsilon_0 \epsilon_{\mathrm{r}} \frac{\partial^2 \mathcal{E}}{\partial t^2} \,. \qquad (A.23)$$

The left-hand side can be simplified by using the vector identity

$$\boldsymbol{\nabla} \times (\boldsymbol{\nabla} \times \mathcal{E}) = \boldsymbol{\nabla} (\boldsymbol{\nabla} \cdot \mathcal{E}) - \nabla^2 \mathcal{E} \,. \qquad (A.24)$$

Equation A.6 with $\varrho = 0$ tells us that $\boldsymbol{\nabla} \cdot \boldsymbol{\mathcal{E}} = 0$. Therefore we obtain the final result:

$$\nabla^2 \boldsymbol{\mathcal{E}} = \mu_0 \mu_r \epsilon_0 \epsilon_r \frac{\partial^2 \boldsymbol{\mathcal{E}}}{\partial t^2} . \tag{A.25}$$

Equation A.25 is of the same form as the wave equation:

$$\frac{\partial^2 y}{\partial x^2} = \frac{1}{v^2} \frac{\partial^2 y}{\partial t^2}, \tag{A.26}$$

where v is the velocity of the wave. We therefore identify eqn A.25 as describing electromagnetic waves with a phase velocity v given by

$$\frac{1}{v^2} = \mu_0 \mu_r \epsilon_0 \epsilon_r. \tag{A.27}$$

In free space $\epsilon_r = \mu_r = 1$ and the velocity of the wave is c, so we have:

$$c = \frac{1}{\sqrt{\mu_0 \epsilon_0}} = 2.998 \times 10^8 \text{ m s}^{-1}. \tag{A.28}$$

At the same time, we see from eqns A.27 and A.28 that the velocity in a medium is given by

$$v = \frac{1}{\sqrt{\epsilon_r \mu_r}} c . \tag{A.29}$$

We define the **refractive index** n of the medium as the ratio of the velocity of light in free space to the velocity in the medium:

$$n = \frac{c}{v} . \tag{A.30}$$

At optical frequencies we can set $\mu_r = 1$, and thus conclude:

$$n = \sqrt{\epsilon_r}. \tag{A.31}$$

This allows us to relate the propagation constants of electromagnetic waves in a medium to the dielectric constant.

The solutions to eqn A.25 are of the form:

The use of complex solutions of the type given in eqn A.32 simplifies the mathematics and is used extensively throughout this book. Physically measurable quantities are obtained by taking the real part of the complex wave. In some texts i is replaced by −j, but this makes no physical difference.

$$\boldsymbol{\mathcal{E}}(z, t) = \boldsymbol{\mathcal{E}}_0 \, e^{i(kz - \omega t)} , \tag{A.32}$$

where $\boldsymbol{\mathcal{E}}_0$ is the amplitude of the wave, z is the direction of propagation, k is the wave vector, and ω is the angular frequency. The magnitude of the wave vector k is given by:

$$k = \frac{2\pi}{\lambda} = \frac{\omega}{v} = \frac{n\omega}{c} , \tag{A.33}$$

where λ is the wavelength inside the medium. The first equality is the definition of k, the second follows by substitution of eqn A.32 into eqn A.25 with v given by eqn A.27, and the third follows from the definition of n given in eqn A.30.

Note that the word 'polarization' is used both for the dielectric polarization $\boldsymbol{P}$ and for the direction of the electric field in an electromagnetic wave. It is usually obvious from the context which meaning is intended.

The direction of the electric field in an electromagnetic wave is called the **polarization**. Several different types of polarization are possible.

- **Linear**: the electric field vector points along a constant direction.
- **Circular**: the electric field vector rotates as the wave propagates, mapping out a circle for each cycle of the wave.
- **Elliptical**: this is similar to circular polarization, except that the rotating electric field vector maps out an ellipse rather than a circle as the wave propagates.
- **Unpolarized**: the light is randomly polarized.

In free space the polarization of a wave is constant as it propagates. However, in anisotropic or chiral materials, the polarization can change as the wave propagates. (See Sections 2.5 and 2.6.)

Figure A.1 depicts a linearly polarized wave propagating along the z axis with the polarization along the x axis. If the beam is travelling parallel to a horizontal optical bench with the x axis perpendicular to the surface, the x-polarized wave shown in Fig. A.1(a) is said to be **vertically polarized**. A y-polarized beam would similarly be called **horizontally polarized**. It is apparent from eqn A.21 or A.22 that the magnetic field must be perpendicular to the electric field. $\mathcal{E}$, $\boldsymbol{H}$, and $\boldsymbol{k}$ therefore form a right-handed system as depicted in Fig. A.1(a). The orthogonal electric and magnetic fields vary sinusoidally in space, as shown in Figure A.1(b).

For the x-polarized wave propagating along the z direction shown in Fig. A.1(a), the components of the complex fields are of the form:

$$\mathcal{E}_x(z, t) = \mathcal{E}_{x0}\, e^{i(kz - \omega t)}$$
$$\mathcal{E}_y(z, t) = 0$$
$$H_x(z, t) = 0 \tag{A.34}$$
$$H_y(z, t) = H_{y0}\, e^{i(kz - \omega t)},$$

where k is the magnitude of the wave vector defined in eqn A.33 and ω is the angular frequency. On substituting the fields from eqn A.34 into eqn A.21, we find that:

$$k\,\mathcal{E}_{x0} = \mu_0 \mu_{\mathrm{r}} \omega\, H_{y0}, \tag{A.35}$$

and hence that:

$$H_{y0} = \frac{\mathcal{E}_{x0}}{Z}, \tag{A.36}$$

where

$$Z = \frac{\mu_0 \mu_{\mathrm{r}} \omega}{k} = \sqrt{\frac{\mu_0 \mu_{\mathrm{r}}}{\epsilon_0 \epsilon_{\mathrm{r}}}} = \frac{1}{c \epsilon_0 n}. \tag{A.37}$$

The second equality in eqn A.37 follows from eqns A.33 and A.27, while the third follows from eqns A.28 and A.31 with $\mu_{\mathrm{r}} = 1$. The quantity Z is called the **wave impedance**, and takes the value of $377\,\Omega$ in free space.

In general, an electromagnetic wave propagating in the z direction will have electric field components along both the x and y axes, with:

$$\mathcal{E}_x(z, t) = \mathcal{E}_{x0}\, e^{i(kz - \omega t)},$$
$$\mathcal{E}_y(z, t) = \mathcal{E}_{y0}\, e^{i(kz - \omega t + \Delta\phi)}. \tag{A.38}$$

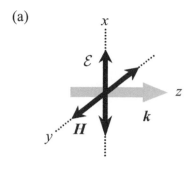

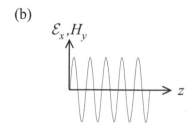

Fig. A.1 (a) The electric and magnetic fields of an electromagnetic wave form a right-handed system. The figure shows the directions of the fields in a wave polarized along the x axis and propagating in the z direction. (b) Spatial variation of the electric and magnetic fields.

Table A.1 Relative amplitudes and phases for the orthogonal components of the electric field for various types of polarization. $\mathcal{E}_{x0}$, $\mathcal{E}_{y0}$, and $\Delta\phi$ are defined in eqn A.38.

Polarization		Relative amplitude	Relative phase $\Delta\phi$
Linear		any	0, π
Positive circular	σ^+	$\mathcal{E}_{x0} = \mathcal{E}_{y0}$	$+\pi/2$
Negative circular	σ^-	$\mathcal{E}_{x0} = \mathcal{E}_{y0}$	$-\pi/2$
Elliptical		$\mathcal{E}_{x0} \neq \mathcal{E}_{y0}$	$\pm\pi/2$
		$\mathcal{E}_{x0} = \mathcal{E}_{y0}$	$\neq 0$, $\pm\pi/2$, or π
Unpolarized		random	random

This can be represented in short-hand form as:

$$\mathcal{E}_0 = \mathcal{E}_{x0}\,\hat{\mathbf{x}} + \mathcal{E}_{y0}\,e^{i\Delta\phi}\,\hat{\mathbf{y}}\,. \tag{A.39}$$

Table A.1 summarizes the effect of varying the relative amplitudes and phases of the two orthogonal components.

For the case of linearly polarized light, the direction of the polarization is given by the resultant of $(\mathcal{E}_{x0}\,\hat{\mathbf{x}} \pm \mathcal{E}_{y0}\,\hat{\mathbf{y}})$, where the + sign is used when $\Delta\phi = 0$ and the − sign when $\Delta\phi = \pi$.

For circularly polarized light, we can either have **right circular polarization** or **left circular polarization** depending on whether the electric field vector rotates to the right (clockwise) or left (anti-clockwise) in a fixed plane as the observer looks towards the light source. In σ^+ circular polarization, the electric field rotates clockwise as seen from the source, making it equivalent to left circular light, and vice versa for σ^- polarization. Since the phase difference between the two orthogonal linear components is $\pm\pi/2$, we can represent σ^+ and σ^- light in the following form:

$$\begin{aligned} \sigma^+ &= \mathcal{E}_0(\hat{\mathbf{x}} + i\hat{\mathbf{y}})/\sqrt{2}\,, \\ \sigma^- &= \mathcal{E}_0(\hat{\mathbf{x}} - i\hat{\mathbf{y}})/\sqrt{2}\,. \end{aligned} \tag{A.40}$$

These can easily be inverted to demonstrate that linearly polarized light can be considered to consist of two opposite circular polarizations. For example, for x-polarized light we have:

$$\mathcal{E}_x \equiv \mathcal{E}_0\hat{\mathbf{x}} = (\sigma^+ + \sigma^-)/\sqrt{2}\,. \tag{A.41}$$

The energy flow in an electromagnetic wave can be calculated from the **Poynting vector**:

In evaluating the Poynting vector for complex fields of the form given in eqn A.32, the real parts must be taken before the multiplication is performed.

$$\boldsymbol{I} = \boldsymbol{\mathcal{E}} \times \boldsymbol{H}\,. \tag{A.42}$$

This gives the power flow per unit area in $\mathrm{W\,m^{-2}}$, which is equal to the **intensity** of the light wave. The intensity is defined as the energy crossing a unit area in unit time, and is therefore given by:

$$I = vu_\nu\,, \tag{A.43}$$

where v is the velocity of the wave and u_ν is the energy density per unit volume of the beam. For a linearly polarized wave, the Poynting vector can easily be evaluated by substituting eqns A.34–A.37 into eqn A.42 to obtain:

$$I = \frac{\langle \mathcal{E}(t)^2 \rangle_{\text{rms}}}{Z} = \frac{1}{2} c \epsilon_0 n \mathcal{E}_0^2, \qquad (A.44)$$

where $\langle \mathcal{E}(t)^2 \rangle_{\text{rms}}$ represents the root-mean-square time average. This shows that the intensity of a light wave is proportional to the square of the amplitude of the electric field. The relationship can be generalized for all light waves irrespective of the particular polarization of the beam.

In many topics covered in this book, it will be necessary to treat the refractive index as a complex number. A well-known example of how such a situation arises occurs when treating the propagation of electromagnetic waves through a conducting medium such as a metal. In a conductor, the current density is related to the electric field through the electrical conductivity σ according to:

$$\boldsymbol{j} = \sigma \boldsymbol{\mathcal{E}}. \qquad (A.45)$$

By using this relationship to substitute for $\boldsymbol{j}$ in eqn A.13, and eliminating $\boldsymbol{D}$, $\boldsymbol{B}$, and $\boldsymbol{H}$ in the same way that led to eqn A.25, we obtain:

$$\nabla^2 \boldsymbol{\mathcal{E}} = \sigma \mu_0 \mu_{\text{r}} \frac{\partial \boldsymbol{\mathcal{E}}}{\partial t} + \mu_0 \mu_{\text{r}} \epsilon_0 \epsilon_{\text{r}} \frac{\partial^2 \boldsymbol{\mathcal{E}}}{\partial t^2}. \qquad (A.46)$$

We now look for plane wave solutions of the type given by eqn A.32. Substitution of eqn A.32 into eqn A.46 gives:

$$k^2 = \sigma \mu_0 \mu_{\text{r}} \omega \, \mathrm{i} + \mu_0 \mu_{\text{r}} \epsilon_0 \epsilon_{\text{r}} \omega^2. \qquad (A.47)$$

This can be made compatible with the usual relationship between ω and k given in eqn A.33 by allowing n to be a complex number. The complex refractive index is usually written $\tilde{n}$, and is defined by

$$k = \tilde{n} \frac{\omega}{c}. \qquad (A.48)$$

By combining eqns A.47 and A.48 we obtain:

$$\tilde{n}^2 = \frac{\mu_{\text{r}} \sigma}{\epsilon_0 \omega} \mathrm{i} + \mu_{\text{r}} \epsilon_{\text{r}}, \qquad (A.49)$$

It is shown in Chapter 7 that eqn A.49 is only valid at low frequencies. This is because the AC conductivity at high frequencies is not the same as the DC conductivity that enters eqn A.45.

where we have made use of eqn A.28. This of course reduces to eqn A.31 if we set $\sigma = 0$ and $\mu_{\text{r}} = 1$. The physical significance of the complex refractive index implied by eqn A.49 is developed in more detail in Section 1.3.

The Maxwell equations also allow us to treat the transmission and reflection of light at an interface between two materials. This situation is depicted in Fig. A.2. Part of the beam is transmitted into the medium and the rest is reflected. The solution for an arbitrary angle of incidence

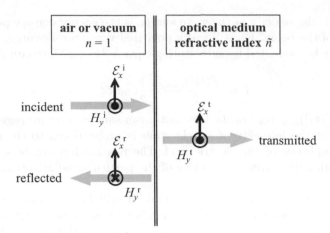

Fig. A.2 Transmission and reflection of light at an interface between air and a medium of refractive index $\tilde{n}$. The incident, transmitted and reflected rays are shown displaced from each other for clarity. All rays are normal to the interface. The symbol $\odot$ for the magnetic fields of the incident and transmitted rays indicates that the field direction is out of the page, while the symbol $\otimes$ for the reflected wave indicates that the field is pointing in to the page.

was treated by Fresnel, and the resulting formulæ are known as **Fresnel's equations**. We restrict ourselves here to the simpler case when the angle of incidence is zero: that is, normal incidence.

We consider again an x-polarized light beam propagating in the z direction, with the field directions as shown in Fig. A.1(a). The electric and magnetic fields are given by eqn A.34. The beam is incident on a medium with complex refractive index $\tilde{n}$. The fields are related to each other through eqn A.36, with Z given by eqn A.37, although we now have to allow for the possibility that n may be complex.

The boundary conditions at the interface between two dielectrics tell us that the tangential components of the electric and magnetic fields are continuous. On applying this to the situation shown in Fig. A.2, we must have that both $\mathcal{E}_x$ and H_y are conserved across the boundary. Hence we can write:

$$\mathcal{E}_x^i + \mathcal{E}_x^r = \mathcal{E}_x^t, \tag{A.50}$$

and

$$H_y^i - H_y^r = H_y^t, \tag{A.51}$$

where the superscript labels i, r, and t refer to the incident, reflected, and transmitted beams respectively. By making use of the relationship between the magnetic and electric fields given in eqns A.36–A.37, we can rewrite eqn A.51 as:

$$\mathcal{E}_x^i - \mathcal{E}_x^r = \tilde{n}\mathcal{E}_x^t, \tag{A.52}$$

where we have assumed that the light is incident from air with $\tilde{n} = 1$ and that $\mu_r = 1$ at the optical frequencies of interest here. Equations A.50 and A.52 can be solved together to obtain

$$\frac{\mathcal{E}_x^r}{\mathcal{E}_x^i} = -\frac{\tilde{n} - 1}{\tilde{n} + 1}. \tag{A.53}$$

This can be re-arranged to obtain the required result for the reflectivity R:

$$R = \left|\frac{\mathcal{E}_x^r}{\mathcal{E}_x^i}\right|^2 = \left|\frac{\tilde{n} - 1}{\tilde{n} + 1}\right|^2. \tag{A.54}$$

For the more general case where the light is reflected at the interface between two materials with complex refractive indices of $\tilde{n}_1$ and $\tilde{n}_2$ respectively, this becomes:

$$R = \left| \frac{\tilde{n}_2 - \tilde{n}_1}{\tilde{n}_2 + \tilde{n}_1} \right|^2 .$$

(A.55)

These formulæ are used in many examples throughout the book.

Further reading

The subject matter of this appendix is standard electromagnetism, and there are numerous books on the market that cover the material, for example: Bleaney & Bleaney (1976), Duffin (1990), Good (1999), Grant & Phillips (1990), or Lorrain, Corson, & Lorrain (2000). The subject is also covered in many optics texts such as Born & Wolf (1999), Hecht (2001), or Smith, King, & Wilkins (2007).

Appendix B

Quantum theory of radiative absorption and emission

B.1 Einstein coefficients   340

B.2 Quantum transition rates   344

B.3 Selection rules   347

Further reading   349

The discussion of the absorption and emission of light by atoms throughout this book presupposes that the reader is familiar with the basic treatment of these processes found in all introductory quantum physics texts. The purpose of this appendix is to give a brief summary of the main results. We begin by discussing the Einstein coefficients in order to introduce the concepts of absorption and emission, and the connection between them. We then move on to discuss how the absorption and emission rates can be calculated by using quantum mechanics. Finally, we briefly discuss the selection rules that apply to radiative transitions. The reader is encouraged to refer to a quantum physics textbook if any of these concepts are unfamiliar.

B.1 Einstein coefficients

The quantum theory of radiation assumes that light is emitted or absorbed whenever an atom makes a jump between two quantum states. These two processes are illustrated in Fig. B.1. Absorption occurs when the atom jumps to a higher level, while emission corresponds to the process in which a photon is emitted as the atom drops down to a lower level. It is customary to label the upper state as level '2' and the lower one as level '1'. Conservation of energy requires that the frequency ν of the photon satisfies:

$$h\nu = E_2 - E_1\,, \qquad (B.1)$$

where E_2 is the energy of level 2 and E_1 is the energy of level 1.

Statistical physics tells us that atoms in excited states have a natural tendency to de-excite and lose their excess energy. Thus the emission of a photon by an atom in an excited state is a spontaneous process. The radiation of light by atoms in excited states is therefore called **spontaneous emission**. This process is illustrated in Fig. B.1(a). One of the electrons in the atom is in level 2 at the start of the process and drops to level 1 by emitting a photon. The frequency of the photon corresponds to the energy difference of the two levels according to eqn B.1. Hence each type of atom has a characteristic emission spectrum determined by its energy levels.

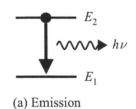

(a) Emission

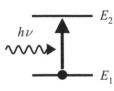

(b) Absorption

Fig. B.1 Optical transitions in an atom: (a) emission, (b) absorption.

The process of **absorption** is illustrated in Fig. B.1(b). The atom is promoted to an excited state by absorbing the required energy from a photon. This promotes an electron from level 1 to level 2. Unlike emission, it is not a spontaneous process. The electron cannot jump to the excited state unless it is stimulated by an incoming photon.

In Section B.2 below we explain how quantum mechanics enables us to calculate the spontaneous emission and absorption rates. At this stage we restrict ourselves to a simpler analysis based on the **Einstein coefficients** for the transition.

Spontaneous emission is governed by the Einstein A coefficient. This gives the probability per unit time that the electron in level 2 will drop to level 1 by emitting a photon. The photon emission rate is therefore proportional to the number of atoms in the excited state and to the A coefficient for the transition. We thus write down the following rate equation for $N_2(t)$, the number of atoms in the excited state:

$$\frac{\mathrm{d}N_2}{\mathrm{d}t} = -A_{21}N_2 \,. \tag{B.2}$$

The subscript '21' on the A coefficient in eqn B.2 makes it plain that the transition starts at level 2 and ends at level 1.

Equation B.2 can be solved for $N_2(t)$ to give:

$$\begin{aligned} N_2(t) &= N_2(0)\ \exp(-A_{21}t) \\ &= N_2(0)\exp(-t/\tau) \,, \end{aligned} \tag{B.3}$$

where

$$\tau = \frac{1}{A_{21}} \,. \tag{B.4}$$

τ is the **natural radiative lifetime** of the excited state. Equation B.4 shows that the number of the atoms in the excited state decays exponentially with a time constant τ due to spontaneous emission. The value of τ for a transition can range from about $1\,\mathrm{ns}$ to several milliseconds. The selection rules that govern whether a particular transition is fast or slow are discussed in Section B.3.

The absorption rate between levels 1 and 2 is governed by the Einstein B coefficient. As mentioned above, this process must be stimulated by the incoming photon field. Following Einstein's treatment, we write the rate of absorption transitions per unit time as:

$$\frac{dN_1}{dt} = -B_{12}N_1 u(\nu) \,, \tag{B.5}$$

where $N_1(t)$ is the number of atoms in level 1 at time t, B_{12} is the Einstein B coefficient for the transition, and $u(\nu)$ is the spectral energy density of the electromagnetic wave at frequency ν in units of $\mathrm{J\,m^{-3}\,Hz^{-1}}$. By writing $u(\nu)$ we are explicitly stating that only the part of the spectrum of the incoming radiation at frequency ν, where $h\nu = E_2 - E_1$, can induce the absorption transitions. Equation B.5 may be considered to be the definition of the Einstein B coefficient.

In the full quantum optical treatment of radiative emission, the photon field is quantized, with a harmonic oscillator energy spectrum given by

$$E_n = (n + \tfrac{1}{2})h\nu \,,$$

n being the number of photons. The factor of $1/2$ corresponds to the zero-point fluctuations of the electromagnetic field. Spontaneous emission is then considered to be a stimulated emission process instigated by the ever-present zero-point fluctuations of the electromagnetic field.

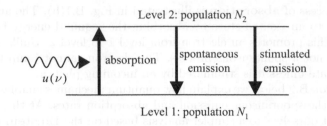

Fig. B.2 Absorption, spontaneous emission, and stimulated emission transitions between two levels of an atom in the presence of electromagnetic radiation with spectral energy density $u(\nu)$.

Stimulated emission is the basis of laser operation. The acronym 'laser' stands for 'light amplification by stimulated emission of radiation'.

The processes of absorption and spontaneous emission that we have described above are fairly intuitive. Einstein realized that the analysis was not complete, and introduced a third type of transition called **stimulated emission**. In this process, the incoming photon field can stimulate downward emission transitions as well as upward absorption transitions. The stimulated emission rate is governed by a second Einstein B coefficient, namely B_{21}. The subscript is now essential to distinguish the B coefficients for the two distinct processes of absorption and stimulated emission.

In analogy with eqn B.5, we write the rate of stimulated emission transitions by the following rate equation:

$$\frac{dN_2}{dt} = -B_{21}N_2u(\nu)\,. \tag{B.6}$$

Stimulated emission is a coherent quantum mechanical effect, in which the photons emitted are in phase with the photons that induce the transition.

The three Einstein coefficients introduced above are not independent parameters: they are all related to each other. If we know one of them, we can work out the other two. To see how this works, we follow Einstein's analysis.

We imagine that the atom is inside a box at temperature T with black walls. The atom will then be bathed in black-body radiation. The black-body radiation will induce both absorption and stimulated emission transitions, while spontaneous emission transitions will also be occurring at a rate determined by the Einstein A coefficient. The three types of transition are indicated in Fig. B.2. If we leave the atom for long enough, it will come to thermal equilibrium with the black-body radiation. In these steady state conditions, the rate of upward transitions due to absorption must exactly balance the rate of downward transitions due to spontaneous and stimulated emission. From eqns B.2, B.5, and B.6, we must therefore have:

The state of equilibrium between the atoms and the radiation occurs whether or not level 1 is the ground state and whether or not transitions take place to and from other levels. The principle of **detailed balance** guarantees that eqn B.7 must hold regardless.

$$B_{12}N_1u(\nu) = A_{21}N_2 + B_{21}N_2u(\nu)\,. \tag{B.7}$$

Since the atoms are in thermal equilibrium with the radiation field at temperature T, the distribution of the atoms among the various energy levels will be governed by the laws of thermal physics. The ratio of N_2

to N_1 will therefore be given by Boltzmann's law:

$$\frac{N_2}{N_1} = \frac{g_2}{g_1} \exp\left(-\frac{h\nu}{k_B T}\right), \tag{B.8}$$

where g_1 and g_2 are the degeneracies of levels 1 and 2 respectively. Now the energy spectrum of a black-body source is given by the Planck formula:

$$u(\nu) = \frac{8\pi h\nu^3}{c^3} \frac{1}{\exp\left(h\nu/k_B T\right) - 1}. \tag{B.9}$$

The only way that eqns B.7–B.9 can be consistent with each other at all temperatures is if:

$$g_1 B_{12} = g_2 B_{21}, \tag{B.10}$$

and

$$A_{21} = \frac{8\pi h\nu^3}{c^3} B_{21}. \tag{B.11}$$

If the atom is embedded within an optical medium with a refractive index n, we replace c by c/n in eqn B.11 to account for the reduced speed of light within the medium.

A moment's thought will convince us that it is not possible to get consistency between the equations without the stimulated emission term. This is what led Einstein to introduce the concept. Equation B.10 tells us that the probabilities for stimulated absorption and emission are the same apart from the degeneracy factors. Furthermore, the interrelationship of the Einstein coefficients tells us that transitions that have a high absorption probability will also have a high emission probability, both for spontaneous processes and stimulated ones.

The relationships between the Einstein coefficients given in eqns B.10 and B.11 have been derived for the case of an atom in equilibrium with black-body radiation. However, once we have derived the interrelationships, they will apply in all other cases as well. This is very useful, because we then only need to know one of the coefficients to work out the other two. For example, we can measure the radiative lifetime to determine A_{21} from eqn B.4, and then work out the B coefficients by using eqns B.11 and B.10.

An important application of the Einstein coefficients is in the analysis of optical amplification in a laser medium. This is achieved when the rate of stimulated emission exceeds the rate of absorption, so that the light intensity increases rather than decreases as it propagates through the medium. We see from eqns B.5 and B.6 that this condition can be written:

$$B_{21} N_2 u(\nu) > B_{12} N_1 u(\nu), \tag{B.12}$$

which, on substituting from eqn B.10, becomes:

$$N_2 > \frac{g_2}{g_1} N_1. \tag{B.13}$$

Equation B.13 describes a non-thermal distribution in which the weighted population of the upper level exceeds that of the lower level. This condition is called **population inversion**, and is a necessary requirement for laser operation.

B.2   Quantum transition rates

The calculation of radiative transition rates by quantum mechanics is based on time-dependent perturbation theory. The light–matter interaction is described by transition probabilities, which can be calculated by using **Fermi's golden rule**. Referring to the radiative processes illustrated in Fig. B.1, we write the rate of transitions as

$$W_{1 \to 2} = \frac{2\pi}{\hbar} |M_{12}|^2 g(h\nu) , \qquad (B.14)$$

where M_{12} is the **matrix element**, and $g(h\nu)$ is the **density of states**.

The density of states is defined so that $g(h\nu)\mathrm{d}E$ is the number of *final* states per unit volume that fall within the energy range E to $E + \mathrm{d}E$, where $E = h\nu$. In the case of radiative transitions between quantized levels in an atom, the initial and final electron states are *discrete*, and the photons are emitted into free space, as illustrated schematically in Fig. B.3. The density of states factor that enters eqn B.14 in this case is therefore the density of *photon* states. The photon density of states in free space is normally written in terms of the frequency rather than the energy, and is given by:

$$g_\nu(\nu) = \frac{8\pi\nu^2}{c^3} . \qquad (B.15)$$

In a medium of refractive index n, c is replaced by c/n. The density of states is thus proportional to the square of the frequency, and therefore the transition rate generally increases with increasing photon frequency.

In solid-state physics, it will frequently be the case that the electron states are no longer discrete, but are broadened into bands. It is then necessary to consider the density of electron states as well as the density of photon states. This electron density of states factor is discussed in detail wherever it occurs in the main text. If both the initial and final levels are in continuous bands, then a suitably weighted joint density of states for both the initial and final bands must be used.

The matrix element that appears in Fermi's golden rule can be written in the compact Dirac notation or explicitly in terms of the overlap integral as:

$$M_{12} = \langle 2|H'|1\rangle = \int \psi_2^*(\boldsymbol{r})H'(\boldsymbol{r})\psi_1(\boldsymbol{r})\,\mathrm{d}^3\boldsymbol{r} , \qquad (B.16)$$

where H' is the perturbation caused by the light wave, $\boldsymbol{r}$ is the position vector of the electron, and $\psi_1(\boldsymbol{r})$ and $\psi_2(\boldsymbol{r})$ are the wave functions of the initial and final states. To evaluate M_{12}, we need to know the wave functions of the states, and also the form of the perturbation due to the light wave.

The perturbation due to the light can be evaluated by calculating the effect of the electromagnetic field on the electron in the atom. From classical electromagnetism we know that the field changes the momentum of a charged particle from $\boldsymbol{p}$ to $(\boldsymbol{p} - q\boldsymbol{A})$, where q is the charge and $\boldsymbol{A}$ is the

The photon density of states in terms of energy is given by:

$$g(h\nu) = g_\nu(\nu)/h^3 .$$

Note that the photon density of states, and hence the radiative emission rate, can be modified by making the atom emit into an optical cavity rather than into free space. (See Fox (2006), chapter 10.)

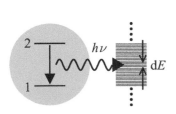

Fig. B.3 Optical transitions between discrete atomic states with photon emission into the continuum of states of free space.

vector potential defined in eqn A.14 of Appendix A. The Hamiltonian for an electron with $q = -e$ in an electromagnetic field is therefore:

$$H = \frac{1}{2m_0}(\boldsymbol{p} + e\boldsymbol{A})^2 + V(\boldsymbol{r})$$

$$= H_0 + \frac{e}{2m_0}(\boldsymbol{p} \cdot \boldsymbol{A} + \boldsymbol{A} \cdot \boldsymbol{p}) + \frac{e^2 \boldsymbol{A}^2}{2m_0}\,. \quad (B.17)$$

Here, H_0 represents the Hamiltonian of the electron before the field was applied, namely:

$$H_0 = \frac{\boldsymbol{p}^2}{2m_0} + V(\boldsymbol{r})\,, \quad (B.18)$$

where $V(\boldsymbol{r})$ is the potential energy of the electron in the atom. The perturbation H' due to the electromagnetic field is thus:

$$H' = \frac{e}{2m_0}(\boldsymbol{p} \cdot \boldsymbol{A} + \boldsymbol{A} \cdot \boldsymbol{p}) + \frac{e^2 \boldsymbol{A}^2}{2m_0}\,. \quad (B.19)$$

For the case of static fields, the first two terms gives rise to the Zeeman effect, while the third causes diamagnetism. We do not consider these effects further here because we are interested in the response of the system to the oscillating fields of a light wave.

The interaction Hamiltonian given by eqn B.19 can be simplified in two ways. Firstly, we can neglect the term in $\boldsymbol{A}^2$ compared to the other two terms on the right-hand side because it is much smaller. Secondly, we can group together the first two terms of H' because the two operators commute. This occurs because $\boldsymbol{\nabla} \cdot \boldsymbol{A} = 0$ in the Coulomb gauge (cf. eqn A.20), and therefore if $\boldsymbol{p} = -i\hbar\boldsymbol{\nabla}$, it follows that $\boldsymbol{p} \cdot \boldsymbol{A} = \boldsymbol{A} \cdot \boldsymbol{p}$.

With these two simplifications, we can write the perturbation due to the light field as:

$$H' = \frac{e}{m_0}\boldsymbol{p} \cdot \boldsymbol{A}\,. \quad (B.20)$$

This is now in a form that can be evaluated for the vector potential of an electromagnetic wave.

If the electric and magnetic fields $\boldsymbol{\mathcal{E}}$ and $\boldsymbol{B}$ of the light wave both vary in time and space as the real part of $\exp i(\boldsymbol{k} \cdot \boldsymbol{r} - \omega t)$ with $\omega = ck$, it follows from eqns A.14 and A.18 that $\boldsymbol{A}$ must also do so. Hence we may write:

$$\boldsymbol{A}(\boldsymbol{r}, t) = \boldsymbol{A}_0 \left(\exp i(\boldsymbol{k} \cdot \boldsymbol{r} - \omega t) + \text{c.c.}\right), \quad (B.21)$$

where from eqn A.18 we see that $\boldsymbol{A}_0$ must be pointing in the same direction as $\boldsymbol{\mathcal{E}}$, namely along the polarization direction.

The exponential in $\boldsymbol{k} \cdot \boldsymbol{r}$ that appears in eqn B.21 should be understood in terms of its Taylor expansion:

$$e^{i\boldsymbol{k} \cdot \boldsymbol{r}} = 1 + i\boldsymbol{k} \cdot \boldsymbol{r} + \tfrac{1}{2}(i\boldsymbol{k} \cdot \boldsymbol{r})^2 + \cdots\,. \quad (B.22)$$

At optical frequencies the wavelength is around 1 μm, and the dimension of a typical atom is $\sim 10^{-10}$ m. Thus $|\boldsymbol{k} \cdot \boldsymbol{r}| \approx 2\pi|\boldsymbol{r}|/\lambda \sim 10^{-3}$. This means that we need only consider the first term in eqn B.22 and can take

The term in $\boldsymbol{A}^2$ corresponds to two-photon rather than one-photon interactions between the atom and the light field. These two-photon interactions are usually very weak, although they may become important with intense laser light fields. The $\boldsymbol{A}^2$ term also causes Rayleigh scattering.

The time dependence of $\boldsymbol{A}$ does not appear here because it has already been assumed that the perturbation has a time dependence of the form $e^{\pm i\omega t}$ in the derivation of Fermi's golden rule.

$\exp i\boldsymbol{k} \cdot \boldsymbol{r} \sim 1$ to a very good approximation. This is called the **electric-dipole** approximation for reasons that will become clear below. In this approximation we need to evaluate matrix elements of the type:

$$M_{12} = \frac{e}{m_0} \langle 2|\boldsymbol{p} \cdot \boldsymbol{A}_0|1\rangle .$$ (B.23)

In the interaction picture we can write the equation of motion of a time-dependent operator $\hat{O}$ as:

$$\frac{\mathrm{d}}{\mathrm{d}t} \hat{O}(t) = \frac{\mathrm{i}}{\hbar}[H_0, \hat{O}] = \frac{\mathrm{i}}{\hbar}(H_0\hat{O} - \hat{O}H_0) .$$ (B.24)

Therefore, with $\boldsymbol{p} = m_0 \mathrm{d}\boldsymbol{r}/\mathrm{d}t$, we have:

$$\begin{aligned}
\langle 2|\boldsymbol{p}|1\rangle &= m_0 \langle 2\,|\mathrm{d}\boldsymbol{r}/\mathrm{d}t|\,1\rangle \\
&= \frac{\mathrm{i}m_0}{\hbar} \langle 2|[H_0, \boldsymbol{r}]|1\rangle \\
&= \frac{\mathrm{i}m_0}{\hbar}(E_2 - E_1)\langle 2|\boldsymbol{r}|1\rangle \\
&= \mathrm{i}m_0\omega_{21}\langle 2|\boldsymbol{r}|1\rangle ,
\end{aligned}$$ (B.25)

where $\hbar\omega_{21}$ is the transition energy. The intermediate step in the third line of eqn B.25 follows because we know from the Schrödinger equation that $H_0\Psi_i = E_i\Psi_i$. Since the $\boldsymbol{A}_0$ that appears in eqn B.23 is just a simple vector and not an operator, we can use eqn B.25 to rewrite eqn B.23 as follows:

$$M_{12} = \mathrm{i}e\omega_{21}\langle 2|\boldsymbol{r} \cdot \boldsymbol{A}_0|1\rangle .$$ (B.26)

Finally, we note from eqns B.21 and A.18 that $\boldsymbol{\mathcal{E}}_0 = \mathrm{i}\omega\boldsymbol{A}_0$. Therefore, if $\omega = \omega_{21}$ as we have been presuming all along, the electric-dipole matrix element is given by:

$$M_{12} = \langle 2|e\boldsymbol{r}{\cdot}\boldsymbol{\mathcal{E}}_0|1\rangle .$$ (B.27)

On comparing eqns B.16 and B.27 we see that in the electric-dipole approximation, the interaction Hamiltonian is just

$$H' = -\boldsymbol{p}_{\mathrm{e}}{\cdot}\boldsymbol{\mathcal{E}}_0 ,$$ (B.28)

where $\boldsymbol{p}_{\mathrm{e}} = -e\boldsymbol{r}$ is the electric dipole moment of the electron. This is in fact exactly equal to the interaction energy we would expect for a dipole in an electric field, and explains why the transitions are called 'electric-dipole' transitions.

The result given in eqn B.27 enables us to evaluate the matrix elements for particular transitions if the wave functions of the initial and final states are known. We can then use Fermi's golden rule to work out $W_{1\rightarrow 2}$ from eqn B.14 in order to obtain the absorption rate, which must be the same as $B_{12}u(\nu)$ in eqn B.5. The spectral energy density $u(\nu)$ of an electromagnetic wave is proportional to $\mathcal{E}_0^2$ (cf. eqns A.43 and A.44), so that we can work out B_{12} from $W_{1\rightarrow 2}$ and A_{21} from eqn B.11.

The final result for transitions between non-degenerate discrete atomic levels by absorption or emission of unpolarized light is:

$$B_{12} = \frac{\pi e^2}{3\epsilon_0\hbar^2} |\langle 2|\boldsymbol{r}|1\rangle|^2 ,$$ (B.29)

and

$$A_{21} = \frac{e^2\omega_{21}^3}{3\pi\epsilon_0\hbar c^3} \left| \langle 2|\boldsymbol{r}|1\rangle \right|^2 . \qquad (B.30)$$

When the levels are degenerate, we must modify eqns B.29 and B.30 to allow for the different transition pathways. For example, if we consider the transitions between atomic levels with quantum numbers j and i, each of which consists of a manifold of degenerate levels labelled by additional quantum numbers m_j and m_i, then eqn B.30 is modified to:

$$A_{ji} = \frac{e^2\omega_{ji}^3}{3\pi\epsilon_0\hbar c^3} \frac{1}{g_j} \sum_{m_j, m_i} \left| \langle j, m_j|\boldsymbol{r}|i, m_i\rangle \right|^2 , \qquad (B.31)$$

where g_j is the degeneracy of the upper state. In solid-state systems with continuous bands, the summation over discrete levels is replaced by the joint density of states for the initial and final bands.

The matrix element for a transition is directly proportional to the **oscillator strength** f_{ij} introduced in the classical treatment of absorption of light by atoms in Section 2.2.2. For transitions between non-degenerate levels, the relationship between them is given by:

$$f_{ij} = \frac{2m\omega_{ji}}{3\hbar} \left| \langle j|\boldsymbol{r}|i\rangle \right|^2 . \qquad (B.32)$$

The oscillator strength was introduced before quantum theory was developed to explain how some atomic absorption and emission lines are stronger than others. With the hindsight of quantum mechanics, it is easy to understand that this is simply caused by the different matrix elements for the transitions.

To conclude this section we note that the quantum optical treatment of the light–matter interaction tells us that the atoms oscillate between the initial and final states while the light field is present. This effect is called Rabi oscillations. We can reconcile this picture with the notion of discrete transitions by realizing that the oscillations can only be observed if the atoms do not scatter in any way. If scattering does occur, the coherence of the oscillations will be broken, and the atom will end up either in the initial or final state, depending on where it is when the scattering event occurs. This then gives the probability that the atom makes a transition to the final state due to the interaction with the light. In the systems we are studying in this book, it will always be the case that the Rabi oscillations are heavily damped, and therefore do not need to be considered.

Two of the powers of ω in eqns B.30 and B.31 come from the photon density of states. (See eqn B.15.) The third comes from the photon energy.

A discussion of optical Rabi oscillations and their connection with the Einstein coefficients may be found in Fox (2006), chapter 9.

B.3 Selection rules

The electric-dipole matrix element given in eqn B.27 can be easily evaluated for simple atoms with known wave functions. This leads to the notion of electric-dipole **selection rules**. These are rules about the quantum numbers of the initial and final states. If the states do not

Table B.1 Electric–dipole selection rules for single electrons. The z axis is usually defined by the direction of an applied static magnetic or electric field. The rule on Δm for circular polarization applies to absorption. The sign is reversed for emission.

Quantum number	Selection rule	Polarization
Parity	changes	
l	$\Delta l = \pm 1$	
m	$\Delta m = +1$	circular: σ^+
	$\Delta m = -1$	circular: σ^-
	$\Delta m = 0$	linear: $\parallel z$
	$\Delta m = \pm 1$	linear: $\parallel (x, y)$
s	$\Delta s = 0$	
m_s	$\Delta m_s = 0$	

satisfy the selection rules, then the electric-dipole transition rate will be zero.

Transitions that obey the electric-dipole selection rules are called **allowed** transitions, while those which do not are called **forbidden** transitions. Electric-dipole-allowed transitions have high transition probabilities, and therefore short radiative lifetimes, typically in the range 1–100 ns. Forbidden transitions, by contrast, are much slower. The different timescales for allowed and forbidden transitions lead to another general classification of spontaneous emission as **fluorescence** and **phosphorescence** respectively. Fluorescence is a 'prompt' process in which the photon is emitted within a few nanoseconds after the atom has been excited, while phosphorescence gives rise to 'delayed' emission which persists for a substantial time.

The electric-dipole selection rules for a single electron in a hydrogenic system with quantum numbers l, m, s, and m_s are summarized in Table B.1. The origin of these rules is as follows:

See Section A.2 in Appendix A for definitions of σ^+ and σ^- circular polarization.

- The parity change rule follows from the fact that the electric-dipole operator is proportional to $\mathbf{r}$, which is an odd function.
- The rule for Δl derives from the properties of the spherical harmonic functions and is consistent with the parity rule because the wave functions have parity $(-1)^l$.
- The rules on Δm can be understood by realizing that σ^+ and σ^- circularly polarized photons carry angular momenta of $+\hbar$ and $-\hbar$ respectively along the z axis, and hence m must change by one unit to conserve angular momentum. For linearly polarized light along the z axis, the photons carry no z component of momentum, implying $\Delta m = 0$, while x or y polarized light can be considered as an equal combination of σ^+ and σ^- photons, giving $\Delta m = \pm 1$.
- The spin selection rules follow from the fact that the photon does not interact with the electron spin, and so the spin quantum numbers never change in the transition.

These selection rules can be generalized to many-electron atoms with quantum numbers (L, S, J) as follows:

(1) The parity of the wave function must change.

(2) $\Delta l = \pm 1$ for the changing electron.

(3) $\Delta L = 0, \pm 1$, but $L = 0 \rightarrow 0$ is forbidden.

(4) $\Delta J = 0, \pm 1$, but $J = 0 \rightarrow 0$ is forbidden.

(5) $\Delta S = 0$.

The parity rule follows from the odd parity of the dipole operator. The rule on l applies the single electron rule to the individual electron that makes the jump in the transition. The rules on L and J follow from the fact that the photon carries one unit of angular momentum. The final rule is a consequence of the fact that the photon does not interact with the spin.

When electric-dipole transitions are forbidden, other types of processes may be possible. For example, **magnetic-dipole** and **electric-quadrupole** transitions are possible between states of the same parity. These processes arise from the higher order terms that we neglected in eqn B.22, and have smaller transition probabilities, leading to longer radiative lifetimes ranging from about 10^{-6} s upwards. In extreme cases it may happen that all standard types of single-photon radiative transitions are forbidden. In this case, the excited state is said to be **metastable**, and the atom must de-excite by transferring its energy to other atoms in collisions, or by some other low probability mechanism that we have not considered here, such as multi-photon emission.

$J = 0 \rightarrow 0$ transitions are strictly forbidden for single-photon transitions

Further reading

This material is covered in most quantum mechanics textbooks. See, for example: Gasiorowicz (1996) or Schiff (1969). More detailed information on atomic selection rules and transition rates may be found in atomic physics texts such as Corney (1977) or Woodgate (1980).

Appendix

C

C.1 Angular momentum in quantum mechanics 350

C.2 Notation for atomic angular momentum states 351

C.3 Sub-level splitting 352

Further reading 353

Angular momentum in atomic physics

The treatment of angular momentum in atomic physics is rather complicated, and often leads to confusion, especially as regards the notation that is used. Since an understanding of atomic angular momentum states is required in several places in this book, a brief summary of the main points is included here.

C.1 Angular momentum in quantum mechanics

In classical physics, it is possible to know the length of an angular momentum vector $\boldsymbol{J}$ and its three components along the Cartesian axes, namely J_x, J_y, and J_z. This is not possible in quantum mechanics, where we can only know $|\boldsymbol{J}|^2$ and one of the components, usually taken to be J_z. For the eigenstates of the angular momentum operators we have:

$$\langle \hat{\boldsymbol{J}}^2 \rangle = J(J+1)\,\hbar^2 \,, \tag{C.1}$$

$$\langle \hat{J}_z \rangle = M_J\,\hbar \,, \tag{C.2}$$

where J and M_J are the appropriate quantum numbers. J can take positive integer or half-integer values, and M_J can take values from $-J$ to $+J$ in integer steps. In Dirac notation, the states with quantum numbers J and M_J are written $|J, M_J\rangle$.

When two quantized angular momenta are added, we cannot apply the simple vector addition rules of classical mechanics, because the length of the resultant vector has to satisfy eqn C.1. Let $\boldsymbol{J}$ be the resultant of two angular momenta $\boldsymbol{J}_1$ and $\boldsymbol{J}_2$ according to:

$$\boldsymbol{J} = \boldsymbol{J}_1 + \boldsymbol{J}_2 \,. \tag{C.3}$$

The quantum number J corresponding to $\boldsymbol{J}$ can only take the values of:

$$(J_1 + J_2), (J_1 + J_2 - 1), \cdots |J_1 - J_2| \,,$$

where J_1 and J_2 are the quantum numbers corresponding to $\boldsymbol{J}_1$ and $\boldsymbol{J}_2$ respectively. This rule is completely general and can be applied to the addition of different types of angular momenta. Several examples are considered in the next subsection.

C.2   Notation for atomic angular momentum states

Quantum mechanics admits of two different types of angular momentum. The first is called **orbital** and is the quantum mechanical counterpart of classical angular momentum. The second is called **spin** and has no classical counterpart.

We start by considering a single electron. There is a general convention in atomic physics that lower case letters refer to single electrons, while upper case letters refer to the equivalent resultant quantities for the whole atom. The orbital angular momentum l is designated by the quantum number l, which can take positive integer values, according to notation given in Table C.1. The z component of l is specified by the quantum number m_l (sometimes just m), which can take integer values from $-l$ to $+l$.

The two quantum numbers that quantify the magnitude and z component of the spin angular momentum s are labelled s and m_s respectively. Experiments on electrons, protons, and neutrons indicate that they each have $s = 1/2$, and therefore $m_s = \pm 1/2$. The m_s states of $\pm 1/2$ are usually called 'spin-up' and 'spin-down' respectively.

When a single particle (e.g. the electron in a hydrogen atom) possesses both orbital and spin angular momentum, we calculate the total angular momentum j according to:

$$j = l + s\,. \tag{C.4}$$

On applying the rule for addition of angular momenta given after eqn C.3, we find that the quantum number j corresponding to j can take values in integer steps from $|l-s|$ to $(l+s)$. For each value of j, the z-component is specified by m_j, which can take values in integer steps from $-j$ to $+j$.

In a many-electron atom, we have to consider how to combine the various types of angular momenta of the individual electrons to find the resultant angular momentum states of the whole atom. The task of doing this is greatly simplified by the fact that the electrons are arranged in shells around the nucleus, and that the total angular momentum of a filled shell is zero. We therefore only have to consider the valence electrons in unfilled shells.

There are a number of different ways to combine the spin and orbital angular momenta in a many-electron atom. The spin and orbital angular momenta interact through the spin–orbit interaction, and the type of coupling that occurs depends on the relative strength of this interaction compared to the others that the electrons experience. These other interactions may be classified into two general categories.

- Internal effects within the atom. The most important of these is the residual electrostatic interaction, which arises from the non–central part of the Coulomb force between the electrons.

- External perturbations from magnetic or electric fields. In ionic crystals, a particularly important source of such perturbations is

In this text we restrict our attention exclusively to electron spin, and so we do not consider here the nuclear spin and its effects, for example, hyperfine structure.

Table C.1 Spectroscopic notation used to designate orbital states.

l	0	1	2	3	4	$\cdots$
Notation	s	p	d	f	g	$\cdots$

For a single electron, the spin–orbit interaction is of the form $\xi l \cdot s$. In a many-electron atom with Russell–Saunders coupling, this generalizes to an interaction of the form $\xi' L \cdot S$.

the electric fields generated by the other charged atoms within the crystal. (See Section 9.3.1.)

The magnitude of the spin–orbit interaction generally increases in approximate proportion to Z^2, where Z is the atomic number, and so the relative sizes of the different types of interaction can change as Z increases. In all isolated atoms with no external perturbations, the spherical symmetry leads to states with well-defined values of the total angular momentum, and hence the quantum number J.

In many isolated atoms, the residual electrostatic interaction is the dominant perturbation, and this leads to a specific type of coupling called **LS coupling** (also called Russell–Saunders coupling.) Here, the total spin and orbital momenta of the atom are determined from:

$$\boldsymbol{L} = \sum_i \boldsymbol{l}_i,\tag{C.5}$$

$$\boldsymbol{S} = \sum_i \boldsymbol{s}_i,\tag{C.6}$$

where the sum is over the valence electrons, and the vector additions are performed according to the rule given after eqn C.3. Once the values of L and S have been determined, the total angular momentum is then worked out from:

$$\boldsymbol{J} = \boldsymbol{L} + \boldsymbol{S}.\tag{C.7}$$

The end result is that for each electronic configuration of the valence electrons, we obtain a set of states labelled by the quantum numbers L, S and J, with J taking values in integer steps from $|L - S|$ to $L + S$. The LS states are split from each other by the residual electrostatic interaction, and the J states corresponding to each LS term are split by the smaller spin–orbit interaction. The $|L, S, J\rangle$ states are written in spectroscopic notation as:

$$|L, S, J\rangle \equiv {}^{(2S+1)}\mathrm{L}_J,\tag{C.8}$$

where the capital roman letter L indicates the value of L according to the convention given in Table C.1. The factor $(2S+1)$ indicates the spin multiplicity: there are $(2S+1)$ M_S states available for each value of S.

In atoms with two valence electrons, there are two possible values of S, namely 0 and 1. Terms with $S = 0$ and $S = 1$ are called **singlets** and **triplets** respectively. Singlet and triplet terms are split by the exchange energy, which originates from the residual electrostatic interaction. The difference between singlet and triplet states is very important when considering the two electrons that form molecular bonds. (See Section 8.2.1.)

C.3   Sub-level splitting

In spherical symmetry, the sub-levels of a particular angular momentum state are degenerate. These sub-levels can be split when external perturbations (eg magnetic fields) break the symmetry, and it is for this

The upper-case lettering used here contrasts with the lower-case notation used in eqn C.4, which applies to an individual electron. The group of J states with the same LS values are called **atomic terms**, while the individual J states are called **levels**.

reason that the quantum numbers that designate the z-component of the angular momentum (i.e. m_l, m_s, M_J, etc.) are often called *magnetic* quantum numbers.

The splitting of the J states of an atom into sub-levels by a magnetic field is called the **Zeeman effect**. The field direction defines the z axis, and the energy shift is given by:

$$\Delta E = g_J \mu_B B_z M_J, \qquad (C.9)$$

where g_J is the g-factor of the level, and B_z is the magnitude of the field. This splits the otherwise degenerate M_J states into $(2J+1)$ sub-levels with equal separations between them, as shown, for example, in Fig. C.1. If LS coupling applies, g_J is given by the Landé g factor:

$$g_J = 1 + \frac{J(J+1) + S(S+1) - L(L+1)}{2J(J+1)}. \qquad (C.10)$$

Note that $g_J = 1$ when $S = 0$ (pure orbital angular momentum: $J = L$), while $g_J = 2$ when $L = 0$ (pure spin angular momentum: $J = S$).

The shift of the atomic levels in an electric field is called the **Stark effect**. The way in which the levels shift depends on the relative strength of the perturbation. At small fields, the energy shift normally varies quadratically with the field strength $\mathcal{E}_z$, with:

$$|\Delta E| \propto \mathcal{E}_z^2 M_J^2. \qquad (C.11)$$

The energy shift can be either positive or negative, and the degeneracy of the sub-levels is only partially lifted, since states with equal $|M_J|$ experience the same shift. The behaviour changes if the field strength is increased to the point that the interaction energy is larger than the unperturbed energy difference between the particular state and another one with the opposite parity. In this limit, the energy shift varies linearly with $\mathcal{E}_z$. This change from the quadratic to the linear Stark effect can readily be observed in semiconductor quantum wells. (See Section 6.5.)

The angular momentum coupling scheme in a crystal can differ from that of the free atom if the electric fields generated by the other atoms in the crystal cause a strong perturbation. This point is particularly important when considering the optical properties of paramagnetic ions embedded within a host crystal. (See Section 9.3.) The shifting and splitting of the levels caused by the crystal-field interaction is called the crystal-field effect.

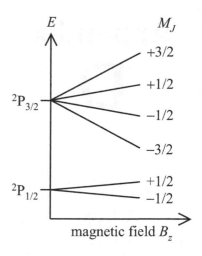

Fig. C.1 Zeeman splitting of the $^2P_{3/2}$ and $^2P_{1/2}$ terms of an alkali atom. These two terms have Landé g factors of 4/3 and 2/3 respectively.

Further reading

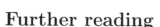

The quantum theory of angular momentum is covered in most quantum mechanics textbooks, for example: Gasiorowicz (1996), Miller (2008), or Schiff (1969). More detailed information on angular momentum states in atoms may be found in atomic physics texts such as Corney (1977) or Woodgate (1980).

Appendix D

Band theory

The electronic states of crystals are described by the band theory of solids. This subject is covered extensively in all solid-state physics texts, and this book presupposes a working knowledge of the main concepts. The purpose of this appendix is to give a brief review of the key points, and to give an overview of the band structure of a few representative materials. The reader is referred to the texts listed under Further Reading for a more extensive treatment.

D.1 Metals, semiconductors, and insulators   354

D.2 The nearly free electron model   356

D.3 Example band structures   359

Further reading   362

D.1 Metals, semiconductors, and insulators

The general concept of band formation is described in Section 1.5.2. The outer orbitals of the atoms in a densely packed solid overlap with each other as the chemical bonds that hold the crystal together are formed. This causes the discrete energy levels of the free atoms to be broadened into bands, as shown schematically in Fig. 1.9. Each band can contain $2N$ electrons, where N is the number of primitive unit cells in the crystal, and the factor of two accounts for the electron spin degeneracy. Electrons fill up the bands to the **Fermi energy** E_F, which is determined by the total electron density.

Figure D.1(a) shows a schematic energy diagram for a monovalent metal such as sodium, or a trivalent metal such as aluminium. These have an odd number of electrons per atom, which means that the highest occupied band will only be half full. The Fermi energy will therefore lie in the middle of the highest occupied band. Electrons with energy just below E_F can easily be excited to empty states just above E_F. This makes it easy to accelerate the electrons with an electric field, and explains why metals are good conductors of electricity.

Figure D.1(b) shows the equivalent energy-level diagram for a semiconductor such as silicon or an insulator such as diamond. These have an even number of electrons per atom, and the highest occupied band is therefore full of electrons. This highest occupied band is called the **valence band**, while the lowest unoccupied band is called the **conduction band**. The Fermi level lies somewhere within the energy gap between the valence and conduction bands. The first empty states for the electrons are in the conduction band, and it requires a minimum amount of energy equal to the **band gap** E_g to excite the electrons to available states. This makes it difficult to accelerate the electrons when

The situation in divalent metals such as magnesium is slightly more complicated. These have an even number of electrons per atom, and so it might be expected that they would behave as semiconductors or insulators. This does not happen because the valence and conduction bands overlap with each other. Electrons below the Fermi level can therefore be excited to empty levels without having to cross an energy gap. This makes them good conductors like the monovalent or trivalent metals.

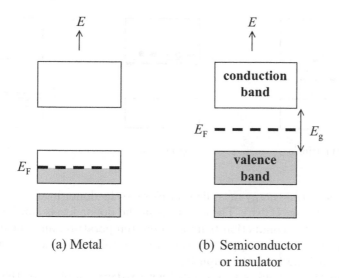

Fig. D.1 Energy level diagrams for (a) a monovalent or trivalent metal, and (b) a semiconductor or insulator. The bands are filled with electrons up to the Fermi level E_F. This is indicated by the shading.

an external electric field is applied, and inhibits the flow of electrical current through the sample. Semiconductors and insulators therefore have much lower electrical conductivities than metals.

The distinction between an insulator and a semiconductor is related to the size of the band gap. Semiconductors have smaller band gaps than insulators. This makes it possible that a significant number of electrons are excited thermally from the valence band to the conduction band at room temperature. The free electrons in the conduction band can conduct electricity easily in the same manner as the free electrons in metals. Semiconductors therefore have a higher conductivity than insulators, but a smaller conductivity than metals because of the smaller number of free electrons.

Figure D.2 shows the conduction and valence bands of a semiconductor in more detail. Figure D.2(a) applies to a pure crystal. This is the same as the case considered in Fig. D.1(b), and the Fermi level lies about half way between the two bands. The thermal excitation of an electron to the conduction band leaves an empty state called a **hole** in the valence band. The hole is equivalent to the absence of an electron, and behaves like a free positive charge. Electrical current can be carried both by the electrons in the conduction band and by the holes in the valence band. The conduction by the thermally excited electrons and holes in a pure crystal is called **intrinsic**.

Figure D.2(b) applies to a crystal with **n-type** doping. In this case, impurities with extra electrons are deliberately introduced into the crystal while it is being grown. For semiconductor crystals such as silicon and germanium that come from group IV in the periodic table, and thus have four valence electrons per atom, this is typically done by adding atoms from group V. The impurities donate one extra electron for each dopant atom. These extra electrons lie in **donor** levels just below the conduction band. The electron states are filled up to the donor levels,

Fig. D.2 Valence and conduction bands for: (a) a pure (intrinsic) semiconductor; (b) a semiconductor with n-type doping; and (c) a semiconductor with p-type doping. The • and ○ symbols represent free electrons and holes respectively. The Fermi energies of the n- and p-type samples lie very close in energy to the impurity levels arising from the donor or acceptor atoms respectively.

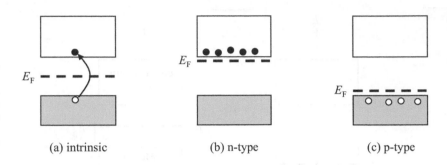

(a) intrinsic   (b) n-type   (c) p-type

and so the Fermi energy must lie very close to the donor level energies, as indicated in Fig. D.2(b). The electrons in the donor levels can be easily excited into the conduction band at room temperature, and the electrical properties of the n-type material are determined by these **extrinsic** electrons arising from the impurity atoms.

Figure D.2(c) applies to a crystal with **p-type** doping. In this case, atoms which have a deficit of one electron per atom are doped into the crystal during its growth. In the case of the group IV elemental semiconductors such as silicon, this is done by doping with atoms from group III. Each of the impurity atoms can accept one electron from the valence band. The **acceptor** levels are just above the top of the valence band, and electrons can easily be excited to these empty states at room temperature. This creates a population of free holes in the valence band that determines the extrinsic electrical conductivity of the sample.

D.2   The nearly free electron model

The motion of free electrons and holes is determined by the E–k dispersion of the solid. If the electrons are completely free, they will only possess kinetic energy. In this case the E–k dispersion is given by:

$$E = \frac{p^2}{2m_0} = \frac{\hbar^2 k^2}{2m_0}, \tag{D.1}$$

where $p = h/\lambda_e = \hbar k$ is the momentum of the electron, λ_e being its de Broglie wavelength.

In a crystal this E–k dispersion relationship is modified because the electrons are not really free. Each atom possesses a certain number of valence electrons. These are the electrons in the outermost atomic shells that determine the chemical properties. The **nearly free electron model** starts by assuming that the valence electrons are released from their atoms and move through the crystal. This leaves behind a regular lattice of positively charged ions. The potential of the ion cores perturbs the motion of the electrons and alters the E–k relationship.

Figure D.3 shows the typical band structure of a simple crystal. The E–k diagram is divided into different **Brillouin zones**, each of which spans a reciprocal lattice vector $\boldsymbol{G}$. This division of $\boldsymbol{k}$-space into Brillouin

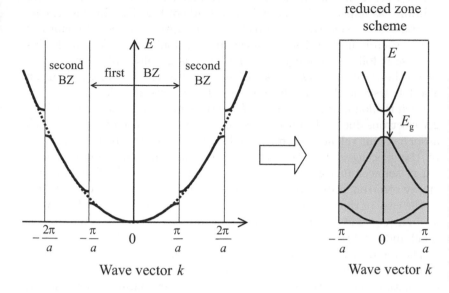

Fig. D.3 Band structure of a simple cubic solid with a lattice constant of a. The left-hand side of the figure shows the E–k relationship across several Brillouin zones, while the right-hand side replots the same band structure in the reduced zone scheme. The dotted line indicates the parabolic dispersion of free electrons given by eqn D.1. Each band within a Brillouin zone can hold $2N$ electrons, where N is the number of unit cells in the crystal. The shading signifies that the band states are filled with electrons. The case shown applies to an atom such as silicon with four valence electrons.

zones reflects the underlying periodicity of the crystal. The reciprocal lattice vectors are defined by

$$e^{i\boldsymbol{G}\cdot(\boldsymbol{r}+\boldsymbol{T})} = e^{i\boldsymbol{G}\cdot\boldsymbol{r}}, \tag{D.2}$$

where $\boldsymbol{T}$ is one of the primitive lattice translation vectors of the crystal.

The E–k relationship shown in Fig. D.3 applies to a simple cubic crystal with a lattice constant of a. In this case eqn D.2 is satisfied with

$$\boldsymbol{G} = \frac{2\pi}{a}(n_x, n_y, n_z), \tag{D.3}$$

where n_x, n_y, and n_z are integers. The band dispersion is drawn for the (100) direction in $\boldsymbol{k}$-space. The central Brillouin zone enclosing the origin runs from $-\boldsymbol{G}/2$ to $+\boldsymbol{G}/2$, namely from $-\pi/a$ to $+\pi/a$. The next zone encloses the section of $\boldsymbol{k}$-space between $\pi/a \to 2\pi/a$ and $-\pi/a \to -2\pi/a$, and so on.

The dotted line in the left-hand side of Fig. D.3 indicates the band dispersion for free electrons that obey eqn D.1. The solid lines indicate the band dispersion for the nearly free electrons. The periodic potential causes a splitting at the zone boundaries, but has a relatively small effect at other points in the Brillouin zone. The band dispersion therefore only departs significantly from that of free electrons near a zone boundary.

The group velocity of the electron wave is given by

$$v_g = \frac{d\omega}{dk} = \frac{1}{\hbar}\frac{dE}{dk}. \tag{D.4}$$

The bands bend over as they approach zone boundaries, as shown in Fig. D.3, so that $dE/dk = 0$ precisely at a zone boundary. This means that the group velocity is zero, which corresponds to a standing wave.

conduction band

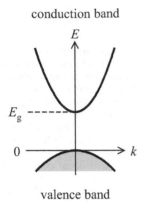

valence band

Fig. D.4 Band dispersion of a semiconductor or insulator near the top of the valence band and the bottom of the conduction band. The energy axis is defined so that $E = 0$ corresponds to the top of the valence band.

Band structure diagrams are usually plotted in the **reduced zone scheme**, as indicated by the right-hand side of Fig. D.3. In this scheme we translate the electron wave vector by an integer number of reciprocal lattice vectors until it lies within the first Brillouin zone. We can do this because it follows from eqn D.2 that there is no physical difference between the wave vectors $\boldsymbol{k}$ and $\boldsymbol{k}+\boldsymbol{G}$ in a periodic crystal. This conclusion also follows from Bloch's theorem, which is discussed below.

Each Brillouin zone contains N $\boldsymbol{k}$-vector states, and can therefore hold $2N$ electrons due to the up–down spin degeneracy of each $\boldsymbol{k}$-state. If each atom has four valence electrons, the first two bands will be full, as shown by the shading in Fig. D.3. This is the situation that applies to four-valent semiconductors such as silicon or germanium. The first available empty electron states are in the next band. This therefore corresponds to the case of the semiconductor or insulator shown in Fig. D.1(b), with an energy gap of E_g between the occupied electron states in the valence band and the first empty states in the conduction band.

The band dispersion near $k = 0$ for a semiconductor or insulator is shown in more detail in Fig. D.4. The top valence band and the lowest conduction band states are shown. The bands are parabolic for small k, and have dispersions given by:

$$E_\mathrm{c}(k) = E_\mathrm{g} + \frac{\hbar^2 k^2}{2m_\mathrm{e}^*}$$
$$E_\mathrm{v}(k) = -\frac{\hbar^2 k^2}{2m_\mathrm{h}^*}, \tag{D.5}$$

where $E = 0$ corresponds to the top of the valence band. The subscripts on the energy identify the conduction and valence bands respectively. Equation D.5 shows that the band dispersion is determined by the **effective mass** m_e^* and m_h^* of the appropriate band.

In general, the effective mass is defined by the curvature of the E–k diagram according to:

$$m^* = \hbar^2 \left(\frac{\mathrm{d}^2 E}{\mathrm{d}k^2} \right)^{-1}. \tag{D.6}$$

The effective mass is therefore a band structure parameter that quantifies the departure of the E–k relationship from the free electron dispersion. It will generally be the case that neither m_e^* nor m_h^* are equal to the free electron mass m_0, and that each material will be different. The negative curvature of the valence band indicates that it is a hole state: hence the 'h' subscript for the effective mass in the valence band. Electrons in the conduction band behave like negatively charged particles of mass m_e^*, while the holes in the valence band behave like positively charged particles of mass m_h^*. Tables D.1 and D.2 give the values of the effective masses for a few important semiconductors.

The nearly free electron approach can be connected to the atomic states of the atoms from which the valence electrons are derived through **Bloch's theorem** (cf. Section 1.5.2):

> *The eigenfunctions of the wave equation for a periodic po-*
> *tential are the product of a plane wave and a function that*
> *has the periodicity of the crystal lattice.*

This implies that the wave function of an electron in a periodic lattice takes the form:

$$\psi(\boldsymbol{r}) = u(\boldsymbol{r})\,\mathrm{e}^{\mathrm{i}\boldsymbol{k}\cdot\boldsymbol{r}}\,. \tag{D.7}$$

where $u(\boldsymbol{r})$ must satisfy $u(\boldsymbol{r}) = u(\boldsymbol{r}+\boldsymbol{T})$. The Bloch functions are therefore modulated plane waves. The envelope function $u(\boldsymbol{r})$ is a wavelike periodic function that relates to the atomic character of the valence electrons. This link is formalized in the tight-binding approach to band structure calculations.

D.3 Example band structures

The band structure of aluminium is shown in Fig. 7.3. Aluminium is a trivalent metal with three valence electrons in a configuration of $3\mathrm{s}^2 3\mathrm{p}^1$. The band structure looks much more complicated than Fig. D.3, but this is mainly a consequence of the way band diagrams are drawn. To understand how the diagram works, we first need to consider the shape of the Brillouin zone in three dimensions.

Aluminium has a **face-centred cubic** (fcc) lattice. The cubic unit cell of an fcc lattice is not the primitive unit cell as it contains four lattice points: one at the origin and three others at the centre of the cube faces with coordinates $(1/2, 1/2, 0)$, $(1/2, 0, 1/2)$, and $(0, 1/2, 1/2)$. The Brillouin zone of the fcc lattice is therefore not a cube, but rather has the shape shown in Fig. D.5. The dispersion of the bands shown in Fig. 7.3 begins by plotting the energy for increasing $\boldsymbol{k}$ outwards from the origin to the X point. We then move across to the L point via the W point, and back to the origin. Finally we go out again from the origin to the X point via the K point.

The departure from the free electron dispersion is actually very small in aluminium. Most of the band diagram can be explained by taking the parabolic dispersion of free electrons shown by the dotted line in Fig. D.3 and folding it back into the complicated shape of the fcc Brillouin zone. The changes in the curvature of the bands at the zone boundaries are then merely caused by the change of direction in which we are moving around the Brillouin zone. Note however that there are small gaps between most bands at the zone edges. These are the band gaps introduced by the lattice potential.

The band structure of copper is shown in Fig. 7.5. Copper has the fcc crystal structure and is a transition metal with an electronic configuration of $3\mathrm{d}^{10}4\mathrm{s}^1$. The band structure is more complicated than that of aluminium because of the need to include the dispersion of both the 3d and 4s bands, which overlap in energy. The 4s bands are approximately parabolic, but the 3d bands are fairly flat. This is a consequence of the strong localization of the d electrons, which means that their orbitals do not overlap much in the crystal. The low dispersion d bands have a

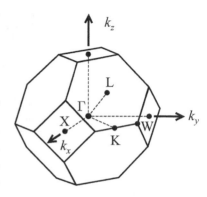

Fig. D.5 Brillouin zone of a face-centred cubic lattice. a is the size of the cubic unit cell in the crystal. High symmetry points within the Brillouin zone are given symbolic names derived from group theory. The origin at $\boldsymbol{k} = (0,0,0)$ is called the Γ point. The X point at $(2\pi/a)(1,0,0)$ identifies the zone edge along the six equivalent (100) directions. The L point at $(\pi/a)(1,1,1)$ is at the zone edge along the eight equivalent (111) directions. An arbitrary point along the $\Gamma \to$ X direction is labelled Δ, while an arbitrary point along the $\Gamma \to$ L direction is given the symbol Λ. The K point corresponds to the zone edge along the (110) direction, while the W point is at the intersection of the square and hexagonal faces of the polygon. The X point corresponds to a wave vector of $2\pi/a$ rather than π/a as Fig. D.3 would suggest because the cubic unit cell of the fcc lattice is not primitive.

Table D.1 Band structure parameters for the group IV elements diamond, silicon, and germanium. All three materials crystallize with the diamond structure and have indirect gaps. E_g^{ind}: indirect band gap; k_{min}: position of the conduction band minimum within the Brillouin zone; valley degeneracy: number of equivalent conduction band minima within the Brillouin zone; $m_e^*(\parallel)$: longitudinal electron effective mass; $m_e^*(\perp)$: transverse electron effective mass; E_g^{dir}: direct band gap at the Γ point; Δ: spin–orbit splitting at the Γ point; m_{hh}^*: heavy-hole effective mass; m_{lh}^*: light-hole effective mass; m_{so}^*: split-off hole effective mass. The effective masses are expressed in units of the free electron mass m_0. The valence band parameters refer to the four-band model shown in Fig. 3.5. After Madelung (1996).

Property	Diamond	Silicon	Germanium
E_g^{ind} (eV) (300 K)	5.47	1.12	0.66
E_g^{ind} (eV) (0 K)	5.5	1.17	0.74
k_{min}	0.76 X	0.85 X	L
Valley degeneracy	6	6	4
$m_e^*(\parallel)$	1.4	0.92	1.58
$m_e^*(\perp)$	0.36	0.19	0.08
E_g^{dir} (eV) (300 K)	6.5	4.1	0.805
Δ (eV)	0.006	0.044	0.29
m_{hh}^*	1.08	0.54	0.3
m_{lh}^*	0.36	0.15	0.04
m_{so}^*	0.15	0.23	0.095

high density of states within a relatively narrow range of energies. These states are very important for both the optical and magnetic properties.

The band structure of the semiconductor silicon is shown in Fig. 3.13. Silicon has four valence electrons and crystallizes with the **diamond structure**. The diamond structure consists of two identical interlocking fcc lattices displaced from each other by $(a/4, a/4, a/4)$. The structure is fcc with a basis of two atoms attached to each lattice point: one at the lattice point itself, and the other displaced by $(a/4, a/4, a/4)$ with respect to it. Silicon therefore has an fcc Brillouin zone, as shown in Fig. D.5.

The real band structure of silicon can be compared to the schematic band dispersions shown in Figs D.3 and D.4. We see that the real material does show the general behaviour indicated by the schematic diagrams, although the actual band diagram is more complicated. One significant difference is the 'camel-back' shape of the conduction band, which means that the minimum of the conduction band occurs near the X point rather than at the Γ point. The band gap of silicon is therefore indirect. This has very important consequences for the optical properties, as discussed in Chapters 3 and 5. Another important difference is the degeneracy of the valence band states at the Γ point. This is usually described in terms of the four-band model shown in Fig. 3.5. The parameters needed to describe the valence band of silicon in this way are listed in Table D.1. The spin–orbit splitting of silicon is too small to be apparent in the low scale band diagram given in Fig. 3.13. Table D.1 also lists the effective masses that describe the conduction band minima near the X point. Note that we must use two separate effective masses to parameterize the anisotropy of the conduction band minimum.

The band structure of germanium is given in Fig. 3.10. Germanium

Table D.2 Band structure parameters for selected direct gap III–V semiconductors with the zinc-blende structure. The parameters listed refer to the four-band model shown in Fig. 3.5. E_g: band gap; Δ: spin–orbit splitting; m_e^*: electron effective mass; m_{hh}^*: heavy-hole effective mass; m_{lh}^*: light-hole effective mass; m_{so}^*: split-off hole effective mass. The effective masses are expressed in units of the free electron mass m_0. After Madelung (1996) and Madelung (1982).

Crystal	E_g (eV) (0 K)	E_g (eV) (300 K)	Δ (eV)	m_e^*	m_{hh}^*	m_{lh}^*	m_{so}^*
GaAs	1.519	1.424	0.34	0.067	0.5	0.08	0.15
GaSb	0.81	0.75	0.76	0.041	0.28	0.05	0.14
InP	1.42	1.34	0.11	0.077	0.6	0.12	0.12
InAs	0.42	0.35	0.38	0.022	0.4	0.026	0.14
InSb	0.24	0.18	0.85	0.014	0.4	0.016	0.47

Table D.3 Structure and band gap data for a number of common semiconductors. E_g is the band gap at 300 K, and the i/d label indicates whether the gap is indirect or direct. SiC crystallizes in more than 200 different modifications, and the data listed here is for the 6H polytype, which has a hexagonal unit cell. ZnS, ZnSe, CdS and CdSe can form stable crystals with either hexagonal or cubic unit cells, and the band gap may be slightly different for the two structural variations. The negative band gap of HgTe signifies that it is a semimetal: the top of the valence band is at higher energy than the bottom of the conduction band. Note that the band gap of InN was originally thought to be around 2 eV. Recent work has now confirmed the value given in this table. After Madelung (1996).

Compound	Crystal structure	E_g (eV)	Type
SiC	6H polytype	2.9	i
AlN	wurtzite	6.2	d
AlP	zinc blende	2.41	i
AlAs	zinc blende	2.15	i
AlSb	zinc blende	1.62	i
GaN	wurtzite	3.44	d
GaP	zinc blende	2.27	i
InN	wurtzite	0.7	d
ZnO	wurtzite	3.4	d
ZnS	wurtzite or zinc blende	3.8 or 3.7	d
ZnSe	wurtzite or zinc blende	2.8 or 2.7	d
ZnTe	zinc blende	2.3	d
CdS	wurtzite or zinc blende	2.5	d
CdSe	wurtzite or zinc blende	1.8	d
CdTe	zinc blende	1.5	d
HgTe	zinc blende	−0.14	semimetal
CuCl	zinc blende	3.17	d
Cu_2O	cuprite	2.2	d

lies one line below silicon in the periodic table, and, like silicon, has the diamond crystal structure. It is not surprising, therefore, that the band structures are fairly similar. There are, however, a number of important differences. Most prominent among these is the fact that the conduction band minimum is at the L point, rather than near the X point. Moreover, the minimum at the Γ point is only just above the one at the L point. The band gap is therefore still indirect, but the optical transitions soon become direct as the photon energy is increased above E_g. The principal band structure parameters of germanium are listed in Table D.1.

The band structure of the III–V compound semiconductor gallium arsenide, which has the **zinc-blende structure**, is given in Fig. 3.4. The zinc-blende structure is similar to the diamond structure, except that the atom at $(a/4, a/4/a/4)$ is different to the one at $(0,0,0)$. The band structure is quite similar to that of germanium, except that the conduction band minimum now lies at the Γ point. GaAs therefore has a direct band gap. This means that GaAs crystals can emit light efficiently when excited, as discussed in Chapter 5. Table D.2 lists the most significant band parameters for GaAs, together with those of a few other direct gap III–V materials.

Less detailed band structure data on other important compound semiconductors are given in Table D.3. Some of these crystals have the zinc-blende structure, while a number of others have the **wurtzite structure**, which has hexagonal symmetry. Several of the II–VI compounds can form stable crystals with either structure, and the band gap may be slightly different between the cubic and hexagonal forms. Cu_2O has its own particular structure, not surprisingly named the cuprite structure. The cuprite structure has cubic symmetry.

Further reading

An introductory treatment of band theory is given in Rosenberg (1988). More detailed accounts are given in Ashcroft and Mermin (1976), Burns (1985), Ibach and Luth (2003), Kittel (2005), or Singleton (2001), and many other texts on solid-state physics.

Semiconductor p–i–n diodes

The p–i–n structure is used extensively in semiconductor opto-electronic devices such as photodiodes, solar cells, light-emitting diodes and optical modulators. The structure is shown schematically in Fig. E.1. It consists of a standard semiconductor p–n diode with a thin undoped i-region of thickness l_i inserted at the junction. This i-region is the optically active part of the diode. The purpose of the p–n junction is to control the number of electrons and holes injected into the active region, and to permit the application of strong electric fields.

In this appendix we discuss the band alignment and electrostatics of the p–i–n structure when an external bias voltage V_0 is applied to the device. The formation of the **depletion region** at the junction is an essential feature of the physics of the p–n diode. The external bias is dropped across the depletion region because it has a very high resistance compared to the heavily doped p- and n-regions. The width of the depletion region at a given voltage is determined by the doping levels in the p- and n-regions, with higher doping giving thinner depletion widths. In a p–i–n structure the residual doping level in the i-region is very small, and so the depletion region can extend across the whole i-region. The extension of the depletion region into the p- and n-regions is very small in comparison to l_i because of the heavy doping level in the contacts. This means that any external voltages that are applied will be dropped almost entirely across the i-region.

Figure E.2 shows a band diagram of a p–i–n structure. Figure E.2(a) shows the band alignments at zero bias, while Fig. E.2(b) applies to the situation with an external voltage applied. At zero bias the Fermi levels of the p- and n- regions align with each other. The energy difference between the Fermi energies and the conduction or valence band edges is small compared to the band gap, and so it is apparent from Fig. E.2(a) that there is a voltage drop of magnitude E_g/e across the i-region. This is equivalent to the built-in voltage V_{bi} that is important for the functioning of solar cells as discussed in Section 3.7. When the bias is applied, the energy difference between the Fermi levels of the p- and n-regions will be equal to $|eV_0|$. This is illustrated in Fig. E.2(b) for the case of reverse bias, that is, when a negative voltage is applied to the p-region with respect to the n-region. Reverse bias tends to increase the voltage drop across the i-region, while forward bias tends to reduce it.

In order to calculate the electric field across the i-region, we need to

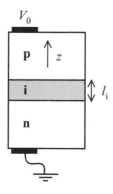

Fig. E.1 Schematic diagram of a p–i–n diode. The bias voltage V_0 is applied to the p-region, so that positive and negative V_0 correspond to forward and reverse bias respectively. The dimensions are not drawn to scale. The i-region is typically only a few microns thick.

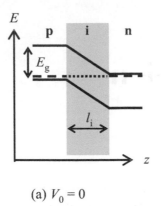
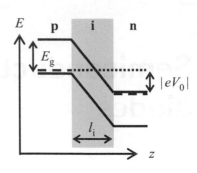

Fig. E.2 Band alignments in a p–i–n diode structure with an i-region thickness of l_i. (a) Bias = 0. (b) Reverse bias voltage V_0 applied. The thick dashed lines indicate the Fermi levels of the doped layers, which lie just above the valence band or just below the conduction band in the p- and n-regions respectively. E_g is the band gap of the semiconductor used for the p- and n-regions.

(a) $V_0 = 0$

(b) Reverse bias V_0 applied

solve Poisson's equation (Appendix A, eqn A.5):

$$\nabla^2 V = -\frac{\varrho}{\epsilon_r \epsilon_0}, \tag{E.1}$$

where V is the voltage and ϱ is the electric charge density. We set up axes so that z is the direction normal to the plane of the diode. We know from the symmetry that the derivatives in the x and y planes must be zero. Poisson's equation therefore reduces to:

$$\frac{\partial^2}{\partial z^2} V(z) = -\frac{\varrho(z)}{\epsilon_r \epsilon_0}. \tag{E.2}$$

We assume that $\varrho = 0$ in the i-region because it is undoped and is fully depleted of all free carriers. The solution to eqn E.2 in the i-region is therefore $V(z) = C_1 z + C_2$, where C_1 and C_2 are constants. The electric field strength can be calculated from eqn A.7 in Appendix A. This gives $\mathcal{E} = -\mathrm{d}V/\mathrm{d}z = -C_1$, which implies that the electric field is constant in the i-region.

We can see from Fig. E.2(b) that the magnitude of the voltage drop across the i-region is approximately equal to $(V_{bi} - V_0)$. The value of the constant electric field in the i-region is therefore given by:

$$\mathcal{E} = \frac{V_{bi} - V_0}{l_i}. \tag{E.3}$$

This shows that negative bias increases the field across the i-region, while a small forward bias reduces $\mathcal{E}$. The field is zero for a forward bias of V_{bi}. At zero bias the field across the i-region is equal to V_{bi}/l_i. This can be a large value. For example, with a GaAs diode we might have $V_{bi} = 1.5\,\mathrm{V}$ and $l_i = 1\,\mu\mathrm{m}$. The field at zero bias is therefore $1.5 \times 10^6\,\mathrm{V\,m^{-1}}$.

Reverse-biased p–i–n structures are also used in semiconductor photodiodes and solar cells, as discussed in Section 3.7. Forward-biased p–i–n structures are used in electroluminescent devices: see Section 5.4.

In this book we frequently come across reverse-biased p–i–n structures in the context of studying the effects of electric fields on the optical properties of semiconductors. Section 3.3.5 discusses the effect of an electric field on the band edge absorption of a bulk semiconductor, while Sections 4.3.1 and 6.5 discuss the effects on the excitons in bulk semiconductors and quantum wells, respectively. In all of these examples, the calibration of the electric field strength from the applied bias is carried out by using eqn E.3.

Further reading

The physics of the p–n junction is discussed in most solid-state physics texts. See, for example: Bleaney and Bleaney (1976), Rosenberg (1988), or Sze (1985). The p–i–n structure is described in detail in Sze (1981).

Solutions to exercises

Chapter 1

(1.1) $R = 0.041$ and $T = 0.92$, assuming that $\alpha = 0$ because the glass is transparent.

(1.2) 2.1.

(1.3) $v = 9.97 \times 10^7\,\mathrm{m\,s^{-1}}$, $\alpha = 9.6 \times 10^6\,\mathrm{m^{-1}}$, and $R = 25.6\%$.

(1.4) $18\,\mu\mathrm{m}$.

(1.5) $T = 0.034$, optical density $= 1.1$.

(1.6) $\tilde{\epsilon}_\mathrm{r} = 1.77 + \mathrm{i}\,9.2 \times 10^{-8}$.

(1.7) Absorbing for blue light, but not for red and green light.

(1.8) (a) $T = (1-R_1)(1-R_2)\mathrm{e}^{-\alpha l} \sum_{k=0}^{\infty}(R_1 R_2 \mathrm{e}^{-2\alpha l})^k$.
(b)(i) -10%, (ii) -1%, (iii) -0.6%.
(c) It is valid to neglect multiple reflections when $\alpha l \gtrsim 1$, and for transparent materials with low refractive indices.

(1.9) (a) The transmitted field amplitude is given by:

$$\mathcal{E}^\mathrm{t} = tt'x\mathcal{E}_0 \sum_{k=0}^{\infty}(x^2 r'^2)^k \,,$$

where r (r') and t (t') are amplitude reflection and transmission coefficients for air to the medium (medium to air) respectively, and $x = \mathrm{e}^{-\alpha l/2}\,\mathrm{e}^{\mathrm{i}\Phi/2}$. The transmission is given by $T = |\mathcal{E}^\mathrm{t}|^2/|\mathcal{E}_0|^2$, and the result is obtained by using $r = -r'$, $r^2 = r'^2 = R$, and $tt' = 1 - r^2 = 1 - R$.
(b) With the same definitions as in part (a), the reflected field amplitude is given by:

$$\mathcal{E}^\mathrm{r} = \mathcal{E}_0 \left(r + \frac{r'tt'x^2}{1 - x^2 r'^2} \right).$$

On inserting $r' = -r$ and $tt' = 1 - r^2 = 1 - R$, we find:

$$\mathcal{E}^\mathrm{r} = \mathcal{E}_0\, r \left(\frac{1 - x^2}{1 - x^2 R} \right).$$

The reflectivity is then found from the ratio $|\mathcal{E}^\mathrm{r}|^2/|\mathcal{E}_0|^2$.

(c) Put $\alpha = 0$ into the transmissivity and reflectivity and add them together.
(d) $(1 - R)^2\,\mathrm{e}^{-\alpha l}$.
(e) Thin-film interference fringes with transmission peaks when $2nl = m\lambda$, m being an integer.

(1.10) Fabry–Perot fringes above band edge with transmission peaks at 875, 933, and 1000 nm. Exponentially decreasing transmission below band edge, with reflectivity roughly constant at 31%.

(1.11) Substitute eqn 1.29 with $\kappa = 0$ into eqn 1.9.

(1.12) 18%, 6%, and 4%.

(1.13) From eqn 1.8 we see that:

$$-\log_{10}(T) = -2\log_{10}(1 - R) + \alpha l/\ln(10)\,.$$

The optical density is found by comparing this to eqn 1.11. If the medium is transparent at λ', and the incoherent limit applies, the transmission will be given by eqn 1.9, which enables R to be determined from a measurement of the transmission. The optical density at λ can then be deduced from a measurement of the transmission at that wavelength. This result will only hold if λ' is close to λ because we are assuming that R does not vary significantly with the wavelength.

(1.14) 99.6%.

(1.15) 14 nm.

(1.16) $\tilde{\epsilon}_\mathrm{r} \approx 1$.

(1.17) (a) 0.294 eV; (b) 8 W and 2 W; (c) 4 W and 6 W.

(1.18) 521 nm.

(1.19) 81%. The scattering cross-section is 11 times larger at 850 nm than at 1550 nm.

(1.20) 3.5 m, 0.22 m.

(1.21) Ice is a uniaxial crystal, but water is a liquid with no preferential axes.

Chapter 2

(2.1)
$$m_1\ddot{x}_1 = -K_s(x_1 - x_2)$$
$$m_2\ddot{x}_2 = -K_s(x_2 - x_1).$$

Divide these equations by m_1 and m_2 respectively and subtract them to obtain the equation of motion for an oscillator of frequency $(K_s/\mu)^{1/2}$:

$$\frac{d^2}{dt^2}(x_1 - x_2) = -K_s\left(\frac{1}{m_1} + \frac{1}{m_2}\right)(x_1 - x_2).$$

(2.2) $-\tan^{-1}[\omega\gamma/(\omega_0^2 - \omega^2)]$.

(2.3) 6.3×10^{-4}.

(2.4) $270\,\mathrm{m}^{-1}$.

(2.5) $\alpha(\omega_0) = Ne^2/n\epsilon_0 m_0 \gamma c$.

(2.6) (a) 5.9; (b) $5.0 \times 10^{12}\,\mathrm{Hz}$; (c) 23 N; (d) $3.0 \times 10^{28}\,\mathrm{m}^{-3}$; (e) about $6 \times 10^{12}\,\mathrm{s}^{-1}$; (f) about $1 \times 10^6\,\mathrm{m}^{-1}$. In part (c), work out the spring constant K_s using eqn 2.2, remembering to use the correct reduced mass ($2.3 \times 10^{-26}\,\mathrm{kg}$). The answers to parts (e) and (f) are only approximate because the data does not exactly follow a simple Lorentzian line shape. The damping rate depends strongly on the frequency, which is why the resonance line is highly asymmetric.

(2.7) The relationship in terms of ω is derived by taking the reciprocal of eqn 2.25 and using $k = n\omega/c$ to obtain:

$$\frac{1}{v_g} = \frac{1}{c}\left(n + \omega\frac{dn}{d\omega}\right).$$

Equation 2.26 follows after substituting $v = c/n$. The relationship in terms of λ is derived by substituting $\lambda = 2\pi c/\omega$ so that $dn/d\omega = -(\lambda^2/2\pi c)dn/d\lambda$.

(2.8) The dispersion in this undamped system is always normal, so that from eqn 2.26 we see that $v_g < v$. If $\omega < \omega_0$, $\epsilon_r > 1$ and therefore $v = c/n = c/\sqrt{\epsilon_r} < c$. Hence $v_g < c$ if $\omega < \omega_0$. For $\omega > \omega_0$ we must work out v_g explicitly:

$$v_g = nc\left(1 + \frac{Ne^2}{\epsilon_0 m_0}\frac{\omega_0^2}{(\omega_0^2 - \omega^2)^2}\right)^{-1}.$$

The denominator is greater than unity, and $n < 1$, so $v_g < c$.

(2.9) This derivation is given in many solid-state or electromagnetism textbooks. See, for example, Kittel, C. (2005), *Introduction to solid state physics* (8th edn), Wiley, New York.

(2.10) Either when the density of absorbing atoms is small or when the frequency is far away from any resonance.

(2.11) $\chi_a = 2.2 \times 10^{-29}\,\mathrm{m}^3$. The two field strengths are $0.8 \times 10^{11}\,\mathrm{Vm}^{-1}$ and $1.4 \times 10^{11}\,\mathrm{Vm}^{-1}$ respectively. It is not surprising that these values are of similar magnitude because the external field must work against the Coulomb forces in the molecule to induce a dipole.

(2.12) Find $\kappa(E)$ from $\alpha(E)$ by using eqn 1.19, and then use eqn 2.36 to obtain $n(E)$ from:

$$n(E) = 1 + \frac{2}{\pi}\int_{E_1}^{E_2}\frac{E'\kappa(E')}{E'^2 - E^2}\,dE'.$$

(2.13) (a) $\lambda_j = 2\pi c/\omega_{0j}$, $A_j = Ne^2 f_j \lambda_j^2/4\pi^2\epsilon_0 m_0 c^2$. (b) $C_1 = (1 + A_1)^{1/2}$; $C_2 = A_1\lambda_1^2/2(1 + A_1)^{1/2}$; $C_3 = A_1(4 + 3A_1)\lambda_1^4/8(1 + A_1)^{3/2}$.

(2.14) (a) $C_1 = 1.5255$, $C_2 = 4824.7\,\mathrm{nm}^2$. (b) 1.5493 and 1.5369. (c) $1.26°$.

(2.15) 14 ps if we assume a time–bandwidth product $\Delta\nu\Delta t = 1$.

(2.16) Use $\epsilon_{11}/\epsilon_0 = \epsilon_{22}/\epsilon_0 = n_o$, $\epsilon_{33}/\epsilon_0 = n_e$, $x = 0$, $z/n(\theta) = \sin\theta$, and $y/n(\theta) = \cos\theta$ to derive the result.

(2.17) $37.1° - 42.3°$.

(2.18) Define axes with z along the optic axis, and x along the direction of propagation, so that the input polarization is $\cos\theta\hat{z} + \sin\theta\hat{y}$. (a) Output polarization = $\cos\theta\hat{z} - \sin\theta\hat{y}$, i.e. rotated by 2θ. (b) Output polarization = $1/\sqrt{2}(\hat{z} + i\hat{y})$, i.e. circular polarization. (c) Elliptically polarized light.

(2.19) $14\,\mu\mathrm{m}$.

(2.20) (a) No (face-centred cubic); (b) no (cubic); (c) yes (hexagonal); (d) yes (hexagonal); (e) no (cubic); (f) no (face-centred cubic); (g) yes (orthorhombic). Sulphur is biaxial.

(2.21) (a) Set $\Delta n\,d = \lambda/2$, with Δn given by eqn 2.51, and solve for $\mathcal{E}$. (b) 85 kV.

(2.22) (a) Split the linear input polarization into two in-phase left- and right-circular beams and recombine them at the output with a relative phase difference ϕ equal to $2\pi(n_R - n_L)\,d/\lambda$. The polarization rotation angle is equal to $\phi/2$. (b) $21.7°/\mathrm{mm}$.

(2.23) (a) The Faraday rotation is negative above and below ω_0, and positive near ω_0. The rotation decays as the frequency is tuned away from resonance.

(b) The magnetic circular dichroism follows a dispersive lineshape, with a negative signal below ω_0 that peaks at $\omega_0 - \mu_B B/\hbar$, and a positive signal above ω_0 that peaks at $\omega_0 + \mu_B B/\hbar$. The signal precisely at ω_0 is zero.

(2.24) 17 cm.

Chapter 3

(3.1) $\boldsymbol{k} = (2\pi/L)(n_x, n_y, n_z)$, where n_x, n_y, and n_z are integers. Each allowed $\boldsymbol{k}$ state occupies a volume of $\boldsymbol{k}$ space equal to $(2\pi/L)^3$, which implies that the number of states in a unit volume of $\boldsymbol{k}$ space is $L^3/(2\pi)^3$. Hence a unit volume of the material would have $1/(2\pi)^3$ states per unit volume of $\boldsymbol{k}$ space.

(3.2) $\mathrm{d}E/\mathrm{d}k = \hbar^2 k/m^*$. Use this in eqn 3.14, and then substitute for k.

(3.3) (a) The parity of a wave function is equal to ± 1 depending on whether $\psi(-\boldsymbol{r}) = \pm\psi(\boldsymbol{r})$. Atomic wave functions have well defined parities because atoms have inversion symmetry about $\boldsymbol{r} = 0$, and hence we must have that $|\psi(-\boldsymbol{r})|^2 = |\psi(\boldsymbol{r})|^2$.

(b) $\boldsymbol{r}$ is an odd function, and so the integral will be zero unless the two wave functions have different parities.

(c) For z polarized light we have:

$$M \propto \int_{\phi=0}^{2\pi} \mathrm{e}^{\mathrm{i}m'\phi}\,(r\cos\theta)\,\mathrm{e}^{\mathrm{i}m\phi}\,\mathrm{d}\phi\,,$$

which is zero unless $m' = m$. For x or y polarized light we have:

$$M \propto \int_{\phi=0}^{2\pi} \mathrm{e}^{\mathrm{i}m'\phi}\,\left(r\sin\theta(\mathrm{e}^{\mathrm{i}\phi} \pm \mathrm{e}^{-\mathrm{i}\phi})\right)\,\mathrm{e}^{\mathrm{i}m\phi}\,\mathrm{d}\phi\,,$$

which is zero unless $m' = m \pm 1$.

(d) $\mathcal{E}^{\pm} \propto \mathrm{e}^{\pm\mathrm{i}\phi}$, so

$$M^{\pm} \propto \int_{\phi=0}^{2\pi} \mathrm{e}^{\mathrm{i}m'\phi}\,\mathrm{e}^{\pm\mathrm{i}\phi}\,\mathrm{e}^{\mathrm{i}m\phi}\,\mathrm{d}\phi\,,$$

which gives $m' = m + 1$ for $\mathcal{E}^+$ and $m' = m - 1$ for $\mathcal{E}^-$.

(3.4) Same as Fig. 3.15, but with a scanning monochromator and an InSb detector.

(3.5) Plot α^2 and $\alpha^{1/2}$ against $\hbar\omega$. Also investigate the temperature dependence of α.

(3.6) Indirect band gap at 2.2 eV. Direct band gap at ~ 2.75 eV.

(3.7) $\alpha \approx 1.2 \times 10^6\,\mathrm{m}^{-1}$.

(3.8) (a) $5.3 \times 10^8\,\mathrm{m}^{-1}$ and $4.1 \times 10^8\,\mathrm{m}^{-1}$. (b) $3.0 \times 10^7\,\mathrm{m}^{-1}$. This is more than an order of magnitude smaller than the electron wave vector. (c) 2.1. (d) 704 nm.

(3.9) $|M|^2 = 3C$ and C respectively for heavy- and light-hole transitions.

(3.10) This follows from eqn A.41.

(3.11) Decreases above $E_g + \Delta$.

(3.12) (a) 4.1 eV. This corresponds to transitions from the p-like valence band to the s-like conduction band. (b) The discussion of the atomic character of bands given in Section 3.3.1 only applies at the Γ point. This means that electric-dipole transitions can be allowed at the zone edges, even though they are forbidden at $k = 0$.

(3.13) 0.75 eV.

(3.14) $\mathcal{E} \approx 1.8 \times 10^6\,\mathrm{V\,m}^{-1}$. This value is obtained by working out the field at which α drops by a factor e^{-1} between E_g and $(E_g - 0.01)$ eV.

(3.15) The first part is easily derived by equating the central force for circular motion with the Lorentz force: $m\omega^2 r = e\omega r B$. The $\Delta n = 0$ selection rule follows from the orthogonality of harmonic oscillator wave functions φ_n, after using Landau level wave functions of the form $\psi_n(\boldsymbol{r}) \propto u(\boldsymbol{r})\,\varphi_n(x, y)\,\mathrm{e}^{\mathrm{i}k_z z}$.

(3.16) (a) $g_{1D}(E) = (2m/Eh^2)^{-1/2}$, where m is the particle mass. (b) $\alpha \propto (\hbar\omega - E_g)^{-1/2}$. (c) The magnetic field quantizes the motion in two dimensions. The absorption coefficient for transitions between Landau levels varies as $(\hbar\omega - E_n)^{-1/2}$, where $E_n = E_g + (n + 1/2)(e\hbar B/\mu)$. This follows from the 1-D density of states and the $\Delta n = 0$ selection rule. $\alpha(\hbar\omega)$ diverges each time the frequency crosses the threshold for a new value of n. These divergences are broadened by scattering. We therefore see dips in the transmission at each value of $\hbar\omega$ that satisfies eqn 3.32. (d) $m_e^* \approx 0.035m_0$ and $E_g = 0.80$ eV. These values refer to the Γ point of the Brillouin zone.

(3.17) $0.46\,\mathrm{A\,W}^{-1}$ at $1.55\,\mu\mathrm{m}$ and $1.05\,\mathrm{A\,W}^{-1}$ at

1.30 μm.

(3.18) (a) The p- and n-regions are good conductors, whereas the i-region is depleted of free carriers and therefore acts like an insulator. (b) 10 pF. (c) 60 ps for the electrons and 200 ps for the holes. (d) 0.2 V.

Chapter 4

(4.1) This is a standard result for any two-particle system.

(4.2) (a) Kinetic energy + Coulomb energy. (b) $E = -\mu e^4/8\epsilon_r^2\epsilon_0^2 h^2 = -(\mu/m_0\epsilon_r^2)R_H$, $a_0 = \epsilon_0\epsilon_r h^2/\pi\mu e^2 = (\epsilon_r m_0/\mu)a_H$, $C = \pi^{-1/2}a_0^{-3/2}$.

(4.3) a_0. $\langle r\rangle = (3/2)a_0$.

(4.4) (a) This is a spherically symmetric function, with a maximum value at $r = 0$. (b) $\langle E\rangle = \hbar^2/2\mu\xi^2 - e^2/4\pi\epsilon_r\epsilon_0\xi$. (c) $\xi_{\min} = 4\pi\epsilon_0\epsilon_r\hbar^2/\mu e^2$, $\langle E\rangle_{\min} = -\mu e^4/8h^2\epsilon_0^2\epsilon_r^2$. (d) $\xi_{\min}$ and $\langle E\rangle_{\min}$ are the same as a_0 and E from Exercise 4.2. The variational method gives the energy and wave function exactly here because our 'guess' wave function had the correct functional form.

(4.5) (b) $E(n)$ and r_n are equal to $-\mu e^4/8h^2\epsilon_0^2\epsilon_r^2 n^2$ and $4\pi\epsilon_0\epsilon_r\hbar^2 n^2/\mu e^2$ respectively. (c) $E(n)$ is identical to the solution of the hydrogen Schrödinger equation. (d) r_1 corresponds to the peak in the radial probability density for the ground state 1s wave function.

(4.6) $E(1) = -39.1$ meV, $r_1 = 2.3$ nm, stable. $E(2) = -9.8$ meV, $r_2 = 9.3$ nm, unstable.

(4.7) 2.2 nm.

(4.8) The refractive index has a maximum value of 3.60 at 1.5146 eV.

(4.9) 394 μm.

(4.10) Substitute $|E(n)| = \mu e^4/8(\epsilon_0\epsilon_r hn)^2$ and $r_n = 4\pi\epsilon_0\epsilon_r\hbar^2 n^2/\mu e^2$ into $\mathcal{E} = e/4\pi\epsilon_0\epsilon_r r^2$ to obtain the result, using $|E(1)| = R_X$ and $r_1 = a_X$.

(4.11) 1.5 meV and 31 nm. $V_0 = +0.55$ V.

(4.12) 1.8 T.

(4.13) $\boldsymbol{B} = \boldsymbol{\nabla}\times\boldsymbol{A} = (0,0,B)$. $\hat{H}' = e^2\boldsymbol{A}^2/2m_0 = e^2B^2(x^2+y^2)/8m_0$. $\langle E\rangle = \langle\psi|\hat{H}|\psi\rangle$. Equation 4.7 follows by adding the contributions of the electron and hole, and remembering that spherical symmetry implies that $\langle x^2\rangle = \langle y^2\rangle = \langle z^2\rangle = \frac{1}{3}\langle r^2\rangle$.

(4.14) $\delta E = +4.9\times 10^{-5}$ eV, $\delta\lambda = -0.026$ nm.

(4.15) 8.1×10^{24} m^{-3} and 1.3×10^{23} m^{-3}.

(4.16) 0.50.

(4.17) 17.2 K.

(4.18) $r_1 = 0.85$ nm: invalid. $r_2 = 3.4$ nm: valid.

Chapter 5

(5.1) See Section 5.2.2.

(5.2) The relaxation within the bands is faster than the radiative recombination.

(5.3) $A_{2p\to 1s} = 6.27\times 10^8$ s^{-1}. $\tau_R = 1.6$ ns.

(5.4) Faster non-radiative recombination at higher temperatures due to phonon emission. $\eta_R(300\,K) = 79\%$, $\eta_R(350\,K) = 56\%$.

(5.5) ZnTe.

(5.6) (a) This follows directly from the definition of α given in eqn 1.3. (b) Set $\dot{N} = I\alpha/h\nu - N/\tau$ equal to zero. (c) 6.6×10^{20} m^{-3}.

(5.7) (a) 1.9×10^{24} m^{-3}. (b) 0.62 ns. (c) 3.5×10^{10} photons.

(5.8) The emission rate is proportional to the probability that the upper level is occupied and that the lower level is empty, that is, $f_e \times f_h$. In the classical limit, $f_{e,h} \propto \exp(-E_{e,h}/k_BT)$, so

$$f_e f_h = \exp(-(E_e + E_h)/k_BT)$$
$$= \exp(-(h\nu - E_g)/k_BT).$$

(5.9) (a) $E_F = -0.216$ eV $= -8.4k_BT$, valid. (b) $E_F = +0.021$ eV $= +0.83k_BT$, invalid.

(5.10) Use $f(E) = 1$ for $E < E_F$ and $f(E) = 0$ for $E > E_F$ to derive the result.

(5.11) Electrons: (a) 0.36 meV, degenerate for $T \ll 4.2$ K; (b) 36 meV, degenerate for $T \ll 420$ K. Holes: (a) 0.073 meV, degenerate for $T \ll 0.9$ K; (b) 7.3 meV, degenerate for $T \ll 85$ K.

(5.12) $k_F = (3\pi^2 N)^{1/3}$.

(5.13) (a) Solid angle $\Omega = 0.049$. (b) 4.2×10^{-4}. (c) $0.53\eta_R$ mW. (d) $0.22\eta_R$ μW.

(5.14) (a) 0.14 eV. (b) 0.012 eV. (c) Electrons degenerate, but not holes. (d) $E_g + E_F^c$ may be read from the spectrum at about 0.94 eV as the point where the luminescence falls to 50% of its peak value. This agrees well with the estimate of E_F^c from the carrier density. (e) Read E_F^c from the data to find $N_e \approx 3 \times 10^{23}$ m^{-3}. $\tau \approx 0.13$ ns.

(5.15) The result follows by working out the relative σ^+ and σ^- transition intensities for an initial population with three times as many electrons in one of spin sub-level as the other, and with the holes equally distributed between their sub-levels.

(5.16) $\tau = (\hbar/g_e\mu_B B_{1/2}) [P_0/P(0)]$, $\tau_S = (\hbar/g_e\mu_B B_{1/2})[1 - P(0)/P_0]^{-1}$.

(5.17) (a) 610 nm. (b) $x = 0.316$.

(5.18) (a) 31%. (b) 4.3×10^{10} Hz. (c) 610 m^{-1}.

(5.19) (a) 150 mW. (b) 26%. (c) 0.77 W A^{-1}, 51%.

(5.20) Set the generation rate equal to the decay rate to derive the result.

Chapter 6

(6.1) About 0.01 K.

(6.2) 9.3 nm and 30 nm.

(6.3) The k vector must satisfy $k = $ integer $\times 2\pi/L$, and the areal density of states in k space is therefore $(1/2\pi)^2$. The result for $g_{2D}(E)$ is derived by writing $g_{2D}(k)dk = 2\pi k dk \times (1/2\pi)^2$, and then using $g_{2D}(E) = 2g_{2D}(k)dk/dE$, with $dk/dE = m/\hbar^2 k$.

(6.4) Use Born–von Karman boundary conditions to deduce that $g_{1D}(k)dk = 1/2\pi dk$. Then apply eqn 3.14 with $E = \hbar^2 k^2/2m$ to derive the result.

(6.5) The function on the right-hand side of eqn 6.26 decreases to zero at $x = \sqrt{\xi}$. It will therefore always cross the $x \tan x$ function between 0 and $\pi/2$, no matter how small ξ is.

(6.6) 7.5 meV for the finite well, 11 meV for the infinite one.

(6.7) (a) This result follow from the orthonormality of the wave functions. (b) The initial and final states must have opposite parities.

(6.8) First step at 1.679 eV due to the $n = 1$ heavy-hole transition. Second step at 1.837 eV due to the light-hole transition. The height of the two steps is in proportion to the reduced masses, that is 0.059:0.036.

(6.9) (a) The transition energies would be lower. Transitions such as hh3 $\rightarrow$ e1 would be weakly allowed. (b) Peaks would appear below the steps due to excitonic absorption.

(6.10) (a) Direct substitution gives:

$$\int_{r=0}^{\infty} \int_{\phi=0}^{2\pi} \Psi^* \Psi \, r dr d\phi = 1 .$$

(b) The result is obtained by working out that

$$\hat{H}\Psi = \left(-\frac{\hbar^2}{2\mu\xi^2} + \frac{\hbar^2}{2\mu\xi^2 r} - \frac{e^2}{4\pi\epsilon_0\epsilon_r r}\right)\Psi ,$$

and then evaluating the integral.
(c) $E_{\min} = -\mu e^4/8(\pi\epsilon_0\epsilon_r\hbar)^2$. This is four times larger than the bulk exciton binding energy found in Exercise 4.4.
(d) $\xi_{\min} = 2\pi\hbar^2\epsilon_0\epsilon_r/\mu e^2 = a_X/2$, where a_X is defined in eqn 4.2.

(6.11) At $d = \infty$ we have bulk GaAs, while at $d = 0$ we have bulk AlGaAs. As d is reduced from ∞, the binding energy increases from 4 meV, going through a peak, and then dropping to 6 meV. The height of the peak would be about 17 meV, that is, four times larger than the binding energy of bulk GaAs.

(6.12) (a) See Section 5.3.5. (b) Heavy-hole exciton and continuum absorption, followed by light-hole exciton and continuum absorption. (c) The heavy-hole continuum starts at 1.592 eV. This implies $d = 9.3$ nm in the infinite well model. The true width would be smaller, because the infinite well model overestimates the confinement energy. (d) 11 meV and 12 meV respectively. A perfect 2-D GaAs quantum well would have $E_b = 4R_X = 16.8$ meV. The experimental binding energies are lower because a real quantum well is not a perfect 2-D system.

(6.13) $\Pi = -100\%$ for the photons in the range given by eqn 6.43. Π decreases for $\hbar\omega > E_g + E_{e1} + E_{lh1}$, and drops to zero for $\hbar\omega > E_g + \Delta + E_{e1} + E_{so1}$.

(6.14) (a) z is an odd function, while $\varphi^*\varphi$ is even. (b) The result follows by taking just the first term in the perturbation, namely $|\langle 1|H'|2\rangle|^2/(E_1 - E_2)$, and substituting the wave functions and energies from eqns 6.11 and 6.13.

(6.15) (a) The experimental shift is smaller, mainly due to the smaller value of d. (b) 3.4 nm, assuming a quadratic Stark shift. (c) $\langle \delta z \rangle \approx 1.6$ nm.

(6.16) The model works quite well for sample A, but

not for sample B. The model breaks down when the size of the Stark shift becomes comparable to the energy splitting of the unperturbed hh1 and hh2 levels. This is essentially the same criterion as for the transition from the quadratic to the linear Stark effect in atomic physics. In sample B, we are in this regime at all the fields quoted.

(6.17) At finite $\mathcal{E}_z$ the inversion symmetry of the quantum well is broken. The states no longer have definite parities, and selection rules based on parity no longer hold.

(6.18) The shift is about 0.02 eV, which is comparable to the linewidth observed in the data. A $\pm 5\%$ variation in d corresponds more or less to a fluctuation of one atomic layer.

(6.19) 14 nm, assuming infinite barriers.

(6.20) (a) z is an odd function, and so the integral will be zero unless $\varphi_n^* \varphi_{n'}$ is also an odd function, which requires that the wave functions must have different parities.
(b) The matrix elements are given by:

$$\langle 1|z|2 \rangle = \frac{2}{d} \int_0^d \sin(\pi z/d) \, z \, \sin(2\pi z/d) \, dz$$
$$= -(16/9\pi^2)d$$
$$\langle 1|z|4 \rangle = \frac{2}{d} \int_0^d \sin(\pi z/d) \, z \, \sin(4\pi z/d) \, dz$$
$$= -(4/45\pi^2)d.$$

The transition strength is proportional to $|M^2|$, and so the $1 \to 4$ transition is weaker than the

$1 \to 2$ transition by a factor $(1/20)^2 = 2.5 \times 10^{-3}$. $\lambda_{1\to2} = 29\,\mu m$.

(6.21) The electric field of the light wave in the medium is maximum for grazing incidence with $\theta = 90°$, when the fractional power of the z component is $1/n^2$. The maximum possible fractional absorption is therefore equal to 9% if $n = 3.3$.

(6.22) The quantized levels occur at energies of 3, 6, 9, 11, 12, 14, 17, in units of $h^2/8m^*d^2$. The degeneracies are, respectively: 1, 3, 3, 3, 1, 6, 3.

(6.23) Substitute $R(r)$ into eqn 6.57 with $l = 0$ and $V(r) = -V_0$ to show that $E = -V_0 + \hbar^2k^2/2m^*$ within the dot. Then set $\sin(kR_0/R_0) = 0$ to find that $k = n\pi/R_0$, where n is an integer, and hence that the energy relative to the bottom of the well is $\hbar^2 n^2 \pi^2 / 2m^* R_0^2$.

(6.24) Larger for cubic dots by a factor $3(3/4\pi)^{2/3} = 1.15$.

(6.25) (a) $E = (n_x + 1/2)\hbar\omega_0 + (n_y + 1/2)\hbar\omega_0$; degeneracy $= n$.
(b) Separate the variables to show that the angular part of the wave function $\Phi(\phi)$ must satisfy $d^2\Phi/d\phi^2 = -m^2\Phi$, where m^2 is the separation constant. The solutions are of the form $\exp im\phi$, and must be single valued. Hence m must be an integer.
(c) $\psi_{0,0}(r,\phi) = \psi_0(x)\psi_0(y)$,
$\psi_{1,\pm1}(r,\phi) \propto \psi_1(x)\psi_0(y) \pm i\psi_0(x)\psi_1(y)$,
$\psi_{2,0}(r,\phi) \propto \psi_2(x)\psi_0(y) + \psi_0(x)\psi_2(y)$,
$\psi_{2,\pm2}(r,\phi) \propto -\psi_2(x)\psi_0(y) \pm \sqrt{2}i\psi_1(x)\psi_1(y) + \psi_0(x)\psi_2(y)$.

Chapter 7

(7.1) $E_F^3 = (9\epsilon_0^2\hbar^2/8m_0)(\pi\hbar\omega_p/e)^4$.

(7.2) $N \sim 10^{11}\,\text{m}^{-3}$.

(7.3) $\delta \sim 0.5\,\text{m}$. To obtain a strong signal in a submerged submarine it is necessary to use much lower frequencies. The data rate would then be very low due to the small carrier frequency.

(7.4) $m_e^* = 1.6\,m_0$.

(7.5) $R = 99.6\%$.

(7.6) $T = 0.16$.

(7.7) The drop in the reflectivity for $\lambda < 600\,\text{nm}$ is caused by interband transitions. The energy gap between the d bands and the Fermi energy can be read from the data as $\sim 2.4\,\text{eV}$. The low reflectivity for green and blue light causes the char-

acteristic yellowish colour.

(7.8) $\epsilon_r = 1$.

(7.9) m_e^* increases from $0.020\,m_0$ at $3.5 \times 10^{23}\,\text{m}^{-3}$ to $0.048\,m_0$ at $4\times10^{24}\,\text{m}^{-3}$. The increase in m_e^* with N_e is caused by the non-parabolicity in the conduction band of InSb.

(7.10) $\tau \sim 1\,\text{ps}$.

(7.11) Calculate the carrier density as in Exercise 5.6, and the free carrier absorption using eqn 7.28. Then add the separate contributions of the electrons and holes together. The final answer is about $200\,\text{m}^{-1}$.

(7.12) (a) $E_F = 0.032\,\text{eV}$. k_F is $6.5 \times 10^8\,\text{m}^{-1}$ and $2.6\times10^8\,\text{m}^{-1}$ for the heavy and light holes respectively. (b) (1): 0.03–0.17 eV, (2): 0.32–0.34 eV,

(3): 0.34–0.42 eV.

(7.13) (a) $m_e^* = 0.85 m_0$. (b) $R_0^* \approx 45$ meV, $R_\pm^* \approx 25$ meV.

(7.14) $R^* = (m_e^*/m_0\epsilon_r^2) \times R_H$. $m_e^* = 0.036 m_0$.

(7.15) Acceptor energy $E_A \sim 8$ meV.

(7.16) Raman scattering from plasmon modes: $N = 4.2 \times 10^{24}$ m^{-3}.

(7.17) 7.2×10^{23} m^{-3}.

(7.18) 0.2%.

(7.19) Aluminium: surface and bulk plasmons.

(7.20) (a) $l_z^d = 390$ nm and $l_z^m = 22$ nm. (b) 28 μm.

(7.21) Resonance at $\omega_p/\sqrt{3}$ for air.

(7.22) (a) 517 nm. (b) 294 nm. The difference is mainly caused by the effect of interband transitions.

Chapter 8

(8.1) $E_1 = (1/2)\hbar\Omega$, $E_2 = (3/2)\hbar\Omega$, $E_3 = (5/2)\hbar\Omega$, $a = (\hbar/m\Omega)^{1/2}$.

(8.2) $d \approx 6.7 \times 10^{-10}$ m, which corresponds to about six carbon–carbon bonds.

(8.3) 4×10^{-2}, 1.6×10^{-3}, 1.4×10^{-5}.

(8.4) 10.2 eV. The ground state of the molecule is more strongly bound than the excited state, and hence the transition energy is larger.

(8.5) (a) The van der Waals interaction energy varies as r^{-6}. (b) Find the point where $dU/dr = 0$. (c) $U(r) = U(r_0) + d^2U/dr^2(r - r_0)^2/2 + \cdots$, where the derivative is evaluated at $r = r_0$. $\Omega^2 = (18B^2/A\mu)(B/2A)^{1/3}$.

(8.6) (a) This follows directly from the Franck–Condon principle: sum over eqn 8.12 for each vibronic sub-level. (b) (i) Zero-phonon line only. (ii) and (iii): Poisson distributions with means of 1 and 5 respectively.

(8.7) S_1 at 4.64 eV, $\Omega/2\pi = 3 \times 10^{13}$ Hz.

(8.8) The configuration diagram is similar to Fig. 8.7, but with two excited states. S_1 state: energy = 5.7 eV, vibrational splitting = 0.11 eV, turning point of $n = 6$ level aligned with Q_0. S_2 state: energy 7.3 eV, vibrational splitting 0.13 eV, turning point of $n = 5$ level aligned with Q_0.

(8.9) Spin–orbit coupling mixes S and L, so that the triplet states contain a small admixture of singlet character through mixing with common L states.

(8.10) Phosphorescence from a triplet state at 1.6 eV.

(8.11) Both give $\hbar\Omega \approx 0.17$ eV.

(8.12) There will be a vibronic band of width ~ 1 eV extending from 3.1 eV downwards, with three or four peaks at energies $(3.1 - n\hbar\Omega)$, where $\hbar\Omega \approx 0.17$ eV.

(8.13) Broad vibronic band from 1.9 eV down to about 1.0 eV. Peaks at 1.9 eV and 1.7 eV.

(8.14) 1.1 eV.

(8.15) 693 nm. (There will be other Raman lines in addition to this one.)

(8.16) Optical excitation creates only singlets, whereas electrical injection creates both singlets and triplets with a probability determined by their statistical weights, namely 1:3. Only singlets emit efficiently, and the population of these is lower by a factor of four in the case of electrical injection.

(8.17) (a) See Exercise 8.16. (b) 5.6 mW. (c) 11%. The efficiency of a real device would be much lower, mainly due to the difficulty of collecting the photons, which are emitted in all directions.

(8.18) (a) $|a_1| = |a_2| = 2 \times a \cos 30°$.
(b) Find $|c|$ by evaluating $c \cdot c$.
(c) $\tan\theta = (n_2 a_0 \sin 60°)/(n_1 a_0 + n_2 a_0 \cos 60°)$.

(8.19) (a) Symmetry requires that the electron wave function should be single valued on rotating the tube by 2π. The result follows by considering the phase change on going round the circumference of the tube. (b) Insert $\mathbf{k} = (\mathbf{k_1} - \mathbf{k_2})/3$ into the result of part (a), and use the definition $\mathbf{a}_i \cdot \mathbf{k}_j = 2\pi\delta_{ij}$.

(8.20) Apply periodic boundary condition to obtain $g(k) = 1/2\pi$, and then use eqn 3.14 with an extra factor of two to account for the fact that the $+k$ and $-k$ velocity states are degenerate.

(8.21) 6.7×10^{-4}.

Chapter 9

(9.1) The solution for a 1-D infinite potential well is given in Section 6.3.2. In a cube the motion is quantized in three dimensions, and the energies for the x, y, and z directions just add together.

(9.2) Equation 9.4 predicts $E = 0.28/a^2$. The experimental energies are lower because a real F-centre is not a rigid cubic box.

(9.3) Either calculate $h\nu = 2.6\,\text{eV}$ from eqn 9.5, or just read $h\nu \approx 2\,\text{eV}$ from Fig. 9.4.

(9.4) $E = (\hbar^2\pi^2/2m_0b^2)(n_x^2 + n_y^2 + n_z^2/4)$ and $h\nu = 3h^2/32m_0b^2$. The frequency is half the value given in eqn 9.5 when $b = \sqrt{2}a$, which is appropriate for an F_2^+ centre. The experimental ratio is about 0.4, which is good agreement considering the simplicity of the model.

(9.5) (a) $N_0 : N_{\pm 1} = 1 : 1.87$; (b) 0.07 K.

(9.6) 0.064 eV.

(9.7) (a) $\langle r\rangle_{3d}/\langle r\rangle_{4f} = (7/12)(Z_{4f}/Z_{3d}) \sim 1.5$ for $Z_{4f} \sim 64$ and $Z_{3d} \sim 25$.
(b) Transition-metal ions have lost the outermost 4s electrons, whereas the 4f orbitals of the rare earths are inside the filled 5s and 5p orbitals.

(9.8) (a) The x, y, and z directions are all equivalent, and so the p_x, p_y, and p_z orbitals must all experience the same interaction energy with the crystal.
(b) The z direction is now different, and so the p_z orbits will have a different energy to the p_x and p_y states.
(c) The singlet is at higher energy because of the greater repulsion from the closer negative charges.

(9.9) (a) The result follows from:
$$\int_0^{2\pi} e^{-im\phi}e^{im''\phi}e^{im'\phi}\,d\phi = 2\pi\delta_{m,(m'+m'')}.$$
(b) This follows from diagonalizing the crystal-field Hamiltonian, which is given by:
$$H_{cf} = \begin{pmatrix} A & 0 & 0 & 0 & D \\ 0 & B & 0 & 0 & 0 \\ 0 & 0 & C & 0 & 0 \\ 0 & 0 & 0 & B & 0 \\ D & 0 & 0 & 0 & A \end{pmatrix}.$$
(c) $2z^2 - x^2 - y^2 \propto |0\rangle$,
$x^2 - y^2 \propto (|2\rangle + |-2\rangle)/\sqrt{2}$,
$xy \propto (|2\rangle - |-2\rangle)/\sqrt{2}$,
$yz \propto (|1\rangle - |-1\rangle)/\sqrt{2}$,
$zx \propto (|1\rangle + |-1\rangle)/\sqrt{2}$.

(d) The dγ states have high probability density along the crystal axes and so the electron of a d^1 configuration experiences strong repulsion, while the hole in a d^9 configuration experiences strong attraction.

(9.10) The relative populations of the 11502 cm^{-1} and 11414 cm^{-1} levels of the $^4F_{3/2}$ term are proportional to $\exp(-\Delta E/k_BT)$. The relative population of the 11502 cm^{-1} level therefore increases from 0.19 at 77 K to 0.66 at 300 K, and the emission intensity increases in proportion to these factors.

(9.11) The stimulated emission rate exceeds the absorption rate if population inversion is present: see Section B.1.

(9.12) (a) Population inversion must occur, which implies that the population of level 2 must exceed that of level 0. (b) 0.3 J. The laser stops working when 10% of the atoms in the upper level have transferred to the lower level.

(9.13) (a) The spectrum is proportional to $|\mathcal{E}(\omega)|^2$, where
$$\mathcal{E}(\omega) = \frac{1}{\sqrt{2\pi}}\int_{-\infty}^{+\infty} \mathcal{E}(t)\,e^{i\omega t}\,dt,$$
and $\mathcal{E}(t) = \exp(-t^2/2\tau^2)\,e^{-i\omega_0 t}$.
(b) $\Delta t = 2\sqrt{\ln 2}\,\tau$. $\Delta\nu = \sqrt{\ln 2}/\pi\tau$.

(9.14) Inhomogeneities in the glass cause local variations in the environment leading to line broadening through the coupling of the laser levels to the local crystal field. $\Delta t = 60\,\text{fs}$ for Gaussian pulses.

(9.15) The transition is parity forbidden. Phosphorescence.

(9.16) The probability for phonon-assisted non-radiative decay increases with T. Equation 5.5 gives $\eta_R(77) = 0.78$ and $\eta_R(300) = 0.03$. The radiative efficiency is too low at 300 K to allow lasing.

(9.17) $P = 3.2\,\text{W}$, assuming that all the pump power is absorbed, and the radiative quantum efficiency is unity. The remaining 1.8 W goes as heat in the crystal.

(9.18) (a) 54%, (b) 69%.

Chapter 10

(10.1) (a) Yes, (b) No, (c) No, (d) Yes, (e) Yes. Germanium and argon are non-polar materials.

(10.2) Solve eqn 10.15 with $\epsilon_r = 1$.

(10.3) 15–$33\,\mu$m.

(10.4) (a) 98%; (b) 84%.

(10.5) (a) $\nu_{TO} = 9.5\,$THz, $\nu_{LO} = 10\,$THz. (b) $\epsilon_\infty = 9.5$, $\epsilon_{st} = 11.8$. (c) About 15 ps. Lyddane–Sachs–Teller predicts $\nu_{LO}/\nu_{TO} = 1.11$, but the experimental ratio is slightly smaller. This is not significant, given that the broadening causes some uncertainty in the experimental values.

(10.6) (a) $1.1 \times 10^7\,$m^{-1}; (b) $3.3 \times 10^6\,$m^{-1}.

(10.7) The phonon lifetime decreases with T as the probability for anharmonic decay increases.

(10.8) $7.8 \times 10^5\,$m^{-1}.

(10.9) (a) $m^{**} = 0.097\,m_0$. (b) $m^* = 0.092\,m_0$.

(10.10) The diamond crystal would have only one peak in the Stokes or anti-Stokes spectrum.

(10.11) Stokes and anti-Stokes peaks from the optical phonon at 15.5 THz. $I(501.2\,\text{nm})/I(528.6\,\text{nm}) = 0.08$.

(10.12) IR active, but not Raman active.

(10.13) GaAs: $h\nu_{TO} = 32.5\,$meV, $h\nu_{LO} = 35.5\,$meV; InP: $h\nu_{TO} = 37.1\,$meV, $h\nu_{LO} = 42.3\,$meV; AlSb: $h\nu_{TO} = 38.7\,$meV, $h\nu_{LO} = 41.1\,$meV; GaP: $h\nu_{TO} = 45.1\,$meV, $h\nu_{LO} = 50.0\,$meV. The small shift of a few wave numbers compared to the infrared data for GaAs in Fig. 10.5 is caused by the slight decrease of the optical phonon frequencies between 4 K and 300 K.

(10.14) Apply conservation of momentum, with $k_1 = k_2 = n\omega/c$.

(10.15) $v_s = 810\,$m s^{-1}.

(10.16) (a) The negative term is the total Coulomb attraction, with the Madelung constant accounting for the contributions of the positive and negative ions from the whole crystal. The positive term represents the short range repulsive force due to the Pauli exclusion principle when the electron wave functions overlap.
(b) r_0 is the value for which $dU/dr = 0$.
(c) The Taylor series about r_0 is:

$$U(r) = U(r_0) + (1/2)(\mathrm{d}^2 U/\mathrm{d}r^2)_{r=r_0}(r - r_0)^2 \\ + (1/6)(\mathrm{d}^3 U/\mathrm{d}r^3)_{r=r_0}(r - r_0)^3 + \cdots .$$

Take $x = r - r_0$ to put this in the form of eqn 10.33, with $U(x)$ defined relative to the minimum at r_0. $C_3 = -22\alpha e^2/3\pi\epsilon_0 r_0^4$.

(10.17) 6 ps, assuming a Lorentzian line shape.

Chapter 11

(11.1) $\mathcal{E} = Ze/4\pi\epsilon_0 r_n^2$. For the outer 3s and 3p electrons in silicon use $Z = 4$ and $n = 3$ to obtain a value of $\sim 5 \times 10^{11}\,$V/m.

(11.2) (a) $6.2 \times 10^7\,$V/m, (b) $1.6 \times 10^5\,$V/m.

(11.3) Only with the field applied.

(11.4) (a) No, (b) yes, (c) no, (d) no, (e) yes, (f) yes. The second-order nonlinear susceptibility is zero if the material has an inversion centre.

(11.5) (a) N_2 cannot increase beyond $N_0/2$ because there is no net absorption when the populations are equal. (b) The rate equations are $\dot{N}_1 = -B_{12}u_\nu g(\nu)(N_1 - N_2)$ and $\dot{N}_2 = B_{12}u_\nu g(\nu)(N_1 - N_2)$. Subtract these to obtain $\mathrm{d}\Delta N/\mathrm{d}t = -2B_{12}u_\nu g(\nu)\Delta N$ where $\Delta N = N_1 - N_2$. Then integrate with $\Delta N(0) = N_0$ to obtain the required result, which implies that the populations will eventually equalize no matter how weak the laser beam is. This misleading conclusion arises from neglecting spontaneous emission and transitions to other levels.

(11.6) $P_x^{(2)} = d_{14}2\mathcal{E}_y\mathcal{E}_z = 0$, and $P_y^{(2)} = d_{25}2\mathcal{E}_z\mathcal{E}_x = 0$. Assume that the beam makes an angle θ with the x axis and then maximize $P_z^{(2)} = d_{36}2\mathcal{E}_x\mathcal{E}_y$.

(11.7) $52°$.

(11.8) (a) $n_o(\mathcal{E}) = n_o - n_o^3 r_{13}\mathcal{E}/2$, $n_e(\mathcal{E}) = n_e - n_e^3 r_{33}\mathcal{E}/2$. (b) $\Delta\Phi(\mathcal{E}) = -\pi n_e^3 r_{33}\mathcal{E}L/\lambda$. (c) The phase change is proportional to the electric field, and hence to the applied voltage.

(11.9) (a) $\Delta\Phi_{x'} = -\Delta\Phi_{y'} = (2\pi L/\lambda)(n_0^3 r_{41}\mathcal{E}_z/2)$, where L is the length of the crystal. $\Delta\Phi = \Delta\Phi_{x'} - \Delta\Phi_{y'}$ gives the result with $\mathcal{E}_z L = V$. (b) 44 kV.

(11.10) 4.2 kV.

(11.11) $\Delta\alpha = (\alpha_0/I_s)I \propto \Delta\epsilon_2$, and $\Delta\epsilon_2 \propto \text{Im}(\chi^{(3)})I$. Hence $(\alpha_0/I_s) \propto \text{Im}(\chi^{(3)})$.

(11.12) Choose z as the direction of propagation and x as the polarization vector so that $\mathcal{E}_y = \mathcal{E}_z = 0$. The only non-zero term is $P_x^{(3)} = \epsilon_0\chi_{xxxx}\mathcal{E}_x^3$, which implies that $\boldsymbol{P}$ is parallel to $\boldsymbol{\mathcal{E}}$.

(11.13) 76 W.

(11.14) 0.06 eV.

(11.15) Follow Example 2.1 to work out the magnitude of the local maximum in n below the absorption line. We then find $|\Delta n| = 0.027$ if we assume that this local maximum is completely saturated.

(11.16) (a) $1.8 \times 10^{23}\,\text{m}^{-3}$. (b) I_s is the intensity required to produce this carrier density, namely $4 \times 10^7\,\text{W}\,\text{m}^{-2}$.

Bibliography

Chapter 1: Introduction

Bach, H. and Neuroth, N. (1995). *The properties of optical glass*. Springer-Verlag, Berlin.

Bleaney, B.I. and Bleaney, B. (1976). *Electricity and magnetism* (3rd edn). Clarendon Press, Oxford. Reissued in two volumes in 1989.

Born, M. and Wolf, E. (1999). *Principles of optics* (7th edn). Cambridge University Press, Cambridge.

Buckley, A.R., Rahn, M.D., Hill, J., Cabanillas-Gonzalez, J., Fox, A.M., and Bradley, D.D.C. (2001). Energy transfer dynamics in polyfluorene-based polymer blends. *Chem. Phys. Lett.*, **339**, 331–6.

Driscoll, W.G. and Vaughan, W. (1978). *Handbook of optics*. McGraw-Hill, New York .

Hecht, Eugene (2001). *Optics* (4th edn). Addison-Wesley, Reading, MA.

Kaye, G.W.C. and Laby, T.H. (1986). *Tables of physical and chemical constants* (15th edn). Longman Scientific, Harlow, UK.

Krause, D. (2005). Glasses. In *Springer handbook of condensed matter and materials data* (eds W. Martienssen and H. Warlimont). Springer-Verlag, Berlin, pp. 523–72.

Lide, D.R. (1996). *CRC handbook of chemistry and physics* (77th edn). CRC Press, Boca Raton, FL.

Lorrain P., Corson D.R., and Lorrain F. (2000). *Fundamentals of electromagnetic phenomena*. W.H. Freeman, Basingstoke.

McCarthy, D.E. (1967). Transmittance of optical materials from $0.17\,\mu$ to $3.0\,\mu$. *Applied Optics*, **6**, 1896–8.

Madelung, O. (1996). *Semiconductors, basic data* (2nd edn). Springer-Verlag, Berlin.

Nye, J.F. (1985). *Physical properties of crystals*. Clarendon Press, Oxford.

Chapter 2: Classical propagation

Bleaney, B.I. and Bleaney, B. (1976). *Electricity and magnetism* (3rd edn). Clarendon Press, Oxford. Reissued in two volumes in 1989.

Born, M. and Wolf, E. (1999). *Principles of optics* (7th edn). Cambridge University Press, Cambridge.

Dressel, M. and Grüner, G. (2002). *Electrodynamics of solids*. Cambridge University Press, Cambridge.

Driscoll, W.G. and Vaughan, W. (1978). *Handbook of optics*. McGraw-Hill, New York.

Hecht, Eugene (2001). *Optics* (4th edn). Addison-Wesley, Reading, MA.

Hoffmann, H.-J. (1995). Differential changes of the refractive index. In *The properties of optical glass* (eds H. Bach and N. Neuroth). Springer-Verlag, Berlin, pp. 96–123.

Kaye, G.W.C. and Laby, T.H. (1986). *Tables of physical and chemical constants* (15th edn). Longman Scientific, Harlow, UK.

Klein, M.V. and Furtak, T.E. (1986). *Optics* (2nd edn). Wiley, New York.

Nye, J.F. (1985). *Physical properties of crystals*. Clarendon Press, Oxford.

Palik, E.D. (1985). *Handbook of the optical constants of solids*. Academic Press, San Diego.

Smith, D.Y., Shiles, E., and Inokuti, M. (2004). Refraction and dispersion in optical glass. *Nuclear instruments and methods in physics research B: beam interactions with materials and atoms*, **218**, 170–5.

Smith, F.G., King, T.A., and Wilkins, D. (2007). *Optics and photonics* (2nd edn). Wiley, Chichester.

Chapter 3: Interband absorption

Aspnes, D.E. (1980). Modulation spectroscopy/electric field effects on the dielectric function of semiconductors. In *Handbook on Semiconductors*, vol. 2 (ed. M. Balkanski). North Holland, Amsterdam, pp. 109–54.

Aspnes, D.E. and Studna, A.A. (1983). Dielectric functions and optical parameters of Si, Ge, GaP, GaAs, GaSb, InP, InAs, and InSb from 1.5 to 6.0 eV. *Phys. Rev. B*, **27**, 985–1009.

Bhattacharya, P. (1997). *Semiconductor optoelectronic devices* (2nd edn). Prentice Hall, New Jersey.

Burns, G. (1985). *Solid state physics*. Academic Press, San Diego.

Chelikowsky J.R. and Cohen, M.L. (1976). Nonlocal pseudopotential calculations for the electronic structure of eleven diamond and zinc-blende semiconductors. *Phys. Rev. B*, **14**, 556–82.

Chuang, S.L. (1995). *Physics of optoelectronic devices*. Wiley, New York.

Cohen, M.L. and Chelikowsky, J. (1988). *Electronic structure and optical properties of semiconductors*. Springer-Verlag, Berlin.

Corney, Alan (1977). *Atomic and laser spectroscopy*. Clarendon Press, Oxford.

Dash, W.C. and Newman, R. (1955). Intrinsic optical absorption in single-crystal germanium and silicon at 77°K and 300°K. *Phys. Rev.*, **99**, 1151–5.

Hamaguchi, C. (2001). *Basic semiconductor physics*. Springer-Verlag, Berlin.

Harrison, W. (1999). *Elementary electronic structure*. World Scientific, Singapore.

Hecht, Eugene (2001). *Optics* (4th edn). Addison-Wesley, Reading, MA.

Ibach, H. and Luth, H. (2003). *Solid-state physics* (3rd edn). Springer-Verlag, Berlin.

Kane, E.O. (1957). Band structure of indium antimonide. *J. Phys. Chem. Solids*, **1**, 249–61.

Kittel, Charles (2005). *Introduction to solid state physics* (8th edn). Wiley, New York.

Klingshirn, C.F. (1995). *Semiconductor optics*. Springer-Verlag, Berlin.

MacFarlane, G.G. and Roberts, V. (1955). Infrared absorption of germanium near the lattice edge. *Phys. Rev.*, **97**, 1714–6.

Madelung, O. (1996). *Semiconductors, basic data* (2nd edn). Springer-Verlag, Berlin.

Palik, E.D. (1985). *Handbook of the optical constants of solids*. Academic Press, San Diego.

Pankove, J.I. (1971). *Optical processes in semiconductors*. Dover, New York.

Seeger, K. (1997). *Semiconductor physics* (6th edn). Springer-Verlag, Berlin.

Singleton, J. (2001). *Band structure and electrical properties of solids*. Clarendon Press, Oxford.

Sze, S.M. (1985). *Semiconductor devices*. Wiley, New York.

Wilson, J. and Hawkes, J. (1998). *Optoelectronics* (3rd edn). Prentice Hall Europe, London.

Woodgate, G.K. (1980). *Elementary atomic structure* (2nd edn). Clarendon Press, Oxford.

Yariv, Amnon (1997). *Optical electronics in modern communications* (5th edn). Oxford University Press, New York.

Yu, P.Y. and Cardona, M. (1996). *Fundamentals of semiconductors*. Springer-Verlag, Berlin.

Zwerdling, S., Lax, B., and Roth, L.M. (1957). Oscillatory magneto-absorption in semiconductors. *Phys. Rev.*, **108**, 1402–8.

Chapter 4: Excitons

Burns, G. (1985). *Solid state physics*. Academic Press, San Diego.

Butov, L.V. (2007). Cold exciton gases in coupled quantum well structures. *J. Phys.: Condens. Matter*, **19**, 295202.

Dexter, D.L. and Knox, R.S. (1965). *Excitons*. Wiley, New York.

Fehrenbach, G.W., Schäfer, W., and Ulbrich, R.G. (1985). Excitonic versus plasma screening in highly excited gallium arsenide. *J. Luminescence*, **30**, 154–61.

Griffin, A., Snoke, D.W., and Stringari, S. (1995). *Bose–Einstein condensation*. Cambridge University Press, Cambridge.

Kavokin, A.V., Baumberg, J.J., Malpuech, G., and Laussy, F.P. (2007). *Microcavities*. Oxford University Press, Oxford.

Kasprzak, J., Richard, M., Kundermann, S., Baas, A., Jeambrun, P., Keeling, J.M.J., et al. (2006). Bose–Einstein condensation of exciton polaritons. *Nature*, **443**, 409–14.

Kittel, Charles (2005). *Introduction to solid state physics* (8th edn). Wiley, New York.

Klingshirn, C.F. (1995). *Semiconductor optics*. Springer-Verlag, Berlin.

Mandl, F. (1988). *Statistical physics* (2nd edn). Wiley, Chichester.

Matsui, A. and Nishimura, H. (1980). Luminescence of free and self trapped excitons in pyrene. *J. Phys. Soc. Jap.*, **49**, 657–63.

Moskalenko, S.A. and Snoke, D.W. (2000). *Bose–Einstein condensation of excitons and biexcitons*. Cambridge University Press, Cambridge.

Palik, E.D. (1985). *Handbook of the optical constants of solids*. Academic Press, San Diego.

Pankove, J.I. (1971). *Optical processes in semiconductors*. Dover, New York.

Rashba E.I. and Sturge, M.D. (1982). *Excitons*. North Holland, Amsterdam.

Reynolds, D.C. and Collins, T.C. (1981). *Excitons: their properties and uses*. Academic Press, New York.

Seeger, K. (1997). *Semiconductor physics* (6th edn). Springer-Verlag, Berlin.

Song, K.S. and Williams, R.T. (1993). *Self–trapped excitons*. Springer-Verlag, Berlin.

Sturge, M.D. (1962). Optical absorption of gallium arsenide between 0.6 and 2.75 eV. *Phys. Rev.*, **127**, 768–73.

Yu, P.Y. and Cardona, M. (1996). *Fundamentals of semiconductors*. Springer-Verlag, Berlin.

Chapter 5: Luminescence

Awschalom, D.D., Loss, D., and Samarth, N. (2002). *Semiconductor spintronics and quantum computation*. Springer-Verlag, Berlin.

Bhattacharya, P. (1997). *Semiconductor optoelectronic devices* (2nd edn). Prentice Hall, New Jersey.

Chuang, S.L. (1995). *Physics of optoelectronic devices*. Wiley, New York.

Dyakonov, M.I. (2008). *Spin physics in semiconductors*. Springer-Verlag, Berlin.

Elliott, R.J. and Gibson, A.F. (1974). *An introduction to solid state physics and its applications*. Macmillan, New York.

Gustafsson, A., Pistol, M.-E., Montelius, L., and Samuelson, L. (1998). Local probe techniques for luminescence studies of low-dimensional semiconductor structures. *J. Appl. Phys.*, **84**, 1715–75.

Kash K. and Shah J. (1984). Carrier energy relaxation in $In_{0.53}Ga_{0.47}As$ determined from picosecond luminescence studies. *Appl. Phys. Lett.*, **45**, 401–3.

Kusrayev, Y. and Landwehr, G. (2008). Special issue on optical orientation. *Semicond. Sci. Technol.*,

23, 110301–114018.

Landsberg, P.T. (1991). *Recombination in semiconductors*. Cambridge University Press, Cambridge.

Madelung, O. (1996). *Semiconductors, basic data* (2nd edn). Springer-Verlag, Berlin.

Meier, F. and Zakharchenya, B.P. (1984). *Optical orientation*. North Holland, Amsterdam.

Nakamura, F., Pearton S., and Fasol, G. (2000). *The blue laser diode* (2nd edn). Springer-Verlag, Berlin.

Pankove, J.I. (1971). *Optical processes in semiconductors*. Dover, New York.

Schubert, E.F. (2006). *Light-emitting diodes* (2nd edn). Cambridge University Press, Cambridge.

Shah, J. (1999). *Ultrafast spectroscopy of semiconductors and semiconductor nanostructures* (2nd edn). Springer-Verlag, Berlin.

Silfvast, W.T. (2004). *Laser fundamentals* (2nd edn). Cambridge University Press, Cambridge.

Svelto, O. (1998). *Principles of lasers* (4th edn). Plenum Press, New York.

Sze, S.M. (1981). *Physics of semiconductor devices* (2nd edn). Wiley, New York.

Sze, S.M. (1985). *Semiconductor devices*. Wiley, New York.

Voos, M., Leheney, R.F., and Shah, J. (1980). Radiative recombination. In *Handbook on semiconductors*, vol. 2 (ed. M. Balkanski). North Holland, Amsterdam, pp. 329–416.

Wilson, J. and Hawkes, J. (1998). *Optoelectronics* (3rd edn). Prentice Hall Europe, London.

Yacobi, B.G. and Holt, D.B. (1990). *Cathodoluminescence microscopy of inorganic solids*. Plenum, New York.

Yariv, Amnon (1997). *Optical electronics in modern communications* (5th edn). Oxford University Press, New York.

Yu, P.Y. and Cardona, M. (1996). *Fundamentals of semiconductors*. Springer-Verlag, Berlin.

Chapter 6: Quantum confinement

Awschalom, D.D., Loss, D., and Samarth, N. (2002). *Semiconductor spintronics and quantum computation*. Springer-Verlag, Berlin.

Bastard, G. (1990). *Wave mechanics applied to semiconductor heterostructures*. Wiley, New York.

Bimberg, D., Grundmann M., and Ledentsov, N.N. (1999). *Quantum dot heterostructures*. Wiley, Chichester.

Blood. P (1999). Visible-emitting quantum well lasers. In *Semiconductor quantum electronics* (eds A. Miller, M. Ebrahimzadeh, and D.M. Finlayson). Institute of Physics, Bristol, pp. 193-211.

Burns, G. (1985). *Solid state physics*. Academic Press, San Diego.

Chuang, S.L. (1995). *Physics of optoelectronic devices*. Wiley, New York.

Davies, A.G., Linfield, E.H., and Pepper, M. (2004). Proceedings of the Discussion Meeting on 'The terahertz gap: the generation of far-infrared radiation and its applications'. *Phil. Trans. R. Soc. Lond. A*, **362**, pp. 197–414.

De Giorgi, M., Tarì, Manna, L., Krahne, R., and Cingolani, R. (2005). Optical properties of colloidal nanocrystal spheres and tetrapods. *Microelectronics Journal*, **36**, 552–4.

Dyakonov, M.I. (2008). *Spin physics in semiconductors*. Springer-Verlag, Berlin.

Esaki, L. and Tsu, R. (1970). Superlattice and negative differential conductivity in semiconductors. *IBM Journal of Research and Development*, **14**, 61–5.

Fox A.M. (1996). Optoelectronics in quantum well structures. *Contemporary Physics*, **37**, 111–25.

Fry, P.W., Itskevich, I.E., Mowbray, D.J., Skolnick, M.S., Finley, J.J., Barker, J.A., et al. (2000). Inverted electron-hole alignment in InAs-GaAs self-assembled quantum dots. *Phys. Rev. Lett.*, **84**, 733–6.

Gasiorowicz, Stephen (1996). *Quantum physics* (2nd edn). Wiley, New York.

Harrison, P. (2005). *Quantum wells, wires and dots* (2nd edn). Wiley, Chichester.

Helm, M. (2000). *Long wavelength infrared emitters based on quantum wells and superlattices.* Gordon and Breach, Amsterdam.

Jaros, M. (1989). *Physics and applications of semiconductor microstructures.* Clarendon Press, Oxford.

Kagan, C.R., Murray, C.B., and Bawendi, M.G. (1996). Long-range resonance transfer of electronic excitations in close-packed CdSe quantum-dot solids. *Phys. Rev. B*, **54**, 8633–43.

Kelly, M.J. (1995). *Low-dimensional semiconductors.* Clarendon Press, Oxford.

Liu, H.C. and Capasso, F. (2000a). *Intersubband transitions in quantum wells: physics and device applications I, Semiconductors and Semimetals*, vol. 62 (series eds R.K. Willardson and E.R. Weber). Academic Press, San Diego.

Liu, H.C. and Capasso, F. (2000b). *Intersubband transitions in quantum wells: physics and device applications II, Semiconductors and Semimetals*, vol. 66 (series eds R.K. Willardson and E.R. Weber). Academic Press, San Diego.

Michler, P. (2003). *Single quantum dots.* Springer-Verlag, Berlin.

Miller, D.A.B., Chemla, D.S., Eilenberger, D.J., Smith, P.W., Gossard, A.C. and Tsang, W.T. (1982). Large room-temperature optical nonlinearity in GaAs/Ga$_{1-x}$Al$_x$As multiple quantum well structures. *Appl. Phys. Lett.*, **41**, 679–81.

Miller, D.A.B. (2008). *Quantum mechanics for scientists and engineers.* Cambridge University Press, New York.

Murray, C.B., Kagan, C.R., and Bawendi, M.G. (2000). Synthesis and characterization of monodisperse nanocrystals and close-packed nanocrystal assemblies. *Annu. Rev. Mater. Sci.*, **30**, 545–610.

Oulton R., Finley J.J., Ashmore A.D., Gregory I.S., Mowbray D.J., Skolnick M.S., et al. (2002). Manipulation of the homogeneous linewidth of an individual In(Ga)As quantum dot. *Phys. Rev. B*, **66**, 045313.

Schiff, L.I. (1969). *Quantum mechanics.* McGraw-Hill, New York.

Singh, J. (1993). *Physics of semiconductors and their heterostructures.* McGraw-Hill, New York.

Singleton, J. (2001). *Band structure and electrical properties of solids.* Clarendon Press, Oxford.

Viña, L. (1999). Spin relaxation in low-dimensional systems. *J. Phys.: Condens. Matter*, **11**, 5929–52.

Weisbuch, C. and Vinter, B. (1991). *Quantum semiconductor structures.* Harcourt, San Diego.

Williams, B.S. (2007). Terahertz quantum-cascade lasers. *Nature Photonics*, **1**, 517.

Woggon, U. (1997). *Optical properties of semiconductor quantum dots.* Springer-Verlag, Berlin.

Woodgate, G.K. (1980). *Elementary atomic structure* (2nd edn). Clarendon Press, Oxford.

Yu, P.Y. and Cardona, M. (1996). *Fundamentals of semiconductors.* Springer-Verlag, Berlin.

Chapter 7: Free electrons

Ashcroft, N.W. and Mermin, N.D. (1976). *Solid state physics.* Saunders College Publishing, Philadelphia.

Barnes, W.L., Dereux, A., and Ebbeson, T.W. (2003). Surface plasmon subwavelength optics. *Nature*, **424**, 824–30.

Bleaney, B.I. and Bleaney, B. (1976). *Electricity and magnetism* (3rd edn). Clarendon Press, Oxford. Reissued in two volumes in 1989.

Born, M. and Wolf, E. (1999). *Principles of optics* (7th edn). Cambridge University Press, Cambridge.

Burns, G. (1985). *Solid state physics.* Academic Press, San Diego.

Ebbeson, T.W., Genet, C., and Bozhevolnyi, S.I. (2008). Surface-plasmon circuitry. *Physics Today*, **61**(5), 44–50.

Ehrenreich, H., Philipp, H.R., and Segall, B. (1963). Optical properties of aluminium. *Phys. Rev.*, **132**, 1918–28.

Givens, M.P. (1958). Optical properties of metals. In *Solid state physics*, vol. 6 (eds F. Seitz and D. Turnbull). Academic Press, New York, pp. 313–52.

Hecht, Eugene (2001). *Optics* (4th edn). Addison-Wesley, Reading, MA.

Jagannath, C., Grabowski, Z.W., and Ramdas, A.K. (1981). Linewidths of the electronic excitation spectra of donors in silicon. *Phys. Rev. B*, **23**, 2023–98.

Kittel, Charles (2005). *Introduction to solid state physics* (8th edn). Wiley, New York.

Lal, S., Link, S., and Halas, N.J. (2007). Nano-optics from sensing to waveguiding. *Nature Photonics*, **1**, 641–8.

Lide D.R. (1996). *CRC handbook of chemistry and physics* (77th edn), CRC Press, Boca Raton.

Maier, Stefan A. (2007). *Plasmonics: fundamentals and applications*. Springer-Verlag, Berlin.

Maier, S.A., and Atwater, H.A. (2005). Plasmonics: localization and guiding of electromagnetic energy in metal/dielectric structures. *J. Appl. Phys.*, **98**, 011101.

Mooradian, A. (1972). Raman spectroscopy of solids. In *Laser handbook* vol. II (eds F.T. Arecchi and E.O. Schulz-duBois). North Holland, Amsterdam, pp. 1409–56.

Moruzzi, V.L., Janak, J.F., and Williams, A.R. (1978). *Calculated electronic properties of metals*. Pergamon Press, New York.

Murray, W.A. and Barnes, W.L. (2007). Plasmonic materials. *Adv. Mater.*, **19**, 3771–82.

Pendry, J.B. (2004). Negative refraction. *Contemp. Phys.*, **45**, 191–202.

Pendry, J.B. and Smith, D.R. (2004). Reversing light with negative refraction. *Physics Today*, **57**(6), 37–43.

Pidgeon, C.R. (1980). Free carrier optical properties of semiconductors. In *Handbook on Semiconductors*, vol. 2 (ed. M. Balkanski). North Holland, Amsterdam, pp. 223–328.

Raether, Heinz (1988). *Surface plasmons*. Springer-Verlag, Berlin.

Ramakrishna, S.A. (2005). Physics of negative refractive index materials. *Rep. Prog. Phys.*, **68**, 449–521.

Segall, B. (1961). Energy bands of aluminium. *Phys. Rev.*, **124**, 1797–806.

Shalaev, V.M. (2007). Optical negative-index metamaterials. *Nature Photonics*, **1**, 41–8.

Singleton, J. (2001). *Band structure and electrical properties of solids*. Clarendon Press, Oxford.

Spitzer, W.G. and Fan, H.Y. (1957). Determination of optical constants and carrier effective mass of semiconductors. *Phys. Rev.*, **106**, 882–90.

Veselago, V.G. (1968). The electrodynamics of substances with simultaneously negative values of ϵ and μ. *Soviet Physics USPEKHI*, **10**, 509–14.

Wyckoff, R.W.G. (1963). *Crystal structures* (2nd edn). Wiley Interscience, New York .

Yu, P.Y. and Cardona, M. (1996). *Fundamentals of semiconductors*. Springer-Verlag, Berlin.

Chapter 8: Molecular materials

Avouris, P. (2009). Carbon nanotube electronics and photonics. *Physics Today*, **62**(1), 34–40.

Banwell, C.N. and McCash, E.M. (1994). *Fundamentals of molecular spectroscopy* (4th edn). McGraw-Hill, London.

Brabec, C., Dyakonov, V., and Scherf, U. (2008). *Organic photovoltaics*. Wiley–VCH, Weinheim.

Castro Neto, A.H., Guinea, F., Peres, N.M.R., Novoselov, K.S., and Geim, A.K. (2009). The electronic properties of graphene. *Rev. Mod. Phys.*, **81**, 109–62

Dresselhaus, M.S., Dresselhaus, G., and Eklund, P.C. (1996). *Science of fullerenes and carbon nanotubes*. Academic Press, San Diego.

Dresselhaus, M.S., Dresselhaus, G., Saito, R., and Jorio, A. (2007). Exciton photophysics or carbon nanotubes. *Annu. Rev. Phys. Chem.*, **58**, 719–47.

Farchioni, R. and Grosso, G. (2001). *Organic electronic materials*. Springer-Verlag, Berlin.

Forrest, S.R. and Thompson, M.E. (2007). Special issue on organic electronics and optoelectronics. *Chemical Reviews*, **107**, 923–1386.

Friend, R.H., Gymer, R.W., Holmes, A.B., Burroughes, J.H., Marks, R.N., Taliani, C., et al. (1999). Electroluminescence in conjugated polymers. *Nature*, **397**, 121–8.

Garbuzov, D.Z., Bulović, V., Burrows, P.E., and Forrest, S.R. (1996). Photoluminescence efficiency and absorption of aluminium-tris-quinolate (Alq$_3$) thin films. *Chem. Phys. Lett.*, **249**, 433–7.

Gasiorowicz, Stephen (1996). *Quantum physics* (2nd edn). Wiley, New York.

Geim, A.K. and MacDonald, A.H. (2007). Graphene: exploring carbon flatland. *Physics Today*, **60**(8), 35–41.

Geim, A.K. and Novoselov, K.S. (2007). The rise of graphene. *Nature Materials*, **6**, 183–91.

Gorman, A.A., Hamblett, I., King, T.A., and Rahn, M.D. (2000). A pulse radiolysis and laser study of the pyrromethene 567 triplet state. *J. Photochem. Photobiol. A: Chemistry*, **130**, 127–132.

Haken, H. and Wolf, H.C. (1995). *Molecular physics and elements of quantum chemistry*. Springer-Verlag, Berlin.

Hartschuh, A., Pedrosa, H.N., Novotny, L., and Krauss, T.D. (2003). Simultaneous fluorescence and Raman scattering from single carbon nanotubes. *Science*, **301**, 1354–6.

Hertel, D., Bässler, H., Scherf, U., and Hörhold, H.H. (1999). Charge carrier transport in conjugated polymers. *J. Chem. Phys.*, **110**, 9214–22.

Kataura, H., Kumazawa, Y., Maniwa, Y., Umezu, I., Suzuki, S., Ohtsuka, Y., and Achiba, Y. (1999). Optical Properties of single-wall carbon nanotubes. *Synthetic Metals*, **103**, 2555–8.

Klessinger, M. and Michl, J. (1995). *Excited states and photochemistry of organic molecules*. VCH Publishers, New York.

Kravets, V.G., Grigorenko, A.N., Nair, R.R., Blake, P., Anissimova, S., Novoselov, K.S., and Geim, A.K. (2010). Spectroscopic ellipsometry of graphene and an exciton-shifted van Hove peak in absorption. *Phys. Rev. B*, **81**, 155413.

Machón, M., Reich, S., Thomsen, C., Sánchez-Portal, D., and Ordejón, P. (2002). *Ab initio* calculations of the optical properties of 4-Å-diameter single-walled nanotubes. *Phys. Rev. B*, **66**, 155410.

Mak, K.F., Shan, J., and Heinz, T.F. (2011). Seeing many-body effects in single- and few-layer graphene: observation of two-dimensional saddle-point excitons. *Phys. Rev. Lett.*, **106**, 046401.

Möller, S. and Weiser, G. (1999). Photoconductivity of polydiacetylene chains in polymer and monomer single crystals. *Chem. Phys.*, **246**, 483-94.

Mueller, G. (2000). *Electroluminescence I & II, Semiconductors and Semimetals*, vols 64 & 65 (series eds R.K. Willardson and E.R. Weber). Academic Press, San Diego.

Nair, R.R., Blake, P., Grigorenko, A.N., Novoselov, K.S., Booth, T.J., Stauber, T., et al. (2008). Fine structure constant defines visual transparency of graphene. *Science*, **320**, 1308.

Pope, M. and Swenberg, C.E. (1999). *Electronic processes in organic crystals and polymers* (2nd edn). Oxford University Press, New York.

Reich, S., Thomsen, C., and Maultzsch, J. (2004). *Carbon nanotubes*. Wiley–VCH, Weinheim.

Ren, S.L., Wang, Y., Rao, A.M., McRae, E., Holden, J.M., Hager, T., et al. (1991). Ellipsometric determination of the optical properties of C$_{60}$ (Buckminsterfullerene) films. *Appl. Phys. Lett.*, **59**, 2678–80.

Saito, S. and Zettl, A. (2008). *Carbon nanotubes: quantum cylinders of graphene*. Elsevier, Amsterdam.

Schlaich, H., Muccini, M., Feldmann, J., Bässler, H., Göbel, E.O., Zamboni, R., et al. (1995). Absorption at the dipole-forbidden optical gap of C$_{60}$. *Chem. Phys. Lett.*, **236**, 135–40.

Scholes, G.D. and Rumbles, G. (2006). Excitons in nanoscale systems. *Nature Materials*, **5**, 683–96.

Slepkov, A.D., Hegmann, F.A., Eisler, S., Elliott, E., and Tykwinski, R.R. (2004). The surprising nonlinear optical properties of conjugated polyyne oligomers. *J. Chem. Phys.*, **120**, 6807–10.

Tang, C.W. and VanSlyke S.A. (1987). Organic electroluminescent diodes. *Appl. Phys. Lett.*, **51**, 913–5.

Watanabe, K. (1954). Photoionization and total cross section of gases. I. Potentials of several

molecules. Cross sections of NH_3 and NO. *J. Chem. Phys.*, **22**, 1564–70.

Wolf, von H.C. (1958). Die niedersten elektronischen Anregungszustände des Anthracen-Kristalls. *Z. Natürforsch.*, **A13**, 414–9.

Wright, J.B. (1995). *Molecular crystals* (2nd edn). Cambridge University Press, Cambridge.

Chapter 9: Luminescence centres

Acosta, V.M., Bauch, E., Ledbetter, M.P., Santori, C., Fu, K.-M.C., Barclay, P.E., et al. (2009). Diamonds with a high density of nitrogen-vacancy centers for magnetometry applications. *Phys. Rev. B*, **80**, 115202.

Ashcroft, N.W. and Mermin, N.D. (1976). *Solid state physics.* Saunders College Publishing, Philadelphia.

Balasubramanian, G., Neumann, P., Twitchen, D., Markham, M., Kolesov, R., Mizuochi, N., et al. (2009). Ultralong spin coherence time in isotopically engineered diamond. *Nature Materials*, **8**, 383–7

Baldacchini G. (1992). Relaxed excited states of color centers. In *Optical properties of excited states in solids* (ed. B. Di Bartolo), NATO ASI Series B, vol. 301. Plenum Press, New York, pp. 255–303.

Blundell, S. (2001). *Magnetism in condensed matter physics.* Clarendon Press, Oxford.

Burns, G. (1985). *Solid state physics.* Academic Press, San Diego.

Di Bartolo, B. (1992). *Optical properties of excited states in solids*, NATO ASI Series B, vol. 301. Plenum Press, New York.

Elliott, R.J. and Gibson, A.F. (1974). *An introduction to solid state physics and its applications.* Macmillan, New York.

Gaebel, T., Popa, I., Gruber, A., Domhan, M., Jelezko, F., and Wrachtrup, J. (2004). Stable single-photon source in the near infrared. *New Journal of Physics*, **6**, 98.

Hayes, W. and Stoneham, A.M. (1985). *Defect and defect processes in nonmetallic solids.* Wiley, New York.

Henderson, B. and Bartram, R.H. (2000). *Crystal-field engineering of solid-state laser materials.* Cambridge University Press, Cambridge.

Henderson, B. and Imbusch, G.F. (1989). *Optical spectroscopy of inorganic solids.* Clarendon Press, Oxford.

Jelezko, F., Popa, I., Gruber, A., Tietz, C., Wrachtrup, J., Nizivtsev, A., and Kilin, S. (2002). Single spin states in a defect center resolved by optical spectroscopy. *Appl. Phys. Lett.*, **81**, 2160–2.

Jelezko, F. and Wrachtrup, J. (2004). Read-out of single spins by optical spectroscopy. *J. Phys.: Condens. Matter*, **16**, R1089–104.

Jelezko, F. and Wrachtrup, J. (2006). Single defect centres in diamond: a review. *Physica Stat. Sol.*(a), **203**, 3207–25.

Kittel, Charles (2005). *Introduction to solid state physics* (8th edn). Wiley, New York.

Koningstein, J.A. and Geusic, J.E.(1964). Energy levels and crystal-field calculations of neodymium in yttrium aluminium garnet. *Phys. Rev*, **136**, A711–6.

Manson, N.B., Harrison, J.P., and Sellars, M.J. (2006). Nitrogen-vacancy center in diamond: Model of the electronic structure and associated dynamics. *Phys. Rev. B*, **74**, 104303.

Mita, Y. (1996). Change of absorption spectra in type-Ib diamond with heavy neutron bombardment. *Phys. Rev. B*, **53**, 11360–4.

Mollenauer, L.F. (1985). Color center lasers. In *Laser handbook*, vol. 4., (eds M.L. Stitch and M. Bass). Elsevier Science Publishers, North Holland, pp. 143–228.

Moulton P.F. (1986). Spectroscopic and laser characteristics of Ti:Al$_2$O$_3$. *J. Opt. Soc. Am. B*, **3**, 125–33.

Mueller-Mach, R., Mueller, G., Krames, M.R., Höppe, H.A., Stadler, F., Schnick, W., et al. (2005). Highly efficient all-nitride phosphor-converted white light emitting diode. *Physica Stat. Sol.* (a), **202**, 1727–32.

Narukawa Y. (2004). White-light LEDs. *Optics & Photonics News*, **15**(4), 24–9.

Schubert, E.F., Kim, J.K., Luo, H., and Xi, J.-Q. (2006). Solid-state lighting—a benevolent technology. *Rep. Prog. Phys.*, **69**, 3069–99.

Shur, M.S. and Žukauskas A. (2005). Solid-state lighting; toward superior illumination. *Proc. IEEE*, **93**, 1691–703.

Silfvast, W.T. (2004). *Laser fundamentals* (2nd edn). Cambridge University Press, Cambridge.

Smets, B. (1992). Advances in sensitization of phosphors. In *Optical properties of excited states in solids* (ed. B. Di Bartolo), NATO ASI Series B, vol. 301. Plenum Press, New York, pp. 349–98.

Svelto, O. (1998). *Principles of lasers* (4th edn). Plenum Press, New York.

Chapter 10: Phonons

Ashcroft, N.W. and Mermin, N.D. (1976). *Solid state physics*. Saunders College Publishing, Philadelphia.

Burns, G. (1985). *Solid state physics*. Academic Press, San Diego.

Hass, M. (1967). Lattice reflection. In *Semiconductors and Semimetals, vol. 3: Optical properties of III–V compounds* (eds R.K. Willardson and A.C. Beer). Academic Press, New York, pp. 3–16.

Henry, C.H. and Hopfield, J.J. (1965). Raman scattering by polaritons. *Phys. Rev. Lett.*, **15**, 964–6.

Houghton, J.T. and Smith, S.D. (1966). *Infra-red physics*. Clarendon Press, Oxford.

Ibach, H. and Luth, H. (2003). *Solid-state physics* (3rd edn). Springer-Verlag, Berlin.

Kittel, Charles (2005). *Introduction to solid state physics* (8th edn). Wiley, New York.

Madelung, O. (1978). *Introduction to solid-state theory*. Springer-Verlag, Berlin.

Madelung, O. (1996). *Semiconductors, basic data* (2nd edn). Springer-Verlag, Berlin.

Maier, Stefan A. (2007). *Plasmonics: fundamentals and applications*. Springer-Verlag, Berlin.

Mooradian, A. (1972). Raman spectroscopy of solids. In *Laser Handbook*, vol. II (eds F.T. Arecchi and E.O. Schulz-duBois). North Holland, Amsterdam, pp. 1409–56.

Pidgeon, C.R. (1980). Free carrier optical properties of semiconductors. In *Handbook on Semiconductors*, vol. 2 (ed. M. Balkanski). North Holland, Amsterdam, pp. 223–328.

Pope, M. and Swenberg, C.E. (1999). *Electronic processes in organic crystals and polymers* (2nd edn). Oxford University Press, New York.

Seeger, K. (1997). *Semiconductor physics* (6th edn). Springer-Verlag, Berlin.

Shah, J. (1999). *Ultrafast spectroscopy of semiconductors and semiconductor nanostructures* (2nd edn). Springer-Verlag, Berlin.

Song, K.S. and Williams, R.T. (1993). *Self-trapped excitons*. Springer-Verlag, Berlin.

Turner, W.J. and Reese, W.E. (1962). Infrared lattice bands of AlSb. *Phys. Rev.*, **127**, 126-31.

Yu, P.Y. and Cardona, M. (1996). *Fundamentals of semiconductors*. Springer-Verlag, Berlin.

Chapter 11: Nonlinear optics

Born, M. and Wolf, E. (1999). *Principles of optics* (7th edn). Cambridge University Press, Cambridge.

Butcher, P.N. and Cotter, D. (1990). *The elements of nonlinear optics*. Cambridge University Press, Cambridge.

Chemla, D.S. (1985). Excitonic optical nonlinearities. *J. Opt. Soc. Am. B*, **2**, 1135–1243.

DeSalvo, R., Said, A.A., Hagan, D.J., Van Stryland, E.W., and Sheik–Bahae, M. (1996). Infrared to ultraviolet measurements of two-photon absorption and n_2 in wide bandgap solids. *IEEE J. Quantum Electron.*, **32**, 1324–33.

Fox, A.M., Maciel, A.C., Shorthose, M.G., Ryan, J.F., Scott, M.D., Davies, J.I., and Riffat, J.R. (1987). Nonlinear excitonic optical absorption in GaInAs/InP quantum wells. *Appl. Phys. Lett.*, **51**, 30–2.

Klein, R.S., Kugel, G.E., Maillard, A. Sifi, A., and Polgár, K. (2003). Absolute non-linear optical coefficients measurements of BBO single crystal and determination of angular acceptance by second harmonic generation. *Optical Materials*, **22**, 163–9.

Kroner, M., Rémi, S., Högele, A., Seidl, S., Holleitner, A.W., Warburton, R.J., et al. (2008). Resonant saturation laser spectroscopy of a single self-assembled quantum dot. *Physica E*, **40**, 1994–6.

Mollenauer, L.F. and Gordon, J.P. (1994). In *Nonlinear spectroscopy of solids* (ed. B. Di Bartolo), NATO ASI Series B, vol. 339. Plenum Press, New York, pp. 451–80.

Nye, J.F. (1957). *Physical properties of crystals*. Clarendon Press, Oxford.

Schmitt-Rink, S., Chemla, D.S., and Miller, D.A.B. (1989). Linear and nonlinear optical properties of semiconductor quantum wells. *Adv. Phys.*, **38**, 89–188.

Sheik–Bahae, M., Hutchings, D.C., Hagan, D.J., and Van Stryland, E.W. (1991). Dispersion of bound electronic nonlinear refraction in solids. *IEEE J. Quantum Electron.*, **27**, 1296–309.

Tang, C.L. (1995). Nonlinear optics. In *Handbook of optics*, vol. II (ed. M. Bass). McGraw-Hill, New York, chapter 38.

Westland, D.J., Fox, A.M., Maciel, A.C., Ryan, J.F., Scott, M.D., Davies, J.I., and Riffat, J.R. (1987). Optical studies of excitons in $Ga_{0.47}In_{0.53}As/InP$ multiple quantum wells. *Appl. Phys. Lett.*, **50**, 839–41.

Yariv, Amnon (1997). *Optical electronics in modern communications* (5th edn). Oxford University Press, New York.

Appendices

Ashcroft, N.W. and Mermin, N.D. (1976). *Solid state physics*. Saunders College Publishing, Philadelphia.

Bleaney, B.I. and Bleaney, B. (1976). *Electricity and magnetism* (3rd edn). Clarendon Press, Oxford. Reissued in two volumes in 1989.

Born, M. and Wolf, E. (1999). *Principles of optics* (7th edn). Cambridge University Press, Cambridge.

Burns, G. (1985). *Solid state physics*. Academic Press, San Diego.

Corney, Alan (1977). *Atomic and laser spectroscopy.* Clarendon Press, Oxford.

Duffin W.J. (1990). *Electricity and magnetism* (4th edn). McGraw-Hill, London.

Fox, M. (2006). *Quantum optics: an introduction.* Clarendon Press, Oxford.

Gasiorowicz, Stephen (1996). *Quantum physics* (2nd edn). Wiley, New York.

Good R.H. (1999). *Classical electromagnetism.* Saunders College Publishing, Fort Worth.

Grant, I.S. and Phillips, W.R. (1990). *Electromagnetism* (2nd edn). Wiley, New York.

r	position vector	ϵ_{st}	static relative dielectric constant
r_{ij}	electro-optic coefficient	ϵ_∞	high-frequency relative dielectric constant
r_p	polaron radius	η	quantum efficiency
R	reflectivity, electrical resistance	η_R	radiative quantum efficiency
R_X	exciton Rydberg constant	θ	angle
S	Huang–Rhys parameter	κ	imaginary part of the complex refractive
$\boldsymbol{S}$	spin angular momentum		index
t	time	λ	wavelength
T	transmissivity, temperature	λ_{deB}	de Broglie wavelength
T_c	critical temperature	μ	reduced mass, chemical potential
T_L	lattice temperature	μ_r	relative magnetic permeability
T_m	melting temperature	ν	frequency
$\boldsymbol{T}$	lattice translation vector	$\bar{\nu}$	wave number
u	envelope function within a Bloch function	ν_{LO}	LO phonon frequency at $q = 0$
$u(\nu)$	energy density of an electromagnetic wave	ν_{TO}	TO phonon frequency at $q = 0$
	at frequency ν	Π	spin polarization
U	potential energy	ρ	density of states
v	velocity of light in a medium	ϱ	electrical charge density
v_g	group velocity	σ	electrical conductivity
v_s	velocity of sound	σ_s	scattering cross-section
$\boldsymbol{v}$	electron velocity	τ	lifetime
V	volume, voltage, Verdet coefficient	τ_{NR}	non-radiative lifetime
V_{bi}	built-in voltage	τ_R	radiative lifetime
W	transition rate	τ_S	spin relaxation time
x	position coordinate	ϕ	azimuthal angle in spherical polar
y	position coordinate		coordinates
z	position coordinate	Φ	optical phase
Z	impedance, atomic number	φ	wave function
α	absorption coefficient, polarizability	χ	electric susceptibility
α_{ep}	electron–phonon coupling constant	χ_a	electric susceptibility per atom
γ	damping rate, fractional loss	χ_M	magnetic susceptibility
γ_ν	gain coefficient	ψ	wave function
γ_{th}	threshold gain coefficient for laser	Ψ	wave function
	oscillation	ω	angular frequency
δ	skin depth	ω_c	cyclotron frequency
Δ	split-off hole band energy	ω_p	plasma frequency
ϵ_r	relative dielectric constant	ω_{sp}	surface plasmon frequency
$\tilde{\epsilon}_r$	complex relative dielectric constant	Ω	phonon angular frequency, Larmor
ϵ_1	real part of the complex relative dielectric		precession angular frequency
	constant	Ω_{LO}	LO phonon angular frequency at $q = 0$
ϵ_2	imaginary part of the complex relative	Ω_{TO}	TO phonon angular frequency at $q = 0$
	dielectric constant		

List of quantum numbers

In atomic physics, lower- and upper-case letters usually refer to individual electrons or whole atoms respectively.

j, J	total angular momentum	m_s, M_S	magnetic (spin angular momentum)
l, L	orbital angular momentum	n	principal
m_j, M_J	magnetic (total angular momentum)	s, S	spin
m_l, M_L	magnetic (orbital angular momentum)		

Index

absorbance, 4
absorption
  coefficient, 3
  definition, 1, 341
  direct, 64–79
  edge, fundamental, 10, 68–82, 154
  excitonic, 73, 98–101, 157–158
  free carrier, 12, 191–196
  impurity, 73, 196–198
  in circularly polarized light, 77, 158
  in electric field, 74–75, 102, 160–164
  in magnetic field, 75
  indirect, 64, 79–82
  infrared, 10, 39, 191–198, 278–280
  interband, 62–91, 152, 188
  intersubband, 166
  intervalence band, 195
  lattice, see phonon absorption
  measurement, 84
  molecular, 32
  nonlinear, 105, 304, 305, 324–326
  quantum dot, 170, 326
  quantum well, 152–160
  saturable, 305, 324–326
  transition metal, 190
  ultraviolet, 10, 39, 42, 46
  vibrational, see phonon absorption
  vibronic, see vibronic absorption
absorption data
  alkali halide F-centre, 251
  Alq3, 229
  ammonia, 225
  C_{60}, 242
  CdSe quantum dot, 171
  CdTe quantum dot, 171
  F_2^+ band in KF, 252
  GaAs, 80, 100–102, 105, 158
  GaAs quantum well, 157, 158, 164, 177
  GaInAs quantum well, 325
  GaN, 117
  GaP, 93
  germanium, 81
  InAs, 73
  InGaAs quantum dot, 326
  LiF, 108
  MeLPPP polymer, 232
  NaCl, 108
  nonlinear, 325, 326
  polydiacetylene, 231
  polyfluorene, 16

polyyne, 226
pyrene, 109
pyrromethene dye, 225
silica glass, 41
silicon, 80, 83
silicon, phosphorus impurity, 197
Ti:sapphire, 261
AC conductivity, 183
acceptor, 191, 356
acetylene (C_2H_2), 226, 230
activity, optical, 55
AgCl, electron–phonon coupling, 283
Al_2O_3, see sapphire
AlGaAs light-emitting diode, 118, 127
AlGaInP quaternary alloy, 128
alkali halide, 32, 285
  F centres, 250
  Frenkel excitons, 108
alkali metal, ultraviolet transparency, 186
Alq3, optical spectra, 229
AlSb
  infrared reflectivity, 293
  Raman scattering, 289
aluminium
  band structure, 189
  interband absorption, 188
  reflectivity, 186
ammonia absorption, 225
amorphous material, 20, 216
Ampere's law, 332
amplifier, optical, 261
angular momentum, 350–353
  coupling schemes, 351
  orbital, 351
  photon, 348
  spin, 351
  total, 351
anharmonicity, 290, 299
anion vacancy, 251
anisotropy
  induced, 53–54
  nonlinear, 297
  optical, 11, 19, 48–54
anthracene ($C_{14}H_{10}$)
  absorption spectrum, 228
  exciton, 109
  structure, 228
anti-Stokes scattering, 286
aromatic hydrocarbon, 109, 227
artificial atom, 144, 167, 326

atomic notation, 351–352
axis, optic, 49

back-scattering geometry, 287
band
  conduction, 354
  electron, 20, 70
  heavy hole, 70
  light hole, 70
  split-off hole, 71
  theory, 354–362
  valence, 354
  vibronic, 21, 228, 247
band edge absorption, 68–82
band filling nonlinearity, 325
band gap, 10, 354
  data, 361
  direct, 63, 116
  indirect, 63, 117
band structure, 356–362
  Γ point, 70, 359
  aluminium, 189
  camel back, 71, 83
  copper, 191
  critical point, 84
  direct gap, 63
  four band model, 70, 77
  GaAs, 69
  germanium, 68, 81
  III–V semiconductor, 69–71
  indirect gap, 63
  L point, 70, 359
  parabolic, 71
  silicon, 84
  X point, 70, 359
BAP mechanism, 124
Beer's law, 3
benzene (C_6H_6) molecule, 15, 214
beryl crystal, 259
biaxial crystal, 49
biexciton, 105
Bir–Aronov–Pikus mechanism, 124
birefringence, 11, 48–53
  circular, 55
  induced, 53–54, 313–316
Bloch's theorem, 20, 358
Boltzmann statistics, 120, 343
Born–Oppenheimer approximation, 220
Bose–Einstein condensation, 105
Bose–Einstein distribution, 81

boundary conditions at an interface, 338
bra-ket notation, 65
Brillouin scattering, 286, 289–290
　stimulated, 321
Brillouin zone, 356
　face-centred cubic, 359
broadening
　by damping, 36
　inhomogeneous, 165, 232
　temporal, 47, 323
buckminsterfullerene, 235, 241–243
built-in voltage, 90, 363

C_2H_2, *see* acetylene
C_2H_4, *see* ethylene
C_6H_6, *see* benzene
$C_{14}H_{10}$, *see* anthracene
$C_{16}H_{10}$, *see* pyrene
C_{60}, *see* carbon nanostructure
calcite crystal, 48, 50, 56
camel back band structure, 71, 83
carbon nanostructure, 235–243
　C_{60} bucky ball, 241–243
　graphene, 236–237
　nanotube, 237–241
carrier, free, *see* free electron
cathodoluminescence, 135–136, 264
Cauchy's dispersion formula, 59
CdSe
　quantum dot, 170
　transmission spectrum, 10
centrosymmetric crystal, 308
charge continuity equation, 199
chirality, 55–56
chromium ion, 17, 259
circular dichroism, 55
circular polarization, 335
classical statistics, 120
classical theories, 23
Clausius–Mossotti relationship, 44
coefficient
　absorption, 3
　electro-optic, 315
　extinction, 6
　nonlinear, 308
　reflection, 3
　transmission, 3
collective excitations, 22–23
colloid, metallic, 206
colloidal quantum dot, 170–172
colour centre, 250–255
　laser, 263
complex dielectric constant, 6–9
complex refractive index, 6–9, 337
conduction band, 354
conductivity
　AC, 183
　DC, 183
　extrinsic, 356
　intrinsic, 355

configuration diagram, 219–221, 249
　colour centre, 251
　molecular, 219
confinement
　electrical, 133
　optical, 133
　quantum, 141–174
conjugated molecule, 15, 109, 227–232
　cyclic, 215
　linear, 215
conjugated polymer, 15, 229–232
copper
　band structure, 191
　density of states, 191
　interband absorption, 190
　reflectivity, 192
Coulomb gauge, 333
covalent bond, polar, 32, 272
Cr^{3+}, *see* chromium ion
$Cr^{3+}:Al_2O_3$, *see* ruby
critical point, 84, 189
cross-section, scattering, 5
crystal
　biaxial, 49
　birefringent, 53
　centrosymmetric, 308
　class, 18
　ionic, 272
　molecular, 227
　nonlinear, 308
　symmetry, 18–20, 49, 52, 256, 308
　uniaxial, 49, 311
crystal-field effect, 20, 255–257
　dynamic, 256
　static, 256
CuCl biexciton, 105
current density, 183, 337
cut glass, 13
cyclotron
　energy, 284
　frequency, 75
　resonance, 284
cyclotron energy, 103

damping
　coefficient, 33
　free carrier, 183
　phonon, 275, 278, 290
de Broglie wavelength, 142, 356
defect, optically-active, 250–255
degeneracy, 122, 130
degenerate four-wave mixing, 319
delayed fluorescence, 229
delocalized state, 22–23
density of states, 21–22
　1-D, 143
　2-D, 155
　electron, 67
　joint, 66, 71–72
　momentum, 67

photon, 344
　quantum dot, 144
　quantum well, 143, 155
　quantum wire, 143
density, optical, 4
depletion region, 363
detailed balance, 342
detector
　infrared, 88, 166
　intersubband, 166
　photoconductive, 89
　quantum well, 163
　responsivity, 87
　semiconductor, 86–91
dextro-rotatory material, 55
diamagnetic shift, excitonic, 103
diamond
　band parameters, 360
　infrared transparency, 11
　NV centre, 253–255
　structure, 360
diatomic molecule, 220
dichroism
　circular, 55
　magnetic circular, 56
dielectric constant, 330
　complex, 6–9
　relative, 7, 331
　static, 35, 275
　tensor, 52
dielectric medium, 330
difference frequency mixing, 306
dimer, 229
diode
　laser, 130–135
　light-emitting, 129–130, 165, 232, 265
　p–i–n, 363–364
　p–n, 126, 130
　photo-, 87–89
dipole
　atomic, 29, 330
　electric, *see* electric dipole
　Hertzian, 30
　molecular, 31
　vibrational, 31, 275
dipole oscillator model, 28–44
　free electron, 180–185
　nonlinear, 299–302
　phonon, 273–276
Dirac delta function, 245
Dirac notation, 65, 344
Dirac point, 236
direct band gap, 63, 116
dispersion, 3, 46–48
　anomalous, 47
　electronic, 356
　glass, 13, 42, 47
　group velocity, 47
　normal, 47
　parameter, 47

phonon, 272
plasmon, 204
polariton, 281
displacement, electric, 330
divergence theorem, 199
donor, 191, 355
levels, 196
doped glass, 16–17
doped semiconductor, 191–198, 355
double refraction, 49
down conversion, 307
Drude–Lorentz model, 33, 180–188
Dyakonov–Perel (DP) mechanism, 124
dye, organic, 225

effective mass, 358
electron, 71
heavy hole, 71
light hole, 71
polaron, 284
reduced, 72
split-off hole, 71
Einstein coefficients, 114, 340–343
electric
displacement, 330
field, 330
permittivity, 331
potential, 331
quadrupole transition, 349
susceptibility, 331
electric dipole
approximation, 346
matrix element, 346
moment, 30, 65, 66, 346
selection rules, 347
transition, 65, 346
electrical confinement, 133
electro-absorption, *see* absorption, in
electric field
electro-optic effect, 75, 160
coefficient, 313
linear, 54, 307, 313–316
modulator, 163, 314, 320
quadratic, 54, 320
electro-reflectance, 75
electroluminescence, 113, 126–135, 232
electromagnetic fields, 330–333
electromagnetic wave, 333–339
electron
bound, 29
free, *see* free electron
inner, 39
nearly free, 356
π, 15, 214
valence, 185, 351
electron–phonon coupling, 119, 247, 257, 282
electron-hole
droplet, 105
overlap, 153, 168

pair, 63, 95
recombination, 115, 126, 164
Elliott–Yafet mechanism, 124
ellipsometry, 86
elliptical polarization, 335
emerald, 259
emission
of light, *see* luminescence
spontaneous, 2, 113, 340
stimulated, 131, 342
energy-loss spectroscopy, 202
epitaxial growth, 127, 144, 172
equilibrium
quasi, 119, 131
thermal, 119, 131, 342
equipartition of energy, 142
erbium optical amplifier, 263
etalon, Fabry–Perot, 4
ethylene (C_2H_4) molecule, 214, 230
evanescent waves, 184
exchange interaction, 352
exciton, 95–109
absorption, 73, 98–101, 157–158
alkali halide, 108
binding energy, 96
Bohr radius, 97
Bose–Einstein condensation, 106
broadening, 100, 104
carbon nanotube, 241
charged, 173
free, 95–106
Frenkel, *see* Frenkel exciton
GaAs, 98, 100–102, 105, 158
in electric field, 102, 160
in magnetic field, 103
indirect semiconductor, 99
interactions, 104
molecular, 108, 231
Mott density, 104
nonlinear absorption, 105, 325
polydiacetylene, 231
quantum dot, 171, 173, 326
quantum well, 157–158, 325
radius, 96
rare gas crystal, 107
Rydberg energy, 97
screening, 99
self-trapped, 285
tightly bound, 95
Wannier, 95
experimental techniques
photoluminescence, 125
Raman scattering, 288
reflectivity, 84
transmission, 84
extinction coefficient, 6
extraordinary ray, 49, 50
EY mechanism, 124

F_2^+ centre, 252

F-centres, 250–253
F8, *see* polyfluorene
Fabry–Perot etalon, 4
face-centred cubic lattice, 359
Faraday effect, 56
Faraday's law, 332
Fermi energy, 122, 354
Fermi's golden rule, 64, 344
Fermi–Dirac statistics, 120
fibre optics, 13, 48, 128, 166, 263, 322–324
field
electric, 330
magnetic, 332
field ionization, 102
filter, coloured glass, 171
fluorescence, 217, 348
delayed, 229
prompt, 229
fluorescent lighting, 264
four-wave mixing, 317
degenerate, 319
Fourier transform spectroscopy, 85
Franck–Condon principle, 221–224, 250
Franz–Keldysh effect, 74–75, 103
oscillations, 74
free electron
absorption, 12, 191–196
conductivity, 183
optical properties, 180–209
oscillator, 32
reflectivity, 14, 180–182, 191–196
free exciton, 95–106
Frenkel exciton, 95, 107–109
alkali halides, 108
molecular crystals, 108, 231
rare gas crystals, 107
self-trapped, 285
frequency
doubling, 300, 306
mixing, 305–308, 317
quadrupling, 306
tripling, 306, 318
Fresnel's equations, 86, 338
fullerene, 241–243
fundamental absorption edge, 10, 68–82, 154

g factor, 353
GaAs
absorption data, 80, 100–102, 105, 158
band parameters, 361
band structure, 69
biexciton, 105
electroluminescence, 130
electron–phonon coupling, 283
exciton, 98, 100–103, 105
infrared reflectivity, 279
laser, 134
light-emitting diode, 127

LO phonon lifetime, 291
p–i–n diode, 102, 130
photoluminescence, 121
plasmon, 202
quantum well, 106, 144, 151, 156, 177
Raman scattering, 202, 289
spin injection, 79
gain coefficient, 132
GaInAs, *see* InGaAs
GaInAsP emitter, 128
GaN
  absorption, 117
  light-emitting diode, 128, 165, 265
  luminescence, 117
  spin injection, 79
GaP
  absorption, 93
  light-emitting diode, 127
  phonon polariton, 282
  Raman scattering, 282, 289
GaSb band parameters, 361
Gauss's law, 331
gemstones, 9
generalized coordinate, 222
germanium
  absorption, 81
  band parameters, 360
  band structure, 68, 81
  detector, 88
  direct band gap, 82
  exciton, 105
  in magnetic field, 76
  indirect band gap, 79, 80
  phonon energy, 82
Glan–Foucault polarizing prism, 50
Glan–Thompson polarizing prism, 51
glass
  coloured, 16, 171, 206
  dispersion, 41
  infrared absorption, 274
  nonlinear refractive index, 322
  optical properties, 12–13
  refractive index, 12
  semiconductor-doped, 16–17, 171
  silica, 12, 42
  stained, 206
  types, 13
gold
  interband absorption, 190
  nanoparticles, 207
  reflectivity, 211
golden rule, *see* Fermi's golden rule
graphene, 235–237
group velocity, 42, 357
  dispersion, 47

Hanle effect, 124
harmonic oscillator, *see* oscillator,
        harmonic
heavy hole

band, 70
  transition, 71
Heisenberg uncertainty principle, 141
Hertzian dipole, 29
heterojunction, 134, 233
heterostructure, 144, 233
HgCdTe detector, 89
hole, 355
HOMO level, 217
hopping transport, 107, 285
horizontal polarization, 335
hot carrier effect, 122
Huang–Rhys parameter, 223, 249
hydrocarbon, aromatic, 227
hydrogen molecule, 220

Iceland Spar, 50
II–VI semiconductor, *see* semiconductor,
        II–VI
III–V semiconductor, *see* semiconductor,
        III–V
impurity
  absorption, 73, 196–198
  level, 196, 355
  paramagnetic, 255–260
InAs
  absorption, 72
  band parameters, 361
  detector, 88
  quantum dot, 172, 326
index ellipsoid, 60, 313
index of refraction, *see* refractive index
indicatrix, 313
indirect band gap, 63, 79–82, 117
indirect transition, 79–82
indium tin oxide (ITO), 233
inelastic light scattering, 285–290
infrared
  absorption, *see* absorption, infrared
  detector, 88, 166
  reflectivity, 193, 273–281
InGaAs
  detector, 88
  photoluminescence, 122
  quantum dot, 326
injection of carriers, 118, 127
InP
  band parameters, 361
  Raman scattering, 289
InSb
  band parameters, 361
  detector, 88
  reflectivity, 193
  spin injection, 79
insulator, 354
  doped, 16–17, 255
  optical properties, summary of, 9–12
  refractive index, 11
  transparency range, 11
intensity, optical, 336

interaction picture, 346
interband absorption, *see* absorption,
        interband
interband luminescence, *see*
        luminescence, interband
intersubband transition, 166
intersystem crossing, 229, 254
intervalence band absorption, 195
ionicity, 272
ionosphere, 182
isotropic medium, 35, 48, 317
  nonlinear, 321
ITO, 233

Jablonski diagram, 227, 242
joint density of states, 66, 71–72

Kane model, 71
Kataura plot, 240
KDP crystal, 297, 309, 312
Kerr effect, 54, 320
  cell, 320
  constant, 54
  gate, 320
  magneto-optical, 56
  optical, 318–321
ket notation, 65
Kleinman symmetry, 322
Kramers–Kronig relationships, 44–46, 86

laevo-rotatory material, 55
Landau levels, 75, 104
Landé g factor, 353
lanthanides, 257
Larmor precession frequency, 124
laser
  cavity, 131
  colour centre, 263
  definition, 342
  diode, 130–135
  efficiency, 133
  four level, 261
  heterojunction, 134
  modes, 132
  Nd:YAG, 261
  oscillation condition, 132
  quantum cascade, 167
  quantum well, 166
  ruby, 269
  solid-state, 261–264
  terahertz, 167
  three level, 261
  Ti:sapphire, 262
  ultrafast, 264
lattice absorption, *see* phonon absorption
lattice matching, 127, 146, 165
lattice vibrations, *see* phonon
LED, *see* light-emitting diode
left-handed material, 208
lens, perfect, 208

Lenz's law, 332
LiF excitonic absorption, 108
lifetime
    non-radiative, 115
    radiative, *see* radiative lifetime
lifting of degeneracies, 19
light-emitting diode, 129–130, 165
    organic, 232–235
    phosphor-converted, 265
    quantum well, 165
    white, 265
light hole
    band, 70
    transition, 71
light polarization, 334
lighting
    fluorescent, 264
    solid-state, 129, 265
linear electro-optic effect, 54, 313–316
linear polarization, 335
local field, 43
localized state, 96, 107, 216, 247
longitudinal cavity mode, 132
longitudinal waves, 200, 276
Lorentz correction, 43
Lorentz model, 29–44, 181
Lorentzian line shape, 36, 303
low-dimensional structures, 141–174
LS coupling, 352
luminescence, 4, 113–136
    cathodo-, 135–136
    centres, 247–266
    efficiency, 115
    electro-, 126–135
    interband, 115–118
    photo-, 118–126
    polarization, 123
    quantum well, 164
luminescence spectra
    Alq3, 229
    C$_{60}$, 242
    carbon nanotube, 241
    F$_2^+$ band in KF, 252
    Ga$_{0.47}$In$_{0.53}$As, 123
    GaAs, 121
    GaAs LED, 130
    GaN, 117
    MeLPPP polymer, 232
    Nd:YAG, 258
    phosphor, 266
    polyfluorene, 234
    pyrromethene dye, 225
    Ti:sapphire, 261
    tricolour fluorescent lamp, 265
    white-light LED, 266
    ZnCdSe quantum well, 165
LUMO level, 217
Lyddane–Sachs–Teller relationship, 276

magnetic

circular dichroism, 56
    dipole transition, 349
    quantum number, 353
magnetization, 332
magneto-absorption, 75
magneto-optics, 56
material dispersion parameter, 47
matrix element, 344
    electric-dipole, 346
    interband, 65, 68
    quantum well, 152
Maxwell's equations, 330–333
MBE, *see* molecular beam epitaxy
medium
    anisotropic, 48–54
    dielectric, 330
    isotropic, 35, 48, 49
    left-handed, 208
MeLPPP polymer, 231
metal–organic chemical vapour
        deposition, *see* MOCVD
metals, 354
    interband absorption, 188
    nanoparticle, 206
    optical properties, 13–14, 180–191
    reflectivity, 14, 185–191
    ultraviolet transparency, 14, 186
metamaterial, 209
metastable state, 349
Mie theory, 207
Miller's rule, 302
mirror symmetry rule, 224, 250, 253, 260
MOCVD, 127, 144
mode, longitudinal cavity, 132
modulator, electro-optic, 163, 314, 320
molecular beam epitaxy, 127, 144
molecular material, 214–243
    aromatic hydrocarbon, 227
    conjugated, 15, 109, 215, 227–232
    crystals, 108, 227, 231
    electronic states, 216–217
    Frenkel excitons, 108
    infrared absorption, 32
    optical properties, 14–16, 216–226
    opto-electronic devices, 232–235
    vibronic coupling, 218–219
molecule
    covalent, 32
    diatomic, 220
    polar, 31
Mollwo–Ivey relationship, 252
momentum scattering time, 183
momentum, electromagnetic, 344
monomer, 229
Mott density, 104
MOVPE, 127, 144
multiple quantum well, 145
multiple resonances, 38–41
multiplicity, spin, 217, 352

n-type semiconductor, *see*
        semiconductor, n-type
NaCl
    absorption spectrum, 108
    infrared refractive index, 58
    ionicity, 32
    Reststrahl band, 280
nanoparticles, 206
nanostructures, 141–174, 235–243
nanotube, 235, 237–241
natural lifetime, *see* radiative lifetime
Nd:YAG, 258
    laser, 261
nearly free electron model, 356–359
negative refraction, 207–209
Neumann's principle, 18, 308
NH$_3$, *see* ammonia
nitride compounds, 128, 165, 265
non-radiative lifetime, 115
non-resonant nonlinearity, 299–302
nonlinear
    absorption, 105, 304, 324–326
    coefficient, 308
    coherence length, 311
    crystal, 308
    frequency mixing, 305–308, 317
    optics, 295–326
    phase shift, 322, 324
    polarization, 296, 306
    refractive index, 319, 325
    susceptibility, 295–298, 302, 321
notation for angular momenta, 351–352
NV centres in diamond, 253–255

O-LED, 232
occupancy factors, 114, 119, 122
octahedral environment, 256, 260
one-dimensional material, 143, 238
open-circuit voltage, 90
optic axis, 49
optical
    activity, 55
    amplification, 131, 261, 343
    anisotropy, 19, 48–54
    chirality, 55–56
    coefficients, 2–5
    confinement, 133
    density, 4
    fibre, 13, 48, 322–324
    intensity, 336
    Kerr effect, 318–321
    limiter, 243
    materials, summary of, 9–17
    modulator, 163, 314, 320
    nonlinearity, 295–326
    orientation, 77, 123–125
    phenomena, definitions, 1–2
    polarization, *see* polarization, optical
    pumping, 254
    rectification, 307

retarder, 51
spin injection, 77, 123, 158
waveguide, 134
opto-electronics
detectors, 86–91
light-emitting devices, 126–135, 164, 232, 265
modulators, 163, 314, 320
organic, 232–235
ordinary ray, 49, 50
organic materials, *see* molecular material
orientation, optical, 77, 123–125
oscillator
anharmonic, 290, 299
atomic, 29–31
bound electron, 28
damped, 31, 33
dipole, 29–38
forced, 33
free electron, 32, 180
harmonic, 33, 341
harmonic (2-D), 170, 178
Lorentz, 33–38
nonlinear, 299–302
strength, 40, 347
vibrational, 31–32, 273
vibronic, 221
output coupler, 132
oxide–confined laser, 134

p–i–n diode, 363–364
p–n diode, 126, 130
P-567 dye, 225
p-type semiconductor, *see*
semiconductor, p-type
parabolic band approximation, 71
parallel band effect, 189
paramagnetic impurity, 255–260
parametric amplification, 307
parity
notation, 241, 260
selection rule, 348
wave function, 148
Pauli exclusion principle, 63, 114
PDA, *see* polydiacetylene
permeability, magnetic, 332
permittivity, relative, 7, 331
perspex, 15
PFO, *see* polyfluorene
phase lag due to scattering, 31
phase matching, 298, 310–312, 318
phonon, 271–291
absorption, 10, 32, 272, 278–280
acoustic, 271, 286, 289
Brillouin scattering, 289–290
dispersion, 272
infrared active, 271–273
lifetime, 275, 278, 290–291
longitudinal optical, 276
optical, 271–273

polariton, 281
Raman scattering, 287–289
scattering, 291
transverse optical, 272
virtual, 283
phonon-assisted transition, 79
phosphor, 136, 264–266
phosphorescence, 217, 229, 348
photoconductivity, 89
photocurrent, 87
spectrum, 103, 163, 231
photodetector, 86–91, 163
photo-elastic effect, 53
photoluminescence, 113, 118–126
excitation spectroscopy (PLE), 125
spectroscopy, 125
time-resolved, 125
photon angular momentum, 348
photovoltaic device, 90–91, 234
π electron, 15, 214
Planck formula, 343
plasma, 180
electron-hole, 104
frequency, 14, 181, 185, 193, 200
oscillations, 198
reflectivity, 180–182, 193
plasmon, 198–207
bulk, 198–202
dispersion, 200
doped semiconductor, 202
metal, 201
Raman scattering, 202
surface, *see* surface plasmon
plasmonics, 202, 206, 287
PLE, 125
plexiglass, 15
Pockels effect, 54, 307, 313–316
point group symmetry, 18
Poisson's equation, 331
polar bond, 32, 272
polariton
exciton, 106
phonon, 281
surface plasmon, 203
polarization, dielectric, 34, 43, 330
nonlinear, 296
polarization, optical, 334
$\sigma^{\pm}$, 77, 336
circular, 335
elliptical, 335
horizontal, 335
linear, 335
p-plane, 86
random, 335
s-plane, 86
vertical, 335
polarization, spin, 78, 158
polarizer, 50
polaron, 282–285
large, 285

mass, 284
radius, 284
small, 285
poly-phenylenevinylene (PPV) LED, 233
polyacetylene molecule, 230
polydiacetylene (PDA)
absorption spectrum, 231
exciton, 109
photocurrent spectrum, 231
structure, 230
polyfluorene
absorption spectrum, 16
luminescence, 234
polymer
conjugated, 15, 109, 229–232
saturated, 230
polythene molecule, 15, 230
polyyne absorption, 226
population inversion, 131, 261, 343
potential
electric, 331
scalar, 333
vector, 332, 345
Poynting vector, 336
PPV, *see* poly-phenylenevinylene
prism, 47
polarizing, 50
propagation of light, 1–2, 28–56
pulse broadening, 47, 323
pyrene ($C_{16}H_{10}$) absorption, 109
pyrromethene 567 dye, 225

quadratic electro-optic effect, 54, 320
quantum
box, *see* quantum dot
cascade laser, 167
confined Stark effect, 160–164, 174
confinement, 141–174
harmonic oscillator, 341
optics, 23, 307, 341, 347
size effect, 16, 141, 226
statistics, 122
transition rate, 344
quantum dot, 142, 167–174, 326
CdSe, 170
colloidal, 170–172
InAs, 172
InGaAs, 326
self-assembled, 172–174, 326
quantum efficiency
detector, 87
laser, 133
radiative, 115
quantum well, 142, 144–167
absorption, 152
exciton, 157–158, 325
finite, 149
GaAs, 144, 151, 156, 158, 163
GaInAs, 165, 325
infinite, 147

laser, 166
  luminescence, 164
  multiple, 145
  quantized levels, 146
  selection rules, 152
  ZnCdSe, 165
quantum wire, 142, 238
quarter-wave plate, 51
quartz, 55
quasi-equilibrium, 120, 131
quaternary alloy, 128
qubit, 254

Rabi oscillations, 347
radiative
  efficiency, 115, 165, 168
  lifetime, 114, 116, 126, 341
  transition, *see* transition, radiative
Raman scattering, 202, 285–289
  experimental arrangement, 288
  GaAs, n-type, 202
  GaP polariton, 282
  III–V semiconductors, 289
  linewidth, 291
  stimulated, 321
  surface-enhanced, 287
  time-resolved, 291
rare-earth ion, 255, 257–259
rare gas crystals, excitons, 107
Rayleigh scattering, 5
reciprocal lattice, 356
rectification, optical, 307
reduced mass, 29, 97, 275
reduced zone scheme, 358
reflection
  coefficient, 3, 7, 278, 339
  measurement, 84
reflectivity
  free carrier, 193
  free electron, 180–182
  infrared, 193, 273–281
  plasma, 180–182
reflectivity spectra
  AlSb, 293
  aluminium, 187
  copper, 192
  GaAs, 279
  gold, 211
  InAs, 279
  n-type InSb, 193
  silver, 14
refractive index
  complex, 6–9, 337
  definition of, 3, 334
  dispersion, 13, 40, 46–48
  ellipsoid, 60, 313
  extraordinary, 50
  imaginary part, 6
  negative, 207–209
  nonlinear, 319, 325

  ordinary, 50
  physical origin, 31
  real part, 6
  relationship to dielectric constant, 7
refractive index data
  fused silica, 12
  glasses, 13
  insulators, 11
  NaCl, 58
  semiconductors, 12
  silica glass, 41, 46
  uniaxial crystals, 51
relative dielectric constant, complex, 7
relative permittivity, 7
relaxation process, 118, 219, 222, 248
resonant nonlinearity, 302–305, 324–326
responsivity, 87
Reststrahl band, 277–278
retarder plate, 51
rotatory power, 60
ruby, 256, 259
  laser, 261, 269
  transmission spectrum, 17
rule of mutual exclusion, 288
Russell–Saunders coupling, 352

sapphire (Al_2O_3), 9, 17, 259
  tranmission, 10
saturable absorber, 305, 324–326
saturated molecule, 15, 230
saturation intensity, 304
scalar potential, 333
scattering
  Brillouin, 286, 289–290
  cross-section, 5
  elastic, 2
  electron–phonon, 119, 194
  inelastic, 2, 201, 285–290
  phonon–phonon, 291
  Raman, 285–289
  Rayleigh, 5
second-harmonic generation, 306
second-order nonlinear optics, 296, 300, 305–316
selection rules, 347–349
  J, 349
  L, 349
  M_J, 77
  S, 349
  $\sigma^{\pm}$, 348
  l, 348
  m, 348
  electric dipole, 348
  intersubband, 166
  IR active phonon, 271
  Landau level, 76
  parity, 154, 348
  quantum well, 152
  Raman, 288
  spin, 227, 348

  wave vector, 66
self-assembled quantum dot, 172–174, 326
self-phase modulation, 322
self-trapping, 107, 285
Sellmeier equation, 59
semi-classical interaction, 23, 65
semiconductor, 354
  absorption, 68–79
  diode, 363–364
  direct gap, 68, 116, 118
  doped, 191–198, 355
  free carrier effects, 191–196
  II–VI, 68, 97, 129, 171, 361
  III–V, 68–71, 97, 128, 289, 361
  indirect gap, 79, 117
  laser, 130–135
  light-emitting diode, 129–130
  luminescence, 115–118
  n-type, 191, 194, 196, 355
  optical properties, summary of, 9–12
  p-type, 129, 191, 195, 356
  photodetector, 86–91
  quantum dot, 167–174, 326
  quantum well, 144–167, 325
  refractive index, 12
  transparency range, 12
semiconductor-doped glass, 16–17, 171
SESAM, 326
short-circuit current, 90
SiC light-emitting diode, 127
$\sigma^{\pm}$ polarization, 77, 336
signal velocity, 42
silica, *see* SiO_2, glass
silicon
  absorption spectrum, 80, 83
  band parameters, 360
  band structure, 84
  effective mass anisotropy, 197
  electron-hole droplet, 105
  exciton, 105
  impurity absorption, 197
  indirect band gap, 79
  photodetector, 88
  solar cell, 91
silver
  interband absorption, 190
  reflectivity, 14
single-photon source, 173, 253
singlet state, 217, 352
SiO_2
  crystalline, 55
  glass, 12, 41, 56, 274
skin depth, 184
slope efficiency, laser, 133
Snell's law, 1, 50, 208
sodium chloride, *see* NaCl
solar cell, 90–91, 234
solid-state laser, 261–264
solid-state lighting, 129, 265

soliton, 322–324
spectral lineshape function, 303
spectroscopic notation, 351–352
spectroscopy
    absorption, 84
    photoluminescence, 125
    reflectivity, 84
    transmissivity, 84
spin
    angular momentum, 351
    injection, 77–79, 123, 158
    multiplicity, 217, 352
    polarization, 78, 123, 158
    pumping, 255
    relaxation, 124
    selection rule, 227, 349
    singlet, 217, 352
    triplet, 217, 352
spin–orbit coupling, 71, 78, 229, 257, 351
spintronics, 77, 158
split-off hole transition, 71
spontaneous emission, 2, 113, 340
stained glass, 206
Stark effect, 259, 353
    linear, 163, 353
    quadratic, 162, 353
    quantum-confined, 160–164
stimulated Brillouin scattering, 321
stimulated emission, 131, 342
stimulated Raman scattering, 321
Stokes scattering, 286
Stokes shift, 4, 219, 248, 253, 260
Stranski–Krastanow crystal growth, 172
substrate material, 126
sum frequency mixing, 306
superlattice, 145
surface plasmon, 202–207
    localized, 207
    polariton, 203
surface wave, 203
surface-enhanced Raman scattering, 287
susceptibility
    electric, 34, 43, 295, 331
    magnetic, 332
    nonlinear, 295–298
    tensor, 49
symmetry
    class, 18
    effects of, 18–20, 308
    Kleinman, 322
    lowering, 19, 53, 148, 153, 163, 256
    translational, 19, 22

terahertz laser, 167
term, atomic, 352
ternary alloy, 128
thermal equilibrium, 119, 131, 342

third-order nonlinear optics, 296, 305, 317–326
threshold current, 132
Ti^{3+} ion, 260
Ti:sapphire
    absorption and emission, 260
    laser, 262
tightly bound exciton, 95
time–bandwidth product, 47, 264
time-resolved luminescence, 122, 125
transition
    allowed, 69, 348
    direct, 64–79
    electric dipole, 65, 346
    electric quadrupole, 349
    forbidden, 348
    free carrier, 195
    indirect, 79
    interband, 62–69, 152
    intersubband, 166
    magnetic dipole, 349
    non-radiative, 115
    phonon-assisted, 79
    probability, 40
    radiative, 114, 340–349
    rate, 64, 344
    selection rules, 347
    singlet–singlet, 217, 228
    singlet–triplet, 217, 229
    two-photon, 345
    vibronic, 216–226, 247–250
    virtual, 302
transition metal, 190
transition-metal ion, 255, 259–260
translational order, 18
transmission
    coefficient, 3
    measurement, 84
transmission spectra
    CdSe, 10
    germanium, in magnetic field, 76
    ruby, 17
    sapphire, 10, 17
transmissivity, 3
transparency range, 9, 11, 12
transverse waves, 200, 276
traps, 115
tricolour lamp, 264
triplet state, 217, 352
tunnelling device, 150
two-dimensional harmonic oscillator, 170, 178
two-dimensional material, 143, 236
two-photon transition, 345
type I and II phase matching, 312

ultraviolet transmission

glass, 13
metals, 186
uniaxial crystal, 49, 311
unpolarized light, 335
Urbach's rule, 198

vacancy, anion, 251
valence band, 354
valence electron, 185, 351
valency, 185
van Hove singularities, 84, 239
vector potential, 332, 345
velocity
    group, 42, 357
    phase, 42, 334
    signal, 42
Verdet coefficient, 56
vertical-cavity surface-emitting laser, 133
vertical polarization, 335
vibrational
    absorption, *see* phonon absorption
    oscillator, 31–32
    relaxation, 219
vibronic
    absorption, 15, 219, 247
    band, 21, 225, 226, 228, 231, 247, 260
    coupling, 218–219, 247–250, 255–257
    emission, 219, 247
    wave function, 223
virtual transition, 302

Wannier–Mott exciton, 95
wave
    electromagnetic, 333–339
    evanescent, 184
    impedance, 335
    surface, 203
    vector, 6, 23, 334
wave function, Bloch, 20, 359
wave number units, 202, 278
wave plate
    half, 51
    quarter, 51
waveguide, 134
Wood's anomaly, 206
wurtzite structure, 79, 362

YAG, *see* Nd:YAG

Zeeman effect, 20, 353
zero-dimensional material, 143, 167
zero-phonon line, 250
zero-point fluctuations, 341
zinc-blende structure, 70, 361
ZnCdTe/ZnSe quantum well, 165
ZnSe, electron–phonon coupling, 283

Printed and bound by CPI Group (UK) Ltd, Croydon, CR0 4YY